Nanocrystals in Nonvolatile Memory

Nanocrystals in Nonvolatile Memory

edited by

Writam Banerjee

PAN STANFORD PUBLISHING

Published by

Pan Stanford Publishing Pte. Ltd.
Penthouse Level, Suntec Tower 3
8 Temasek Boulevard
Singapore 038988

Email: editorial@panstanford.com
Web: www.panstanford.com

British Library Cataloguing-in-Publication Data
A catalogue record for this book is available from the British Library.

Nanocrystals in Nonvolatile Memory

ISBN 978-981-4774-73-4 (Hardcover)
ISBN 978-1-351-20327-2 (eBook)

I dedicate this book to my parents, Mr. Parboti Shankar Banerjee and Mrs. Gopa Banerjee, whose patience, hard work, and impressive attitude toward life and family inspired me to have a strong belief in hope and made me capable of editing this book. On this note, I must share my point of view about the magical word "hope."

I believe,
*Every mortal object in this living planet starts its new sun with the same magical word **HOPE**.*

H stands for Highly,

O stands for Organized or Oriented,

P stands for Positive, and

E stands for Energy.

Contents

Preface   xv
Acknowledgments   xix

1. Nanocrystal Materials, Fabrications, and Characterizations   1

Sujan Chowdhury and Puspendu Barik

1.1   Introduction   2
    1.1.1   Nanomaterials for a Nonvolatile
        Memory Device   4
    1.1.2   Overview of Nonvolatile Memory   5
    1.1.3   Classification of Nanomaterials   7
1.2   Synthesis and Fabrication of Nanocrystals
    for NVM   9
    1.2.1   0D Nanocrystals   11
    1.2.2   1D Nanocrystals   14
    1.2.3   2D Nanocrystals   20
    1.2.4   3D Nanocrystals   25
1.3   Characterization of Nanoparticles   26
    1.3.1   Microscopy Technique   27
        1.3.1.1   Electron microscopy   27
        1.3.1.2   Reflection high-energy
              electron diffraction   32
        1.3.1.3   Scanning probe microscopy   33
    1.3.2   X-Ray-Based Methods   37
        1.3.2.1   X-ray diffraction   37
        1.3.2.2   Small-angle X-ray scattering   38
        1.3.2.3   X-ray photoelectron
              spectroscopy   40
        1.3.2.4   X-ray absorption
              spectroscopy   43
    1.3.3   Light-Based Spectroscopic Techniques   44
        1.3.3.1   Light scattering techniques   45
        1.3.3.2   Ultraviolet/visible
              spectroscopy   46
        1.3.3.3   Photoluminescence
              spectroscopy   47

	1.3.3.4	Raman spectroscopy	49
	1.3.3.5	Fourier transform infrared spectroscopy	50
1.4	Summary		51

2. Modeling and Simulation of Nanocrystal Flash Memory   75

Bikash Sharma and Chandan Kumar Sarkar

2.1	Introduction	76
2.2	Developments in Nanocrystal Memory	79
2.3	Model for Nanocrystal and Nitride-Trap Memory	82
2.4	Memory Device Scaling with the Use of a Silicon Nanocrystal	87
2.5	Modeling of Tunneling Currents	90
2.6	Model for the Charging and Discharging Process	96
2.7	Programming Time Model	101
2.8	Growth of Metal (Au) Nanocrystals in High-κ Dielectrics	104
2.9	Retention Characteristics Model	106
2.10	Tunneling Characteristics of Metal-Nanocrystal- and Semiconductor-Nanocrystal-Based Gate Dielectrics	109
	2.10.1 Fowler–Nordheim Tunneling	114
	2.10.2 Direct Tunneling	115
2.11	Conclusion	120

3. Charge Trapping and High-κ Nanocrystal Flash Memory   127

Meng Chuan Lee and Hin Yong Wong

3.1	Introduction to Charge Storage Nonvolatile Memory	128
3.2	Evolution of Nanocrystal-Based CS-NVM	139
3.3	Reliability Challenges of Nanocrystal-Based CS-NVM	145
3.4	Technical Mitigations	152
3.5	Summary	165

4. Silicon Nanocrystal Flash Memory   173

Lili Zhao, Tiezheng Lv, and Guofeng Fan

| 4.1 | Introduction | 174 |
| 4.2 | Si NCs in Flash Memory | 175 |

4.2.1   Structure Development of a Si NC
         Floating Gate   175
         4.2.1.1   Si NC floating-gate story   176
         4.2.1.2   Preparation of Si NCs for flash
                   memory   180
4.2.2   Electrical Characteristics of Si
         Nanocrystal in Flash Memory   183
4.3   Si Nanocrystal Trap Center Studied by
       Deep-Level Transient Spectroscopy   184
4.4   Engineering for Improved Si Nanocrystal Flash
       Memory   192

5. **Synthesis, Characterization, and Memory Application
   of Germanium Nanocrystals in Dielectric Matrices   199**
   Wee Kiong Choi and Writam Banerjee
5.1   Introduction   200
5.2   Synthesis of Ge Nanocrystals   203
       5.2.1   Ge Atoms for Nanocrystal Growth   203
       5.2.2   Effect of Ge Concentration and
                Annealing Temperature   205
       5.2.3   Effect of Annealing Ambient   210
       5.2.4   Effect of an Oxide Barrier Layer   211
       5.2.5   Influence of Dielectric Matrices   213
5.3   Characterizations of Ge Nanocrystals   216
       5.3.1   Photoluminescence Properties   216
       5.3.2   Electroluminescence Properties   219
       5.3.3   Stress in Ge Nanocrystals Embedded
                in Dielectrics   222
5.4   Ge Nanocrystal-Based Floating-Gate Memory
       Devices   228
       5.4.1   Fabrication of Ge Nanocrystal
                Memory Structures   228
       5.4.2   Control of Nanocrystal Size   231
       5.4.3   Retention Properties   237
       5.4.4   High-κ Dielectrics   242
       5.4.5   Characterization Ge-Nanocrystal-
                Based Transistors   245
5.5   Summary   253

6. Nanographene Flash Memory   **263**

Jianling Meng, Rong Yang, and Guangyu Zhang

6.1   Introduction   264
    6.1.1   Graphene Fundamentals   264
        6.1.1.1   Structure and electronic
                properties of graphene   264
        6.1.1.2   Graphene nanostructures
                and graphene nanosheets   267
6.2   Preparation/Synthesis of Graphene and
    Nanographene   268
    6.2.1   Graphene Thin-Film Preparation   268
        6.2.1.1   Reduced graphene oxide   268
        6.2.1.2   Chemical vapor depositions   273
    6.2.2   Synthetic Strategies for Nanographene   280
        6.2.2.1   Top-down methods   281
        6.2.2.2   Bottom-up methods   287
6.3   Graphene-Based Flash Memory   291
6.4   Graphene Nanostructures Flash Memory   297
    6.4.1   Memory Window   298
    6.4.2   P/E Transient Time   309
    6.4.3   Retention Characteristics   310
    6.4.4   Endurance Cycles   313
6.5   Graphene Memory Hybrids   313
    6.5.1   Flexible Transparent Flash Memory   314
    6.5.2   3D Stacking   317
6.6   Conclusion and Prospects   319

7. Data Recovery of Flash Memory   **327**

Bernard Kasamani Shibwabo and
Ismail Ateya Lukandu

7.1   Introduction   328
7.2   How Computers Store Information   332
    7.2.1   Kinds of Computer Memory   332
    7.2.2   Bits and Memory   333
7.3   Flash Memory   335
    7.3.1   Introduction to Flash Memory   335
    7.3.2   The Features of Flash Memory   337
    7.3.3   Transistors   337
    7.3.4   NAND and NOR Flash Memory   343

7.4	Data Recovery		345
	7.4.1	Introduction to Data Recovery	345
	7.4.2	The Need for Data Recovery	346
	7.4.3	Data Extraction/Acquisition	347
		7.4.3.1 Data extraction tools	347
		7.4.3.2 Physical extraction	348
7.5	Data Recovery in Flash Media		348
	7.5.1	Data Loss on Flash Media	348
		7.5.1.1 Bit flipping	350
		7.5.1.2 Bad block handling	351
		7.5.1.3 Life span/endurance	351
		7.5.1.4 Retention	351
	7.5.2	Bad Blocks	352
	7.5.3	File Systems	353
	7.5.4	File Attributes	354
	7.5.5	Flash Data Recovery Techniques	356
		7.5.5.1 Fundamental concepts	356
		7.5.5.2 The flash translation layer and flash data recovery	358
		7.5.5.3 Data recovery for data loss due to a virus attack	360
		7.5.5.4 Data recovery software	362
		7.5.5.5 Best practice for flash	366
7.6	Windows User Laboratory Activities		366
7.7	Summary		367

8. Nanocrystals in Resistive Random Access Memory   **369**

Writam Banerjee and Qi Liu

8.1	Introduction		370
	8.1.1	Background	370
	8.1.2	Prototype NVM Technologies	373
		8.1.2.1 Ferroelectric random access memory	373
		8.1.2.2 Phase change memory	375
		8.1.2.3 Spin-transfer torque random access memory	376
	8.1.3	Emerging NVM Technologies	379
		8.1.3.1 Emerging FeRAM	379
		8.1.3.2 Carbon memory	380
		8.1.3.3 Mott memory	381

	8.1.3.4	Macromolecular memory	382
	8.1.3.5	Molecular memory	382
	8.1.3.6	Resistive random access memory	383
	8.1.3.7	History of RRAM	386
8.2	Mechanisms and Materials in RRAM		389
8.2.1	Resistive Switching Mechanisms		389
	8.2.1.1	Electrochemical metallization type	390
	8.2.1.2	Valence change memory type	396
	8.2.1.3	Thermochemical reaction type	400
8.2.2	Materials in RRAM		401
	8.2.2.1	Metal electrode layer	401
	8.2.2.2	Insulating layer	405
	8.2.2.3	Defect-related improvement of RRAM performance	406
8.3	Applications of Nanocrystals in RRAM		408
8.3.1	Improvement of Electrical Performance		409
	8.3.1.1	Forming process	409
	8.3.1.2	SET/RESET operation	410
	8.3.1.3	Reliability of RRAM devices	412
8.3.2	Conductive Filament Formation Based on Nanocrystal Migration		412
8.3.3	Charge Trapping Using NC		415
8.3.4	Threshold Switching to Memory Switching		418
8.4	Nanocrystals as the Seed Layer in RRAM		424
8.4.1	Effect of Nanocrystals in the Resistive Switching Layer		424
	8.4.1.1	Colloidal nanocrystals as the switching layer	425
	8.4.1.2	Local electric field enhancement with nanocrystals	427
	8.4.1.3	Formation of homogeneous NCs for RRAM applications	428

		8.4.2	Bottom Electrode Modification		432
			8.4.2.1	Nanocrystal-based bottom electrode	432
			8.4.2.2	Nanopyramid-shaped bottom electrode	435
			8.4.2.3	Arc-shaped bottom electrode	437
	8.5	Summary and Future Scope			438

9. Measurement Aspects of Nonvolatile Memory   449

Alberto Campisi

	9.1	Introduction			450
	9.2	Testing Memory			451
		9.2.1	Memory Tester		452
			9.2.1.1	Digital channel	454
			9.2.1.2	PMU	458
			9.2.1.3	Device power supply	460
			9.2.1.4	Control unit	461
			9.2.1.5	Capture memory: data buffer memory	463
			9.2.1.6	Redundancy analysis processor	465
			9.2.1.7	Other remarks on flash testers	465
		9.2.2	DUT Built-In Test-Oriented Resources		470
			9.2.2.1	DMA	474
			9.2.2.2	Threshold distribution	475
	9.3	Test Flow			480
		9.3.1	Wafer Sort		481
		9.3.2	Final Test		490
	9.4	Brief History of Flash			492
	9.5	Redundancy			493
	9.6	Cycling			493
	9.7	Retention			494
	9.8	Silicon Debug/Design Validation			496
	9.9	Testing Readiness			496
	9.10	Characterization			497
		9.10.1	Shmoo Plot		497
	9.11	Qualification			500
	9.12	Datasheet			500

9.12.1	Product General Description		501
9.12.2	Pin Name and Function		501
9.12.3	Product Conceptual Schematic		502
9.12.4	Command Set		502
9.12.5	DC Characteristics		503
9.12.6	AC Characteristics		503
9.12.7	Endurance Characteristics		504
9.12.8	Package Dimensions		504
9.12.9	Order Code		504
9.13	Datasheet Gray Areas		505
9.14	Error Correction Code		506

Index 513

Preface

Modern electronics means living in the age of quantum physics. The abundant advantages of quantum physics have been realized in quantum dots, quantum wires, quantum wells, and nanocrystals. The advantages of these nanoparticles have been adopted in many areas of applications. This book focuses specifically on the application of nanoparticles in the field of nonvolatile memory. Nanocrystals and nonvolatile memory are two of the most lucrative areas of research, have attracted considerable attention of both companies and academia, and are already introduced at various levels in university education. Needless to say, this encourages writing of a suitable textbook based on the applications of nanocrystals in nonvolatile memory that can fulfill the requirements of the curricula.

It was October 2014. The idea of the book came just after I had delivered an invited talk entitled "Nanocrystals for Nonvolatile Memory Applications" in "Nano S&T 2014 – BIT's 4th Annual World Congress of Nano Science & Technology 2014." The chapters of this book are written by people from various institutes from different countries around the globe. The topics included in this book are as diverse as the fabrication methods of various nanocrystals, nanocrystal-based baseline and emerging nonvolatile memory devices, and the electrical characterization techniques of the same, written by the best-possible experts on the subjects.

This book has three layers. Part 1 introduces a huge set of nanocrystal materials and the fabrication methods and characterization of the nanocrystals. Part 2 looks at a detailed analysis of charge trapping and nanocrystal-based flash memory devices, including some of the new aspects, like nanographene flash memory. Part 3 looks at the advantages of nanocrystals for emerging memory technologies, particularly the fabrication, properties, and performance of nanocrystal-based resistive random access memory technology.

The scientific literature contains a large volume of materials suitable for nanoparticles or nanocrystals, uses of those nanocrystals in nonvolatile memory devices, and also electrical characterizations of these devices. The complete discussion is guided by plenty of illustrations, useful and relevant figures, and related references.

Chapter 1, "Nanocrystal Materials, Fabrications, and Characterizations": As the name suggests, this chapter deals with materials suitable for nanocrystals and their fabrication and characterization using various techniques.

Chapter 2, "Modeling and Simulation of Nanocrystal Flash Memory": The advantages of nanocrystals are used rigorously by nonvolatile flash memory technology, and it is one of the mature technologies. This chapter deals with the modeling and simulation of nanocrystal-based flash memory devices.

Chapter 3, "Charge Trapping and High-κ Nanocrystal Flash Memory"; Chapter 4, "Silicon Nanocrystal Flash Memory"; and Chapter 5, "Synthesis, Characterization, and Memory Application of Germanium Nanocrystals in Dielectric Matrices": These chapters discuss several high-κ and semiconductor nanocrystals for flash memory technology and its uses.

Chapter 6, "Nanographene Flash Memory": This is a chapter on the most recent achievements in the use of graphene for flash memory applications. It considers the uses of various forms of graphene in transparent, flexible electronic device design.

Chapter 7, "Data Recovery of Flash Memory": Data recovery may be necessary because of physical damage to the storage device or logical damage to the file system that stops it from being mounted by the host operating system. This is a suitable chapter for those who are interested in information technology and engineering.

Chapter 8, "Nanocrystals in Resistive Random Access Memory": This chapter discusses the uses of nanocrystals in nonvolatile emerging memory technologies. In particular, the uses of nanocrystals for different purposes have been clearly identified.

Chapter 9, "Measurement Aspects of Nonvolatile Memory": The chapter provides details about electrical measurements of nonvolatile memory devices.

All of the chapters are based on current topics and are internally linked with each other. The chapters may be useful for different disciplines, but all together, they are under one roof: *Nanocrystals in Nonvolatile Memory*. The present version of the book is meant to be a textbook or a reference book for students preparing for their master's degrees in science and also for doctoral students. Apart from science graduate students and students in materials engineering, this book is also expected to be useful to students with an electronics and information technology background.

This book has tried to discuss the subject matter in a clear and concise way, including many explanatory diagrams to discuss all the issues. The hard work will be amply rewarded if the book proves to be useful for students. Any suggestions for further improvement of this book are most welcome.

Dr. Writam Banerjee

Key Laboratory of Microelectronic Devices and Integrated Technology

Institute of Microelectronics, Chinese Academy of Sciences

Beijing, China

University of Chinese Academy of Sciences, Beijing, China

Jiangsu National Synergetic Innovation Center for Advances Materials (SICAM), Nanjing, China

2018

Acknowledgments

At first, I would like to thank all the authors—Dr. Sujan Chowdhury, Dr. Puspendu Barik, Mr. Bikash Sharma, Prof. Chandan Kumar Sarkar, Dr. Meng Chuan Lee, Prof. Hin Yong Wong, Prof. Lili Zhao, Dr. Tiezheng Lv, Dr. Guofeng Fan, Prof. Wee Kiong Choi, Dr. Jianling Meng, Prof. Rong Yang, Prof. Guangyu Zhang, Prof. Bernard Kasamani Shibwabo, Prof. Ismail Ateya Lukandu, Prof. Qi Liu, and Dr. Alberto Campisi— for their valuable contribution to this book.

I am very grateful to Prof. Nicole Herbots, professor of physics, Arizona State University, United States, and founder and chief strategy officer (CSO) of SiO_2 Nanotech LLC and Prof. Vijay Arora, professor of electrical engineering, IEEE-EDS distinguished lecturer, and Leading Educator of the World 2005, from the Department of Electrical Engineering and Physics, Wilkes University, United States, for guiding me at every step of this book.

I would like to thank Prof. Ming Liu, the director of the Institute of Microelectronics of Chinese Academy of Sciences, Beijing. I would like to convey my thanks to all other faculty members, students, and staff members of the Key Laboratory of Microelectronic Devices and Integrated Technology, Institute of Microelectronics, Chinese Academy of Sciences, Beijing, China.

This is a great opportunity to thank the most respected teachers of my life, Prof. Subhas Chandra Samanta, ex-professor of Department of Physical Science, Midnapore College, India, and Prof. Pradipta Panchadhyay, professor at Prabhat Kumar College, Contai, India.

Finally, most importantly, my hearty thanks to my parents, my brother (Mr. Debottam Banerjee), and last but foremost Ms. Ananya Dutta for their support at each and every step.

Chapter 1

Nanocrystal Materials, Fabrications, and Characterizations

Sujan Chowdhury [a],* and Puspendu Barik[b],*

[a]*Department of Chemical Engineering, Universiti Tunku Abdul Rahman, Malaysia*
[b]*Department of Physics, Indian Institute of Science, Bangalore, India*
sujan@utar.edu.my; puspendub@iisc.ac.in

One of the foremost challenges is to develop nanocrystal (NC) floating-gate nonvolatile memory (NVM) devices to overcome the challenge of scaling down, where NCs serve as discrete charge storage elements to make a thinner floating gate without any significant leakage with time. In recent years, significant improvements have been made in NC fabrication and in the prototypes of NC-based NVM. In this chapter, we briefly emphasize the basics of NCs and their general synthesis methods, mostly the wet-chemical method. Subsequently, we focus on those NCs that have been used for NVM devices in the recent past. We also discuss the recent efforts and research activities regarding the fabrication and characterization of NVM devices made of these NCs, which are further classified into four

_____
*Both authors have contributed equally to this work.

Nanocrystals in Nonvolatile Memory
Edited by Writam Banerjee
Copyright © 2018 Pan Stanford Publishing Pte. Ltd.
ISBN 978-981-4774-73-4 (Hardcover), 978-1-351-20327-2 (eBook)
www.panstanford.com

categories: zero-dimensional, one-dimensional, two-dimensional, and three-dimensional.

1.1 Introduction

On December 29, 1959, Richard Feynman gave a classic talk entitled "There Is Plenty of Room at the Bottom" at the annual meeting of the American Physical Society at the California Institute of Technology, and he discussed the consequences of measuring and manipulating materials on the nanoscale for the first time [1]. Over the last decades, there has been an increased interest in nanocrystals (NCs)/ nanomaterials after the pioneering work of Russia's Alexei Ekimov in 1981 on semiconducting quantum dots (QDs) in a glass matrix involving exploring their electronic and optical properties [2]. In the year 1984, Louis Brus [3] discovered colloidal semiconductor NCs (i.e., CdS QDs). After this, the research on nanomaterials increased rapidly and a new area of research was formed called "nanotechnology."

An NC is a material particle that can also be called a nanoparticle (NP), composed of atoms in a single- or polycrystalline arrangement. A variety of metal and semiconductor NPs have been synthesized and proposed as potential building blocks of optical electronic devices owing to a variety of basic reasons: emerging technologies (integrated circuits, quantum electronics, biochips, etc.), precise control over these material, and the fact that the device structure can be formed on very low length scales [4, 5]. In fact, a theoretical estimate of limits of computation, based on the use of the uncertainty principle, shows that irrespective of geometric factors, the optimum localization for computation is on the nanometer length scale (1 nm ≈ 2 nm for operation at room temperature), that is, the molecular length scale [6].

NPs mostly rank as passive nanostructures and represent almost the only part of nanotechnology with a commercial significance. However, it is sometimes questioned whether they can truly represent nanotechnology. This is because they are not new; they have existed for many centuries. In 1449, Flemish glassmaker John Utynam was granted a patent in England for making stained glass combining nanoparticulate gold; in the early 16th century, Swiss medical doctor and chemist von Hohenheim (Paracelsus) synthesized and used Au

NPs for patients suffering from certain diseases [7]. The chemical synthesis of NPs had been well established by the middle of the 19th century (e.g., Thomas Graham's method for making ferric hydroxide NPs) [8]. In general, NPs are material particles whose size remains in the range of 1–100 nm and the behavior of NPs is between that of a macroscopic solid and that of an atomic or molecular system. However, how an NP is observed and defined depends on the specific application. In this regard, Table 1.1 reviews the definitions of NPs and nanomaterials given by various organizations [9].

Table 1.1 Definitions of nanoparticles and nanomaterials by various organizations

Name of the organization	Nanoparticle	Nanomaterial
ISO	A particle spanning 1–100 nm (diameter)	–
CEN	–	A nano-object with all three external dimensions on the nanoscale
OECD	A particle for which one, two, or three dimensions are confined to the nanoscale	A material that is either a nano-object or nanostructured
ASTM International	An ultrafine particle whose length in two or three places is 1–100 nm	–
SCENIHR	A particle for which at least one dimension is on the order of 100 nm or less	A material with more than one external dimension or internal structure on the nanoscale that could exhibit novel characteristics compared to the same material without nanoscale features
NIOSH	A particle with a diameter between 1 and 100 nm or a fiber spanning a range of 1–100 nm	–

(Continued)

Table 1.1 (*Continued*)

Name of the organization	Nanoparticle	Nanomaterial
SCCP	A particle for which at least one side is in the nanoscale range	A material for which at least one side or internal structure is on the nanoscale
BSI	A particle for which all the fields or diameters are in the nanoscale range	A material for which at least one side or internal structure is on the nanoscale

ISO, International Organization for Standardization; CEN, European Committee for Standardization; OECD, Organisation for Economic Co-operation and Development; SCENIHR, Scientific Committee on Emerging and Newly Identified Health Risks; NIOSH, National Institute of Occupational Safety and Health; SCCP, Scientific Committee on Consumer Products; BSI, British Standards Institution.

This chapter provides basic ideas and physical concepts of NCs that provide an understanding of NCs applicable for nonvolatile memory (NVM). Our point of view is that a general introduction of NCs and a short overview of their synthesis method, fabrication, and characterization technique will supplement greatly the ability of professionals to contribute in the various areas of nanotechnology.

1.1.1 Nanomaterials for a Nonvolatile Memory Device

Particles downsized to the nanometer scale have a tendency to perform differently from the bulk solids of the same material because of the effects of the activities of the atoms or molecules themselves. The difference occurs because of the modification of the bonding state of the atoms or molecules constructing the NC. For example, if a cube with a side length of 1 cm is divided into cubes of 1 μm, the number of particles on the surfaces of the cubes will be 10^{12}, and if these cubes are divided further into cubes of 1 nm, the number of particles on the surfaces of the cubes will now be 10^{21} (as we divide a cube, with each successive division the number of surfaces increases, even though the cubes are getting smaller). The number of atoms or molecules to be found on the surface of an NC plays an important role since they are more active than the atoms or molecules inside the solid NC because surface atoms or molecules form bonds easily with the surrounding materials and cause numerous changes in the NC's properties. There are two distinct categories of nanomaterials: (i) nanostructured materials that refer to condensed bulk materials made of grains in the nanometer size range and (ii) nanophase materials that are usually dispersive NPs. Here, the nanometer size range covers a wide range, from as low as 1 nm to as high as 200 nm. However, to distinguish a nanophase or a nanostructured material from its bulk material, it should have at least one important and unique property different from the bulk material.

In metals and semiconductors, quantum size effects mean that the electronic wave functions of conduction electrons delocalize over the entire particle and, therefore, we assume electrons as "particles in a box." Consequently, the densities of state (DOS) and the energies of the particles depend significantly on the size of the box, which at first indicates the size dependence [10]. The overall behavior of bulk crystalline materials changes when the constituent particle's dimension reduces to the nanoscale. The smaller the dimensions of the nanostructure, the wider the separation between the energy levels, leading to a spectrum of discrete energies. Because of quantum confinement in QDs, quantum wires, quantum wells, and bulk, there are changes in the energy spectrum, as shown in Fig. 1.1.

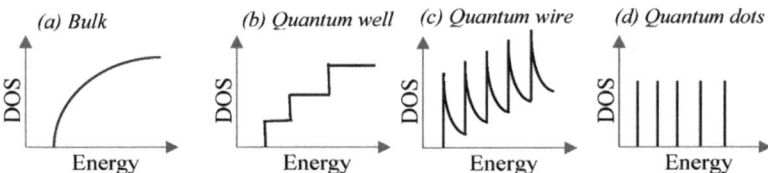

Figure 1.1 Density of states (DOS) (schematically) for (a) a bulk semiconductor, (b) quantum wells, (c) quantum wires, and (d) quantum dots.

1.1.2 Overview of Nonvolatile Memory

In recent years, NVM devices are giving way to high-density memory arrays, which have low costs, low power consumption, high-speed operations, and good reliability [11]. A schematic of the status of research on emerging NVM is shown in Fig. 1.2. One of the emerging NVM devices is an NC memory device that has a multiple-nano-floating-gate (FG) structure. Instead of charge stored in a single FG, charge can be stored in many floating NCs. NC memory can extend the FG scaling by allowing a further decrease in the tunnel oxide thickness. NC memory offers enhanced robustness and fault tolerance of distributed charge storage. Among various NCs, for NVM technology, metallic NCs were extensively investigated over semiconductor NCs because of several benefits, such as enhanced gate control ability (i.e., stronger coupling with the conduction channel), higher density of states (DOS), smaller energy disturbance, and larger work function [12–14].

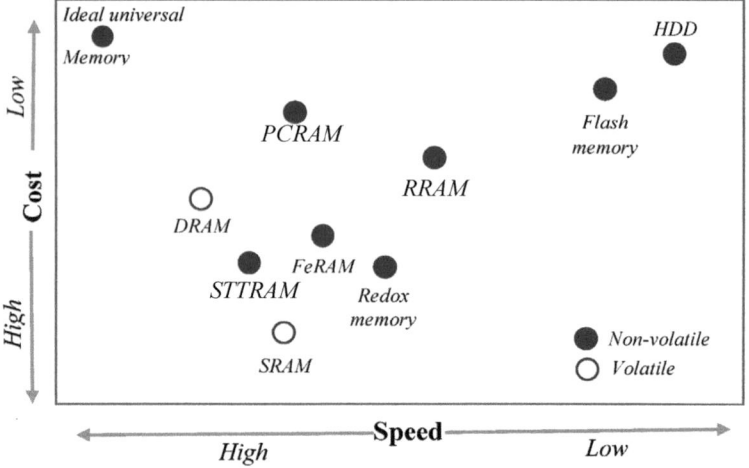

Figure 1.2 Status of research studies on emerging types of nonvolatile memory (HDD, hard disk drive; NVM, nonvolatile memory; FeRAM, ferroelectric random access memory; SRAM, static random access memory; DRAM, dynamic random access memory; RRAM, resistive random access memory; STTRAM, spin-transfer torque random access memory; PCRAM, phase-change random access memory).

According to the prediction from the International Technology Roadmap for Semiconductor (ITRS), silicon metal-oxide-semiconductor field-effect transistors (MOSFETs) are already available in the nanoscale. FG flash memory is comparatively much slower in terms of operation, with poor endurance. For long data retention in a conventional FG memory device, a tunnel oxide layer with a thickness greater than 8 nm is required [15, 16] and this corresponds to a node that is greater than 20 nm. To improve the speed of a FG device, the thickness of the tunnel oxide layer should be reduced to less than the thickness (8 nm ≈ 11 nm) of the layer used in present commercial flash memory devices. A possible solution to this problem is to implement an NC FG, as shown schematically in Fig. 1.3.

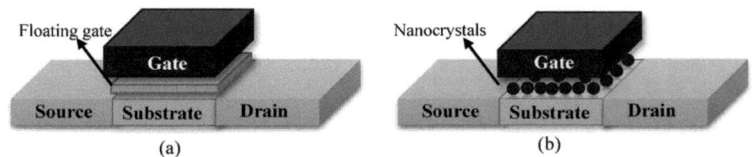

Figure 1.3 (a) Floating-gate NVM structure and (b) nanocrystal NVM structure.

To overcome the problems of the FG memory, NC FGs using discrete NCs as charge-trapping centers have received a lot of attention in recent years. Using the charge-trapping layer structure can lower the operational voltage of the device and further improve the operational speed of the device. Moreover, NCs can develop a thinner tunnel oxide layer without losing nonvolatility of the memory device.

1.1.3 Classification of Nanomaterials

The physical properties of nanoscale materials mainly depend on two crucial nanoscale geometrical parameters: the size of the nanoscale materials and their shape. Early investigations have been focused on the nanoscale size effect; the properties of NCs are influenced by the size of the NCs [17, 18], and the size also influences electron transport properties. Similarly, the shape of NCs plays a crucial role in the determination of their various physical properties [19]. One can classify the shapes of NCs by their dimensionality. Some basic geometrical motifs of NCs (as shown in Fig. 1.4) in different dimensions are as follows: 0D—sphere, hollow sphere, bead, etc.; 1D—rods, wires, tube, ribbon, belt, etc.; 2D—discs, plates, sheet, wall, etc.; and 3D—flowers, cubes, polyhedrons, etc.

In the case of the size restriction in one dimension, we get a 2D structure, the so-called quantum well. In the case of the 2D confinement, the relevant 1D structure would be a quantum wire.

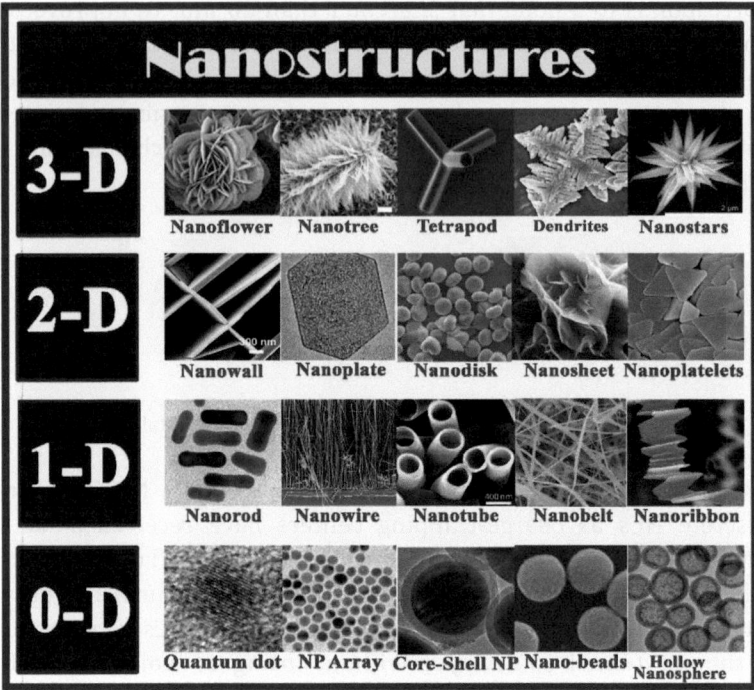

Figure 1.4 Different types of nanostructures. (i) 3D: a nanoflower (Reprinted by permission from Springer Customer Service Center GmbH: Springer Nature, *Nature Nanotechnology*, Ref. [22], 2012), a nanotree [23] (https://creativecommons.org/licenses/by/3.0/), a tetrapod [24] (https://creativecommons.org/licenses/by/3.0/), dendrites (Reprinted with permission from Ref. [25]. Copyright (2008) American Chemical Society), and nanostars [26] (https://creativecommons.org/licenses/by/3.0/); (ii) 2D: a nanowall [27] (https://creativecommons.org/licenses/by/3.0/), a nanoplate (Reprinted from Ref. [28], Copyright (2009), with permission from Elsevier), a nanodisk (Reprinted from Ref. [29], Copyright (2008), with permission from Elsevier), a nanosheet [30] (https://creativecommons.org/licenses/by/3.0/), and nanoplatelets (Reprinted with permission from Ref. [31]. Copyright (2013) American Chemical Society); (iii) 1D: a nanorod and a nanowire (Reprinted by permission from Springer Customer Service Center GmbH: Springer Nature, *Nature*, Ref. [32], 2008), a nanotube (Reprinted with permission from Ref. [33]. Copyright (2010) American Chemical Society), a nanobelt [34] (https://creativecommons.org/licenses/by/3.0/), and a nanoribbon (Reprinted with permission from Ref. [35]. Copyright (2008) American Chemical Society); (iv) 0D: a quantum dot, an NP array, and a core-shell NP [36] (https://creativecommons.org/licenses/by/2.0/); nanobeads [37] (https://creativecommons.org/licenses/by/3.0/); and a hollow nanosphere nanowire (Reprinted by permission from Springer Customer Service Center GmbH: Springer Nature, *Nature Materials,* Ref. [38], 2014).

Finally, if the motion of electrons, holes, and excitons is restricted in all three directions, we come to a quasi-0D system, the so-called "quantum dot." In QDs, the DOS is just a series of delta functions, given that all three dimensions exhibit carrier confinement. With respect to the presence of features at the nanoscale, 3D nanomaterials can contain dispersions of NPs, bundles of nanowires, and nanotubes as well as multinanolayers. In terms of nanocrystalline structure, bulk nanomaterials can be composed of a multiple arrangement of nanosized crystals, most typically in different orientations.

Metallic NCs are potential candidates to satisfy simultaneously fast program/erase speed and long retention time under low-voltage operation. Some major advantages of metal NCs over semiconductor NCs are (i) a higher DOS around the Fermi level, (ii) scalability of the NC size, (iii) a wide range of available work functions, and (iv) smaller energy perturbations due to carrier confinement [20, 21].

1.2 Synthesis and Fabrication of Nanocrystals for NVM

In the last couple of decades, morphology-dependent (like zero, one, two, and three, i.e., hierarchical) nanomaterials have been synthesized through different methods to enable enhancement of fundamental properties and open up the window for the improvement of NVM devices through enormous enhancement [39]. However, several synthetic techniques have been summarized in Fig. 1.5, and some basic things to consider are common to all approaches. In the synthesis of NPs, you need to use a scheme or a process that achieves at least the following conditions among many others: (i) control of particle size, size distribution, shape, crystal structure, and composition distribution; (ii) NPs with the lowest possible amount of impurities or defects; (iii) control of aggregation, agglomeration, and composite formation; (iv) equilibrium of physical properties, structures, and reactants; (v) higher reproducibility and productivity; and (vi) higher mass production, scale-up, and lower costs. For both research and practical purposes, precipitation and chemical reaction are the best methods of forming NCs because it is easier to control the size, shape, and density of the NCs using these methods.

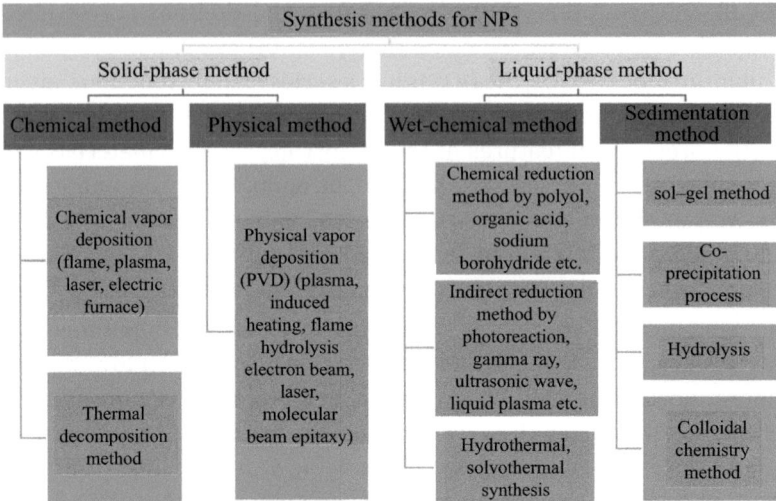

Figure 1.5 Representative synthetic methods for NPs.

Nanomaterials are usually fabricated using either of two known methodologies, the top-down and the bottom-up approach, using the solid phase or liquid phase method [40, 41]. Indeed, colloidal NCs can freely arrange themselves to form larger structures via a self-assembly process to exhibit the unique properties as demonstrated by the constituent building blocks [42]. The crucial goal of NC self-assembly is the fabrication of highly ordered 2D or 3D superlattices or other well-defined structures [43]. Structures created through self-assembly display potential in the areas of photonics, plasmonics, and charge transport; and materials fabricated by self-assembly show promise in magnetic, electronic, photovoltaic, biomedical, sensing, nanoelectronic, and catalytic uses [44–47]. Structures of self-assembled NPs are influenced by many interparticle forces [48]: (i) van der Waals forces; (ii) internal and external magnetic and electrostatic forces; (iii) repulsive steric, confining, or jamming forces; (iv) solvation, structural, and depletion forces; (v) capillary forces; (vi) convective forces; and (vii) friction and lubrication forces. In most cases, NPs do not self-assemble into their thermodynamically lowest energy state, and hence, external energy is required to guide them into particular structures or assemblies. The self-assembly processes to produce desired nanostructured materials, involving interparticle and externally applied forces, are explicitly discussed theoretically and experimentally in a large number of articles published in the past two or three decade [48–50].

1.2.1 0D Nanocrystals

Organic bistable memory devices (OBDs) have attracted attention for memory device application due to the high switching speed, on/off ratio, and fabrication process through a flexible form on a plastic substrate [51–60]. An island-like structure of aluminum NPs (40, 50, and 100 nm) based on organic tris-(8-hydroxyquinoline) aluminum (AlQ$_3$) was fabricated by thermal evaporation at a base pressure of 3.75×10^{-6} torr with a deposition rate of 0.1 nm/s on to indium tin oxide (ITO) electrodes [59]. The thickness of the aluminum layer and the size of the NPs are directly involved in electrical bistability, and this aluminum layer has better charge storage elements than a continuous metal layer when biased to a sufficiently high voltage. Indeed, a MoO$_3$ interlayer with AlQ$_3$ (at a deposition rate of 0.1 nm/s and a vacuum pressure of 1.0×10^{-6} torr) provides an on/off ratio of over 200 at a low reading voltage of 1 V [60]. The use of Au NPs was realized in some OBDs [52, 53, 58, 61–63]. Au NPs are capped with organic self-assembled monolayers (SAMs) through vapor deposition (in a vacuum of 1 or 2×10^{-5} torr) to serve as FGs, which trap carrier charges in the particles [61, 62]. The efficiency with which the charges can be detrapped or kept in the NPs depends on the character and chain length of the SAM-forming molecules as well as the gold particle sizes [61]. Besides, Au NPs (5 nm) are dip-coated with organic cross-linked polyvinyl alcohol (PVA) to form the devices through spin-coating techniques that exhibit clockwise hysteresis in their capacitance–voltage characteristics because PVA consists of a good gate insulator [64]. Onlaor et al. [65] observed that bistability depends on the deposition rate of a copper phythalocyanine (CuPc) layer. The degree of crystallinity and the grain sizes of the CuPc are found to decrease with increasing deposition rates of 0.01, 0.1, 0.5, and 1.0 nm/s by a thermal evaporation technique with a base pressure of 1.5×10^{-6} torr, while the thickness of CuPc films was fixed at 100 nm for an ITO/CuPc/Al device, whereas the dislocation density and strain increased. The device exhibited two distinctive states of conductivity with the maximum on/off current ratio of ~100 at a reading voltage of +3 V. The block copolymer templates in a 0.5% (w/w) solution of reverse micelles of polystyrene-block-poly(vinylpyridine) (PS-b-PVP, MW 80.5 kDa) in *meta*-xylene used to guide the patterning of ZnO exhibit diameters of 38 nm and heights of 14 nm by spin-coating at 3000 rpm [66]. The coated micelles were then subjected to atomic layer deposition (ALD) of ZnO performed using an ALD system at 70°C in pulses as $(C_2H_5)_2Zn$ (17 mtorr, 0.3 s)/N_2 purge (2 mtorr, 2 s)/H_2O (17 mtorr, 0.3 s)/N_2 purge (2 mtorr, 2 s), which offers tunability of the template that guides and controls the geometric attributes of the resulting ZnO nanopatterns. Dip-pen nanolithography (DPN) deposits soft and hard materials in different nanopatterns on various surfaces by using high-resolution atomic force microscopy (AFM) that utilizes both electric force microscopy (EFM) and piezoresponse force microscopy (PFM) [67, 68]. A NiO nanodot (ND) of ~30 nm diameter could converted into two bridging Au nanowires by a DPN technique using nickel carbonate ($Ni_2(CO_3)$ $(OH)_2$) and $[AuCl_4]^-$ complex solutions, where an AFM tip is writing the left nanowire at a speed of ~300 nm/s, DPN time of ~100 ms [68]. Besides, suitably adjusting the dip-pen deposition time can help fabricate nanomaterials like ferroelectric $PbTiO_3$ (PTO) square-shaped NDs of sizes ranging from 37 to 200 nm with a thickness of ~22 nm [69]. In a typical process, low-viscosity-alcohol-modified PTO sol was deposited on a Nb-doped $SrTiO_3$ substrate and dried for a week at room temperature and then annealed under an oxygen atmosphere at 650°C for 1 min. to obtain pervoskite PTO, which shows a maximum storage density of ~0.5 Tbit/in^2. Ferroelectric copolymers such as poly(vinylidene fluoride-ran-trifluoroethylene) (PVDF-TrFE) have been studied because of their low crystallization temperatures, flexibility, and nontoxicity [70]. Indeed, PVDF-TrFE ND-suspended single-walled carbon nanotube (SW-CNT) channels were made by DPN using a procedure in which plasma-enhanced chemical vapor deposition (PECVD) was used for the preparation of SW-CNT, the precursor sol consisting of a randomly copolymerized solution made of 70/30 mol% polyvinylidene fluoride and polytrifluoroethylene was dipped into the SW-CNT, and then it was slowly dried for several days at room temperature and annealed at 140°C for 1 h [70].

The sol–gel process is a wet-chemical synthesis approach for generating NPs by gelation and precipitation techniques. NPs synthesized in aqueous media display a relatively broad size distribution in comparison with NPs synthesized in organic media [71–74]. The highest-quality NPs (i.e., of a uniform size and with high crystallinity) are synthesized using conventional hydrophobic stabilizers that have carboxylic acid or ammonium moieties (e.g., oleic acid, palmitic acid, ammonium fluoride (NH_4F)/ethylene glycol solution, or tetraoctylammonium) in a nonpolar solvent and produce large quantities of a variety of metal (Ag, Au, and Pd) and metal-oxide (Fe_3O_4, MnO_2, and $BaTiO_3$) NPs of a uniform size and high concentrations [71, 74, 75]. Magnetite (Fe_3O_4) NPs, having diameters of 7, 9, 12, and 15 nm, show well-controlled size monodispersity, and *V–I* hysteresis depends on the size of the NP [72]. Spinel-structured NPs (MFe_2O_4, where M = Mn, Co, or Ni) exhibit bipolar switching characteristics, and their size can be controlled by varying the reaction temperature from 200°C to 305°C and reaction time from 30 min. to 2 h in the organic solvents 1,2-hexadecanediol, oleic acid, and oleylamine soluation [72]. Polymer-based nanocomposites exhibit promising applications in flexible memory devices owing to flexibility, solution processability, low cost, better scalability, and 3D-stacking capability [73, 76–78]. Various polymer-based nanocomposites have been utilized as active layers—nanocomposites such as polystyrene/8-hydroxyquinoline/ Au, pentacene/Al, *tris*-(8-hydroxyquinoline) aluminum/Al, polyvinyl pyrrolidone (PVP)/MoS_2, poly(methyl methacrylate) (PMMA)/graphene, PMMA/ZnO, poly(*N*-vinyl carbazole)/Au, and polystyrene/Al_2O_3 and inorganic phases such as Au, TiO_2, ZnO, graphene, graphene oxide (GO), and 6-phenyl-C61 butyric acid methyl ester (PCBM)—owing to the understanding of the thermal stability, chemical resistance, and excellent mechanical strength of memory devices [73, 76–82]. In addition, organic solvents undergo layer-by-layer assembly with amine functionalized or PVP polymer inorganic (PI) nanomaterial composites (including Ag, Au, Pd, Zn, Au-Cu, Fe-Pt, Fe_3O_4, MnO_2, and $BaTiO_3$) while incurring minimal physical and chemical degradation of the inorganic NPs [79–81, 83]. Furthermore, commercial grade or synthesized (ZnO, PbS, Au–Cu, Au–Pt–Ag) nanomaterials are mixed with by ultrasonication or sol–gel techniques a different polymeric solution of PVP, PMMA–ethanol, biphenyltetracarboxylic acid dianhydride p-phenylene diamine (BPDA-PPD), *N*-methyl-2-pyrrolidone (NMP), or polyvinyl alcohol/polyacrylamide co-acrylic acid with glycerol ionic liquid (PVA-PAA-glycerol) at various concentrations to form PI nanomaterial composites [78–82, 84–88]. Ng et al. [89] prepared an organic memory device with embedded Au NPs (FCC structured) by hydrothermal reaction. The size of Au NPs increases with the increase in $HAuCl_4$ concentration in solution, which contributes to the excellent memory effects [89–91]. In the Sn–W–O ternary system, α-$SnWO_4$ is one of the prominent n-type semiconductors with a narrow bandgap of about 1.64 eV. Irregular-shaped α-$SnWO_4$ NCs were synthesized by the hydrothermal method at 200°C (for 24 h) in the presence of cationic surfactant cetyltrimethylammonium bromide (CTAB) in a neutral pH enrironment [92]. Sun et al. [93] obtained $CuWO_4$ NPs with the direct mixture of $Na_2WO_4 \cdot 2H_2O$, $Cu(NO_3)_2 \cdot 6H_2O$, and CTAB, where the mixture was kept at 200°C for 72 h. Indeed, a $Ag/CuWO_4$/fluorine-doped tin oxide (FTO) device was fabricated through spin-coating at 5000 rpm for 10 s and these samples were subsequently dried at 120°C in vacuum overnight. The device exhibits excellent photoelectron double-controlled resistive switching (RS) memory characteristics with an off/on resistance ratio of $\sim 10^3$. Ferroelectric $BiMnO_3$ (BMO) cubic NDs (10 and 20 nm in size) on a Nb-doped $SrTiO_3$ substrate were fabricated by pulsed laser deposition (PLD) at 800°C in the presence of an oxygen partial pressure of 0.2 torr with the nominal deposition rate of 0.005 nm/ pulse and subsequent cooling, after deposition, to room temperature under 300 torr oxygen ambient [94]. The BMO NDs showed bipolar RS as well as a typical piezoresponse characteristic.

1.2.2   1D Nanocrystals

Electrically driven resistance alteration in metal/metal-oxide/ metal junctions, the so-called RS, is a candidate for next-generation universal NVM. Otsuka et al. [95] buried Ni nanorods (NRs) into the nanoholes (diameter 30 nm $\approx$ 70 nm and density 80 G pores/ $inch^2$) of an anodic aluminum oxide (AAO) by pulsed electroplating techniques. The voltage was 3V, the time 5 min., and the pulse width 10 ms in the interval of 990 ms at 53°C. The electrolytes were a mixture of nickel sulfate, nickel chloride, and boric acid. Afterward, the 10 nm NiO layer was attributed treated with 200 W oxygen radio frequency (RF) plasma for 10 min. at room temperature to restrict the formation of too many filaments. Oka et al. [96] fabricated different types of Pt (70-nm-thick)/NiO nanowire (100-nm-thick)/Pt RS junctions, varying the spatial locations of amorphous SiO_2 (50-nm-thick) passivation layer on the junctions by a combined process of electron beam (EB) lithography, RF sputtering, and PLD techniques. A spontaneous chemical vapor transport method at 650°C–850°C for 2 h in a constant gas flow of 120 SCCM (SCCM denotes cubic centimeter per minute at STP) carrier gas of Ar can help grow single-crystalline FeSi nanowires that exhibit strong memory effects [97]. Additionally, the catalyst-assisted vapor transport method has helped synthesize 1D ZnO nanowires that exhibit a high (up to 10^3) on/off ratio at a zero gate voltage [98–102]. Besides, an on/off resistance ratio of over 100 times higher can be achieved by the addition of carbon dots onto a 1D ZnO structure [101–104]. Carbon nanotubes are perfect 1D nanostructures to reveal chemical stability, mechanical robustness, and room-temperature ballistic transport [105–108]. Carbon nanotubes show an extremely high mobility of 9000 cm^2/V.s and are synthesized via chemical vapor deposition (CVD) over metal nanoparticles (MNPs) at a temperature of 900°C, with methane as the feedstock gas, on a conducting substrate like silicon [107–109]. Heterostructured metal-oxide nanowires are potential candidates to incorporate rich functionalities into nanowire-based devices, for example, Au-SiO_2, Au/Ga_2O_3, MgO/$Fe_{3-\delta}O_4$, MgO/TiO_2, Ta/SnO_2, Sb/SnO_2, SnO_2/In_2O_3, CoO_x/MgO, and NiO/MgO [110–123]. Nagashima et al. [110, 111] fabricated 5- and 10-nm-thick cobalt oxide shell layers on a 5 μm MgO core nanowire by using an in situ nanowire template method–like combination of vapor-liquid-solid (VLS) growth and thin film growth [111–119]. Multicomponent VLS oxide nanowires are fabricated using relatively similar vapor pressures of two metal elements [122]. DPN is a nanowriting procedure that employs an AFM tip as a "nanopen" to deposit organic molecules like starburst polyamidoamine dendrimers and polypropylene imine dendrimers [39, 67, 124–127]. Hu et al. fabricated a resistive random access memory (RRAM) device based on the Ag/TiO_2/Ag and graphene/TiO_2/graphene sandwich structure on both SiO_2/Si and flexible polyethylene terephthalate (PET) substrates using photolithography and atmospheric pressure chemical vapor deposition (APCVD), respectively, and the switching behavior was changed from unipolar to bipolar because of the unique physical properties of graphene [128, 129]. N-doped or undoped GO was reduced with dimethylformamide (DMF) or hydrazine using solvothermal or chemical treatment and dried in a vacuum oven at 80°C for 24 h under Ar containing H_2 for 30 s ≈ 90 s and cooled at room temperature [130]. Then, N-doped or undoped GO was spin-coated onto an ITO or Au electrode, which allows the control of the characteristics of polarization-induced rewritable NVM devices.

The sol-gel technique is an attractive synthetic method because of its simplicity and flexibility, which allow for optimization and production of bulk-morphology-controlled nanomaterials. Compared to conventional semiconductor NCs, the core/shell zinc-blende (ZB)/wurtzite (WZ) heterocrystalline structure consists of CdX (X = S, Se, or Te) and can form a new type of superlattice where the bandgaps of WZ Cd compounds are larger than their ZB counterparts' [131–133]. Moreover, a 3D geometric shape makes it impossible for the tetrapod structures to lie flat on the planar substrate surfaces and thus seriously hampers the capacitance coupling between the gating electric field and the tetrapod conduction channels [131]. Ferroelectric insulating films possess a higher dielectric constant for the development of ferroelectric field-effect transistors (FeFETs) with NVM functions [102, 131, 134]. In the typical synthesis of FeFETs, lead zirconate titanate [$Pb(Zr_{0.3},Ti_{0.7})O_3$] (PZT; several microns in length and ~200 nm in diameter) nanowire, lead acetate trihydrate [$Pb(OCOCH_3)_2 \cdot 3H_2O$, 99.5%], zirconium tetra n-propoxide ($Zr(OC_3H_8)_4$, 70%), and titanium (IV) butoxide ($Ti(OC_4H_9)_4$, 98%) were used as a precursor material; methanol/acetic acid was used as a mixed solvent and baked on a hotplate at 70°C in an air atmosphere for 5 min. [135, 136]. Then, PZT was coated on a silicon surface by spin-coating techniques, followed by crystallization of the PZT by conventional thermal annealing in air at 650°C for 15 min. However, lead-free ferroelectric materials have attracted increasing interest due to the need for global environmental protection [137].

A hydrothermal technique efficiently controls the synthesis of morphology-dependent nanomaterials. Compared with other routes, this method has advantages such as simple instrumentation, low cost, relatively low toxicity, abundant natural resources, and easy manipulation.

It will be beneficial to realize large-scale production of nanomaterials [93]. For instance, Senthilkumar et al. [138] used the simple hydrothermal method for the synthesis of an anatase phase TiO_2 NR film on an FTO substrate. The equal-volume ratio of 1.2 ml titanium isopropoxide and 36% hydrochloric acid (HCl) with deionized (DI) water was stirred for 10 min. or till a clear and transparent solution was produced. The cleaned FTO was placed in a glass bottle containing the solution. The bottle was tightly closed with a Teflon cap and kept at a temperature of 165°C for 5 h, after which the substrate was removed and rinsed thoroughly with DI water and finally the film was annealed at 200°C for 1 h in an air atmosphere. The average length and diameter of the TiO_2 NRs were ~650 nm and ~120 nm, respectively, and the oxidation/reduction reaction occurred at the interface of Ti and TiO_2 NRs to exhibit superior endurance (10^5), superior retention (5×10^3 s), and a high on/off ratio (>10^4). Zhang et al. [139] investigated bipolar RS behavior in the single-crystalline rutile phase of TiO_2 NR arrays using an inert Pt top electrode. In this report, authors demonstrated the asymmetric nature of the current hysteresis and bipolar RS behaviors in hydrothermally grown anatase TiO_2 using a reactive Ti top electrode with an improved on/off ratio and endurance. These TiO_2 NR film–based devices have the potential to be used in next-generation NVM applications. Chuang et al. [103] utilized Si-substrate-modified Au metal for ZnO NR synthesis. An 80-nm-thick SiO_2 dielectric layer was first deposited on a Si substrate by electron-beam evaporation. Then, a 2-μm-thick Au metal layer was formed by standard photolithography to serve as a catalyst and metal contact. The patterned substrates were immersed in a hydrothermal solution of zinc nitrate tetrahydrate and hexamethylenetetramine (HMT, $C_6H_{12}N_4$) in an alkaline solution at 90°C for 8 h to grow 0.15 and 2.6 μm ZnO NRs [103, 140]. Zinc nitrate provides Zn^{2+} ions, while HMT provides OH^- ions, and the hydrothermal solution serves as a kinetic buffer for pH of the solution [141]. To obtain NRs with a low aspect ratio, ammonium nitrate is added to the growth solution to maintain the growth solution at a low pH value. According to the work by Richardson and Lange [142, 143], a growth solution with a low pH value produces low-aspect-ratio ZnO NRs while that with a high pH produces high-aspect-ratio ZnO NRs. When ammonium nitrate is added, the presence of NH_4^+ ions reduces the concentration of the OH^- ions, which results in a low pH value in the growth solution. As a result, ZnO NRs with a low aspect ratio are formed. In the absence of

HMT, an aqueous solution containing various concentrations of zinc nitrate hexahydrate and ammonium nitrate heated at 90°C for 2 h directly forms vertically aligned ZnO NRs 2.9 m in length and ~66 nm in diameter [144]. The laterally bridged 1D ZnO-based memory provides an on/off current ratio higher than 1×10^6 and retention up to 1.56×10^6 s [98, 103, 144]. The geometry of a ZnO NR leads to a narrow dispersion of the on/off ratio owing to the formation of straight and extensible conducting filaments along the aligned ZnO NRs. To improve further RS performance, transition metals ions are incorporated onto the 1D NR structure by hydrothermal or solvothermal methods to enhance the memory performance [98, 144–148]. The as-grown ZnO:Mn film has a polycrystalline WZ structure with a weak c axis orientation. Manganese dopants exists in 2+ states and can significantly depress the concentration of intrinsic donors, such as interstitial zincs or oxygen vacancies [147]. To date, only a few reports have been published on the controlled synthesis of ceria (CeO_2) NCs by a hydrothermal or solvothermal process for memristor applications [148–151]. In a typical synthesis, oleic acid and *tert*-butylamine perform the crucial role of forming ceria nanocubes from a cerium(III) nitrate solution in a toluene solvent heated at 200°C for 3 h. It is known that surfactant molecules can play a key role in forming ordered assemblies [149–151]. Oxygen vacancies of ceria nanomaterials with an optimum three-layer dip-coating device (a thickness of ~162 nm) strengthened the switching (off/on) ratio to a value higher than 10^4. The endurance performance was more than 10^2 cycles, the retention time was longer than 10^3 s, and the stability was up to 480 K.

Metal tungstates (MWO_4, where M = Ni, Co, Zn, Fe, Ca, Cu, or Sn) were crystallized with the wolframite type of structure, which is made of hexagonal close-packed oxygen atoms with certain octahedral sites filled by M^{2+} and W^{6+} cations in an ordered way [92, 93, 152–157]. Metal tungstate family compounds of 1D nanostructure have been mostly synthesized in the presence of cationic surfactant CTAB associated with a highly alkaline solution at a higher temperature [92, 152–158]. In a typical hydrothermal synthesis route, iron tungstate ($FeWO_4$) nanowires (6 μm in length and 80 nm in diameter) were formed at 140°C for 72 h in the presence of CTAB [155]. It is obvious that the morphology evolution is a function of hydrothermal treatment conditions (reaction times). In a Ag/FeWO$_4$/Ti device, titanium might act as an oxygen-gathering material to induce oxygen vacancies at the Ti/FeWO$_4$ interface, resulting in a Schottky-like barrier, and titanium plays a similar role as reported in the Ti/ZrO$_2$/Pt device [155]. Additionally, ZnWO$_4$ nanowires ~2 μm in length and 50–70 nm in diameter were synthesized at 180°C for 24 h [154]. In addition, CaWO$_4$ nanostructures, for example, nanospheres, NRs, and nanoplates, were synthesized through an Ostwald ripening process followed by self-assembly at 180°C for 12 h, 48 h, and 96 h, respectively [156].

Molybdenum diselenide (MoSe$_2$), an indirect-bandgap semiconductor with a bandgap of 1.7–1.9 eV, is similar to the MS$_2$ sandwich structure, which is composed of stacked atom layers (Se–M–Se) held together by van der Waals force [159–162]. Yan et al. [160] synthesized 5–10 μm MoSe$_2$ NRs by a hydrothermal method. Sodium molybdate (Na$_2$MoO$_4$·2H$_2$O) and selenium (Se) were kept in the autoclave and the solution was stirred until they completely dissolved. An appropriate amount of hydrazine hydrate (N$_2$H$_4$·H$_2$O) was added to maintain the pH at 12. Then, solid sodium hydroxide (NaOH) was added and the mixture heated in the autoclave at 220°C for 100 h to obtain a black precipitate after washing with absolute ethanol and distilled water. The final product was dried in a vacuum box at 60°C overnight. In addition, 1D single MoSe$_2$ NRs have been fabricated through photolithography techniques and the device demonstrates stable RS memory behaviors with an on/off ratio (memory window) of ~50 at room temperature [159, 160]. Glucose, a soft reducing agent, reduces the silver tellurite (Ag$_2$TeO$_3$) to Ag$_2$Te-Ag-axial-junction nanowires by a solvothermal method at 165°C. Glucose also reduces the aqueous dispersion of Ag$_2$TeO$_3$ to Ag$_2$Te and Ag-incorporated Ag$_2$Te in an alkaline medium with varying pH to obtain Ag$_2$Te nanowires 200–500 nm in diameter and ~10 μm in length, which possess magnetoresistive properties [163]. Bismuth-containing perovskites have attracted much attention for NVM application [164, 165]. In the hydrothermal method, PVP is mixed with Bi(NO$_3$)$_3$·5H$_2$O and CoCl$_2$·6H$_2$O in the highly alkaline sodium hydroxide solution to form BiCoO$_3$ microribbons (width 2 μm) at 120°C for 180 h [165]. BiCoO$_3$, similar to PbTiO$_3$, is isostructural, having a large displacement of the Co^{3+} ions from the center of the octahedron, which supports a pyramidal rather than an octahedral coordination. This coordination leads to the distortion to a high c/a ratio of 1.267 and a calculated saturate ferroelectric polarization of 120 $\mu C/cm^2$. The white-light-controlled resistance switching on 1D composites represents a good rectifying property and bipolar behavior at room temperature and offers potential application for nonvolatile-light-controlled memory applications [154, 166–169]. Indeed, titanium (IV) isopropoxide and $CuCl_2 \cdot 2H_2O$ were dissolved in the aqueous hydrochloric acid solution and transferred to a 50 ml Teflon-lined stainless-steel autoclave. Cleaned, FTO-coated glass substrates (14 Ω per square) were put into the autoclave solution for 3 h at 180°C to form composite NRs. TiO_2/Cu_2O composite NRs (length 3.2 μm and diameter 220 nm) were rinsed with DI water and subsequently annealed at 450°C for 2.5 h in air. The white-light-controlled RS phenomenon results from the light-modulated trapped electrons in the Schottky-like depletion layer, and the resistance in the dark is 240 times more than that under white-light illumination [168].

One-dimensional NCs, particularly NRs and nanowires, can be arranged into a gathering of higher-order assembly structures. One-dimensional NCs are important for photovoltaic and photocatalytic applications and offer significant opportunities for the fabrication of novel electronic, optoelectronic, and sensing devices. The use of NRs and nanowires as building blocks to construct ordered structures definitely has the potential to provide a revolutionary improvement in several areas of technology. Besides, self-assembly of colloidal 1D NCs can take place (i) on substrates by controlling the evaporation of the solvent, the external field, and the use of the template; (ii) at interfaces; and (iii) in solutions by means of chemical bonding, depletion of attraction forces, and linker-mediated interactions [44]. The choice of self-assembly processes has enabled the configuration of the desired properties for many functional device applications.

1.2.3 2D Nanocrystals

Multiferroic $BiFeO_3$ (BFO) has been intensively investigated as an oxide-resistive switch [170–176]. For the PLD process for the formation of a BFO thin film, the nominal laser energy density, the laser repetition rate, the oxygen ambient pressure, and the growth temperature were 2.6 J/cm^2, 10 Hz, 9.7×10^{-3} torr, and 650°C, respectively [177]. In a typical synthesis process, a bulk target was prepared by using a conventional solid-state reaction method [176]. High-purity, commercially available powders of the chemicals, that is Bi_2O_3 and Fe_2O_3, were mixed in particular molar ratios. The mixtures were well ground, pelletized, and fired in an air environment at 400°C for 10 h, at 700°C for 12 h, and finally at 880°C for 5 h, with intermediate grinding. After the PLD process, the BFO thin films were annealed in situ at 390°C with the oxygen ambient pressure of 200 mbar for 60 min. Following the deposition, circular metal (e.g., Au or Pt) top contacts with an area of 0.045 mm^2 and a thickness of 110 nm were prepared by direct current (DC) or RF magnetron sputtering using a metal shadow mask and subsequent annealing was carried out at 400°C for 5 min. in ambient air [176, 177]. Transparent electronics have extensive application in next-generation electronic circuitry and memory structures for fabricating transparent devices [178–182]. A ZnO thin film sandwiched by ITO electrodes reveals unipolar switching characteristics for transparent memristors [183]. In addition, wide-bandgap magnesium-doped ZnO, Gd_2O_3, HfO_x, and (La,Sr)MnO_3 thin films have extensively focused on resistance switching properties of transparent memristors [178–189]. To create memory devices, transparent ZnO, Al_2O_3, Al:ZnO, and HfO_x have been grown by ALD or plasma-enhanced atomic layer deposition (PEALD) and RF sputtering [179–181, 190–194].

A transition metal oxide (e.g., NiO, TiO_2, CuO_x, or MgO) is included as the resistive oxide in the metal/resistive oxide/metal (MRM) sandwich structure, which constitutes the core of the capacitor-like memory element [185, 195–208]. Lee et al. [209] have grown a NiO thin film by ALD on Pt (111)/Ti/SiO_2/Si (001) substrates using Ni(EtCp)$_2$—where EtCp = ethylcyclopentadienyl, $(C_2H_5)C_5H_4$)—as the Ni precursor and oxygen gas (O_2) as the oxygen source at a working temperature of 300°C under a working pressure of 5×10^{-3} torr with an oxygen content of 5%–30%, where Pt top and bottom electrodes were deposited using DC magnetron sputtering. Growing techniques of NiO thin films by ALD at 300°C and EB deposition at 40°C on Si, Ni, Pt, W, and TiN substrates exhibit roughness in 1.2–6.2 nm range and the NiO electron density is 1.35–1.96 e^- $Å^{-3}$ spread around the nominal value of 1.83 e^- $Å^{-3}$ for bulk cubic polycrystalline NiO [207, 208]. $Pr_{0.7}Ca_{0.3}MnO_3$ (PCMO) films were grown using a PLD method on a Pt/Ti/SiO_2/Si (Pt–Si) substrate at a temperature of

650°C and the Pt/PCMO/Pt–Si device showed unipolar RS behavior. Metal nitride films also exhibit stable RS behaviors [210–217]. Kim et al. [210–216] demonstrated the feasibility of silicon nitride (Si_3N_4, deposited via PECVD) dielectrics as an RS material and utilized Cu, Si, Ti, Au, and Ag as the top and bottom electrodes to obtain high performance during operation. Besides, conductive properties of polymers have the ability to transport positive charges (holes) [218–220]. Poly(1,3,5-trimethyl-1,3,5-trivinyl cyclotrisiloxane) (pV3D3)-based RRAM arrays are fabricated via the solvent-free technique called the initiated chemical vapor deposition (iCVD) process for flexible memory application [221]. Spinel oxide Co_3O_4, $NiFe_2O_4$, $CoFe_2O_4$, and $ZnFe_2O_4$ and ferromagnet $La_{0.7}Sr_{0.3}MnO_3$ [222] thin films are prepared in a chemical solution techniques in the presence of 2-methoxyethanol, assisted by spin-coating at around 3000 rpm for 30 s and then baked at 300°C for 5 min. and annealed at 600°C for 1 h in an air ambient atmosphere; these are promising materials for NVM application because of their excellent RS characteristics [223–227]. Sun et al. [228] fabricated a metal oxide (ZnO or NiO) and ZnO/NiO diode–based charge-trapping layer of ~20 nm thickness by e-beam evaporation at room temperature on a Si substrate. This layer was annealed at 500°C to obtain crystallized ZnO and NiO to form a n- and p-type oxide. Indeed, they found that the ZnO/NiO charge-trapping layer demonstrates a performance superior to that of other charge-trapping layers such as ZrON, ZrO_2, multipleTa_2O_5, Tb_2O_3, and $SrTiO_3$ owing to the comparable memory window achieved for the lower operation time like 1 ms. RS behavior has been observed in a variety of materials, including binary, ternary, and complex oxides as prepared by RF sputtering [147, 199, 203, 229–245]. In a typical synthesis, ZrO_2 films on $Pt/Ti/SiO_2/Si$ substrates were formed at around 200°C by a RF magnetron sputter by employing a ceramic ZrO_2 target to obtain a metal/ZrO_2/metal device [199, 231]. The working pressure to sputter ZrO_2 was 10 mtorr, which was maintained by a gas mixture of oxygen and argon at a mixing ratio of 1:2 with a total flow rate 18 sccm that operated over 10,000 RS cycles by a sweeping DC voltage, and the on/off memory states showed good stability at a stress voltage of under 0.3 V to retain the memory state of data for over 10^5 s [199].

Considering the transition layer with excess Zr, high-voltage stress causes ionization of metallic Zr and generates positive Zr ions. The positive charges in ZrO cause band bending, which in turn enhances current flow through the transition layer at the critical bias (or set voltage, over 1.6 V) to enable the ZrO film to the low-resistance state. If we apply bias above the critical value, we expect electron accumulation at the transition layer because of the limited current flow at the top oxide layer. It might cause recombination of electrons with positive Zr ions, which in turn reduces the electric field and the current flow. Indeed, NiO_x was deposited at a working pressure of 5 mtorr and a substrate temperature of 300°C by reactive RF magnetron sputtering in an oxygen environment of ~20% of oxygen in the O_2/Ar mixture in the flowing chamber [237]. The oxygen partial pressure during sputtering and the post-thermal process are crucial to form Pt/Ni–O/Pt, Pt/TaO_x/Pt, Pt/Co–O/Pt, Pt/TaO_x/TaON/Pt, Ag/ZnS-Ag/$CuAlO_2$/Pt, and Cu/TaO_x/TiN trilayers [184, 230, 239, 246–252]. A 30-nm-thick NiO film was deposited on a flexible copper substrate by the RF method using a Ni target in the Ar and O_2 mixture (Ar:O_2 = 2:3) at room temperature, and the top electrode of a Ti layer about 100-nm-thick was obtained by using e-beam evaporation with a shadow mask 0.22 nm in diameter [253]. A (111)-oriented BFO thin film was deposited on the $SrRuO_3$ (SRO)/$SrTiO_3$(STO) (111) substrate using the RF deposition method at a base pressure of 3.0×10^{-6} torr and a deposition pressure of 10 mtorr with Ar and O_2 at a ratio of 4:1 at a substrate temperature of 650°C [240]. Circular Au top electrodes of 0.10 mm diameter were sputtered on the thin film surfaces by using a shadow mask.

Rare earth (RE) metal oxides are attractive for NVM devices because of their dielectric constants, larger energy bandgaps, and predicted chemical and thermal stability [241, 254–257]. Indeed, an Yb_2O_3 thin film of 20-nm-thickness was deposited on a TiN/Si substrate by reactive magnetron RF sputtering of an Yb_2O_3 target in Ar (30 sccm) ambient at room temperature [254]. The RF sputtering power and pressure of the sputter system were set to 150 W and 4 mtorr. Kao et al. [241] compared Ti-doped Gd_2O_3 with that of a Gd_2O_3 and Ti-doped Gd_2O_3 trapping layers in metal-oxide high-κ-oxide silicon (MOHOS) flash memory structure. A Gd_2O_3 film was deposited by the RF sputtering process with 10 mtorr at room temperature with precursors of O_2 (3 sccm) and Ar (21 sccm) at 150 W, followed by rapid thermal annealing (RTA) in O_2 ambient for 30 s two different temperatures of 800°C and 900°C. The Gd_2TiO_5

trapping layer annealed at 900°C had a higher window of 3.8 V in the capacitance–voltage (C–V) hysteresis loop and demonstrated good endurance with superior reliability. A trapping layer mixture of RE metal oxide and Ti can result in an increase in the dielectric constant and barrier height from the trapping layer to the tunneling oxide [241–245, 255, 256]. An improvement in RS characteristics of CeO_2-based devices has been reported by charge transfer through Al metal as a dopant and fabricated by RF sputtering at around 100 W, 10 mtorr, and room temperature [257]. The Ti/CeO_2:Al/Pt sandwich structure exhibits significantly better switching characteristics, including lower forming voltage, improved and stable SET/RESET voltages, enhanced endurance of more than 10^4 repetitive switching cycles, and large memory window ($R_{OFF}/R_{ON} > 10^2$) as compared to an undoped Ti/CeO_x/Pt device owing to the effect on both electronic charge transfer from valence to conduction bands of Al and Al_2O_3 and the formation stability of oxygen vacancies in the conductive filament.

Metal, transition metal oxides (e.g., Ag, VO_2, TiO_2, NiO, GaZnO, Nb_2O_5, ZrO, and $SrTiO_x$), and perovskites (e.g., $La_{1-x}Ca_xMnO_3$ and $SrZrO_3$) have been widely used as RS materials for NVM devices owing to consisting thermal stability and operating durability in air [74, 258–265]. In a typical sol-gel reaction to synthesize a transition metal oxide like TiO_2 or Nb_2O_5, first the precursor solutions after a hydrolysis and condensation reaction with hydrochloric acid were thermally annealed at above 400°C for 2 h in a nitrogen atmosphere and finally annealed at the same temperature for 4.5 h in air. The initial thermal annealing process in nitrogen can induce oxygen vacancies accompanied by an increase in local electron concentration. Layers created through this process can be used as efficient active layers in RS memory devices [259, 266, 267]. Anatase TiO_2 films have been obtained at 450°C, and residual carboxylate or carbonate groups are known to prevent crystallization of the Nb_2O_5 precursor below 580°C [259, 268–271]. Furthermore, a sol-gel-derived TiO_2, Nb_2O_5, or InGaZnO film with an amorphous structure could display the RS memory property [259, 260, 272, 273]. Besides, perovskites such as $La_{1-x}Ca_xMnO_3$ and $CaCu_3Ti_4O_{12}$ and doped perovskite oxides such as Mo-doped $SrZrO_3$ were prepared by solvents like ethylene glycol, acetic acid, and acetylacetone in a sol–gel solution process [261, 262, 274, 275]. Ferroelectric thin films such as $PbZrTiO_3$ (PZT), $BaTiO_3$

(BTO), barium calcium titanate (BCT), barium zirconate titanate (BZT), barium strontium titanate (BST), $Ba(Ti_{0.89}Sn_{0.11})O_3$ (BTS), and some piezoelectric ceramics have been studied extensively for applications in nonvolatile ferroelectric random access memory (FeRAM), capacitor equipment, piezoelectric device, infrared (IR) sensors, etc., and were typically synthesized in the presence of glacial acetic acid and ethylene glycol solution [137, 261, 262, 274]. In addition, a conjugated rod-coil diblock copolymer was fabricated onto ITO glass using a spin-coating technique at a speed of 1500 rpm for 60 s and exhibited an on/off ratio up to 1×10^7 [276].

1.2.4   3D Nanocrystals

NP-modified self-assembled block copolymers as charge-trapping elements could be realized by utilizing the saturation of programmed/erased states [276–282]. A hierarchical supercluster of flowerlike Au NPs (~11.7 nm) with block copolymer PS-b-PVP in *meta*-xylene solutions of 0.5% (w/w) concentration was self-assembled on a Si surface by spin-coating at 6000 rpm, where the polymer template was then removed using O_2 plasma reactive ion etch for 10 min. at 30 W, 65 mtorr, and 20 sccm O_2 [283]. Kim et al. [284] fabricated a 3D structure of nitride-based (Si_3N_4) RRAM and observed that the high- and low-resistance state can be effectively modulated by the film thickness and compliance current, respectively. In a typical synthesis, at first, $SiO_2/Si/SiO_2/Si$ layers were sequentially deposited on the Si substrate by ion implantation at an acceleration energy of 40 KeV and a dose of 1×10^{15} cm^{-2} at 1000°C. Next, patterning and dry etching were performed to make room for resistive material (RM) and vertical electrode (VE) deposition. Subsequently, Si_3N_4 layers of different thicknesses (7–15 nm) were deposited as a RM by the PECVD method at 280°C using 5% SiH_4/N_2 (800 sccm), NH_3 (10 sccm), and N_2 (1200 sccm) as the precursors in the operating pressure of 580 mtorr at 60 W. Finally, a nickel top electrode of 100 μm was deposited by a thermal evaporator. Sun et al. [285] synthesized the multiferroic $BiCoO_3$ nanoflowers by the hydrothermal process. The typical current-voltage (I-V) characteristics of the $Ag/BiCoO_3/Ag$ structures exhibit extreme change in resistance with an on/off ratio of $\sim 10^5$.

1.3   Characterization of Nanoparticles

In this section, we will restrict ourselves to an overall more detailed sketch of the common characterization techniques employed for the investigation of nanostructures. Commonly, there exist some essential primary probes, for example, electrons, X-rays, ions, atoms, light (including visible [Vis], ultraviolet [UV], and IR), neutrons, and sound, which may be used to excite secondary effects such as electrons, X-rays, ions, light, neutrons, sound, and heat from the excited sample. The preferred secondary effects may be monitored as a function of one or more variables, for example, energy, temperature, mass, intensity, time, angle, and phase. At present, various techniques are available for detecting, measuring, and characterizing NPs precisely. However, a particular method cannot be the "best" method; rather a suitable method is considered to be the best one taking into consideration several real-world limitations, for example, the type of sample, the information required, and time and the cost of the analysis. A straightforward technique may detect the presence of NPs, but other specific techniques may give the quantity, crystal structure, shape, size distribution, or surface area of the NPs. Some specific measurement techniques can assess the chemical contents of NPs, the reactions on the surface of the NPs, the interactions with other chemical species, etc. Sometimes, to get more information in a single measurement, more than one measurement techniques can be combined.

Moreover, different measurement techniques will be suitable for diverse types of samples and the amount of sample required can also differ according to the choice of technique. In some experimental techniques, a suitable mathematical model is required to obtain results. Finally, an important restriction must be kept in mind that the choice of technique from among the available techniques could vary extensively according to the costs of calibration and maintenance, the instrument's accuracy, the time required for measurement, etc. Measurement techniques have been continuously developing with pinpoint accuracy in recent times, and upgraded by scientific and industrial research. The correctness of the different methods is not always determined, but one can compare the results of the same sample by different techniques to identify any errors as instruments will vary in their accuracy, sensitivity, and application ability depending on the manufacturer. In this chapter, we are only interested in the size distribution, structure, and chemical compositions of NPs for their usage in NC NVM, and several common techniques are described next.

1.3.1 Microscopy Technique

Microscopic imaging techniques are the most developed characterization tools to create surface images of a specimen. When NPs have all three external dimensions in the nanoscale, the optical microscopy technique (theoretical spatial resolution limited to $\sim\lambda/2$) is not able to form an image of such NPs. To overcome the limitation, electron microscopy techniques are commonly used to characterize the NPs because of their intrinsic subnanometer scale resolution. However, some optical techniques, such as near-field scanning optical microscopy (NSOM), can be used for the imaging of NPs because of a typical resolution between 50 and 100 nm, reaching 20 nm and 2–5 nm of lateral and vertical spatial resolution, respectively [286, 287].

1.3.1.1 Electron microscopy

The electron microscope uses a beam of electrons to create an image of the specimen, utilizing the same principles as an optical microscope but rather than photons exploiting focused electrons. This technique offers much higher magnifications and greater resolving power than a light microscope, allowing it to observe much smaller objects in finer detail. All electron microscopes use electromagnetic (EM) and/or electrostatic lenses to control the path of electrons. Electron microscopy is further divided into several subcategories according to the imaging technique used, as shown in Fig. 1.6.

1.3.1.1.1 *Transmission electron microscopy*

Transmission electron microscopy (TEM) is used to check the cross section of NCs embedded between two layers of an oxide, which is the critical region in the NC memory device. Plan-view and cross-sectional TEM were explored to characterize the morphology, distribution, and crystal structure of NPs and also NPs in the device structure [83]. TEM is a microscopy technique where a beam of electrons transmits through an ultrathin specimen and an image forms because of the interaction of the electrons transmitted through the specimen; the image is magnified and focused onto an imaging device. It can be used to determine a lot of information about the NC, for example, its size, its shape and the distance between adjacent NCs. Alternate modes of use allow TEM to observe several properties, for example, compositional information, crystal orientation, and electronic structure. The resolution of a modern TEM is about 0.1 nm, which is the typical separation between two atoms in a solid. The ability to determine the positions of atoms inside materials has been made possible by high-resolution TEM (HRTEM), which is an essential tool for nanotechnology research and development in many fields. Kim et al. [288] used field-emission TEM (FETEM) to study V_3Si NPs, which are ~4 nm in size and have a spherical shape. Yang et al. [289] used HRTEM to determine the thickness of the tunnel oxide and the top control oxide along with the NPs in SiO_2/Co NPs/SiO_2 (3 nm/4.5 nm/20 nm) structure on a Si substrate. Banerjee et al. [290] investigated IrO_x NDs 1.3 nm in diameter with a high density of 1×10^{13}/cm^2 by HRTEM, as shown in Fig. 1.7. A TEM image of Ru NCs formed by a hybrid CVD/ALD reaction reveals the diameter distribution to be 1–4 nm [291]. Amouroux et al. [292] used energy-filtered TEM (EFTEM) along with HRTEM to capture the location of crystalline Si dots within the amorphous SiO_2 matrix.

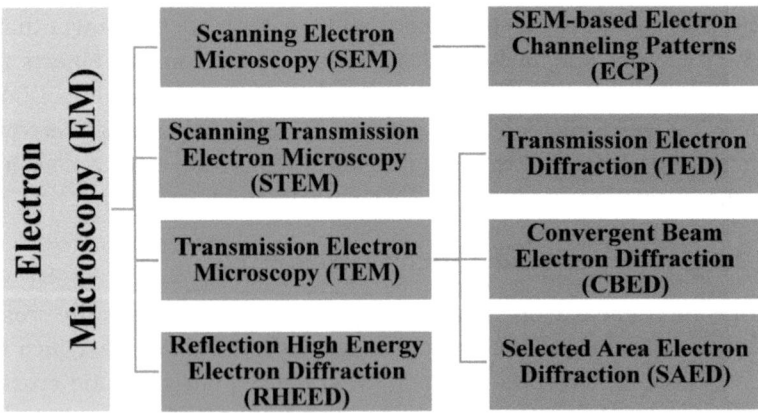

Figure 1.6   Different forms of the electron microscopy technique.

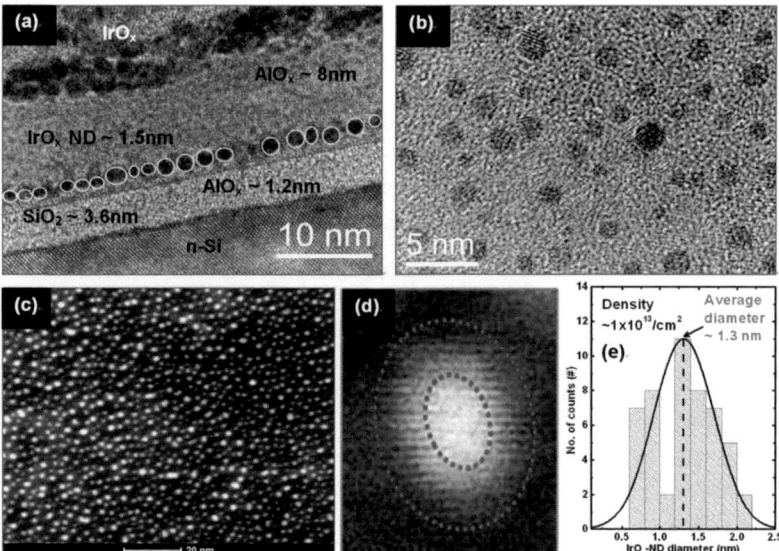

Figure 1.7 (a) Cross-sectional HRTEM of IrO_x NDs in an $IrO_x/Al_2O_3/IrO_x$ NDs/$Al_2O_3/SiO_2/n$-Si metal-insulator-semiconductor (MIS) structure, (b) plan-view TEM image, (c) STM image, (d) a single core-shell IrO_x ND, and (e) a histogram of IrO_x NDs with an average size of 1.3 nm and a density of $10^{13}/cm^2$ [290] (https://creativecommons.org/licenses/by/2.0/).

1.3.1.1.2 *Scanning electron microscopy*

Scanning electron microscopy (SEM) uses a focused beam of high-energy electrons to produce a range of signals at the surface of solid specimens. SEM can measure the morphology, density, and average size of NCs. Unlike TEM, where electrons in the primary beam transmitted through the sample are measured, SEM produces images by detecting secondary electrons that are emitted from the surface because of excitation by the primary EB. The high-resolution, 3D images formed by SEM provide topographical, morphological, and compositional information [120, 293–297] (as shown in Fig. 1.8), making them important in a diverse field of science and industry applications, such as life science, biology, gemology, medical and forensic science, semiconductor inspection, production line of miniscule products and assembly of microchips for computers, and metallurgy [298, 299].

Figure 1.8 (a, upper panel) Tilted SEM images of a TiW/SiO$_x$/TiW (MIM) device; (a, lower panel) magnified cross-sectional SEM images of the device structure. (b) Cross-sectional FESEM image of a thick GO in an Al/GO/Al device where the charging layer shows the formation of Al$_2$O$_3$ in the top Al electrode and GO interface. (c) Cross-sectional SEM images of electrochemically deposited Ni nanodots. (d) TEM micrographs of a phase change memory 1T1R test cell. (e) TEM image of the cross section of a ferrocene molecular flash memory device. (f) SEM image of a dendrite bridging two electrodes. (a) Reproduced from Ref. [293] with permission from The Royal Society of Chemistry. (b) Reprinted from Ref. [294] (under a Creative Commons CC-BY license). (c) Reprinted from Ref. [295] (under a Creative Commons CC-BY license). (d) Reprinted from Ref. [297], copyright (2016), with permission from Elsevier. (e) Reprinted with permission from Ref. [120]. Copyright (2015) American Chemical Society. (f) Reproduced from Ref. [296] with permission from The Royal Society of Chemistry.

Ren et al. [300] determined temperature-dependent Si NCs' nucleation and growth behavior taking SEM images, which indicates a clear trend of decreasing dot density and size, segregation, and NC density variation. Zou and Michael [301] calculated the thickness of Cu$_x$O ($\sim$2.3 nm) by using a cross-sectional view of a SEM image of a Cu/Cu$_x$O/Ag metal-oxide-metal NVM device fabricated on a Kapton substrate. Ren et al. [302] also fabricated a NiSi NC memory device with a single triangular-shaped Si NW channel on a silicon on insulator (SOI) substrate, and they characterized the memory device with SEM (tilted image), TEM, and X-ray photoelectron spectroscopy (XPS). Amouroux et al. [292] used a technique called critical dimension SEM (CDSEM) to determine the size distribution of Si

NCs in order to characterize "in line" the Si NCs (i.e., a size of 13 nm, a density of 5×10^{11} cm^{-2}, and a coverage of 55%) instantaneously after the growth process. They also used other "off-line" imaging techniques, like topographic AFM and TEM.

1.3.1.1.3 *Scanning transmission electron microscopy*

Secondary (backscattered) electrons are used for imaging in scanning transmission electron microscopy (STEM) as in SEM. A scanning transmission electron microscope is a conventional transmission electron microscope equipped with additional scanning coils, detectors, and circuitry to produce images, as does a scanning electron microscope, with enhanced spatial resolution (by a TEM setup). This technique can have multiple detectors working concurrently to collect different but complementary pieces of information and has use in elemental analysis of samples. Modern aberration-corrected scanning transmission electron microscopes have a depth of field of only a few nanometers, and so it becomes possible to reconstruct the set of images into a 3D illustration of the specimen's structure [303]. A scanning transmission electron microscope is capable of microanalysis by two analysis methods: energy dispersive X-ray (EDX) analysis and electron energy loss spectroscopy (EELS). STEM has the ability to detect single atoms in the high-angle annular dark-field STEM (HAADF-STEM) mode, where images are constructed using large-angle scattered electrons [304–306], as shown in Fig. 1.9a–b. Parreira et al. [307] studied amorphous ZrO_2 resistive memory devices sandwiched between Pt and Ti electrodes and mapped the Zr, O, and Ti profiles by using EELS data, whereas they used EDX data to map the Pt in the memory device. Schamm et al. [308] described the method to visualize all Si NPs within SiO_2 film; they performed reliable size and density measurements by using EELS along with TEM and STEM imaging. Siles et al. [306] analyzed the amorphous or nanocrystalline nature of TiO_x in the memory device using STEM-EELS spectra and HAADF-STEM. A typical STEM-EELS spectrum is shown in Fig. 1.9e–g, where it maps for the second- and third-order resonant modes after switching.

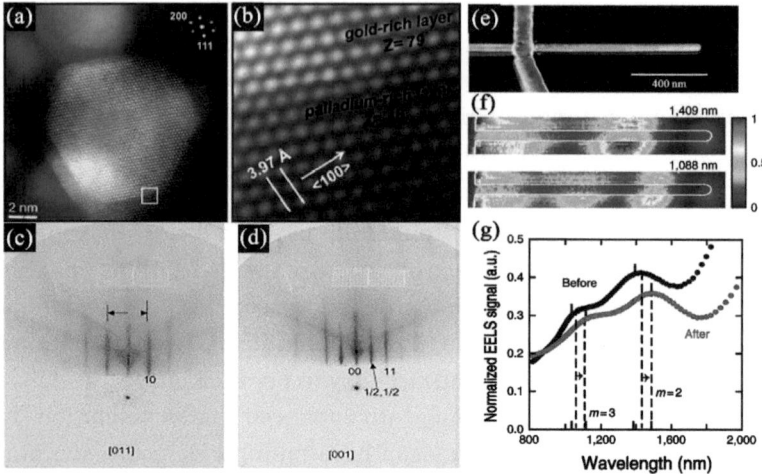

Figure 1.9 (a) Atomic-resolution HAADF-STEM image of a cuboctahedral Au/Pd NP; (b) enlarged image of a small part of the exterior layer of an NP; (c, d) RHEED patterns for a Au(100)-(1×1) electrode covered with one multilayer Pd at [011] and [001] azimuths; (e) STEM image of a memristive optical antenna after breakdown (Ag nanowire shown in longitudinal direction with a Au electrode in the vertical); (f) STEM-EELS maps for the second- and third-order resonant modes after switching; (g) STEM-EELS spectra showing a red shift in resonant frequencies before and after breakdown. (a, b) Reproduced from Ref. [305] with permission from The Royal Society of Chemistry. (c, d) Reprinted from Ref. [309], Copyright (2006), with permission from Elsevier. (e–g) Reprinted from Ref. [310] (under a Creative Commons CC-BY license).

1.3.1.2   Reflection high-energy electron diffraction

Reflection high-energy electron diffraction (RHEED) measures the angular intensity distribution of electrons "reflected" from a crystalline surface under bombardment by high-energy electrons at a grazing incidence of 1–5 degrees. Typical RHEED patterns are shown in Fig. 1.9c–d, which depict that the Au(100)-(1×1) electrode is covered with one multilayer Pd at [011] and [001] azimuths [309]. This analytical tool is superior for characterizing thin films during growth by molecular beam epitaxy (MBE) since it is very sensitive to surface structure and morphology [311]. RHEED can determine several properties of materials; it can be used for qualitative assessment of the structural perfection of the surface, determination of the reciprocal lattice of the surface from the diffraction pattern geometry, determination of the atomic structure of the surface, and control of the by-layer growth of epitaxial films with atomic precision [312]. With the RHEED oscillation measurement technique, one can measure the epitaxial growth of a film up to a less-than-one-monolayer precision [313, 314]. Schilling et al. [315] studied especially the texture formation and epitaxy of the specimens of the FePt NPs (7.6 nm) by RHEED. They applied in situ RHEED even in between annealing steps of the FePt NPs and averaged the structural information over a much larger area.

1.3.1.3   Scanning probe microscopy

Scanning probe microscopy (SPM) is commonly used for imaging and characterizing a surface structure down to a few nanometers with atomic structure details. A scanning probe microscope cannot replace an electron microscope, but this tool "feels" the surface and constructs an image to represent it. SPM can target and manipulate just one atom or molecule as well as perform high-resolution imaging of surfaces under ambient conditions [316, 317]. It represents the development of the concept of a profilometer. In principle, one can apply various fundamental interactions between the probe (atoms of the tip) and the sample atoms for practical operation of the SPM. Depending on the particular choice of interactions, there are three main types of SPM [317]: scanning tunneling microscopy (STM), atomic force microscopy (AFM), and NSOM.

1.3.1.3.1   *Scanning tunneling microscopy*

STM senses the surface topography via tunneling the current flowing between tip and sample as they are a very short distance apart. STM relies on the electrical conductivity of the sample, so at least some electrical features on the sample surface must be present to some extent. STM is an imaging technique and it also modifies the sample surface. Recently, the field of STM lithography has become more significant. Moreover, STM can be used to obtain atomic-scale structural (shape of the NP crystal) and spectroscopic information on surfaces and the real-time control of the optical-field distribution at a molecular scale [318].

There are mainly two STM techniques—the constant-current method and the constant-height method. By scanning the tip over the surface and simultaneously monitoring the tunneling current at a given tip height and bias, an "atomic" image of the surface is obtained. On the other hand, the scanning tunneling spectroscopy (STS) technique measures the tunneling current as a function of the bias, keeping the tip position fixed at a particular position, which means the local density of electronic states of the conductor can be measured as a function of the energy [316, 317]. The STS technique creates the image of small molecules or NCs attached to the atomically flat substrate and their energy band structure. In a recent review, Swart et al. elaborately described how one can use the STM and the STS to obtain information about the confined electronic orbitals and related energy levels of individual semiconductor QDs [316]. Potapenko et al. [319] reported the study of nanocrystalline phases of TiO_2 by STM, and they described elaborately the influence of the reactive layer thickness on the areal and size distributions of the NPs. Tunneling spectroscopy analyzes the energy levels (DOS) of a QD, which can be altered by quantum coupling. Hence, tunneling spectra may reveal direct information on the local strength of quantum coupling in an array of NCs [316] as studied for CdSe QDs [320]. Osváth et al. [321] explained the Au-NP-modulated local density of electronic states in graphene by STM and STS measurements (typical STM images are shown in Fig. 1.10c–e).

1.3.1.3.2 *Atomic force microscopy*

AFM provides high-resolution characterization of local topographic, mechanical, and chemical composition, mapping of thermal properties of the NC, and resolutions well beyond the optical diffraction limit, below 50 nm, by combining the strength of AFM and IR spectroscopy, a new probe-based measurement technique called AFM-IR [317]. AFM is a suitable technique more commonly applicable to conductive and insulating materials, including inorganic and synthetic materials, polymers and polymer matrix, and biological structures. There are numerous variations of these techniques. AFM may operate in several modes that differ according to the force between the tip and the surface: contact mode, noncontact mode, intermittent contact mode (tapping mode), lateral force mode, magnetic force, and thermal scanning [322]. Figure 1.10a–b shows a typical tapping-mode AFM image of graphene transferred onto Au NPs [321]. Conductive AFM (CAFM) provides topographic and electrical information of the sample simultaneously and has been used to investigate the properties of gate oxides since the 1990s [323–325]. CAFM can provide the precise size distribution of the NPs. Using this technique, one can realize the localized nature of RS based on nanometric conductive filaments with diameters of ~1 nm [325]. A combination of AFM and SEM images can determine the NP size as well as the concentration of NPs on the surface. By using these methods, Yang et al. [289] observed the average diameter (6–8 nm), the center-to-center (8–10 nm) distance, and the surface density (10^{12}/cm^2) of Co NPs prepared by the laser irradiation method. Using a similar measurement technique, Kanoun et al. [326] observed Ge NCs' diameter (8.5 nm) and density (6 × 10^{11}/cm^2) grown by CVD. Prakash et al. [327] used AFM to determine the roughness and surface morphology of IrO$_x$/high-κ_x/W structures containing AlO$_x$, GdO$_x$, HfO$_x$, and TaO$_x$ switching materials.

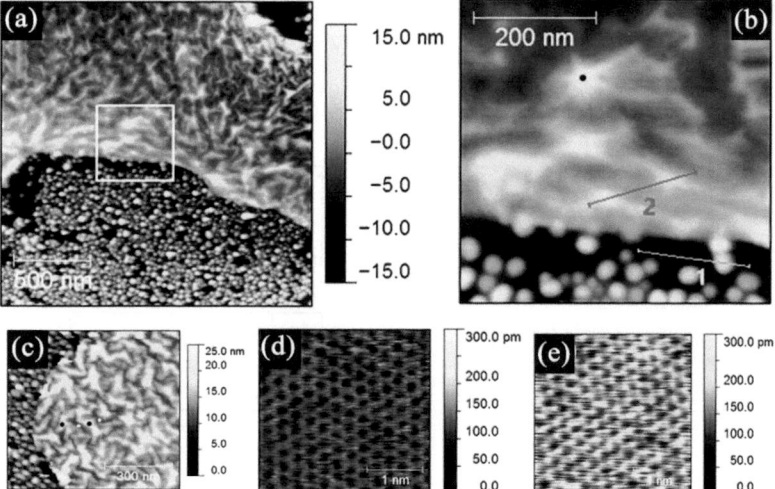

Figure 1.10 (a) Tapping-mode AFM image of graphene transferred onto Au nanoparticles. (b) Higher-magnification image of the area marked by a white square in (a). (c) STM image of graphene/Au NPs where the left part represents uncovered Au NPs. The black-and-white dots in (c) represent NP-supported and NP-suspended graphene regions, respectively. Atomic-resolution STM images of (d) NP-supported and (e) NP-suspended graphene. Reproduced from Ref. [321] with permission from The Royal Society of Chemistry.

1.3.1.3.3 *Near-field scanning optical microscopy*

The main principle of NSOM is the EM interaction of two very closely positioned nano-objects, which represent a probe and a sample. NSOM is a local probe technique with a resolving power of 10–50 nm, not being limited by diffraction, which unwraps new viewpoints for optical characterization, particularly in biology, microelectronics, and materials science [328, 329]. NSOM scans a very small light source very close to the sample and is an example of an extended AFM as it records EM information (light) in addition to the topography. Detection of this light energy forms an image of the sample and it offers a resolution below the diffraction limit.

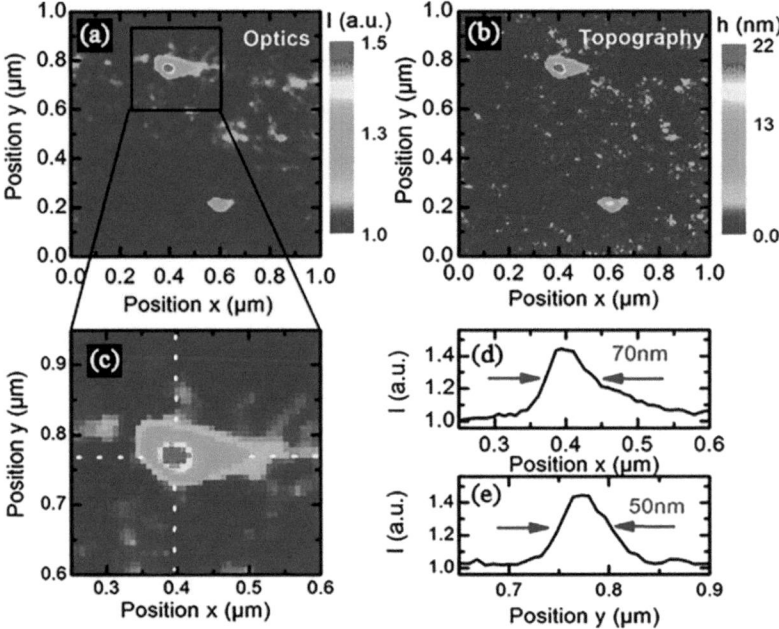

Figure 1.11 (a) 2D optical s-NSOM image of two Au NPs with radius <30 nm, (b) the corresponding shear-force topographical image, (c) zoomed-in image indicated by the square in (a), and (d, e) cross sections of the optical intensity along the *x* and *y* directions, marked by dashed lines in (c). Reprinted from Ref. [330]. Copyright (2011) American Chemical Society.

Oshikane et al. [286] reached a lateral spatial resolution of 20 nm and a vertical spatial resolution of 5–2 nm with a noise level of photomultiplier tube (PMT) by the NSOM system. In aperture NSOM, a small aperture is used for local illumination of a sample in which the imaging resolution depends on the size of the aperture and the skin depth of the metal coating [287, 331]. On the other hand, apertureless NSOM uses a sharp metallic tip or a metal-coated tip. The scattered light is collected in the far field, and high spatial resolution is obtained due to the local field enhancement [287, 332]. The detection sensitivity and spatial resolution of apertureless NSOM can be enhanced by means of localized plasmonic resonance, that is, coupling free-space light to the propagating surface plasmon polaritons (SPPs) [333]. So, tip-based NSOM can be used to study the local field distribution on a variety of samples and provides important nanoscale information inaccessible by other optical microscopy methods [287]. Sadiq et al. [330] explored the scattering-type NSOM (s-NSOM) technique for individual Au NPs on a glass substrate and demonstrated sub-30 nm resolution imaging of localized SPP fields of spherical and elliptical NPs (a typical image shown in Fig. 1.11).

1.3.2 X-Ray-Based Methods

X-ray-based methods, such as X-ray diffraction (XRD), X-ray scattering (XRS), X-ray absorption spectroscopy (XAS), and XPS, tend to be greatly surface specific and can offer significant information on surface properties and coatings, crystallographic structure, or elemental composition. In some cases, X-ray spectroscopy estimates the elemental composition and quantitative analysis of NPs/NCs along with SEM and TEM. X-ray-based characterization methods can be simply explained as follows [334]: XRD—getting to know the arrangement of atoms; XRS—learning about particle shape and morphology; XAS—exploring the chemical composition and local structure; and XPS—learning about the chemical nature and elemental composition of the surface.

1.3.2.1 X-ray diffraction

XRD patterns have been widely used in nanotechnology research as a basic characterization tool for obtaining primary physical characteristics, such as the crystal structure, the crystallite size, the microstrain of a lattice, and the dislocation structure. A typical diffraction pattern of CuO, Cu_2O, and Cu_2O/CuO phases is shown in Fig. 1.14a [335]. The polycrystals in nanocrystalline materials are the source of the broadening of diffraction peaks. Moreover, an inhomogeneous lattice strain and structural faults produce the broadening of diffraction peaks. There are several techniques to analyze the XRD line profiles, such as Scherrer [336], Williamson–Hall [337], and Warren–Averbach [338] methods. The naivest and most extensively used method for estimating the average crystallite size is the full width at half maximum (FWHM) of a diffraction peak using the Scherrer equation [336]. For a single-crystalline NP, the crystallite size and the particle size are the same but they are not the same for polycrystalline materials. Therefore, the particle size estimated from either XRD or TEM for single-crystalline materials is more or less same. When NPs are outlined by well-defined boundaries or they are dispersed NPs, particle size estimations from XRD and TEM values are in good agreement. Liu et al. [83] observe no change in FWHM of the XRD peaks, which leads them to conclude that there is no obvious coalescence or sintering of FePt NCs during the annealing process. Hall et al. [339] have studied Au NPs 4 nm in size by TEM and XRD. They have suggested a Fourier analysis method as an alternative to the widely used Scherrer formula to determine the size from XRD data. Rana et al. [340] determined the formation of a TiO interfacial layer in between CeO_2 and Ti top electrode by XRD peaks related to the monoclinic structure of TiO.

1.3.2.2 Small-angle X-ray scattering

In small-angle scattering, experiments are intended to measure scattered intensity at very small scattering vectors, with 2θ ranging from a few microradians to tens of radians, in order to explore systems with characteristic sizes ranging from crystallite sizes (a few angstroms) to colloidal sizes (up to a few microns). Small-angle X-ray scattering (SAXS) measurement can give many characteristic parameters of the sample, including molecular weight, excluded particle volume, maximum particle dimension, and the radius of gyration [341]. Shortage of best contrast or overlap particle boundaries often complicates the analysis of TEM images. In contrast, SAXS and XRD are indirect methods but provide more reliable information from the statistical point of view [342]. SAXS in combination with X-ray crystallographic data can be very powerful for the analysis of large varieties of multicomponent systems. A combination of characterization methods is essential to reveal sufficient information, including mean size, particle volume fraction, number density, and the type of distribution (Gaussian, Lifshitz–Slyozov–Wagner [LSW], or log-normal), to understand the NP formation process [343].

Individually, UV/Vis spectroscopy, SAXS, and TEM do not provide adequate information to follow in situ process of nucleation and growth of NPs, for example, Au NPs. UV/Vis spectroscopy provides insufficient information on the number of particles, TEM introduces many artifacts, and SAXS does not offer a unique solution for the shape of the NPs' distribution. Koerner et al. [344] used a combination of UV/Vis spectroscopy, TEM, and SAXS (in situ) to reveal a change in the mechanism of the growth process. They provided new insights into the impact of stabilization additives on NP formation, the impact of thermodynamic factors on evolution of the size distribution, molecular details of stabilizers, and chemical interactions at the NP surface [344]. Rieker et al. [345] studied and compared the particle size distribution of monodisperse Si spheres (90–400 nm) by using SAXS and TEM. Yu et al. [346] observed an ordering transition for very small octadecanethiol-capped Au NCs of 1.7 nm (when heated to moderate temperature) by using grazing incidence small-angle XRS (GISAXS) and SAXS measurements. These measurements reveal the average diameter and the reasonably polydisperse size distribution of Au NCs in toluene. The degree of ordering in NC and block copolymer composite films can be estimated by using GISAXS [347]. Wang et al. [348] developed a technique—combination of both SAXS and wide-angle X-ray scattering (WAXS)—that allows in situ exploration of the atomic structure and nanoscale superstructure (composed of 5 nm Fe_3O_4 and 10 nm Au NPs and other cubic and hexagonal NPs) phase relations under pressure. Slyusarenko et al. [349] calculated the radii of both Au rods and spheres by SAXS data and confirmed the results by TEM and optical absorption spectra (as shown in Fig. 1.12).

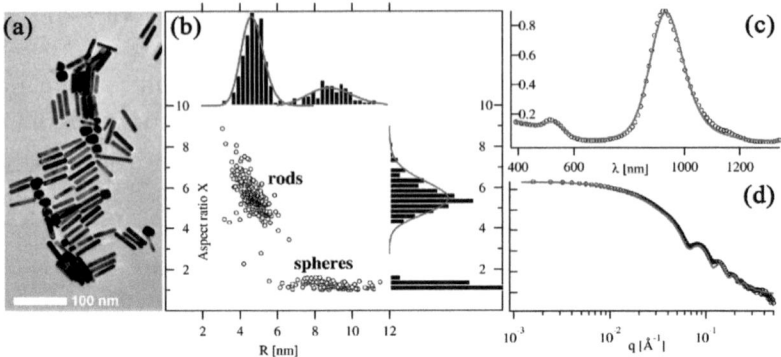

Figure 1.12 Three techniques at a glance for a Au NR. (a) TEM image; (b) TEM analysis: aspect ratio versus radius for all analyzed particles, with histograms of both parameters; (c) UV-vis absorption data; (d) SAXS data. Reproduced from Ref. [349] with permission from The Royal Society of Chemistry.

1.3.2.3 X-ray photoelectron spectroscopy

XPS is used to check the NCs' chemical nature, the chemical or electronic state of each element on the surface, and the elemental composition of the area that exists within a short distance below the material surface [350]. XPS spectra are obtained by irradiating a material with a beam of X-rays while simultaneously measuring the kinetic energy and number of electrons that escape from the top; 1 to 10 nm of the material is analyzed, which requires ultrahigh vacuum conditions. Many XPS databases [351, 352] give easy access to the binding energies of many photoelectron and Auger-electron spectral lines of many elements and their composite compounds, so one can easily identify unknown measured lines by matching to previous measurements. To measure XPS spectra for metal nanoparticles (MNPs), one needs a proper sample preparation method as MNPs easily spoil and can alter properties over time or with exposure to air, light, or any other environment, chemical or otherwise.

XPS can determine many useful properties of NPs: (i) surface and bulk composition (and sometimes functional groups) of NPs, for example, V_3Si [353], SeO_2/TiO_2 [354], and Pd-Cu alloy [355], (ii) layer thickness and structure, for example, of multiwalled carbon nanotubes [356], SiO_2/Si [357], and Ge and GeO_2, (iii) particle coatings, for example, of silane- and *N*-isopropylacrylamide-coated

Fe_2O_3 [358], citrate-stabilized Au NPs (10–80 nm) [359], and Al_xO_y@ SiO_2 [360], (iv) particle size (sometimes when it cannot be obtain by other methods), for example, of Pd NPs [361], and (v) surface segregation or enrichment, for example, of Cu-Cr alloy [362] and Mn and Sb aggregation [363]. Kim et al. [353] have studied the charge loss mechanism of a NVM device with V_3Si NPs embedded in a SiO_2 dielectric layer, and they used XPS measurement to confirm the existence of V_3Si phases. Miller et al. [355] investigated the composition of a Pd-Cu alloy by two physical probes, XPS and low-energy ion scattering spectroscopy (LEISS) (near-surface and top layer of a Pd-Cu alloy annealed to temperatures in the 400–1000 K range). Mohai and Bertóti [356] developed an algorithm based on XPS intensity data and described how to determine the quantitative information of the modification of N_2-RF-plasma-modified multiwalled carbon nanotubes; the algorithm has proved to provide reasonable quantitative data of the modification. Preisler et al. [364] studied the oxidation and reduction of species at the interface of silicon and cerium oxide using XPS, for example, the reduction of CeO_2 to Ce_2O_3 in the presence of Si. Tabet et al. [365] used high-resolution XPS to analyze thin oxide layers grown on germanium substrates and observe the presence of a high density of electron states situated at the oxide/germanium interface, which leads to ~0.6 eV band bending in XPS spectra. Jablonski and Zemek [357] analyzed the system SiO_2/Si and suggested that the uncertainty of the XPS approach in determining layer thickness will be 0.4 nm in the range of thicknesses between 2.5 and 7.8 nm, while the uncertainty will be 0.25 nm for thicknesses below 2.5 nm. The positions, integrated areas, and FWHM of XPS spectra offer insight into the geometrical structure of the Al_xO_y monolayer, for example, larger FWHM of the protonated site compared with the neutral one means a larger distribution in bond lengths and angles and an extensive range of H-bonding environments at the protonated site [360]. The thickness of the NP shell can be determined from energy-dependent spectra using different photon energies of the synchrotron source and calculating the integrated Si 2p/Al 2p ratios [360]. Wojcieszak et al. [361] applied different XPS models to calculate the average particle size of Pd NPs for a Pd/TiO_2(H_2O) sample prepared by a microemulsion method and were able to determine a size as small as 1 nm by using the Davis model [366] and

0.9 nm by using the Kerkhof–Moulijn [367] model. Wojcieszak et al. [361] have concluded that the Davis model is a more appropriate model to determine the size of all particles, small as well as large, as compared with the Kerkhof–Moulijn model.

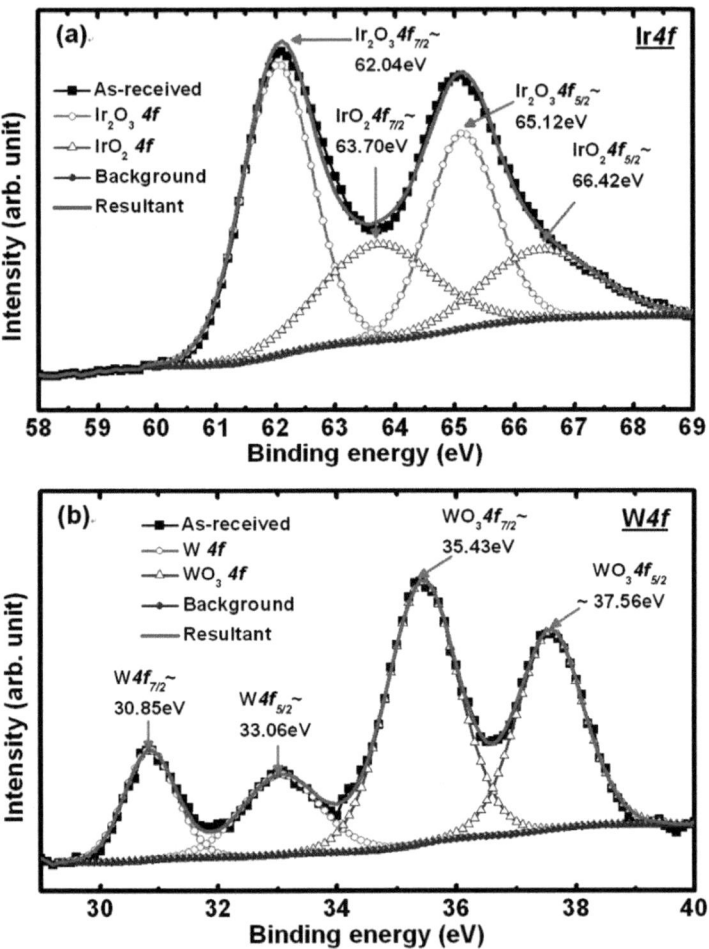

Figure 1.13 XPS spectra of the Ir_4f core levels with Ir_2O_3 and IrO_2 compositions. (a) XPS of the Ir_4f electrons from IrO_x NDs and (b) W_4f core-level electrons in the WO_x layer. Peak fitted using Shirley background subtraction and Gaussian–Lorentzian functions [290] (https://creativecommons.org/licenses/by/2.0/).

The variation of the Ce^{3+} concentration and the thermal stability of ceria NPs can be monitored by analysis of the XPS peak's line shape of the Ce 3d spectrum, as shown in Ref. [368]. Suzuki et al. [362] used angle-resolved X-ray photoelectron spectroscopy (AR-XPS) to study the surface segregation of the Cr-Cu alloy, measuring the spectra at different take-off angles, that is, the angle between analyzer and detector. Caruso et al. [363] also studied the Mn segregates at the surface of the Ni-Mn-Sb alloy and Sb segregates at the surface of the Ti-Co-Sb alloy using AR-XPS measurements. Banerjee et al. [290] investigated the formation of polarity-dependent RRAM devices using $IrO_x/Al_2O_3/IrO_x$ NDs/Al_2O_3/WO_x/W structure and they confirmed the IrO_x NDs, Al_2O_3, and WO_x layers by XPS analysis, as shown by Fig. 1.13.

1.3.2.4 X-ray absorption spectroscopy

X-ray absorption spectroscopy (XAS), comprising an X-ray absorption near-edge spectroscopy (XANES) region and an extended X-ray absorption fine structure (EXAFS) region, is a widely used well-established analytical technique for determining the local geometric and/or electronic structure of matter, extensively used for the characterization of semiconductors in solid/liquid, crystalline/amorphous, and bulk/nanoscale form [369, 370]. The structural information delivered by XAS contains average interatomic distances and the number and chemical identities of nearest neighbors within 5–6 Å of a selected atom species. XRD cannot clearly distinguish between two closely related crystal phases (e.g., Fe_3O_4 and γ-Fe_2O_3) with almost the same "d" value, and it requires a more powerful local structural probe, for example, XAS, to detect and eventually distinguish between crystallographically similar phases [371]. XAS is also used to study NPs' formation mechanisms [372, 373], chemical transformation of bimetallic NPs to ternary metallic NPs by a cation redox reaction [372], and the local electronic structures of NPs and NCs and their composite structures, as shown in Fig. 1.14b [335].

Hwu et al. [374] used XAS to identify the electronic structure and crystal structure of nanocrystalline TiO_2 prepared by gas condensation. Near-edge X-ray absorption fine structure (NEXAFS) and X-ray-excited optical luminescence (XEOL) spectroscopy are the characterization techniques to obtain information on the electronic and optical properties of luminescent materials, such as Si NCs [375]. NEXAFS and XEOL are frequently used in the field of Si nanomaterials to characterize the luminescent species in porous Si, Si nanowires, Si NCs, and Si heterostructures [375]. Chang et al. [376] studied the quantum confinement effect in nanocrystalline diamond (size range 3.5–5 μm) using XAS spectra. Hamad et al. [377] investigated size-dependent structural disorder by XANES in InAs and CdSe NCs (size 1.7–8 nm). Polte et al. [378–380] reported temporal regimes (or phases) of metallic NP growth conducted with in situ monitoring by XANES and SAXS, which allowed real-time determination of NP mean size and number density (by SAXS) and mean metal oxidation state (by XANES) [381]. Polte et al. [380] developed a method to analyze the formation of NPs (i.e., coalescence of nuclei and further monomer attachment) via in situ SAXS and XANES using synchrotron radiation to overcome difficulties arising during time-resolved measurements.

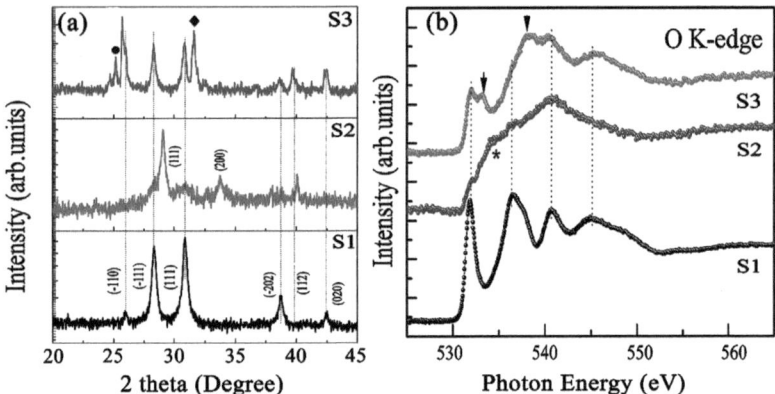

Figure 1.14 (a) XRD patterns of samples S1 (CuO NPs; only CuO peaks are indexed); S2 (Cu$_2$O/CuO NCs; only Cu$_2$O peaks are indexed); and S3 (CuO/TiO$_2$ NCs), where ● represents (001)TiO$_2$/(112)Ti$_2$O$_3$ and ◆ represents (004)TiO$_2$/(11-1)Ti$_2$O$_3$. (b) O K-edge XANES spectra of samples S1 (CuO NPs), S2 (Cu$_2$O/CuO NCs), and S3 (CuO/TiO$_2$ NCs). Reproduced from Ref. [335] with permission from The Royal Society of Chemistry.

1.3.3 Light-Based Spectroscopic Techniques

In a spectroscopic study, the interactions of EM radiation with matter are thoroughly investigated to obtain information about a sample. Matter emits EM radiation stimulated by the application of energy in the form of heat, electrical energy, light, particles, or a chemical reaction. Spectroscopic techniques provide information about elemental ᵢ composition, structural information, and concentration (e.g., UV-vis absorption, luminescence, and X-ray emission). Spectroscopic techniques such as UV/Vis spectroscopy are used to quantify the light that is absorbed and scattered by a sample. Spectroscopy provides information about NP size, aggregation, structure, stabilization, surface chemistry, etc. Dynamic light scattering (DLS), also called photon correlation spectroscopy or quasi-elastic light scattering, fluctuations in the scattered light that explicitly depend on particle diffusion, is frequently used to determine particle size and size distribution.

1.3.3.1 Light scattering techniques

In this technique, scattered light is used to measure diffusion rates (Brownian motion) of particles in stable suspensions to determine a size on the basis of Stokes–Einstein equation [382, 383]. The equation is based on the assumption that particles should be hard spheres, which may not represent many NPs. However, one can *assume* NPs to be hard spheres. Hence, one measures the hydrodynamic diameters of NPs and NP aggregates. The light source could be laser light, X-rays, or neutrons as per the requirement to probe the different size ranges and particle compositions. Additionally, light scattering techniques are a nondestructive, sensitive, and powerful analytical tool to characterize macromolecules and QD/NP colloids in solution [384]. The main limitation of this technique is that even a tiny fraction of micrometer-size particles can destroy the signal from the NPs because this technique measures the hydrodynamic diameter rather than the actual size of the NPs [385]. Using this technique, one can detect the size of monodisperse particles as low as 0.1 nm and as high as 10 μm provided that one knows the viscosity of the dispersed liquid. DLS data can determine the NP's size distribution and stability in solution over a time interval to check the agglomeration and/or precipitation of the NPs [386]. Nishioka et al. used DLS and UV absorption data to characterize uniform size distribution of Ag NPs, and steady production with almost equivalent-quality yield of Ag NPs [386]. Zimbone et al. estimated an average radius of 32 nm for the spherical Au NPs in an as-prepared solution by depolarized DLS (DDLS) data and determined the concentration of Au NPs in the solution from the correlation between DDLS and UV/Vis spectroscopy [387].

1.3.3.2 Ultraviolet/visible spectroscopy

Optical absorption spectra of NPs are highly sensitive to size, shape, concentration, agglomeration state, and refractive index of the adjacent area of the NP surface [388–390], which makes UV/Vis/IR spectroscopy a convenient tool for identifying, characterizing, and studying these materials. Plasmonic NPs (gold, silver, etc.) strongly interact with specific wavelengths of light [390], as shown by a typical absorption spectra of Au NRs in Fig. 1.12c. For example, the information acquired on the bandgap energy (E_g) is very useful to estimate the dispersion and local structure of NPs formed by transition metal oxides, sulfides, and selenides [391]. Many methods have been proposed to estimate E_g of NPs by using optical absorption spectroscopy. Davis and Mott [392] investigated the behavior of the optical absorption coefficient near the absorption edge and derived a power law that relates the absorption coefficient with the photon energy. The absorption spectra of semiconductor QDs provide information about E_g (indirectly the size of QDs) and the purity and homogeneity of the QDs. UV/Vis/IR spectroscopy is a primary tool to characterize semiconductor QDs.

Like semiconductors, absorption and emission in metals and MNPs is influenced by the size, shape, and surrounding medium of the particle. The principal feature in the visible absorption spectrum of MNPs is the surface plasma resonance (SPR). For MNPs, the deviations from spherical particle geometry affect SPR as well [389]. Spheroidal, rod, nanowire, and gold nanoshells have two plasma resonances, one for transverse and one for longitudinal vibrational mode, as shown in Fig. 1.12c. In general, the transverse plasma resonance (TPR) maintains its peak position; on the other hand, the longitudinal plasma resonance (LPR) is red-shifted (e.g., more than 100 nm for gold NRs). The position of LPR depends on the aspect ratio of the rod structure (i.e., the ratio between the length and diameter of the rod) [377, 389]. Semiconducting NPs with gold nanoshells show a similar behavior, where the LPR red shift depends on the thickness of the shell and the diameter of the dielectric core [393]. Electronic relaxation in NRs is very comparable to that in spherical NPs [389]. The relaxation dynamics of surface plasmons' (SPs') electronic oscillations in spherical Au and Ag NPs and NRs can provide a lot of valuable information, like the polydispersity of that system, estimation of the aspect ratio of the rod, and the shape of the NPs [394].

1.3.3.3 Photoluminescence spectroscopy

Photoluminescence (PL) spectroscopy is a contactless, nondestructive spectroscopic method of probing photoexcited species (bound exciton, free carriers, localized exciton on defect/doping state, etc.) and their lifetime. PL spectra reveal several important characteristics of nanomaterials: (i) bandgap determination [395, 396], (ii) impurity level and defect detection [397], (iii) recombination mechanisms [398], and (iv) surface structure and excited states [396]. Time-resolved spectroscopic methods—transient absorption (TA) and time-resolved photoluminescence (TR-PL) spectroscopy—can be employed to study the dynamical processes in semiconductor nanostructures over a large set of time scales and wavelengths [399]. These measurement techniques explore exciton-exciton and carrier-carrier interactions. Using the PL technique, very low concentrations of optical centers can be detected, but it can't give us quantitative information. The PL spectra of a semiconductor can be analyzed to determine the electronic bandgap [395]. PL spectra provide information to quantify the elemental composition of compound semiconductor. The amount of PL emitted from a certain material is directly proportional to the relative amount of radiative and nonradiative recombination process. Among these process, the nonradiative rates usually depend on impurities. On the other hand, the amount of PL and its dependence on the level of photoexcitation and temperature are directly associated with the radiative recombination process. Thus, PL can monitor the fluctuations in impurity levels and defect states of a material as a function of growth and processing conditions and help understand the underlying physics of the recombination mechanism. However, PL is very sensitive to surface effects or adsorbed species of semiconductor NPs, and thus PL is a probe of electron-hole surface processes like other detection techniques, such as XRD, IR spectroscopy, and Raman spectroscopy.

Bruhn et al. [400] studied the carrier multiplication dynamics in silicon NCs in an oxide matrix, employing PL as a characterizing technique. Park et al. [401] assessed the separate effects of plasmonic absorption and emission enhancement by a systematic analysis of the pump fluence dependence of low-temperature PL resulting from individual CdSe/CdS deposited on Ag films. Harbold and Wise [402] investigated the low-energy electronic states of PbSe QDs using PL and have explained the possible origins of the broad emission lines observed despite size-selective excitation and the splitting of the emission spectrum. Cassette et al. [399] used a PL-based time-resolved spectroscopic technique to study the dynamics of excitons in colloidal semiconductor nanostructures below 100 fs; energy and charge transfer; carrier multiplication; vibrational and electronic coherences; coupling, coherence, and ultrafast dynamics of excitons and charge carriers, etc. Klimešová et al. [403] investigated Li doping in Si NCs and explained the lattice expansion due to Li doping using the PL spectroscopic method. Boer et al. [404] studied the spectral shift and enhanced quantum efficiency in phonon-free PL from Si NCs, as shown in Fig. 1.15.

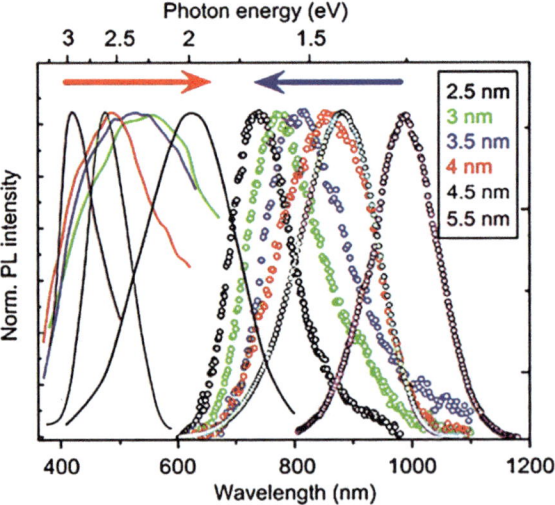

Figure 1.15 Hot PL and excitonic PL bands for all Si NC samples with average diameters of 2.5, 3, 3.5, 4, 4.5, and 5.5 nm. PL spectra show a blue shift on the long-wavelength side and a red shift on the short-wavelength side with a decrease in nanocrystal size. Reprinted by permission from Macmillan Publishers Ltd: *Nature Nanotechnology* (Ref. [404]) copyright (2010).

1.3.3.4   Raman spectroscopy

Raman spectroscopy, in combination with other techniques, can be used to get information (composition, defect chemistry, phonon interactions, etc.) about a nanostructure [405]. Raman spectroscopy is possibly the most effective tool for acquiring information on vibrational and electronic structures [405] and is based on the inelastic scattering of visible light by matter.

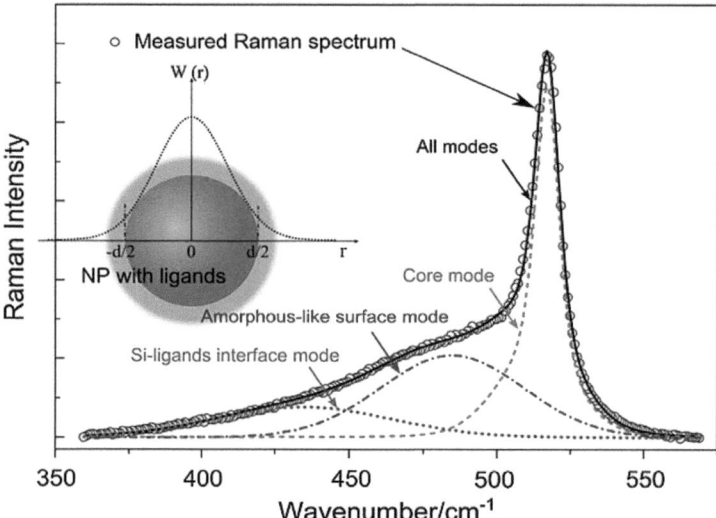

Figure 1.16   Scheme of a Si NP and its typical measured Raman spectrum. $W(r)$ is the Gaussian confinement function used in the phonon confinement model (PCM); the measured Raman spectrum of 3.8 nm Si NPs can be accurately fitted with three modes (dotted lines). From Ref. [406]. Copyright © 2015. Reproduced with permission of John Wiley & Sons, Inc.

In the Raman spectra, both Stokes and anti-Stokes radiations closely depend on molecular vibrations of the nanostructure under investigation. Therefore, information obtained by measuring the Raman shift provides direct information at the molecular level of the nature of the chemical bonds and symmetry of the nanostructure. The red shift of the phonon peak as well as a broadening of its line width can be used to confirm the formation of NCs and to estimate their sizes in various matrices by using an appropriate model [405, 407]. Gao et al. [407] developed a quantitatively accurate model of the size dependence of the Raman spectra of NCs, and they compared the models for diamond, Si, CdS, CdSe, InP, and InAs NCs. Zhang et al. [406] studied size-dependent evolution of phonon confinement in Si NPs via Raman spectroscopy, as shown in Fig. 1.16. Zhang et al. [408] observed the formation of graphene oxide in the C:SiO$_x$ layer by analyzing Raman and Fourier transform infrared (FTIR) spectra (shown in Fig 1.17).

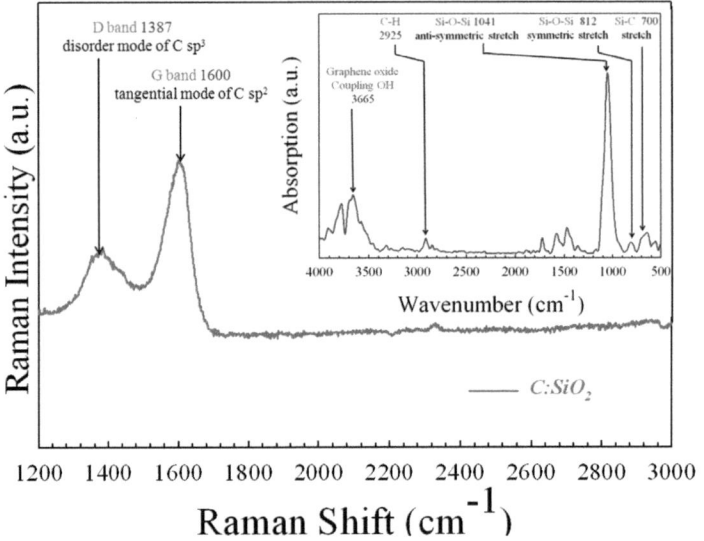

Figure 1.17 Raman spectra of a C sp^2 and C sp^3 in a C:SiO$_x$ film, which confirm graphene oxide (GO) formation. The inset shows the corresponding FTIR spectra in which −OH coupling with GO is clearly observed at the wavenumber 3665 cm^{-1} [408] (https://creativecommons.org/licenses/by/2.0/).

1.3.3.5 Fourier transform infrared spectroscopy

FTIR is a spectroscopic technique to get an IR spectrum of absorption or emission of a solid, liquid, or gas and is a very useful tool for surface characterization of NP surfaces. Under certain conditions, FTIR spectroscopy identifies the chemical composition of the NP surface. Typical FTIR spectra are shown in the inset of Fig. 1.17, and the spectra confirm the formation of graphene oxide [408].

A FTIR spectrometer simultaneously collects high-spectral-resolution data over a wide spectral range as an interferometer is involved in the measurement system. The attenuated total reflectance Fourier transform infrared (ATR-FTIR) spectroscopic technique is able to demonstrate in situ characterization of the liquid/solid interface to probe surface adsorption on NP surfaces [409]. Khomenkova et al. [410] fabricated high-dielectric-constant (high-κ) stacked dielectric structures based on the combination of Hf-based SiN_x materials, and their properties versus deposition conditions were studied by ATR-FTIR and HRTEM techniques. Lehninger et al. [411] have discovered that the phase separation process occurs during annealing of Ge-ZrO_2 composite films by using spectroscopic ellipsometry, Raman scattering, and FTIR spectroscopy. Analyzing FTIR and PL spectra, Van Duy et al. [412] have shown that the richest Si material incorporates numerous nonbridging oxygen hole centers (defect sources) and a rich Si phase in the base material along with a different bonding character of Si.

1.4 Summary

In this chapter, we reviewed the status of current research on NPs, which have use in NVM, and focused on nanocrystalline materials—their fabrication and characterization methods—to reconnoiter what is pertinent to the state-of-the-art NC memory. Nowadays, various kinds of NPs have attracted much attention as an alternative for NVM devices because of scalability, fast write/erase speeds, low operating voltages, and long retention times, especially metal, semiconductor, and oxides NCs. Metal NCs are advantageous over others as they have higher densities of state (DOS), an extensive range of work functions, and lesser energy perturbation. One can realize the enhancement of performance in memory through the integration of metal NCs with high-κ dielectric materials. We have presented most commonly used methods of fabricating nanocrystalline materials with different dimensional structures: chemical, hydrothermal, self-assembly, deposition techniques, etc. We suggest the chemical method as the superior one because it helps form uniform and easily controllable discrete NC structures. By improving the fabrication methods of low-dimension nanocrystalline materials, we are able to find their applications in semiconductor electronics, sensors, special polymers, magnetics, advanced ceramics, etc. To develop NC-based

NVM with high efficiency, it is essential to study how to control the density of NCs and we must improve our current practice of size control and fabrication methodologies for different types of NPs and optimize their characterization techniques.

References

1. Feynman, R.P. (1959). There's Plenty of Room at the Bottom. http://www.zyvex.com/nanotech/feynman.html.

2. The article "Nanotechnology Timeline" under the section "Nanotechnology 101" by United States National Nanotechnology Initiative. http://www.nano.gov/timeline, http://www.nano.gov/timeline.

3. Rossetti, R., Ellison, J.L., Gibson, J.M., and Brus, L.E. (1984). *J. Chem. Phys.*, **80**, 4464.

4. Talapin, D.V., Lee, J.-S., Kovalenko, M.V., and Shevchenko, E.V. (2010). *Chem. Rev.*, **110**, 389.

5. Grim, J.Q., Manna, L., and Moreels, I. (2015). *Chem. Soc. Rev.*, **44**, 5897.

6. Korkin, A., Gusev, E., Labanowski, J., and Luryi, S. (eds.) (2007). *Nanotechnology for Electronic Materials and Devices*, Springer US.

7. Ramsden, J. (2011). *Nanotechnology: An Introduction*, Elsevier.

8. Ramsden, J.J. (2011). What is nanotechnology?, in *Nanotechnology*, Elsevier, pp. 1–14.

9. Horikoshi, S., and Serpone, N. (2013). Introduction to nanoparticles, in *Microwaves in Nanoparticle Synthesis*, Wiley-VCH Verlag GmbH & Co. KGaA, Weinheim, Germany, pp. 1–24.

10. Roduner, E. (2006). *Chem. Soc. Rev.*, **35**, 583.

11. Lai, S. (2008). *IEEE International Electron Devices Meeting*, 1.

12. Shalchian, M., Grisolia, J., Assayag, G. Ben, Coffin, H., Atarodi, S.M., and Claverie, A. (2005). *Appl. Phys. Lett.*, **86**, 163111.

13. Chen, W.-R., Chang, T.-C., Yeh, J.-L., Sze, S.M., and Chang, C.-Y. (2008). *Appl. Phys. Lett.*, **92**, 152114.

14. Chen, W.-R., Chang, T.-C., Liu, P.-T., Lin, P.-S., Tu, C.-H., and Chang, C.-Y. (2007). *Appl. Phys. Lett.*, **90**, 112108.

15. Lu, C.Y. (2012). *J. Nanosci. Nanotechnol.*, **12**, 7604.

16. Wang, L., Yang, C.-H., and Wen, J. (2015). *Electron. Mater. Lett.*, **11**, 505.

17. Alivisatos, A.P. (1996). *Science*, **271**, 933.

18. El-Sayed, M.A. (2004). *Acc. Chem. Res.*, **37**, 326.

19. Hu, J., Li Ls, Yang, W., Manna, L., Wang Lw, and Alivisatos, A.P. (2001). *Science*, **292**, 2060.

20. Liu, Z., Lee, C., Narayanan, V., Pei, G., and Kan, E.C. (2002). *IEEE Trans. Electron Devices*, **49**, 1614.

21. Liu, Z., Lee, C., Narayanan, V., Pei, G., and Kan, E.C. (2002). *IEEE Trans. Electron Devices*, **49**, 1606.

22. Ge, J., Lei, J., and Zare, R.N. (2012). *Nat. Nanotechnol.*, **7**, 428.

23. Ko, S.H., Lee, D., Kang, H.W., Nam, K.H., Yeo, J.Y., Hong, S.J., Grigoropoulos, C.P., and Sung, H.J. (2011). *Nano Lett.*, **11**, 666.

24. Modi, G. (2015). *Adv. Nat. Sci. Nanosci. Nanotechnol.*, **6**, 33002.

25. Zhou, L., Wang, W., and Xu, H. (2008). *Cryst. Growth Des.*, **8**, 728.

26. Sajanlal, P.R., and Pradeep, T. (2009). *Nano Res.*, **2**, 306.

27. Pradhan, D., Sindhwani, S., and Leung, K.T. (2010). *Nanoscale Res. Lett.*, **5**, 1727.

28. Kumari, L., Li, W.Z., Vannoy, C.H., Leblanc, R.M., and Wang, D.Z. (2009). *Ceram. Int.*, **35**, 3355.

29. Wang, M., Hahn, S.H., Kim, J.S., Chung, J.S., Kim, E.J., and Koo, K.-K. (2008). *J. Cryst. Growth*, **310**, 1213.

30. Zhao, J., Sun, M., Liu, Z., Quan, B., Gu, C., and Li, J. (2015). *Sci. Rep.*, **5**, 16019.

31. Lee, Y.-I., Kim, S., Jung, S.-B., Myung, N. V., and Choa, Y.-H. (2013). *ACS Appl. Mater. Interfaces*, **5**, 5908.

32. Hochbaum, A.I., Chen, R., Delgado, R.D., Liang, W., Garnett, E.C., Najarian, M., Majumdar, A., and Yang, P. (2008). *Nature*, **451**, 163.

33. Qu, X., and Komatsu, T. (2010). *ACS Nano*, **4**, 563.

34. Li, L., Yang, S., Han, F., Wang, L., Zhang, X., Jiang, Z., and Pan, A. (2014). *Sensors*, **14**, 7332.

35. Campos-Delgado, J., Romo-Herrera, J.M., Jia, X., Cullen, D.A., Muramatsu, H., Kim, Y.A., Hayashi, T., Ren, Z., Smith, D.J., Okuno, Y., Ohba, T., Kanoh, H., Kaneko, K., Endo, M., Terrones, H., Dresselhaus, M.S., and Terrones, M. (2008). *Nano Lett.*, **8**, 2773.

36. Nomoev, A.V., Bardakhanov, S.P., Schreiber, M., Bazarova, D.G., Romanov, N.A., Baldanov, B.B., Radnaev, B.R., and Syzrantsev, V.V. (2015). *Beilstein J. Nanotechnol.*, **6**, 874.

37. Kang, J.S., Lim, J., Rho, W.-Y., Kim, J., Moon, D.-S., Jeong, J., Jung, D., Choi, J.-W., Lee, J.-K., and Sung, Y.-E. (2016). *Sci. Rep.*, **6**, 30829.

38. Wang, G.-H., Hilgert, J., Richter, F.H., Wang, F., Bongard, H.-J., Spliethoff, B., Weidenthaler, C., and Schüth, F. (2014). *Nat. Mater.*, **13**, 293.

39. Siddiqui, G.U.D., Ali, J., Doh, Y.H., and Choi, K.H. (2016). *Mater. Lett.*, **166**, 311.

40. Biswas, A., Bayer, I.S., Biris, A.S., Wang, T., Dervishi, E., and Faupel, F. (2012). *Adv. Colloid Interface Sci.*, **170**, 2.

41. Yu, H.-D., Regulacio, M.D., Ye, E., and Han, M.-Y. (2013). *Chem. Soc. Rev.*, **42**, 6006.

42. Nie, Z., Petukhova, A., and Kumacheva, E. (2010). *Nat. Nanotechnol.*, **5**, 15.

43. Wang, L., Xu, L., Kuang, H., Xu, C., and Kotov, N.A. (2012). *Acc. Chem. Res.*, **45**, 1916.

44. Zhang, S.-Y., Regulacio, M.D., and Han, M.-Y. (2014). *Chem. Soc. Rev.*, **43**, 2301.

45. Glotzer, S.C., and Solomon, M.J. (2007). *Nat. Mater.*, **6**, 557.

46. Kim, J.K., Yang, S.Y., Lee, Y., and Kim, Y. (2010). *Prog. Polym. Sci.*, **35**, 1325.

47. Nozik, A.J., Beard, M.C., Luther, J.M., Law, M., Ellingson, R.J., and Johnson, J.C. (2010). *Chem. Rev.*, **110**, 6873.

48. Min, Y., Akbulut, M., Kristiansen, K., Golan, Y., and Israelachvili, J. (2008). *Nat. Mater.*, **7**, 527.

49. Ozin, G.A., Hou, K., Lotsch, B.V., Cademartiri, L., Puzzo, D.P., Scotognella, F., Ghadimi, A., and Thomson, J. (2009). *Mater. Today*, **12**, 12.

50. Ross, C.A., Berggren, K.K., Cheng, J.Y., Jung, Y.S., and Chang, J.-B. (2014). *Adv. Mater.*, **26**, 4386.

51. Lauters, M., McCarthy, B., Sarid, D., and Jabbour, G.E. (2005). *Appl. Phys. Lett.*, **87**, 231105.

52. Wang, H.P., Pigeon, S., Izquierdo, R., and Martel, R. (2006). *Appl. Phys. Lett.*, **89**, 183502.

53. Tu, C.-H., Kwong, D.-L., and Lai, Y.-S. (2006). *Appl. Phys. Lett.*, **89**, 252107.

54. Lai, Y.-S., Tu, C.-H., Kwong, D.-L., and Chen, J.S. (2005). *Appl. Phys. Lett.*, **87**, 122101.

55. Kanwal, A., and Chhowalla, M. (2006). *Appl. Phys. Lett.*, **89**, 203103.

56. Ikeda, H., Sakata, J., Hayakawa, M., Aoyama, T., Kawakami, T., Kamata, K., Iwaki, Y., Seo, S., Noda, Y., Nomura, R., and Yamazaki, S. (2006). *SID Symp. Dig. Tech. Pap.*, **37**, 923.

57. Chen, J., Xu, L., Lin, J., Geng, Y., Wang, L., and Ma, D. (2006). *Appl. Phys. Lett.*, **89**, 83514.

58. Prakash, A., Ouyang, J., Lin, J.-L., and Yang, Y. (2006). *J. Appl. Phys.*, **100**, 054309.

59. Reddy, V.S., Karak, S., Ray, S.K., and Dhar, A. (2009). *Org. Electron.*, **10**, 138.

60. Yook, K.S., Jeon, S.O., Joo, C.W., Lee, J.Y., Kim, S.H., and Jang, J. (2009). *Org. Electron.*, **10**, 48.

61. Tseng, C., and Tao, Y. (2009). *J. Am. Chem. Soc.*, **131**, 12441.

62. Tseng, R.J., Huang, J., Ouyang, J., Kaner, R.B., and Yang, Y. (2005). *Nano Lett.*, **5**, 1077.

63. Ouyang, J., Chu, C.-W., Szmanda, C.R., Ma, L., and Yang, Y. (2004). *Nat. Mater.*, **3**, 918.

64. Kim, H.J., Jung, S.M., Kim, Y.H., Kim, B.J., Ha, S., Kim, Y.S., Yoon, T.S., and Lee, H.H. (2011). *Thin Solid Films*, **519**, 6140.

65. Onlaor, K., Tunhoo, B., Keeratithiwakorn, P., Thiwawong, T., and Nukeaw, J. (2012). *Solid State Electron.*, **72**, 60.

66. Suresh, V., Huang, M.S., Srinivasan, M.P., and Krishnamoorthy, S. (2013). *ACS Appl. Mater. Interfaces*, **5**, 5727.

67. Salaita, K., Wang, Y., and Mirkin, C.A. (2007). *Nat. Nanotechnol.*, **2**, 145.

68. Son, J.Y., Shin, Y.S., and Shin, Y.H. (2011). *Appl. Surf. Sci.*, **257**, 9885.

69. Jong, Y.S., Shin, Y.H., Sangwoo, R., Hyungjun, K., and Jang, H.M. (2009). *J. Am. Chem. Soc.*, **131**, 14676.

70. Son, J.Y., Ryu, S., Park, Y.C., Lim, Y.T., Shin, Y.S., Shin, Y.H., and Jang, H.M. (2010). *ACS Nano*, **4**, 7315.

71. Ko, Y., Baek, H., Kim, Y., Yoon, M., and Cho, J. (2013). *ACS Nano*, **7**, 143.

72. Kim, T.H., Jang, E.Y., Lee, N.J., Choi, D.J., Lee, K.J., Jang, J.T., Choi, J.S., Moon, S.H., and Cheon, J. (2009). *Nano Lett.*, **9**, 2229.

73. Choi, D.J., Kim, J.-K., Seong, H., Jang, M.-S., and Kim, Y.-H. (2015). *Org. Electron.*, **27**, 65.

74. Sandouk, E.J., Gimzewski, J.K., and Stieg, A.Z. (2015). *Sci. Technol. Adv. Mater.*, **16**, 45004.

75. Gao, X., She, X.-J., Liu, C.-H., Sun, Q.-J., Liu, J., and Wang, S.-D. (2013). *Appl. Phys. Lett.*, **102**, 23303.

76. Chou, Y.H., You, N.H., Kurosawa, T., Lee, W.Y., Higashihara, T., Ueda, M., and Chen, W.C. (2012). *Macromolecules*, **45**, 6946.

77. Tondelier, D., Lmimouni, K., Vuillaume, D., Fery, C., and Haas, G. (2004). *Appl. Phys. Lett.*, **85**, 5763.

78. Kao, P.C., Liu, C.C., and Li, T.Y. (2015). *Org. Electron.*, **21**, 203.

79. Kim, Y., Lee, C., Shim, I., Wang, D., and Cho, J. (2010). *Adv. Mater.*, **22**, 5140.

80. Onlaor, K., Thiwawong, T., and Tunhoo, B. (2014). *Org. Electron.*, **15**, 1254.

81. Ayesh, A.I., Qadri, S., Baboo, V.J., Haik, M.Y., and Haik, Y. (2013). *Synth. Met.*, **183**, 24.

82. Gupta, R.K., Ying, G., Srinivasan, M.P., and Lee, P.S. (2012). *J. Phys. Chem. B*, **116**, 9784.

83. Liu, X.-J., Gao, M.-Y., Li, A.-D., Cao, Y.-Q., Li, X.-F., Guo, B.-L., Cao, Z.-Y., and Wu, D. (2014). *J. Alloys Compd.*, **588**, 103.

84. Haik, M.Y., Ayesh, A.I., Abdulrehman, T., and Haik, Y. (2014). *Mater. Lett.*, **124**, 67.

85. Yoo, D., Song, Y., Jang, J., Hwang, W.T., Jung, S.H., Hong, S., Lee, J.K., and Lee, T. (2015). *Org. Electron.*, **21**, 198.

86. Onlaor, K., Tunhoo, B., Thiwawong, T., and Nukeaw, J. (2013). *Appl. Phys. A*, **112**, 495.

87. Nenna, G., Masala, S., Bizzarro, V., Re, M., Pesce, E., Minarini, C., and Di Luccio, T. (2012). *J. Appl. Phys.*, **112**, 044508.

88. Sarma, S. (2016). *J. Polym. Eng.*, **36**, 293.

89. Ng, S.A., Razak, K.A., Cheong, K.Y., and Aw, K.C. (2016). *Thin Solid Films*, **615**, 84.

90. Naka, K., Itoh, H., Tampo, Y., and Chujo, Y. (2003). *Langmuir*, **19**, 5546.

91. Ahmad, Z., Ooi, P.C., Aw, K.C., and Sayyad, M.H. (2011). *Solid State Commun.*, **151**, 297.

92. Han, P., Sun, B., Cheng, S., Yu, F., Jiao, B., and Wu, Q. (2016). *J. Alloys Compd.*, **681**, 516.

93. Sun, B., Jia, X.J., Wu, J.H., and Chen, P. (2015). *Solid State Commun.*, **223**, 1.

94. Ahn, Y., Son, J.Y., and Jang, J. (2016). *J. Phys. Chem. C*, **120**, 11739.

95. Otsuka, S., Furuya, S., Takeda, R., Shimizu, T., Shingubara, S., Watanabe, T., Takano, Y., and Takase, K. (2012). *Microelectron. Eng.*, **98**, 367.

96. Oka, K., Yanagida, T., Nagashima, K., Kanai, M., Kawai, T., Kim, J.S., and Park, B.H. (2011). *J. Am. Chem. Soc.*, **133**, 12482.

97. Hung, S.W., Wang, T.T.J., Chu, L.W., and Chen, L.J. (2011). *J. Phys. Chem. C*, **115**, 15592.

98. Chiang, Y.-D., Chang, W.-Y., Ho, C.-Y., Chen, C.-Y., Ho, C.-H., Lin, S.-J., Wu, T.-B., and He, J.-H. (2011). *IEEE Trans. Electron Devices*, **58**, 1735.

99. Hong, W.-K., Sohn, J.I., Hwang, D.-K., Kwon, S.-S., Jo, G., Song, S., Kim, S.-M., Ko, H.-J., Park, S.-J., Welland, M.E., and Lee, T. (2008). *Nano Lett.*, **8**, 950.

100. Yoon, J., Hong, W.K., Jo, M., Jo, G., Choe, M., Park, W., Sohn, J.I., Nedic, S., Hwang, H., Welland, M.E., and Lee, T. (2011). *ACS Nano*, **5**, 558.

101. Wang, X., Xu, J., Shi, S., Zhang, X., Zhang, X., Shi, X., Li, S., Li, L., Liu, X., and Li, L. (2016). *Phys. Lett. A*, **380**, 262.

102. Liao, L., Fan, H.J., Yan, B., Zhang, Z., Chen, L.L., Li, B.S., Xing, G.Z., Shen, Z.X., Wu, T., Sun, X.W., Wang, J., and Yu, T. (2009). *ACS Nano*, **3**, 700.

103. Chuang, M.Y., Chen, Y.C., Su, Y.K., Hsiao, C.H., Huang, C.S., Tsai, J.J., and Yu, H.C. (2014). *ACS Appl. Mater. Interfaces*, **6**, 5432.

104. Wang, F., Xie, Z., Zhang, H., Liu, C., and Zhang, Y. (2011). *Adv. Funct. Mater.*, **21**, 1027.

105. Fu, W., Xu, Z., Bai, X., Gu, C., and Wang, E. (2009). *Nano Lett.*, **9**, 921.

106. Liao, A.D., Araujo, P.T., Xu, R., and Dresselhaus, M.S. (2014). *Nat. Commun.*, **5**, 5673.

107. Yao, J., Jin, Z., Zhong, L., Natelson, D., and Tour, J.M. (2009). *ACS Nano*, **3**, 4122.

108. Sakurai, T., Yoshimura, T., Akita, S., Fujimura, N., and Nakayama, Y. (2006). *Jpn. J. Appl. Phys.*, **45**, L1036.

109. Fuhrer, M.S., Kim, B.M., Dürkop, T., and Brintlinger, T. (2002). *Nano Lett.*, **2**, 755.

110. Nagashima, K., Yanagida, T., Oka, K., Taniguchi, M., Kawai, T., Kim, J.-S.S., and Park, B.H. (2010). *Nano Lett.*, **10**, 1359.

111. Nagashima, K., Yanagida, T., Oka, K., Kanai, M., Klamchuen, A., Kim, J.S., Park, B.H., and Kawai, T. (2011). *Nano Lett.*, **11**, 2114.

112. Klamchuen, A., Yanagida, T., Kanai, M., Nagashima, K., Oka, K., Kawai, T., Suzuki, M., Hidaka, Y., and Kai, S. (2010). *Appl. Phys. Lett.*, **97**, 73114.

113. Klamchuen, A., Yanagida, T., Kanai, M., Nagashima, K., Oka, K., Seki, S., Suzuki, M., Hidaka, Y., Kai, S., and Kawai, T. (2011). *Appl. Phys. Lett.*, **98**, 53107.

114. Klamchuen, A., Yanagida, T., Nagashima, K., Seki, S., Oka, K., Taniguchi, M., and Kawai, T. (2009). *Appl. Phys. Lett.*, **95**, 53105.

115. Marcu, A., Yanagida, T., Nagashima, K., Oka, K., Tanaka, H., and Kawai, T. (2008). *Appl. Phys. Lett.*, **92**, 173119.

116. Nagashima, K., Yanagida, T., Tanaka, H., Seki, S., Saeki, A., Tagawa, S., and Kawai, T. (2008). *J. Am. Chem. Soc.*, **130**, 5378.

117. Huang, Y.-T., Huang, C.-W., Chen, J.-Y., Ting, Y.-H., Lu, K.-C., Chueh, Y.-L., and Wu, W.-W. (2014). *ACS Nano*, **8**, 9457.

118. Wang, Y., Lew, K., Ho, T., Pan, L., Novak, S.W., Dickey, E.C., Redwing, J.M., and Mayer, T.S. (2005). *Nano Lett.*, **5**, 2139.

119. Dugaiczyk, L., Ngo-Duc, T.T., Gacusan, J., Singh, K., Yang, J., Santhanam, S., Han, J.W., Koehne, J.E., Kobayashi, N.P., Meyyappan, M., and Oye, M.M. (2013). *Chem. Phys. Lett.*, **575**, 112.

120. Zhu, H., Pookpanratana, S.J., Bonevich, J.E., Natoli, S.N., Hacker, C.A., Ren, T., Suehle, J.S., Richter, C.A., and Li, Q. (2015). *ACS Appl. Mater. Interfaces*, **7**, 27306.

121. Oka, K., Yanagida, T., Nagashima, K., Tanaka, H., and Kawai, T. (2009). *J. Am. Chem. Soc.*, **131**, 3434.

122. Meng, G., Yanagida, T., Nagashima, K., Yoshida, H., Kanai, M., Klamchuen, A., Zhuge, F., He, Y., Rahong, S., Fang, X., Takeda, S., and Kawai, T. (2013). *J. Am. Chem. Soc.*, **135**, 7033.

123. Hsu, C.-W., and Chou, L. (2012). *Nano Lett.*, **12**, 4247.

124. Piner, R.D. (1999). *Science*, **283**, 661.

125. Son, J.Y., Lee, J.H., Song, S., Shin, Y.H., and Jang, H.M. (2013). *ACS Nano*, **7**, 5522.

126. McKendry, R., Huck, W.T.S., Weeks, B., Fiorini, M., Abell, C., and Rayment, T. (2002). *Nano Lett.*, **2**, 713.

127. Kianian, S., Rosendale, G., Manning, M., Hamilton, D., Huang, X.M.H., Robins, K., Kim, Y.W., Rueckes, T., Robinson, K., Kim, Y.W., Rueckes, T., Robins, K., Kim, Y.W., and Rueckes, T. (2010). *Proc. Eur. Solid State Device Res. Conf.*, 404.

128. Hu, Y., Perello, D., Yun, M., Kwon, D.H., and Kim, M. (2013). *Microelectron. Eng.*, **104**, 42.

129. Han, G.H., Güneş, F., Bae, J.J., Kim, E.S., Chae, S.J., Shin, H.J., Choi, J.Y., Pribat, D., and Lee, Y.H. (2011). *Nano Lett.*, **11**, 4144.

130. Seo, S., Yoon, Y., Lee, J., Park, Y., and Lee, H. (2013). *ACS Nano*, **7**, 3607.

131. Fu, W., Qin, S., Liu, L., Kim, T.H., Hellstrom, S., Wang, W., Liang, W., Bai, X., Li, A.P., and Wang, E. (2011). *Nano Lett.*, **11**, 1913.

132. Wang, W., and Bai, F. (2005). *Appl. Phys. Lett.*, **87**, 193109.

133. Wang, W., and Bai, F. (2003). *ChemPhysChem*, **4**, 761.

134. Kartawidjaja, F.C., Zhou, Z., and Wang, J. (2006). *J. Electroceram.*, **16**, 425.

135. Shen, Z.K., Chen, Z.H., Qiu, Z.J., rui Lu, B., Wan, J., Deng, S.R., Jiang, A.Q., Qu, X.P., Liu, R., and Chen, Y. (2010). *Microelectron. Eng.*, **87**, 869.

136. Shen, Z.K., Chen, Z.H., Li, H., Qu, X.P., Chen, Y., and Liu, R. (2011). *Appl. Surf. Sci.*, **257**, 8820.

137. Zhu, Z., Guo, X., Wang, Z., Yuan, G., Wang, Y., Cai, Z., Chen, L., Chen, J., Zhang, Y., and Liu, Z. (2015). *Chem. Phys. Lett.*, **638**, 168.

138. Senthilkumar, V., Kathalingam, A., Kannan, V., Senthil, K., and Rhee, J.K. (2013). *Sens. Actuators, A*, **194**, 135.

139. Zhang, F., Gan, X., Li, X., Wu, L., Gao, X., Zheng, R., He, Y., Liu, X., and Yang, R. (2011). *Electrochem. Solid-State Lett.*, **14**, H422.

140. Xiao, J., Ong, W.L., Guo, Z., Ho, G.W., and Zeng, K. (2015). *ACS Appl. Mater. Interfaces*, **7**, 11412.

141. Ashfold, M.N.R., Doherty, R.P., Ndifor-Angwafor, N.G., Riley, D.J., and Sun, Y. (2007). *Thin Solid Films*, **515**, 8679.

142. Richardson, J.J., and Lange, F.F. (2009). *Cryst. Growth Des.*, **9**, 2576.

143. Richardson, J.J., and Lange, F.F. (2009). *Cryst. Growth Des.*, **9**, 2570.

144. Chang, W.Y., Lin, C.A., He, J.H., and Wu, T.B. (2010). *Appl. Phys. Lett.*, **96**, 242109.

145. Lim, J.-H., and Park, S.-J. (2006). Contacts to ZnO, in *Zinc Oxide Bulk, Thin Films and Nanostructures*, Vol. 287, Elsevier, pp. 267–283.

146. Chu, D., Younis, A., and Li, S. (2012). *J. Phys. D: Appl. Phys.*, **45**, 355306.

147. Yang, Y.C., Pan, F., Liu, Q., Liu, M., and Zeng, F. (2009). *Nano Lett.*, **9**, 1636.

148. Younis, A., Chu, D., Lin, X., Yi, J., Dang, F., and Li, S. (2013). *ACS Appl. Mater. Interfaces*, **5**, 2249.

149. Younis, A., Chu, D., Li, C.M., Das, T., Sehar, S., Manefield, M., and Li, S. (2014). *Langmuir*, **30**, 1183.

150. Younis, A., Chu, D., Mihail, I., and Li, S. (2013). *ACS Appl. Mater. Interfaces*, **5**, 9429.

151. Dang, F., Kato, K., Imai, H., Wada, S., Haneda, H., and Kuwabara, M. (2011). *Cryst. Growth Des.*, **11**, 4129.

152. Sun, B., Zhao, W., Wei, L., Li, H., and Chen, P. (2014). *Chem. Commun.*, **50**, 13142.

153. Sun, B., Li, H., Wei, L., and Chen, P. (2014). *CrystEngComm*, **16**, 9891.

154. Zhao, W.X., Sun, B., Liu, Y.H., Wei, L.J., Li, H.W., and Chen, P. (2014). *AIP Adv.*, **4**, 77127.

155. Sun, B., and Li, C.M. (2014). *Chem. Phys. Lett.*, **604**, 127.

156. Sun, B., Jia, X.J., Wu, J.H., and Chen, P. (2015). *J. Alloys Compd.*, **653**, 95.

157. Sun, B., Li, X., Liang, D., and Chen, P. (2016). *Chem. Phys. Lett.*, **643**, 66.

158. Zhou, G., Zhao, W., Ma, X., and Zhou, A.K. (2016). *J. Alloys Compd.*, **679**, 2.

159. Dong, H., Zhang, X., Zhao, D., Niu, Z., Zeng, Q., Li, J., Cai, L., Wang, Y., Zhou, W., Gao, M., and Xie, S. (2012). *Nanoscale*, **4**, 2571.

160. Yan, Y., Sun, B., and Ma, D. (2015). *Chem. Phys. Lett.*, **638**, 103.

161. Zhang, X., Wen, F., Xiang, J., Wang, X., Wang, L., Hu, W., and Liu, Z. (2015). *Appl. Phys. Lett.*, **107**, 103109.

162. Han, P., Sun, B., Cheng, S., Yu, F., Jiao, B., and Wu, Q. (2016). *J. Alloys Compd.*, **664**, 619.

163. Batabyal, S.K., and Vittal, J.J. (2008). *Chem. Mater.*, **20**, 5845.

164. Yan, Y., Sun, B., and Ma, D.J. (2016). *J. Mater. Sci. Mater. Electron.*, **27**, 512.

165. Sun, B., Zhao, W., Li, H., Wei, L., and Chen, P. (2014). *Chem. Phys. Lett.*, **613**, 100.

166. Park, J., Lee, S., Lee, J., and Yong, K. (2013). *Adv. Mater.*, **25**, 6423.

167. Sun, B., and Li, C.M. (2015). *Phys. Chem. Chem. Phys.*, **17**, 6718.

168. Sun, B., Liu, Y., Lou, F., and Chen, P. (2015). *Chem. Phys.*, **457**, 28.

169. Sun, B., Wu, J., Jia, X., Lou, F., and Chen, P. (2015). *J. Sol-Gel Sci. Technol.*, **75**, 664.

170. Bhatnagar, A., Roy Chaudhuri, A., Heon Kim, Y., Hesse, D., and Alexe, M. (2013). *Nat. Commun.*, **4**, 1840.

171. Pintilie, L., Dragoi, C., Chu, Y.H., Martin, L.W., Ramesh, R., and Alexe, M. (2009). *Appl. Phys. Lett.*, **94**, 232902.

172. Qu, T.L., Zhao, Y.G., Xie, D., Shi, J.P., Chen, Q.P., and Ren, T.L. (2011). *Appl. Phys. Lett.*, **98**, 173507.

173. Yin, K., Li, M., Liu, Y., He, C., Zhuge, F., Chen, B., Lu, W., Pan, X., and Li, R.-W. (2010). *Appl. Phys. Lett.*, **97**, 42101.

174. Wang, C., Jin, K., Xu, Z., Wang, L., Ge, C., Lu, H., Guo, H., He, M., and Yang, G. (2011). *Appl. Phys. Lett.*, **98**, 192901.

175. You, T., Du, N., Slesazeck, S., Mikolajick, T., Li, G., Bürger, D., Skorupa, I., Stöcker, H., Abendroth, B., Beyer, A., Volz, K., Schmidt, O.G., and Schmidt, H. (2014). *ACS Appl. Mater. Interfaces*, **6**, 19758.

176. Kim, W., Son, J.Y., and Jang, H.M. (2014). *ACS Appl. Mater. Interfaces*, **6**, 6346.

177. You, T., Ou, X., Niu, G., Bärwolf, F., Li, G., Du, N., Bürger, D., Skorupa, I., Jia, Q., Yu, W., Wang, X., Schmidt, O.G., and Schmidt, H. (2015). *Sci. Rep.*, **5**, 18623.

178. Wu, S., Ren, L., Qing, J., Yu, F., Yang, K., Yang, M., Wang, Y., Meng, M., Zhou, W., Zhou, X., and Li, S. (2014). *ACS Appl. Mater. Interfaces*, **6**, 8575.

179. Mundle, R., Carvajal, C., and Pradhan, A.K. (2016). *Langmuir*, **32**, 4983.

180. Liu, K.C., Tzeng, W.H., Chang, K.M., Chan, Y.C., and Kuo, C.C. (2010). *Thin Solid Films*, **518**, 7460.

181. Mourey, D.A., Zhao, D.A., Sun, J., and Jackson, T.N. (2010). *IEEE Trans. Electron Devices*, **57**, 530.

182. Shi, L., Shang, D., Sun, J., and Shen, B. (2009). *Appl. Phys. Express*, **2**, 101602.

183. Seo, J.W., Park, J.-W., Lim, K.S., Yang, J.-H., and Kang, S.J. (2008). *Appl. Phys. Lett.*, **93**, 223505.

184. Liu, K.C., Tzeng, W.H., Chang, K.M., and Wu, C.H. (2010). *Surf. Coat. Technol.*, **205**, S379.

185. Liu, D., Cheng, H., Wang, G., Zhu, X., and Shao, Z. (2013). *IEEE Electron Device Lett.*, **34**, 1506.

186. Liu, K.-C., Tzeng, W.-H., Chang, K.-M., Chan, Y.-C., and Kuo, C.-C. (2011). *Microelectron. Eng.*, **88**, 1586.

187. Wu, S., Luo, X., Turner, S., Peng, H., Lin, W., Ding, J., David, A., Wang, B., Van Tendeloo, G., Wang, J., and Wu, T. (2014). *Phys. Rev. X*, **3**, 1.

188. Liu, K.C., Tzeng, W.H., Chang, K.M., Chan, Y.C., Kuo, C.C., and Cheng, C.W. (2010). *Microelectron. Reliab.*, **50**, 670.

189. Acharya, S.K., Nallagatla, R.V., Togibasa, O., Lee, B.W., Liu, C., Jung, C.U., Park, B.H., Park, J.Y., Cho, Y., Kim, D.W., Jo, J., Kwon, D.H., Kim, M., Hwang, C.S., and Chae, S.C. (2016). *ACS Appl. Mater. Interfaces*, **8**, 7902.

190. Privitera, S., Bersuker, G., Lombardo, S., Bongiorno, C., and Gilmer, D.C. (2015). *Solid State Electron.*, **111**, 161.

191. Calka, P., Martinez, E., Lafond, D., Dansas, H., Tirano, S., Jousseaume, V., Bertin, F., and Guedj, C. (2011). *Microelectron. Eng.*, **88**, 1140.

192. Privitera, S., Bersuker, G., Butcher, B., Kalantarian, A., Lombardo, S., Bongiorno, C., Geer, R., Gilmer, D.C., and Kirsch, P.D. (2013). *Microelectron. Eng.*, **109**, 75.

193. Lian, X., Miranda, E., Long, S., Perniola, L., Liu, M., and Suñé, J. (2014). *Solid State Electron.*, **98**, 38.

194. Cabout, T., Buckley, J., Cagli, C., Jousseaume, V., Nodin, J.F., De Salvo, B., Bocquet, M., and Muller, C. (2013). *Thin Solid Films*, **533**, 19.

195. Deng, T., Ye, C., Wu, J., He, P., and Wang, H. (2016). *Microelectron. Reliab.*, **57**, 34.

196. Wu, J., Ye, C., Zhang, J., Deng, T., He, P., and Wang, H. (2016). *Mater. Sci. Semicond. Process.*, **43**, 144.

197. Hsu, C.-C., Wu, C.-H., and Wang, S.-Y. (2016). *J. Alloys Compd.*, **663**, 262.

198. Luo, Y.Y., Pan, S.S., Xu, S.C., Zhong, L., Wang, H., and Li, G.H. (2016). *J. Alloys Compd.*, **664**, 626.

199. Wang, S.Y., Tsai, C.H., Lee, D.Y., Lin, C.Y., Lin, C.C., and Tseng, T.Y. (2011). *Microelectron. Eng.*, **88**, 1628.

200. Yang, J.-B., Chang, T.-C., Huang, J.-J., Chen, Y.-T., Yang, P.-C., Tseng, H.-C., Chu, A.-K., Sze, S.M., and Tsai, M.-J. (2013). *Thin Solid Films*, **528**, 26.

201. Yang, J.-B., Chang, T.-C., Huang, J.-J., Chen, S.-C., Yang, P.-C., Chen, Y.-T., Tseng, H.-C., Sze, S.M., Chu, A.-K., and Tsai, M.-J. (2013). *Thin Solid Films*, **529**, 200.

202. Yang, W.-Y., Kim, W.-G., and Rhee, S.-W. (2008). *Thin Solid Films*, **517**, 967.

203. Lin, C.C., Tu, B.C., Lin, C.C., Lin, C.H., and Tseng, T.Y. (2006). *IEEE Electron Device Lett.*, **27**, 725.

204. Yoshida, C., Kurasawa, M., Lee, Y.M., Aoki, M., and Sugiyama, Y. (2008). *Appl. Phys. Lett.*, **92**, 113508.

205. Shima, H., Takano, F., Muramatsu, H., Akinaga, H., Inoue, I.H., and Takagi, H. (2008). *Appl. Phys. Lett.*, **92**, 43510.

206. Chen, A., Haddad, S., Wu, Y.C., Lan, Z., Fang, T.N., and Kaza, S. (2007). *Appl. Phys. Lett.*, **91**, 123517.

207. Park, G.-S., Li, X.-S., Kim, D.-C., Jung, R.-J., Lee, M.-J., and Seo, S. (2007). *Appl. Phys. Lett.*, **91**, 222103.

208. Lamperti, A., Spiga, S., Lu, H.L., Wiemer, C., Perego, M., Cianci, E., Alia, M., and Fanciulli, M. (2008). *Microelectron. Eng.*, **85**, 2425.

209. Lee, M.J., Han, S., Jeon, S.H., Park, B.H., Kang, B.S., Ahn, S.E., Kim, K.H., Lee, C.B., Kim, C.J., Yoo, I.K., Seo, D.H., Li, X.S., Park, J.B., Lee, J.H., and Park, Y. (2009). *Nano Lett.*, **9**, 1476.

210. Kim, H.-D.D., An, H.-M.M., and Kim, T.G. (2012). *Microelectron. Eng.*, **98**, 351.

211. Kim, H.-D., An, H.-M., Kim, K.C., Seo, Y., Nam, K.-H., Chung, H.-B., Lee, E.B., and Kim, T.G. (2010). *Semicond. Sci. Technol.*, **25**, 65002.

212. Kim, H.-D., An, H.-M., and Kim, T.G. (2011). *J. Appl. Phys.*, **109**, 016105.

213. Kim, S., Jung, S., Kim, M.-H., Cho, S., and Park, B.-G. (2015). *Solid State Electron.*, **114**, 94.

214. Kim, S., Kim, M.-H., Kim, T.-H., Cho, S., and Park, B.-G. (2016). *J. Alloys Compd.*, **686**, 479.

215. Kim, S., and Park, B. (2016). *J. Alloys Compd.*, **663**, 256.

216. Kim, S., and Park, B.-G. (2016). *Appl. Phys. Lett.*, **108**, 212103.

217. Jiang, X., Ma, Z., Xu, J., Chen, K., Xu, L., Li, W., Huang, X., and Feng, D. (2015). *Sci. Rep.*, **5**, 15762.

218. Chen, H., Cheng, N., Ma, W., Li, M., Hu, S., Gu, L., Meng, S., and Guo, X. (2016). *ACS Nano*, **10**, 436.

219. Lim, S.L., Ling, Q., Teo, E.Y.H., Zhu, C.X., Chan, D.S.H., Kang, E.T., and Neoh, K.G. (2007). *Chem. Mater.*, **19**, 5148.

220. Das, S., and Appenzeller, J. (2011). *Nano Lett.*, **11**, 4003.

221. Jang, B.C., Seong, H., Kim, S.K., Kim, J.Y., Koo, B.J., Choi, J., Yang, S.Y., Im, S.G., and Choi, S.-Y. (2016). *ACS Appl. Mater. Interfaces*, **8**, 12951.

222. Moreno, C., Munuera, C., Valencia, S., Kronast, F., Obradors, X., and Ocal, C. (2010). *Nano Lett.*, **10**, 3828.

223. Hu, W., Zou, L., Chen, R., Xie, W., Chen, X., Qin, N., Li, S., Yang, G., and Bao, D. (2014). *Appl. Phys. Lett.*, **104**, 143502.

224. Hu, W., Qin, N., Wu, G., Lin, Y., Li, S., and Bao, D. (2012). *J. Am. Chem. Soc.*, **134**, 14658.

225. Hu, W., Chen, X., Wu, G., Lin, Y., Qin, N., and Bao, D. (2012). *Appl. Phys. Lett.*, **101**, 63501.

226. Hu, W., Zou, L., Lin, X., Gao, C., Guo, Y., and Bao, D. (2016). *Mater. Des.*, **103**, 230.

227. Wu, W., Shan, B., Feng, K., and Nan, H. (2016). *Mater. Sci. Semicond. Process.*, **44**, 18.

228. Sun, C.E., Chen, C.Y., Chu, K.L., Shen, Y.S., Lin, C.C., and Wu, Y.H. (2015). *ACS Appl. Mater. Interfaces*, **7**, 6383.

229. Carcia, P.F., McLean, R.S., Reilly, M.H., and Nunes, G. (2003). *Appl. Phys. Lett.*, **82**, 1117.

230. Hu, W., Zou, L., Gao, C., Guo, Y., and Bao, D. (2016). *J. Alloys Compd.*, **676**, 356.

231. Lin, C.-Y., Wu, C.-Y., Wu, C.-Y., Lee, T.-C., Yang, F.-L., Hu, C., and Tseng, T.-Y. (2007). *IEEE Electron Device Lett.*, **28**, 366.

232. Kim, J., Inamdar, A.I., Jo, Y., Woo, H., Cho, S., Pawar, S.M., Kim, H., and Im, H. (2016). *ACS Appl. Mater. Interfaces*, **8**, 9499.

233. Tararam, R., Joanni, E., Savu, R., Bueno, P.R., Longo, E., and Varela, J.A. (2011). *ACS Appl. Mater. Interfaces*, **3**, 500.

234. Yoo, I.K. (2005). *IEEE Electron Device Lett.*, **26**, 719.

235. Qi, J., Olmedo, M., Ren, J., Zhan, N., Zhao, J., Zheng, J.G., and Liu, J. (2012). *ACS Nano*, **6**, 1051.

236. Son, J.Y., Shin, Y.H., Kim, H., and Jang, H.M. (2010). *ACS Nano*, **4**, 2655.

237. Chowdhury, M., Long, B., Jha, R., and Devabhaktuni, V. (2012). *Solid State Electron.*, **68**, 1.

238. Hasan, M., Dong, R., Lee, D.S., Seong, D.J., Choi, H.J., Pyun, M.B., and Hwang, H. (2008). *J. Semicond. Technol. Sci.*, **8**, 66.

239. Shima, H., Takano, F., Akinaga, H., Tamai, Y., Inoue, I.H., and Takagi, H. (2007). *Appl. Phys. Lett.*, **91**, 12901.

240. Wu, J., Wang, J., Xiao, D., and Zhu, J. (2011). *Mater. Res. Bull.*, **46**, 2183.

241. Kao, C.H., Chen, C.C., and Lin, C.J. (2015). *Vacuum*, **118**, 74.

242. Starykov, O., and Sakurai, K. (2005). *Vacuum*, **80**, 117.

243. Dargis, R., Fissel, A., Schwendt, D., Bugiel, E., Krügener, J., Wietler, T., Laha, A., and Osten, H.J. (2010). *Vacuum*, **85**, 523.

244. Pan, T.-M., Yu, T.-Y., and Hsieh, Y.-Y. (2007). *J. Appl. Phys.*, **102**, 074111.

245. Wang, Y.Q., Chen, J.H., Yoo, W.J., Yeo, Y.-C., Kim, S.J., Gupta, R., Tan, Z.Y.L., Kwong, D.-L., Du, A.Y., and Balasubramanian, N. (2004). *Appl. Phys. Lett.*, **84**, 5407.

246. Mikhaylov, A.N., Belov, A.I., Guseinov, D. V., Korolev, D.S., Antonov, I.N., Efimovykh, D. V., Tikhov, S. V., Kasatkin, A.P., Gorshkov, O.N., Tetelbaum, D.I., Bobrov, A.I., Malekhonova, N. V., Pavlov, D.A., Gryaznov, E.G., and Yatmanov, A.P. (2015). *Mater. Sci. Eng. B*, **194**, 48.

247. Rosário, C.M.M., Gorshkov, O.N., Kasatkin, A.P., Antonov, I.N., Korolev, D.S., Mikhaylov, A.N., and Sobolev, N.A. (2015). *Vacuum*, **122**, 293.

248. Singh, B., and Mehta, B.R. (2014). *Thin Solid Films*, **569**, 35.

249. Chen, M.-C., Chang, T.-C., Chiu, Y.-C., Chen, S.-C., Huang, S.-Y., Chang, K.-C., Tsai, T.-M., Yang, K.-H., Sze, S.M., and Tsai, M.-J. (2013). *Thin Solid Films*, **528**, 224.

250. Mähne, H., Slesazeck, S., Jakschik, S., Dirnstorfer, I., and Mikolajick, T. (2011). *Microelectron. Eng.*, **88**, 1148.

251. Zhang, L., Xu, H., Wang, Z., Yu, H., Ma, J., and Liu, Y. (2016). *Appl. Surf. Sci.*, **360**, 338.

252. Jeon, H.H., Park, J., Jang, W., Kim, H.H., Song, H., Kim, H.H., Seo, H., and Jeon, H.H. (2015). *Curr. Appl. Phys.*, **15**, 1005.

253. Wang, H., Zou, C., Zhou, L., Tian, C., and Fu, D. (2012). *Microelectron. Eng.*, **91**, 144.

254. Tseng, H.-C., Chang, T.-C., Huang, J.-J., Chen, Y.-T., Yang, P.-C., Huang, H.-C., Gan, D.-S., Ho, N.-J., Sze, S.M., and Tsai, M.-J. (2011). *Thin Solid Films*, **520**, 1656.

255. Kao, C., Chen, H., Tsung, Y., Sing, J., and Lu, T. (2012). *Solid State Commun.*, **152**, 504.

256. Kao, C.H., Chen, H., Chen, S.Z., Hung, S.-H., Chen, C.Y., He, Y.-Y., Lin, S.-R., Hsieh, K.-M., and Lin, M.-H. (2015). *Vacuum*, **118**, 69.

257. Ismail, M., Ahmed, E., Rana, A.M., Hussain, F., Talib, I., Nadeem, M.Y., Panda, D., and Shah, N.A. (2016). *ACS Appl. Mater. Interfaces*, **8**, 6127.

258. Liu, C.-Y., Wu, P.-H., Wang, A., Jang, W.-Y., Young, J.-C., Chiu, K.-Y., and Tseng, T.-Y. (2005). *IEEE Electron Device Lett.*, **26**, 351.

259. Lee, C., Kim, I., Choi, W., Shin, H., and Cho, J. (2009). *Langmuir*, **25**, 4274.

260. Baek, H., Lee, C., Choi, J., and Cho, J. (2013). *Langmuir*, **29**, 380.

261. Liu, C.-Y., and Tseng, T.-Y. (2007). *J. Phys. D: Appl. Phys.*, **40**, 2157.

262. Zhang, T., Su, Z., Chen, H., Ding, L., and Zhang, W. (2008). *Appl. Phys. Lett.*, **93**, 172104.

263. Das, S., Majumdar, S., and Giri, S. (2010). *J. Phys. Chem. C*, **114**, 6671.

264. Shi, Q., Huang, W., Zhang, Y., Yan, J., Zhang, Y., Mao, M., Zhang, Y., and Tu, M. (2011). *ACS Appl. Mater. Interfaces*, **3**, 3523.

265. Kim, A., Song, K., Kim, Y., and Moon, J. (2011). *ACS Appl. Mater. Interfaces*, **3**, 4525.

266. Yang, J.J., Pickett, M.D., Li, X., Ohlberg, D.A.A., Stewart, D.R., and Williams, R.S. (2008). *Nat. Nanotechnol.*, **3**, 429.

267. Barman, S., Deng, F., and McCreery, R.L. (2008). *J. Am. Chem. Soc.*, **130**, 11073.

268. Narendar, Y., and Messing, G.L. (1997). *Chem. Mater.*, **9**, 580.

269. Knauth, P., and Tuller, H.L. (1999). *J. Appl. Phys.*, **85**, 897.

270. Frohlich, K. (2013). *Mater. Sci. Semicond. Process.*, **16**, 1186.

271. Lai, C.H., Chen, C.H., and Tseng, T.Y. (2013). *Surf. Coat. Technol.*, **231**, 399.

272. Hu, W., Zou, L., Chen, X., Qin, N., Li, S., and Bao, D. (2014). *ACS Appl. Mater. Interfaces*, **6**, 5012.

273. Senthilkumar, V., Kathalingam, A., Kannan, V., and Rhee, J.-K. (2012). *Microelectron. Eng.*, **98**, 97.

274. Lin, C.-Y., Lin, C.-C., Huang, C.-H., Lin, C.-H., and Tseng, T.-Y. (2007). *Surf. Coat. Technol.*, **202**, 1319.

275. Chang, L.C., Lee, D.Y., Ho, C.C., and Chiou, B.S. (2007). *Thin Solid Films*, **516**, 454.

276. Lian, S.-L., Liu, C.-L., and Chen, W.-C. (2011). *ACS Appl. Mater. Interfaces*, **3**, 4504.

277. Rosa, C. De, Auriemma, F., Girolamo, R. Di, Pepe, G.P., Napolitano, T., and Scaldaferri, R. (2010). *Adv. Mater.*, **22**, 5414.

278. Leong, W.L., Mathews, N., Tan, B., Vaidyanathan, S., Dötz, F., and Mhaisalkar, S. (2011). *J. Mater. Chem.*, **21**, 8971.

279. Leong, W.L., Mathews, N., Mhaisalkar, S., Lam, Y.M., Chen, T., and Lee, P.S. (2009). *J. Mater. Chem.*, **19**, 7354.

280. Leong, W.L., Lee, P.S., Lohani, A., Lam, Y.M., Chen, T., Zhang, S., Dodabalapur, A., and G. Mhaisalkar, S. (2008). *Adv. Mater.*, **20**, 2325.

281. Lee, J.-S., Kim, Y.-M., Kwon, J.-H., Sim, J.S., Shin, H., Sohn, B.-H., and Jia, Q. (2011). *Adv. Mater.*, **23**, 2064.

282. Chiu, Y.-C., Chen, T.-Y., Chen, Y., Satoh, T., Kakuchi, T., and Chen, W.-C. (2014). *ACS Appl. Mater. Interfaces*, **6**, 12780.

283. Suresh, V., Kusuma, D.Y., Lee, P.S., Yap, F.L., Srinivasan, M.P., and Krishnamoorthy, S. (2015). *ACS Appl. Mater. Interfaces*, **7**, 279.

284. Kim, S., Kim, H., Jung, S., Kim, M.-H., Lee, S.-H., Cho, S., and Park, B.-G. (2016). *J. Alloys Compd.*, **663**, 419.

285. Sun, B., Li, Q., Liu, Y., and Chen, P. (2015). *Funct. Mater. Lett.*, **8**, 1550001.

286. Oshikane, Y., Kataoka, T., Okuda, M., Hara, S., Inoue, H., and Nakano, M. (2007). *Sci. Technol. Adv. Mater.*, **8**, 181.

287. Johnson, T.W., Lapin, Z.J., Beams, R., Lindquist, N.C., Rodrigo, S.G., Novotny, L., and Oh, S.-H. (2012). *ACS Nano*, **6**, 9168.

288. Kim, D., Lee, D.U., Lee, H.J., and Kim, E.K. (2012). *Thin Solid Films*, **521**, 94.

289. Yang, J.Y., Yoon, K.S., Choi, W.J., Do, Y.H., Kim, J.H., Kim, C.O., and Hong, J.P. (2007). *Curr. Appl. Phys.*, **7**, 147.

290. Banerjee, W., Maikap, S., Lai, C.-S., Chen, Y.-Y., Tien, T.-C., Lee, H.-Y., Chen, W.-S., Chen, F.T., Kao, M.-J., Tsai, M.-J., and Yang, J.-R. (2012). *Nanoscale Res. Lett.*, **7**, 194.

291. Farmer, D.B., and Gordon, R.G. (2007). *J. Appl. Phys.*, **101**, 124503.

292. Amouroux, J., Faivre, E., Boivin, P., Muller, C., Deleruyelle, D., Fares, L., Maillot, P., Putero, M., and Jalaguier, E. (2012). *Int. Semicond. Conf. Dresden-Grenoble*, 169.

293. Fowler, B.W., Chang, Y.-F., Zhou, F., Wang, Y., Chen, P.-Y., Xue, F., Chen, Y.-T., Bringhurst, B., Pozder, S., and Lee, J.C. (2015). *RSC Adv.*, **5**, 21215.

294. Pradhan, S.K., Xiao, B., Mishra, S., Killam, A., and Pradhan, A.K. (2016). *Sci. Rep.*, **6**, 26763.

295. Song, J.-M., and Lee, J.-S. (2016). *Sci. Rep.*, **6**, 18967.

296. Di Mauro, E., Carpentier, O., Yáñez Sánchez, S.I., Ignoumba Ignoumba, N., Lalancette-Jean, M., Lefebvre, J., Zhang, S., Graeff, C.F.O., Cicoira, F., and Santato, C. (2016). *J. Mater. Chem. C*, **4**, 9544.

297. Xu, Z., Liu, B., Chen, Y., Zhang, Z., Gao, D., Wang, H., Song, Z., Wang, C., Ren, J., Zhu, N., Xiang, Y., Zhan, Y., and Feng, S. (2016). *Solid State Electron.*, **116**, 119.

298. Vernon-Parry, K.D.D. (2000). *III-Vs Rev.*, **13**, 40.

299. Kourkoutis, L.F., Plitzko, J.M., and Baumeister, W. (2012). *Annu. Rev. Mater. Res.*, **42**, 33.

300. Ren, J., Hu, H., Liu, F., Chu, S., and Liu, J. (2012). *J. Appl. Phys.*, **112**, 54311.

301. Zou, S., and Michael, C. (2014). *IEEE 64th Electron. Components Technol. Conf.*, 441.

302. Ren, J., Yan, D., Chu, S., and Liu, J. (2013). *Appl. Phys. A*, **111**, 719.

303. Pennycook, S.J., Lupini, A.R., Varela, M., Borisevich, A.Y., Peng, Y., Oxley, M.P., van Benthem, K., and Chisholm, M.F. (2007). Scanning transmission electron microscopy for nanostructure characterization, in *Scanning Microscopy for Nanotechnology Techniques and Applications* (eds. Zhou, W., and Wang, Z.L.), Springer US, New York, pp. 152–191.

304. Tanaka, N. (2014). Introduction, in *Scanning Transmission Electron Microscopy of Nanomaterials*, Imperial College Press, pp. 1–8.

305. Ferrer, D., Blom, D.A., Allard, L.F., Mejía, S., Pérez-Tijerina, E., and José-Yacamán, M. (2008). *J. Mater. Chem.*, **18**, 2442.

306. Siles, P.F., Archanjo, B.S., Baptista, D.L., Pimentel, V.L., Joshua, J., Neves, B.R.A., and Medeiros-Ribeiro, G. (2011). *J. Appl. Phys.*, **110**, 24511.

307. Parreira, P., Paterson, G.W., McVitie, S., and MacLaren, D.A. (2016). *J. Phys. D: Appl. Phys.*, **49**, 95111.

308. Schamm, S., Bonafos, C., Coffin, H., Cherkashin, N., Carrada, M., Ben Assayag, G., Claverie, A., Tencé, M., and Colliex, C. (2008). *Ultramicroscopy*, **108**, 346.

309. Pinheiro, A.L.N., Zei, M.S., Luo, M.F., and Ertl, G. (2006). *Surf. Sci.*, **600**, 641.

310. Schoen, D.T., Holsteen, A.L., and Brongersma, M.L. (2016). *Nat. Commun.*, **7**, 12162.

311. Oura, K., Lifshits, V.G., Saranin, A.A., Zotov, A. V., and Katayama, M. (2003). *Surface Science: An Introduction*, Springer Berlin Heidelberg, Berlin.

312. Ichimiya, A., and Cohen, P.I. (2004). *Reflection High-Energy Electron Diffraction*, Cambridge University Press, Cambridge.

313. Mitsuishi, K., Hashimoto, I., Sakamoto, K., Sakamoto, T., and Watanabe, K. (1995). *Phys. Rev. B*, **52**, 10748.

314. Henini, M. (2012). *Molecular Beam Epitaxy: From Research to Mass Production*, Elsevier, Amsterdam.

315. Schilling, M., Ziemann, P., Zhang, Z., Biskupek, J., Kaiser, U., and Wiedwald, U. (2016). *Beilstein J. Nanotechnol.*, **7**, 591.

316. Swart, I., Liljeroth, P., and Vanmaekelbergh, D. (2016). *Chem. Rev.*, **116**, 11181.

317. Voigtländer, B. (2015). *Scanning Probe Microscopy*, Springer Berlin Heidelberg, Berlin, Heidelberg.

318. Riedel, D., Delattre, R., Borisov, A.G., and Teperik, T. V. (2010). *Nano Lett.*, **10**, 3857.

319. Potapenko, D. V., Hrbek, J., and Osgood, R.M. (2008). *ACS Nano*, **2**, 1353.

320. Liljeroth, P., Overgaag, K., Urbieta, A., Grandidier, B., Hickey, S.G., and Vanmaekelbergh, D. (2006). *Phys. Rev. Lett.*, **97**, 96803.

321. Osváth, Z., Deák, A., Kertész, K., Molnár, G., Vértesy, G., Zámbó, D., Hwang, C., and Biró, L.P. (2015). *Nanoscale*, **7**, 5503.

322. Haugstad, G. (2012). *Atomic Force Microscopy*, John Wiley & Sons, Inc., Hoboken, NJ, USA.

323. Murrell, M.P., Welland, M.E., O'Shea, S.J., Wong, T.M.H., Barnes, J.R., McKinnon, A.W., Heyns, M., and Verhaverbeke, S. (1993). *Appl. Phys. Lett.*, **62**, 786.

324. O'Shea, S.J. (1995). *J. Vac. Sci. Technol. B*, **13**, 1945.

325. Lanza, M. (2014). *Materials (Basel).*, **7**, 2155.

326. Kanoun, M., Baron, T., Gautier, E., and Souifi, A. (2006). *Mater. Sci. Eng. C*, **26**, 360.

327. Prakash, A., Maikap, S., Banerjee, W., Jana, D., and Lai, C.-S. (2013). *Nanoscale Res. Lett.*, **8**, 379.

328. Pohl, D.W., and Courjon, D. (eds.) (1993). *Near Field Optics*, Springer Netherlands, Dordrecht.

329. Moreno-Flores, S., and Toca-Herrera, J. L. (2012). What brings scanning near-field optical microscopy: the eyes at the nanoscale, in *Hybridizing Surface Probe Microscopies: Toward a Full Description of the Meso- and Nanoworlds*, CRC Press, Boca Raton, FL, pp. 119–152.

330. Sadiq, D., Shirdel, J., Lee, J.S., Selishcheva, E., Park, N., and Lienau, C. (2011). *Nano Lett.*, **11**, 1609.

331. Hecht, B., Sick, B., Wild, U.P., Deckert, V., Zenobi, R., Martin, O.J.F., and Pohl, D.W. (2000). *J. Chem. Phys.*, **112**, 7761.

332. Zenhausern, F., O'Boyle, M.P., and Wickramasinghe, H.K. (1994). *Appl. Phys. Lett.*, **65**, 1623.

333. Zhang, Z., Ahn, P., Dong, B., Balogun, O., and Sun, C. (2013). *Sci. Rep.*, **3**, 2803.

334. Modrow, H. (2005). X-ray methods for the characterization of nanoparticles, in *Nanofabrication Towards Biomedical Applications*, Wiley-VCH Verlag GmbH & Co. KGaA, Weinheim, FRG, pp. 163–196.

335. Sharma, A., Varshney, M., Park, J., Ha, T.-K., Chae, K.-H., and Shin, H.-J. (2015). *RSC Adv.*, **5**, 21762.

336. Scherrer, P. (1918). *Math. Klasse*, **2**, 98.

337. Williamson, G., and Hall, W. (1953). *Acta Metall.*, **1**, 22.

338. Warren, B.E., and Averbach, B.L. (1952). *J. Appl. Phys.*, **23**, 497.

339. Hall, B.D., Zanchet, D., and Ugarte, D. (2000). *J. Appl. Crystallogr.*, **33**, 1335.

340. Rana, A.M., Ismail, M., Ahmed, E., Talib, I., Khan, T., Hussain, M., and Nadeem, M.Y. (2015). *Mater. Sci. Semicond. Process.*, **39**, 211.

341. Kikhney, A.G., and Svergun, D.I. (2015). *FEBS Lett.*, **589**, 2570.

342. Borchert, H., Shevchenko, E. V., Robert, A., Mekis, I., Kornowski, A., Grübel, G., and Weller, H. (2005). *Langmuir*, **21**, 1931.

343. Li, T., Senesi, A.J., and Lee, B. (2016). *Chem. Rev.*, **116**, 11128.

344. Koerner, H., MacCuspie, R.I., Park, K., and Vaia, R.A. (2012). *Chem. Mater.*, **24**, 981.

345. Rieker, T., Hanprasopwattana, A., Datye, A., and Hubbard, P. (1999). *Langmuir*, **15**, 638.

346. Yu, Y., Jain, A., Guillaussier, A., Voggu, V.R., Truskett, T.M., Smilgies, D.-M., and Korgel, B.A. (2015). *Faraday Discuss.*, **181**, 181.

347. Ong, G.K., Williams, T.E., Singh, A., Schaible, E., Helms, B.A., and Milliron, D.J. (2015). *Nano Lett.*, **15**, 8240.

348. Wang, Z., Chen, O., Cao, C.Y., Finkelstein, K., Smilgies, D.-M., Lu, X., and Bassett, W.A. (2010). *Rev. Sci. Instrum.*, **81**, 93902.

349. Slyusarenko, K., Abécassis, B., Davidson, P., and Constantin, D. (2014). *Nanoscale*, **6**, 13527.

350. Hofmann, S. (2013). Introduction and outline, in *Auger- and X-Ray Photoelectron Spectroscopy in Materials Science*, 1st ed., Springer Berlin Heidelberg, Berlin, pp. 1–10.

351. Thompson, A.C., Attwood, D.T., Gullikson, E.M., Howells, M.R., Kortright, J.B., Robinson, A.L., Underwood, J.H., Kim, K.-J., Kirz, J., Lindau, I., Pianetta, P., Winick, H., Williams, G.P., and Scofield, J.H. (2009). X-ray data from the X-ray data booklet online, http://xdb.lbl.gov/.

352. Website link for XPS database for different elements and composite structures: (i) http://srdata.nist.gov/xps/intro.aspx, (ii) http://techdb.podzone.net/eindex.html, (iii) http://techdb.podzone.net/xps-e/.

353. Kim, D., Uk Lee, D., Kyu Kim, E., and Cho, W.-J. (2012). *Appl. Phys. Lett.*, **101**, 233510.

354. Rajamanickam, D., Dhatshanamurthi, P., and Shanthi, M. (2015). *Spectrochim. Acta Part A*, **138**, 489.

355. Miller, J.B., Matranga, C., and Gellman, A.J. (2008). *Surf. Sci.*, **602**, 375.

356. Mohai, M., and Bertóti, I. (2012). *Surf. Interface Anal.*, **44**, 1130.

357. Jablonski, A., and Zemek, J. (2009). *Surf. Interface Anal.*, **41**, 193.

358. Rahimi, M., Yousef, M., Cheng, Y., Meletis, E.I., Eberhart, R.C., and Nguyen, K. (2009). *J. Nanosci. Nanotechnol.*, **9**, 4128.

359. Belsey, N.A., Shard, A.G., and Minelli, C. (2015). *Biointerphases*, **10**, 19012.

360. Olivieri, G., and Brown, M.A. (2016). *Top. Catal.*, **59**, 621.

361. Wojcieszak, R., Genet, M.J., Eloy, P., Ruiz, P., and Gaigneaux, E.M. (2010). *J. Phys. Chem. C*, **114**, 16677.

362. Suzuki, S., Ishikawa, Y., Isshiki, M., and Waseda, Y. (2002). *Mater. Trans.*, **43**, 2523.

363. Caruso, A., Borca, C., Ristoiu, D., Nozières, J., and Dowben, P. (2003). *Surf. Sci.*, **525**, L109.

364. Preisler, E.J., Marsh, O.J., Beach, R.A., and McGill, T.C. (2001). *J. Vac. Sci. Technol. B*, **19**, 1611.

365. Tabet, N., Faiz, M., Hamdan, N., and Hussain, Z. (2003). *Surf. Sci.*, **523**, 68.

366. Davis, S. (1989). *J. Catal.*, **117**, 432.

367. Kerkhof, F.P.J.M., and Moulijn, J.A. (1979). *J. Phys. Chem.*, **83**, 1612.

368. Spadaro, M.C., D'Addato, S., Gasperi, G., Benedetti, F., Luches, P., Grillo, V., Bertoni, G., and Valeri, S. (2015). *Beilstein J. Nanotechnol.*, **6**, 60.

369. Newville, M. (2014). *Rev. Mineral. Geochem.*, **78**, 33.

370. Schnohr, C.S., and Ridgway, M.C. (2015). Introduction to X-ray absorption spectroscopy, in *X-Ray Absorption Spectroscopy of Semiconductors* (eds. Schnohr, C.S., and Ridgway, M.C.), Springer Berlin Heidelberg, Berlin, pp. 1–26.

371. Balasubramanian, C., Joseph, B., Gupta, P., Saini, N.L., Mukherjee, S., Di Gioacchino, D., and Marcelli, A. (2014). *J. Electron. Spectrosc. Relat. Phenom.*, **196**, 125.

372. Chen, C.-H., Sarma, L.S., Chen, J.-M., Shih, S.-C., Wang, G.-R., Liu, D.-G., Tang, M.-T., Lee, J.-F., and Hwang, B.-J. (2007). *ACS Nano*, **1**, 114.

373. Hwang, B.-J., Tsai, Y.-W., Sarma, L.S., Tseng, Y.-L., Liu, D.-G., and Lee, J.-F. (2004). *J. Phys. Chem. B*, **108**, 20427.

374. Hwu, Y., Yao, Y.D., Cheng, N.F., Tung, C.Y., and Lin, H.M. (1997). *Nanostruct. Mater.*, **9**, 355.

375. Kelly, J.A., Henderson, E.J., Clark, R.J., Hessel, C.M., Cavell, R.G., and Veinot, J.G.C. (2010). *J. Phys. Chem. C*, **114**, 22519.

376. Chang, Y.K., Hsieh, H.H., Pong, W.F., Tsai, M.-H., Chien, F.Z., Tseng, P.K., Chen, L.C., Wang, T.Y., Chen, K.H., Bhusari, D.M., Yang, J.R., and Lin, S.T. (1999). *Phys. Rev. Lett.*, **82**, 5377.

377. Hamad, K.S., Roth, R., Rockenberger, J., van Buuren, T., and Alivisatos, A.P. (1999). *Phys. Rev. Lett.*, **83**, 3474.

378. Polte, J., Erler, R., Thünemann, A.F., Sokolov, S., Ahner, T.T., Rademann, K., Emmerling, F., and Kraehnert, R. (2010). *ACS Nano*, **4**, 1076.

379. Polte, J., Tuaev, X., Wuithschick, M., Fischer, A., Thuenemann, A.F., Rademann, K., Kraehnert, R., and Emmerling, F. (2012). *ACS Nano*, **6**, 5791.

380. Polte, J., Ahner, T.T., Delissen, F., Sokolov, S., Emmerling, F., Thünemann, A.F., and Kraehnert, R. (2010). *J. Am. Chem. Soc.*, **132**, 1296.

381. Wang, F., Richards, V.N., Shields, S.P., and Buhro, W.E. (2014). *Chem. Mater.*, **26**, 5.

382. Jadzinsky, P.D., Calero, G., Ackerson, C.J., Bushnell, D.A., and Kornberg, R.D. (2007). *Science*, **318**, 430.

383. Grassian, V.H. (2008). *J. Phys. Chem. C*, **112**, 18303.

384. Kumar, N., and Kumbhat, S. (2016). Characterization tools for nanomaterials, in *Essentials in Nanoscience and Nanotechnology*, John Wiley & Sons, Inc., New Jersey, pp. 77–148.

385. Tomaszewska, E., Soliwoda, K., Kadziola, K., Tkacz-Szczesna, B., Celichowski, G., Cichomski, M., Szmaja, W., and Grobelny, J. (2013). *J. Nanomater.*, **2013**, 1.

386. Nishioka, M., Miyakawa, M., Kataoka, H., Koda, H., Sato, K., and Suzuki, T.M. (2011). *Nanoscale*, **3**, 2621.

387. Zimbone, M., Calcagno, L., Messina, G., Baeri, P., and Compagnini, G. (2011). *Mater. Lett.*, **65**, 2906.

388. Roy, D., and Fendler, J. (2004). *Adv. Mater.*, **16**, 479.

389. Norman Jr, T., Grant, C., and Zhang, J. (2005). Optical and dynamic properties of gold metal nanomaterials, in *Nanoparticle Assemblies and Superstructures*, CRC Press, Boca Raton, FL, pp. 193–206.

390. van Dijk, M.A., Tchebotareva, A.L., Orrit, M., Lippitz, M., Berciaud, S., Lasne, D., Cognet, L., and Lounis, B. (2006). *Phys. Chem. Chem. Phys.*, **8**, 3486.

391. Herrera, J., and Sakulchaicharoen, N. (2009). Microscopic and spectroscopic characterization of nanoparticles, in *Drug Delivery Nanoparticles Formulation and Characterization*, CRC Press, Boca Raton, FL, pp. 239–251.

392. Davis, E.A., and Mott, N.F. (1970). *Philos. Mag.*, **22**, 903.

393. Oldenburg, S., Averitt, R., Westcott, S., and Halas, N. (1998). *Chem. Phys. Lett.*, **288**, 243.

394. Link, S., and El-Sayed, M.A. (1999). *J. Phys. Chem. B*, **103**, 8410.

395. Ullrich, B., Singh, A.K., Bhowmick, M., Barik, P., Ariza-Flores, D., Xi, H., and Tomm, J.W. (2014). *AIP Adv.*, **4**, 123001.

396. Gilliland, G.D. (1997). *Mater. Sci. Eng. R*, **18**, 99.

397. Djurišić, A.B., Leung, Y.H., Tam, K.H., Hsu, Y.F., Ding, L., Ge, W.K., Zhong, Y.C., Wong, K.S., Chan, W.K., Tam, H.L., Cheah, K.W., Kwok, W.M., and Phillips, D.L. (2007). *Nanotechnology*, **18**, 95702.

398. Linnros, J., and Grivickas, V. (2002). Carrier lifetime: free carrier absorption, photoconductivity, and photoluminescence, in *Characterization of Materials*, John Wiley & Sons, Inc., Hoboken, NJ, USA.

399. Cassette, E., Dean, J.C., and Scholes, G.D. (2016). *Small*, **12**, 2234.

400. Bruhn, B., Limpens, R., Chung, N.X., Schall, P., and Gregorkiewicz, T. (2016). *Sci. Rep.*, **6**, 20538.

401. Park, Y.-S., Ghosh, Y., Xu, P., Mack, N.H., Wang, H.-L., Hollingsworth, J.A., and Htoon, H. (2013). *J. Phys. Chem. Lett.*, **4**, 1465.

402. Harbold, J.M., and Wise, F.W. (2007). *Phys. Rev. B*, **76**, 125304.

403. Klimeov, E., Vack, J., Holy, V., and Pelant, I. (2013). *Nanomater. Nanotechnol.*, 1.

404. de Boer, W.D.A.M., Timmerman, D., Dohnalová, K., Yassievich, I.N., Zhang, H., Buma, W.J., and Gregorkiewicz, T. (2010). *Nat. Nanotechnol.*, **5**, 878.

405. Zhang, S.-L. (2012). *Raman Spectroscopy and Its Application in Nanostructures*, John Wiley & Sons, Ltd, Chichester, UK.

406. Zhang, P., Feng, Y., Anthony, R., Kortshagen, U., Conibeer, G., and Huang, S. (2015). *J. Raman Spectrosc.*, **46**, 1110.

407. Gao, Y., Zhao, X., Yin, P., and Gao, F. (2016). *Sci. Rep.*, **6**, 20539.

408. Zhang, R., Chang, K.-C., Chang, T.-C., Tsai, T.-M., Chen, K.-H., Lou, J.-C., Chen, J.-H., Young, T.-F., Shih, C.-C., Yang, Y.-L., Pan, Y.-C., Chu, T.-J., Huang, S.-Y., Pan, C.-H., Su, Y.-T., Syu, Y.-E., and Sze, S.M. (2013). *Nanoscale Res. Lett.*, **8**, 497.

409. Mudunkotuwa, I.A., Al Minshid, A., and Grassian, V.H. (2014). *Analyst*, **139**, 870.

410. Khomenkova, L., Normand, P., Gourbilleau, F., Slaoui, A., and Bonafos, C. (2016). *J. Nano Res.*, **39**, 121.

411. Lehninger, D., Khomenkova, L., Roder, C., Gartner, G., Abendroth, B., Beyer, J., Schneider, F., Meyer, D.C., and Heitmann, J. (2015). *ECS Trans.*, **66**, 203.

412. Van Duy, N., Jung, S., Nga, N.T., Son, D.N., Cho, J., Lee, S., Lee, W., and Yi, J. (2010). *Mater. Sci. Eng. B*, **175**, 176.

Chapter 2

Modeling and Simulation of Nanocrystal Flash Memory

Bikash Sharma* and Chandan Kumar Sarkar*

NanoDevice Simulation Lab, Department of Electronics and Telecommunication Engineering, Jadavpur University, 188 Raja S. C. Mallick Road, Kolkata 700032, West Bengal, India
ju.bikash@gmail.com

"Memory" in a simple dictionary meaning is "the ability to remember information, experiences, etc.," and in a computer it stands for the part where data is stored in a temporary or permanent manner.

All information processing can be viewed as consisting of the sequential actions of directives on the data or information. Steps taken for operation of any data or information consist of sensing, processing/interpreting, and acting.

This chapter provides a comprehensive review of modeling and simulation results for nanocrystal-based nonvolatile memory (flash) memory devices. The various models and simulation results have been duly depicted, along with the progress made in the development of such memory devices. The chapter showcases the

---

*Both authors have contributed equally to this work.

Nanocrystals in Nonvolatile Memory
Edited by Writam Banerjee
Copyright © 2018 Pan Stanford Publishing Pte. Ltd.
ISBN 978-981-4774-73-4 (Hardcover), 978-1-351-20327-2 (eBook)
www.panstanford.com

initial development of nanocrystal-based flash memory devices followed by various theoretical and mathematical models in use. The different current models for nitride-trap-based memory are depicted. Then scaling properties are reviewed with respect to the use of semiconductor (Si) nanocrystals. Model and simulation results are discussed for tunneling currents and programming time in such devices. Then there is a brief discussion to bring about an idea about the formation of nanocrystals. As the size of such nanocrystals is very small and they are difficult to grow, a view about their growth is quite essential. Next the retention characteristics of nanocrystal-based flash memory and use of high-dielectric-constant (high-κ) materials to enhance the scaling of such devices are reviewed. Finally, there is a comparative study of tunneling currents for nonvolatile memory devices based on semiconductor and those based metal nanocrystals. Hence, the chapter has tried to effectively bring out the progress made in the modeling and simulations done for nanocrystal-based nonvolatile (flash) memory with various parameters of such devices. The chapter reviews various parameters, like retention, growth, charging (programming) and discharging (erase), tunneling currents, and scaling models, for nonvolatile (flash) memory.

2.1 Introduction

Archiving of information and data has been done since the dawn of human civilization, as far back as in 2000 BC, during the Egyptian civilization, through hieroglyphics—a system in which pictures are used instead of words. Hieroglyphs were written on mud or stone tablets. Archiving information by this mechanism is very long-lasting (nonvolatile), but the speed of writing is very slow. And keeping in mind the amount of data or information that we have today we need a mechanism to have a repository that's fast, long-lasting, and miniaturized enough to fit a large amount of data in a small amount of space. Hence a number of methods and technology have evolved and undergone tremendous change and upgradation to address this challenge and provide a suitable device and storage mechanism. Memory devices have evolved from mud or stone tablets, cog on wheels, punched cards, and capacitor or rotating drums to electromechanical relays, vacuum tubes, cathode-ray tubes (CRTs), magnetic drums, magnetic tapes, magnetic core, magnetic disks, transistor (bipolar), transistor (metal-oxide-semiconductor field-effect transistors [MOSFETs]), ferroelectric memory, flash memory, nanocrystal memory, and other latest technologies.

In this chapter we will discuss the various modeling and simulations methods deployed in different literatures to enhance some specific parameters and characteristics of the nanocrystal flash memory.

A flash memory operates on the principle of storing charges in the floating gate of a MOSFET, as depicted in Fig. 2.1a [1]. A floating gate stores the charge carriers with the application of proper biases and voltages across gate, source, and drain. The charge gets transported via tunnel oxide under the influence of quantum mechanical tunneling. The charge is stored even after the voltages are removed. Subsequently, alteration to two distinct values, 0 for erased state and 1 for programmed state (Fig. 2.1b), evaluates the amount of charge stored in a floating gate.

For a metal-oxide-semiconductor (MOS) transistor, the threshold voltage can be written from the basic theory as

$$V_{th} = 2\phi_f + \phi_b - \frac{Q_i}{C_i} - \frac{Q_d}{C_i} - \frac{Q_t}{\varepsilon_i} d_i ,\qquad(2.1)$$

where

ϕ_b = the work-function difference between the gate and the bulk material,

ϕ_f = the Fermi potential of the semiconductor at the surface,

Q_i = the fixed charge at the substrate/tunnel oxide interface,

Q_d = the charge in the silicon depletion layer,

Q_t = the charge stored in the control oxide at a distance d_i from the gate,

C_i = the capacitance between the gate and the substrate, and

ε_i = the dielectric constant of the oxide.

The charge stored Q_t causes the threshold shift, given by

$$\Delta V_{th} = -\frac{Q_t}{\varepsilon_i} d_i .\qquad(2.2)$$

We can read the state of a flash memory by the application of a gate voltage V_{read} with a threshold voltage having two values. The transistor conducts current in one state while there is no current flow in the other state. The stored charge retains the memory states even after the biases are withdrawn.

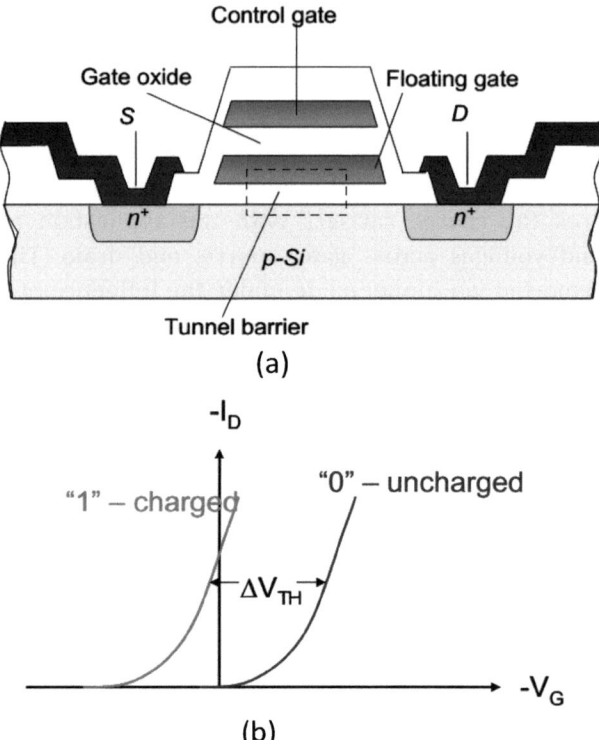

Figure 2.1 (a) Schematic representation of a floating-gate memory and (b) influence of charge in the dielectric on the threshold of a p-channel transistor. Reprinted from Ref. [1], with the permission of AIP Publishing.

Electrons generated in the silicon substrate get stored in the floating gate and travel across the tunnel oxide in a flash memory as indicated in Fig. 2.1a. The tunnel oxide thickness should be optimized such that on application of a high electric field, the electrons should be able to travel across it while when there is no electric field the electrons should remain trapped within it. Also, even after the removal of the applied voltage there should exist a minimum electric field across the tunnel barrier that prevents the charges from tunneling back into the substrate.

In a memory device the overall speed and the charge retention properties are balanced by optimizing the tunnel oxide thickness. As the programming and erasing speeds are high when the tunnel thickness is small, the charge retention becomes a problem and vice versa with thicker tunnel oxide.

This optimization is very important for analyzing the overall performance of a flash device.

The equation for a parallel plate capacitor is given by

$$C = A\kappa\varepsilon_0/t, \tag{2.3}$$

where κ is the material's dielectric constant, ε_0 is the free space permittivity, A is the area of the capacitor area, and t is the dielectric layer thickness.

It is observed that the capacitance of the gate dielectric can be decreased by decreasing the area. Hence, until most recently the thickness t of the gate dielectric was being reduced. But the charge retention and the tunneling problem posed a limit to the minimum thickness that can be used for an effective memory device. Also, we can observe from Eq. 2.3 that the same can be achieved by increasing the dielectric constant while keeping the thickness to the minimum. Hence, high-κ dielectrics have come in the picture and are being used extensively.

2.2 Developments in Nanocrystal Memory [2]

Today flash nonvolatile memory finds application in almost all portable electronic devices, though due to extreme miniaturization of the device, the scaling capability has reached its limit. The reliability of these floating-gate flash memory devices is facing serious issues because the charges stored get leaked through the thin tunneling oxide layer. Hence, the use of discrete nanocrystals in the floating gate provides better reliability, higher scalability, and better operation speed [2].

In 1995, IBM proposed the use of discrete nanocrystals to store charge [3]. The discrete electron storing capability allows the nanocrystal memory to store 2 bits per cell. This facilitates an increase in the memory density [3].

Ting et al. [2] proposed a model for injecting or removing charges that makes nanocrystals store charge (Fig. 2.2). Hence for the program and erase operations an appreciable amount of voltage is required for turning on the transistor (Fig. 2.2c). The drain current is read by application of the read voltage (V_{read}) across the program and erase operations, subsequently allowing the determination of the 0 or 1 memory state. The program operation is mainly adopted by channel hot electron injection (CHEI) and Fowler–Nordheim (FN) tunneling, while the erase operation is mainly adopted by FN tunneling [4].

A high applied electric field induces FN tunneling, with the electron tunneling across a triangular barrier. And at a low applied electric field, direct tunneling is induced, where the electrons tunnel across trapezoidal barriers. In the CHEI mechanism a fraction of energetic electrons traveling at high speeds in the channel get scattered and thereafter change course to tunnel into the oxide defects (as per the lucky electron model).

In an advanced nonvolatile memory cell we look for sufficient threshold voltage shift, good retention characteristics, high reliability, and low power consumption [5–7].

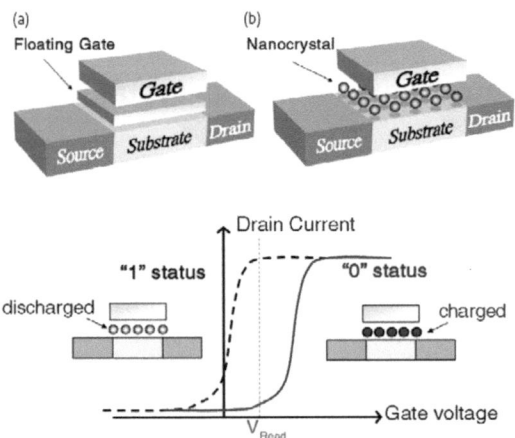

Figure 2.2 (a) Structure of a floating-gate nonvolatile memory, (b) structure of a nanocrystal-based nonvolatile memory, and (c) program and erase modes of the nanocrystal memory device. Reprinted from Ref. [2], Copyright (2011), with permission from Elsevier.

With development in nanotechnology, various metal nanocrystals, such as Ag, Au, Al, Ni, Co, W, TiN, Hf, and NiS_2, have been studied for their respective memory effects. Their effects have been demonstrated through the fabrication and study of their characteristics [8–24]. Also with a higher effective thickness the memory effect is shown to improve [8, 9, 10, 24]. The following is a summary of memory characteristics based on various metal nanocrystals (as depicted in Table 2.1) from various studies [11, 12, 25–32].

Table 2.1 Memory characteristics of various materials

Element	Tunnel dletectric	Sweep voltage (V)	Memory windows (V)
Si	SiO_2	10 to –10	4.6
Ge	SiO_2	5 to –5	0.42
Au	SiO_2	2 to –4	2.3
Ag	SiO_2	2 to –4	2.1
Pt	SiO_2	2 to –4	3.8
W	SiO_2	10 to –10	9
Ni	SiO_2	10 to –10	4
	HfO_2	2 to –2	0.75
NiSi	SiO_2	3 to –3	1.04
	HfO_2	3 to -3	1.38
Co	SiO_2	7 to –7	1.8
$CoSi_2$	SiO_2	3 to –3	1.1
Mo	SiO_2	9 to –11	3.6
Al	AlN	5 to –5	1
$Ni_{1-x}Fe_x$	PI	11 to –6	2
TiN	Al_2O_3	16 to –16	2.9
HfO_2	SiO_2	$V_g = 9$ V, $V_d = 9$ V	2.2
CeO_2	SiO_2	$V_g = 9$ V, $V_d = 9$ V	3.1

Source: [2]

2.3 Model for Nanocrystal and Nitride-Trap Memory [33]

De Salvo et al. [33] proposed a model based on a modified-floating-gate-like approach (as shown in Fig. 2.3) and a trap-like approach. Temperature, bias, and time-based analyses of static and dynamic charging and discharging were carried out, with detailed theoretical and experimental study [33].

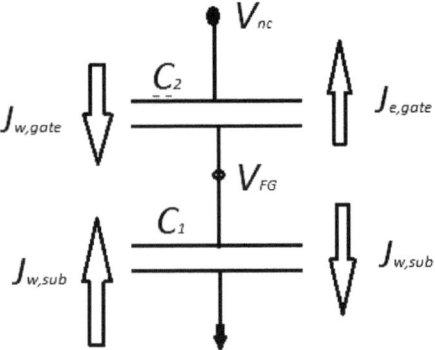

Figure 2.3 Capacitor model for traditional memory used in the floating-gate-like approach [33].

The floating-gate-like approach model was derived from the current-continuity mechanism [34] and floating-gate memory equations [35] for a Si-nanocrystal-based memory. This methodology takes into account the proportion of charges filling into and emptying from the nanocrystals of trapped layer dielectrics. The same is given by

$$\frac{dQ_{nc}}{dt} = J_e - J_w ,\qquad(2.4)$$

where

Q_{nc} = the charge embedded/trapped in the nanocrystals,

J_w = charging of the currents in the nanocrystals by tunneling of charges from the substrate to the trap (floating) layer, and

J_e = discharging of currents from the nanocrystals by tunneling of charges from the trap (floating) layer to the substrate.

The flux of the tunneling (charging) electrons is measured by the FN tunneling current, later extended to direct current as a part of the charging current, given by [39]

$$J_{\text{w,sub}}(V_{\text{nc}}) = \frac{4\pi q^3 m_{\text{Si}} kT}{h^3} \int_0^{\Phi} \Theta_{\text{sub}}(E, V_{\text{nc}}).\ln\left[1 + \exp\left(\frac{E_{\text{sub}} - E}{kT}\right)\right] dE .$$

(2.5)

Similarly, a part of the discharging current, the electrons tunneling from the nanocrystal to the substrate, is given by

$$J_{\text{e}}(V_{\text{nc}}, V_{\text{c}}) = \frac{4\pi q^3 m_{\text{Si}} kT}{h^3} \int_0^{\Phi} \Theta_{\text{nc}}(E, V_{\text{nc}}, V_{\text{c}}).\ln\left[1 + \exp\left(\frac{E_{\text{nc}} - E}{kT}\right)\right] dE ,$$

(2.6)

where

$$\Theta_{\text{nc}}(E, V_{\text{nc}}, V_{\text{c}}) = \Theta_{\text{nc,sub}}(E, V_{\text{nc}}) + \Theta_{\text{nc,gate}}(E, V_{\text{nc}}, V_{\text{c}}),$$

$\phi =$ the barrier height (in electron volts) of the Si/SiO$_2$ layer interface,

$E =$ electron energy in electron volts,

$E_{\text{sub}} =$ the Fermi potential of the substrate, and

$V_{\text{nc}} =$ the silicon nanocrystal potential.

The analytical calculations were done by approximating the above equations replaced by their low-temperature values [36], given by the more generalized equation

$$J_{\text{x,y}}(V) = A.\left(\frac{V}{T_{\text{i}}}\right)^2 \Theta_{\text{y}}(V),$$

(2.7)

where

$A = q^3 m_{\text{Si}}/(16\pi^2 \hbar m_{\text{ox}} \Phi)$,

$x =$ charging or discharging,

$y =$ substrate, floating-gate/trap layer, or nanocrystal,

$\Theta_{\text{y}} =$ corresponding Wentzel–Kramer–Brillouin (WKB) transparency at $E = 0$,

$T_{\text{i}} =$ thickness of the tunnel dielectric, and

$V =$ ad hoc potential drop.

Equation 2.7, which expresses FN current, is basically for metallic electrodes. Hence, for the nonmetallic character of a Si nanocrystal and the inversion charge in the channel the above equations have to be modified. The tunneling current for electron tunneling from the inversion layer is given from Eq. 2.6 as follows:

$$J_{w,sub}(V_{nc}) = f_{sub} \cdot Q_{inv} \cdot \Theta_{sub}(V_{nc}), \qquad (2.8)$$

where

Q_{inv} = the absolute inversion charge and

f_{sub} = the escape frequency ($\sim 10^{13}$ Hz [37]).

Applying Boltzmann's statistics for undoped Si nanocrystals, the FN expression from Eq. 2.8 can be rewritten as follows [33]:

$$J_e(V_{nc}, V_c) \approx \frac{4\pi q^3 m_{Si}(kT)^2}{h^3}$$

$$\exp\left(\frac{E_{nc}}{kT}\right) \cdot \left[\Theta_{nc,sub}(kT, V_{nc}) + \Theta_{nc,gate}(kT, V_{nc}, V_c)\right]. \qquad (2.9)$$

From the above equation we have the relation between the charge of the Si nanocrystal and the volume electron concentration in the Si nanocrystal as follows:

$$J_e(V_{nc}, V_c) = \frac{4\pi q^3 m_{Si} kT}{h^3} \cdot \frac{n_{nc}}{N_c} \cdot \left[\Theta_{nc,sub}(kT, V_{nc}) + \Theta_{nc,gate}(kT, V_{nc}, V_c)\right]$$

$$= f_{nc} \cdot |Q_{nc}| \cdot \left[\Theta_{nc,sub}(kT, V_{nc}) + \Theta_{nc,gate}(kT, V_{nc}, V_c)\right]$$

$$(2.10)$$

where

f_{nc} = ($4\pi^2 m_{Si}(kT)^2/h^3\Phi_{nc}N_c$) is the escape frequency in the Si nanocrystal,

n_{nc} = $N_c \cdot \exp(E_F/kT)$ is the volume electron concentration in the Si nanocrystal, and

N_c = the effective density of states in the Si conduction band.

In the above equations the nanocrystals have been assumed to be large in size, thus allowing us to neglect the quantum or Coulomb blockade effects.

Also, applying Weinberg's escape frequency,

$$f_{nc} = E_0/h, \qquad (2.11)$$

where E_0 = the lowest level in the silicon nanocrystal's conduction band.

Therefore, after applying the single-electron charging effects to the charging and discharging currents in Eqs. 2.8–2.10, accounted for by the Coulomb blockade effect, the current can be calculated by the modified equation

$$J = \frac{J_0 \dfrac{\Delta E}{kT}}{\exp\left(\dfrac{\Delta E}{kT}\right) - 1} \approx J_0 \exp\left(-\frac{\Delta E}{2kT}\right),$$ (2.12)

where

J = current in the presence of the Coulomb blockade effect,

J_0 = current without the Coulomb blockade effect, and

ΔE = the charging energy characteristic of the Si nanocrystal coupled with the substrate.

Also, Yano et al. [38] consider

$$\Delta E \approx n.\Delta E_0,$$ (2.13)

where

$$\Delta E_0 \approx \frac{q T_i}{\varepsilon_{ox} \Phi_{nc}^2}$$ (2.14)

is the basic charging energy. Hence, the value of the current is in integral multiples of the number of electrons with the Coulomb blockade factor (CBF) [38], given by

$$CBF = \exp(-\Delta E_0/2kT),$$ (2.15)

and the current is given by

$$J \approx J_0 CBF^n.$$ (2.16)

Here, a unit increment is carried out by the exponent (n), yielding an integer value for $|Q_{nc}|/q$ each time.

Neglecting the Coulomb blockade effect due to a larger average nanocrystal size, the silicon nanocrystal potential (V_{nc}) is calculated as a classical capacitor model (as shown in Fig. 2.3) [35, 39, 40] for the conventional floating-gate memory devices, given by

$$V_{nc} = \frac{C_2}{C_1 + C_2} V_c + \frac{Q_{nc}}{C_1 + C_2},$$ (2.17)

where

C_1 and C_2 = the control gate capacitor and the substrate coupling capacitor, respectively, and

Q_{nc} = the floating-gate charge.

The drain current transfer characteristics for the formulation of the MOSFET, where gate and threshold voltages with inversion charge are used for faster computations.

Hence, the threshold voltage (V_{th}) is given by

$$V_{th} = V_{th0} - \frac{Q_{FG}R_{nc}}{C_2},$$ (2.18)

where R_{nc} = the density of the silicon nanocrystal.

Here, R_{nc}, which is the density of the silicon nanocrystal, acts as a MOSFET's trapped charge-weighting factors.

Also, the product $Q_{FG}R_{nc}$ = the whole channel surface's effective areal density of charges in the MOSFET. The MOSFET operation is evaluated considering the uniform distribution of the trapped charge (in the nanocrystal) over the channel.

Shockley–Read–Hall statistics are applied in the trap-like approach model. In this model the silicon nanocrystal is treated as a continuum of traps at a distance of t from the substrate in the SiO_2 layer. The tunnel dielectric of Si_3N_4 was used in this model, where a large number of traps were considered to be uniformly distributed over the interface and the trapping layer embedded with silicon nanocrystals. According to the model applied the time evolution is given by

$$\frac{df_t(t,E_t)}{dt} = w_e + r_h - (r_e + w_e + r_h + w_h)f_t$$ (2.19)

where

r_e, r_h = the charge's emission rate (e = electron and h = hole) and
w_e, w_h = the charge's capture rate.

Also, the capture and emission rates depend on the carrier densities, the depth of the trap cross section, and the applied electrical field [41]. The continuity potential relation is estimated by the surface potential, given by

$$V_c(t) = V_{fb} - \Psi_s(t) - \frac{Q_{SC}(\Psi_s,t)}{C_{ox-eq}} + \frac{Q_{ot}(t)}{C_2},$$ (2.20)

where

V_{fb} = the flat-band voltage of a chargeless dielectric,

Ψ_S = surface potential,

Q_{SC} = the charge of the semiconductor,

Q_{ot} = the trapped charge in the oxide, and

$C_{ox\text{-}eq}$ = the capacitance of the oxide, given by $C_1.C_2/(C_1 + C_2)$.

Characteristics of the device can be evaluated by the time-dependent gate voltage application. The self-consistent solution of Eqs. 2.19 and 2.20 facilitates the computation of the inversion charge and trapped charge time evolution (assuming $t = 0$ and the traps are at equilibrium). As only the flux of the current through the bottom dielectric is considered for the computation in this model at $J_{in/out,sub} = 0$, a steady state is reached. Here, the current from and into the gate dielectric is neglected. Also, by applying Schulz's formulation [42] and Yano's method [38] the Coulomb blockade effect can be accounted for in this model. The type of trap and the charge state determine the modification of the capture and emission rates calculated by the application of these formulations.

2.4 Memory Device Scaling with the Use of a Silicon Nanocrystal [43]

Steimle et al. [43] proposed a model for scaling nonvolatile memory using a silicon nanocrystal, as depicted later. The oxide thickness has been scaled to 10 nm in the conventional silicon-oxide-nitride-oxide-silicon (SONOS)-type memory. Read disturb effects were caused during program and erase operations induced by the hot carrier injection and FN tunneling, respectively. Hence, further scaling wasn't easy due to saturation occurring during erase operation [44], as reported by Steimle et al. [43] in Fig. 2.4, which affected the SONOS-type memory. In this regard the introduction of nanocrystals (initially silicon) for charge storage in nanocrystal-based memory provided an advantage over the conventional floating-gate-type memory. The oxide thickness could be scaled down to 5 nm. However, issues like charge confinement in nanocrystals pose some problem in the use of this type of device. Fabricating nanocrystals of appropriate size and density and conserving them during the fabrication process of the device are also critical.

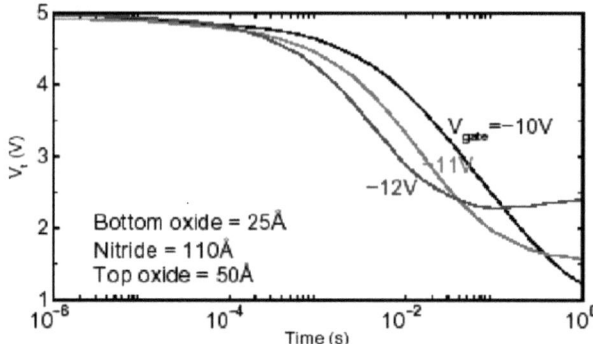

Figure 2.4 Erase saturation in SONOS nonvolatile memory, indicating the reduced memory window as the erase voltage is increased. Reprinted from Ref. [43], Copyright (2007), with permission from Elsevier.

With the storage of electrons in the nanocrystals, the induced threshold voltage shift is given by [45]

$$\Delta V_{th} = \frac{mke}{\varepsilon_{med}}\left(t_{cox} = \frac{\varepsilon_{med}t_{nc}}{2\varepsilon_{Si}}\right), \qquad (2.21)$$

where

m = the density (number) of nanocrystals,

k = the average number of electrons stored per nanocrystal,

e = the charge of an electron,

ε = the permittivity of the medium,

t_{nc} = the diameter of the nanocrystal, and

t_{cox} = the thickness of the control oxide.

To overcome the probability of defects in any nanocrystal, they should be adequately isolated electrically. Hence, a minimum of 4 nm separation has to be incorporated. Also, Coulomb blockade effects become prominent when the size of the nanocrystal becomes lesser than 3 nm and hence adversely affects the charge retention capability. The nanocrystal's self-capacitance is given as

$$C = 2\pi\varepsilon_{med}t. \qquad (2.22)$$

Additional electrons are required to charge the nanocrystal. With the addition of the nth electron, the rise in energy of the nanocrystal is given by [43]

$$\Delta E_{n,n-1} = \frac{e^2}{2C}[n(n-1)-(n-1)(n-2)] = \frac{(n-1)e^2}{C}. \qquad (2.23)$$

Hence, with the addition of each electron the electrochemical potential change is given by

$$\Delta\mu = \Delta E_{n,n-1} - \Delta E_{n-1,n-2} = q^2/C. \qquad (2.24)$$

Electron storage is restricted due to the loss of charge through the bottom oxide, occurring due to tunneling that is highly affected by the increasing of the energy levels. But it enhances the erasing process due to FN tunneling, which becomes dominant at such high energy levels. Deeper electron storage traps (3 eV) enhance electron retention in the nanocrystal memory as compared to nitride traps (1–2 eV). Another advantage of nanocrystal memory is that with an increase in the energy level followed by enhanced FN tunneling, the saturation threshold voltage does not depend on the gate bias. Electrons are tunneled from the nanocrystal to the substrate because of more charged electrons existing in the bottom oxide field as compared to the top oxide field. As fields and FN currents equalize, a steady state is reached. And no effective charge transfer takes place through the nanocrystals in a steady state. Using WKB approximation [46] shows the erase characteristics as reported by Steimle et al. [43] in Fig. 2.5, with Coulomb blockade and gate injection effects. In nanocrystal-embedded gate dielectric memory, data retention is less sensitive to temperature because of the deeper charge traps. In nanocrystal nonvolatile-based memory bit cells, the electrical operation is apparent due to the discrete nature of nanocrystals and due to Coulomb blockade, which limits the storage of charge. While the programming of the nanocrystals is slow in the case of direct tunneling when the tunnel oxide thickness is less than 4 nm, it is possible to operate the nanocrystals at a gate voltage of 6–10 V. However, at higher gate voltages, of 12–14 V, FN tunneling gets dominant. And the programming of the memory cells with smaller nanocrystal sizes becomes difficult due to the absence of capacitive coupling effects and large, localized fields.

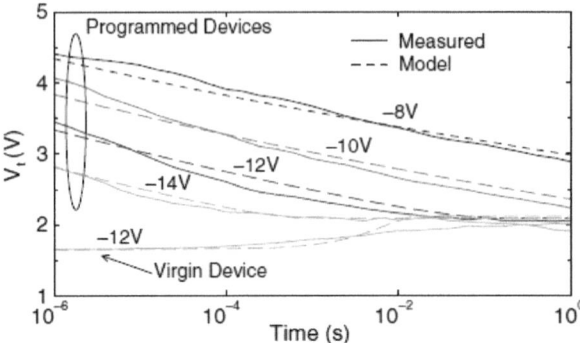

Figure 2.5 Erase characteristics of a nanocrystal device with 50 Å bottom oxide, 70 Å top oxide, and nanocrystals with a mean diameter of approximately 60 Å. Reprinted from Ref. [43], Copyright (2007), with permission from Elsevier.

2.5 Modeling of Tunneling Currents [47]

The study of tunneling currents in semiconductor (Si) nanocrystal–embedded nonvolatile memory was proposed in a model by Chakraborty et al. [47]. The limitations due to the extensive scaling of the dielectrics have been successfully improved by the use of a Si-nanocrystal-based memory device [48]. Low power operations have been realized with efforts in embedding silicon nanocrystals in a silicon dioxide (SiO$_2$) gate dielectric [49, 50]. It's of great interest to study the tunneling current at the interface of a silicon substrate and a composite gate dielectric embedded with Si nanocrystals. For a memory device the issue of tunneling current is very important.

The model considered for investigation by Chakraborty et al. consists of a layered structure where one layer of SiO$_2$ embedded with Si nanocrystals is stacked on top of a pure SiO$_2$ layer.

On application of bias across the gate terminal, the electrons tunnel through the composite gate dielectric from the Si substrate. The model investigates direct tunneling [51] and FN tunneling [52, 53], which account for the gate current in a nanocrystal-embedded gate dielectric composite. This demands descriptions of the effective mass, the dielectric constant, and the bandgap. Further, these effects are dependent upon the nanocrystal size, the volume fraction, and the tunneling barrier height.

The applied field with respect to the barrier height defines the type of tunneling that exists. FN tunneling is prominent when the applied voltage is higher than the barrier height, while direct tunneling exists when the applied voltage is lower than the barrier height. The WKB approximation method is deployed to calculate the FN tunneling probability.

A narrow potential well confines the carriers near the silicon substrate in a MOS structure. Electrons tunnel from the Si substrate across the gate dielectric embedded with Si nanocrystals on application of an electric field. Due to the stacked structure and existence of two different dielectrics of a pure gate SiO_2 layer and Si nanocrystal embedded in a SiO_2 layer, a parallel plate capacitor is formed.

For the composite gate dielectric the equivalent dielectric constants in terms of the oxide thickness is given by

$$\varepsilon_{eqv} = \left[\frac{t_{ox}}{\varepsilon_{ox}t} + \frac{t-t_{ox}}{\varepsilon_{nc\text{-}SiO_2}t} \right]^{-1}, \qquad (2.25)$$

where

ε_{eqv} = the equivalent dielectric constant,

ε_{ox} = the dielectric constant of the pure SiO_2,

$\varepsilon_{nc}\text{-}SiO_2$ = the dielectric constant of the Si-nanocrystal-embedded SiO_2,

t = total gate dielectric thickness, and

t_{ox} = pure SiO_2 layer thickness.

Considering randomly distributed nanocrystals (assumed to be spherical) in the dielectric medium, the dielectric constant following the Maxwell Garnett theory of the composite gate dielectric is given by

$$\varepsilon_{nc-SiO_2} = \frac{\varepsilon_{ox}[2v(\varepsilon_{nc-Si}-\varepsilon_{ox})+(\varepsilon_{nc-Si}+2\varepsilon_{ox})]}{\varepsilon_{nc-Si}+2\varepsilon_{ox}-v(\varepsilon_{nc-Si}-\varepsilon_{ox})}, \qquad (2.26)$$

where

v = the volume fraction of the Si **nanocrystal** and

ε_{nc-Si} = the dielectric constant of the Si nanocrystal, which is given by

$$\varepsilon_{nc-Si} = \varepsilon_{Si} - t_s(\varepsilon_{Si}-1). \qquad (2.27)$$

Here

t_s = dielectric susceptibility whose relative change depends on the size of the nanocrystal,

ϕ_b = the barrier height between the composite gate dielectric and the edge of the conduction band of the silicon substrate and is given by

$$\phi_b = (E_{g_{eqv}} - E_{g_{Si}})/2 \,, \tag{2.28}$$

where

$E_{g_{eqv}}$ = the equivalent bandgap of the gate dielectric and

$E_{g_{Si}}$ = silicon's bandgap.

An average of the bandgaps is considered to calculate the $E_{g_{eqv}}$ of the composite gate dielectric, given by

$$E_{g_{eqv}} = \frac{E_{g_{ox}} + E_{g_{nc\text{-}SiO_2}}}{2} \,, \tag{2.29}$$

where

$E_{g_{ox}}$ = the bandgap of pure SiO$_2$ and

$E_{g_{nc\text{-}SiO_2}}$ = the bandgap of the SiO$_2$-embedded nanocrystals.

The bandgap of embedded dielectrics is calculated assuming the formation of an alloy of two materials, Si nanocrystals and SiO$_2$. As per the virtual crystal approximation $E_{g_{nc\text{-}SiO_2}}$ is given by

$$E_{g_{nc\text{-}SiO_2}} = E_{ox}(1-v) + E_{g_{nc\text{-}Si}}v \,, \tag{2.30}$$

where

$$E_{g_{nc\text{-}Si}} = E_{g_{Si}} + \frac{C}{d_0^n}\left(\frac{d_m}{d_0}\right)^{n(2n+5)/3}. \tag{2.31}$$

Here

d_0 = the mean size of the nanocrystal,

d_m = the maximum size in a log-normal distribution,

n = 1.22 and d_0/d_m = 0.7 [55], and

C = 3.9.

FN tunneling is induced by the discharge of electrons either from the gate electrode or from the inversion layer [56]. It occurs when the applied voltage $V > (\phi_b - E_0)/q$ arises as a triangular potential barrier (Fig. 2.6a [47]).

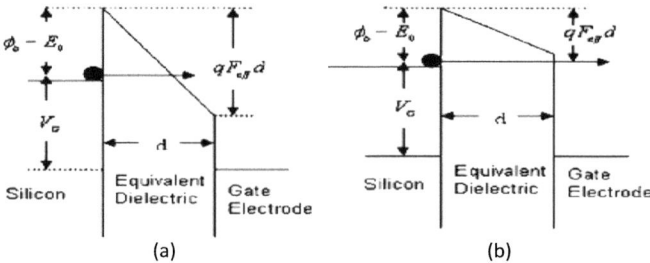

Figure 2.6 Band structure criteria for (a) FN tunneling and (b) direct tunneling. Reprinted from Ref. [47], with the permission of AIP Publishing.

Near the silicon substrate surface, applying the linear potential approximation,

$$\varepsilon_{Si}F_{Si} = F_{eqv}\varepsilon_{eqv} = F, \qquad (2.32)$$

where

F_{Si} = silicon's surface field,

F_{eqv} = the equivalent field within the composite dielectric, and

F = field in the vacuum.

Hence, the electron density equation is given by

$$N_s = \frac{\varepsilon_o \varepsilon_{eqv}}{q} = \frac{\varepsilon_o F}{q}, \qquad (2.33)$$

where

N_s = the electron density at the surface of silicon,

ε_o = the vacuum's dielectric constant, and

q = the charge of an electron.

During the field emission accounted for by the strong electric field, a quantum well is formed near the surface of the Si substrate. At the sub-band, the electron concentration is given by

$$N_n = n_v \left[\frac{m_d^*}{\pi \hbar^2} \right] kT \ln \left[1 + \exp\left(\frac{E_f - E_n}{kT} \right) \right] = \frac{\varepsilon_o F}{q}, \qquad (2.34)$$

where

n_v = valley degeneracy,

$m_d^* = (m_z^* m_y^*)^{1/2}$ is the state density mass,

m_y^* and m_z^* = effective masses for the motion parallel to the interface,

$\hbar$ = the reduced Plank's constant,

k = Boltzmann's constant,

T = the absolute temperature,

E_f = the Fermi level, and

E_n = the quantization level.

When $\qquad \phi_b - E_o < qF_{eqv}t < \phi - E_o,$

$$D(E_o) = \{\sin^2\theta_2 \cosh^2(\theta_3 - \theta_1) + \cos^2\theta_2 \cosh^2[\theta_3 + \theta_1 + \ln(4)]\}^{-1}$$

(2.35)

$$D(E_o) = \exp\left[-\frac{4\sqrt{2m_{eqv}}}{3q\hbar F_{eqv}}(\phi_b - E_o)^{3/2}\right],$$

(2.36)

where m_{eqv} = the equivalent effective mass of the composite dielectric.

Figure 2.6b depicts the energy band diagram for direct tunneling [47]. When $V < (\phi_b - E_o)/q$ in the case of direct tunneling a trapezoidal energy barrier is formed. Also, due to the insulating film between the electrodes a potential barrier is created. The current density due to the direct tunneling is given by Simmons model [57],

$$J_d = \left(\frac{e}{4\pi^2\hbar t^2}\right)\left\{\left(\phi_b - E_0 - \frac{eV}{2}\right)\exp\left[-\frac{2(2m_{eqv})^{1/2}}{\hbar}\alpha(\phi_b - E_0 - \frac{eV}{2})^{1/2}t\right]\right.$$
$$\left. - (\phi_b - E_0 + \frac{eV}{2})\exp\left[-\frac{2(2m_{eqv})^{1/2}}{\hbar}\alpha(\phi_b - E_0 + \frac{eV}{2})^{1/2}t\right]\right\}.$$

(2.37)

For a low field range, it can be rewritten as

$$J_d = \frac{[2m_{eqv}(\phi_b - E_0)]^{1/2}\alpha q^2 V}{\hbar^2 t}\exp\left[\frac{2\alpha\sqrt{2m_{eqv}(\phi_b - E_0)}}{\hbar}t\right], \quad (2.38)$$

where

m_{eqv} = electron effective mass,

ϕ_b = the barrier height,

V = the applied bias, and

α = a unitless adjustable variable introduced to modify the simple triangular barrier model [54].

Figure 2.7 depicts the various potential barrier shapes induced by the tunneling of electrons, also being influenced by the thickness of the oxide [47]. Figure 2.8 depicts the simulated results as reported by Chakraborty et al. [47] for the total gate tunneling current as a function of applied gate voltage. Figure 2.9 depicts the plot for $\ln(J_{FN}/F_2)$ (FN tunneling) at a constant applied field as reported by Chakraborty et al. [47].

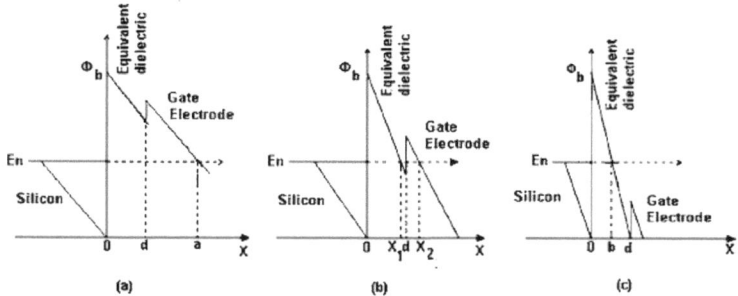

Figure 2.7 Band bending of a composite gate dielectric at different applied electric fields under different conditions. Reprinted from Ref. [47], with the permission of AIP Publishing.

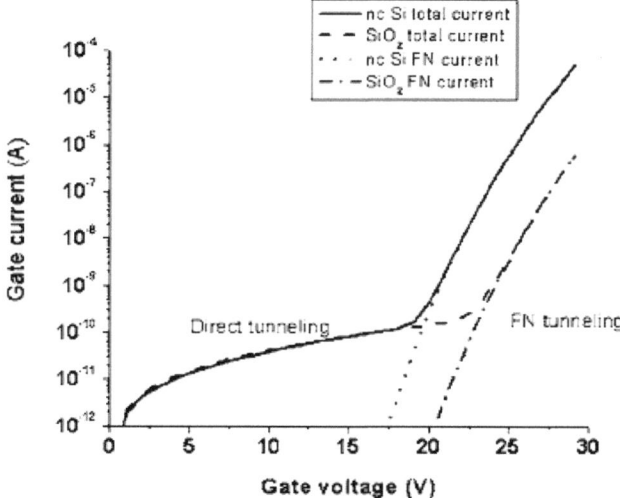

Figure 2.8 Gate current–voltage characteristics of direct and FN tunneling in Si-nanocrystal-embedded SiO_2 of MOS structure and comparison with pure SiO_2 gate oxide. Reprinted from Ref. [47], with the permission of AIP Publishing.

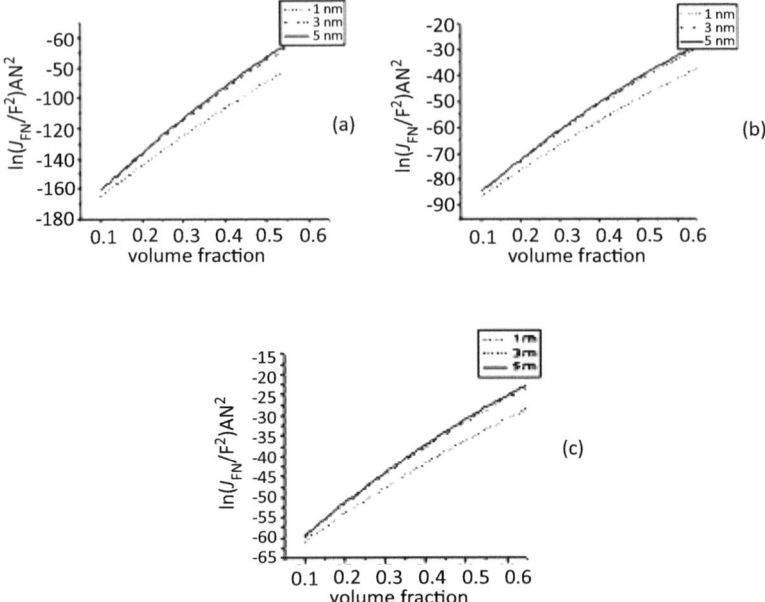

Figure 2.9 Plot of $\ln(J_{FN}/F_2)$ as a function of volume fraction at a fixed gate voltage with different crystallite sizes. Reprinted from Ref. [47], with the permission of AIP Publishing.

2.6 Model for the Charging and Discharging Process [58]

In this model, Campera and Iannaccone [58] adopt a one-dimensional structure for the study where Fig. 2.10 depicts the vertical cross section of a nanocrystal through the z direction (central axis). The simulation of single-electron charging has also been demonstrated by Thean and Leburton [94]. In the previous one-dimensional model proposed [59] for studying MOS capacitors, the generation and recombination processes adhering to the tunneling of charges through the nanocrystals have been modified to suit the characteristics studied. Hence the model deploys the understanding of two different physical possibilities for the study. Firstly, the presence of a large gap between the quantum dot and the silicon dioxide surrounding layer results in electron confinement that defines the quantum dot. The electrochemical potential of the nanocrystal is driven by the bias conditions and the number of electrons in the nanocrystal. Secondly, inside or at the interface of the Si/SiO_2 layers the presence of many localized traps is considered. One electron per trap due to Coulomb repulsion is considered for the analysis of the nanocrystal as a unique system, where the electron to be placed in the trap requires the least electrochemical potential. The model, for simplicity, adopts identical energy traps and the increment of the energies in the trap by a fixed charging energy. The model assumes the nanocrystals to be identical and uniformly distributed. The capacitive coupling among the nanocrystals is not taken into account, and the additional electrons introduced in the nanocrystal induce an electrochemical potential, which is considered to be incremented by the factor

$$\Delta\mu = q^2/2C, \tag{2.39}$$

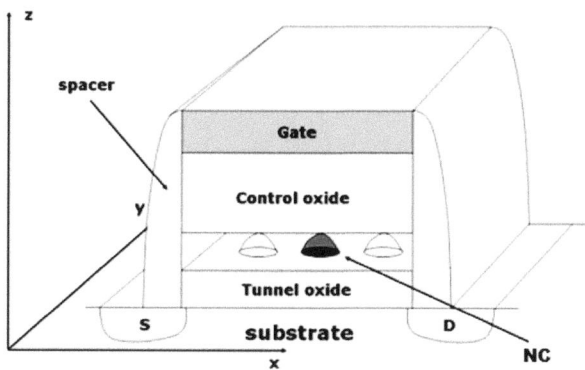

Figure 2.10 Schematic view of the nanocrystal flash memory with nanocrystals deposited by LPCVD. Reprinted from Ref. [58], Copyright (2005), with permission from Elsevier.

where

    q = electron charge and

    C = the total capacitance in the nanocrystal.

    Hence, the total capacitance is given by

$$C = C_1 + C_2 = \frac{\pi\varepsilon_{Si}d^2}{2t_{cox} + \dfrac{\varepsilon_{ox}}{\varepsilon_{Si}}d} + \frac{\pi\varepsilon_{Si}d^2}{2t_{cox} + \dfrac{\varepsilon_{ox}}{\varepsilon_{Si}}d'}, \tag{2.40}$$

where

d = the diameter of the nanocrystal,

t_{tox} = the thickness of the tunnel oxide layer,

t_{cox} = the thickness of the control oxide layer,

ε_{Si} = the dielectric constant of silicon, and

ε_{ox} = the dielectric constant of silicon dioxide.

A one-dimensional structure for the profile of the valence and conduction bands of a nanocrystal depicting the vertical cross section through the z direction (central axis) is represented in Fig. 2.11 [58]. For each applied gate voltage, the self-consistent Poisson-Schrödinger solver is used to compute the band profile. The quantum confinement and the mass anisotropy at the emitter and the silicon conduction band, respectively, have been considered in this model by the use of the self-consistent Poisson-Schrödinger solver. For the computation, the tunneling current is assumed to be low so that two different Fermi energies exist. To separate the local equilibrium that has two regions for consideration of the electrochemical potential (μ_{NC}), quasi-equilibrium approximation is used.

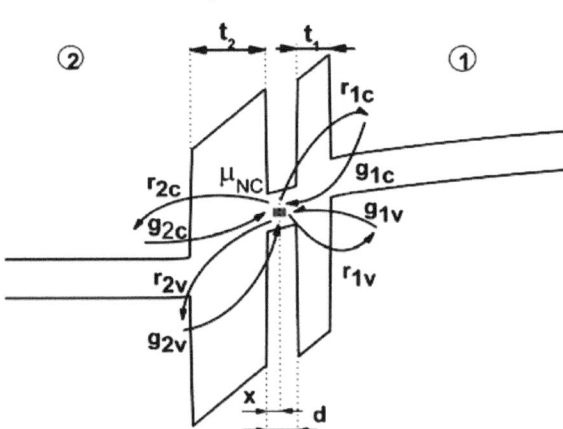

Figure 2.11 Band profile of the structure used in the model and generation and recombination rates. μ_{NC} indicates the electrochemical potential of the nanocrystal. Reprinted from Ref. [58], Copyright (2005), with permission from Elsevier.

The rate of tunneling from the electrode to the nanocrystal is termed as the "generation rate," while the rate of tunneling from the nanocrystal to the electrode is termed as the "recombination rate." Eight different rates of generation and recombination have been depicted in Fig. 2.11 [58]. The transition rate as per the Fermi "golden rule" from one of the band states in the electrode ($|\beta\rangle$) to the contact (Kohn-Sham) state ($|\alpha\rangle$) in the nanocrystal is given by [58]

$$v_{\beta \to \alpha} = \frac{2\pi}{\hbar} |M(\beta,\alpha)|^2 h_\Gamma (E_\alpha - E_\beta), \qquad (2.41)$$

where

$|\alpha\rangle$ = a contact state in the nanocrystal and

$|\beta\rangle$ = a state in one band of the electrode

The transition rate from the contact state in the nanocrystal to the band state in the electrode is given by

$$v_{\beta \to \alpha} = \sigma_{\beta,\alpha} J(\beta, x') = \sigma_{\beta,\alpha} T(E_1) v(E_1). \qquad (2.42)$$

The trap cross section per unit energy is defined by comparing Eqs. 2.40 and 2.42, given by

$$\sigma_{\alpha,\beta} = k.h_\Gamma (E_\alpha - E_\beta), \qquad (2.43)$$

where

k = the unknown constant surface determined from experimental fitting (in m^2).

This model defines the trap cross section differently by considering the variation of the initial and final state energies. The generation and recombination rates are computed using just a multiplying coefficient and are determined by k. The transition state from the band states in the electrode ($|\beta\rangle$) to the contact state ($|\alpha\rangle$) in the nanocrystal is integrated to get the generalized expression for the generation (g) and recombination (r) rates over the occupied and unoccupied states, respectively, as follows:

$$g = 2 \int_\beta v_{\beta \to \alpha} \rho_\beta f_\beta d\beta \qquad (2.44)$$

and

$$r = \int_\beta v_{\beta \to \alpha} \rho_\beta (1 - f_\beta) d\beta. \qquad (2.45)$$

The existence of the two spin states accounts for the factor of 2 as in the equation for the generation of occupied states. On the other hand, in the recombination state the trapped electrons have to have the same spin. Also, the generation state through the substrate's conduction band into the nanocrystal is given by the expression [58]

$$g_{1c} = 2\int_{\beta} \sigma_{\beta,\alpha} T(E_1) v(E_1) \rho_\beta(E) f_\beta$$

$$= 2\int_{E_1}^{\infty} k T(E_1) v(E_1) \rho_1(E_1) \rho_T dE_1 . \int_0^{\infty} h_T(E_1 + E_T) dE_T \tag{2.46}$$

At time t, the probability that n electrons are present in the nanocrystal is given by [58]

$$\frac{dP(n,t)}{dt} = r(n+1)P(n+1,t) + g(n)P(n-1,t) - P(n,t)[r(n) + g(n+1)],$$

$$\tag{2.47}$$

where

$g(n) =$ the total generation rate of the nth electron,

$r(n) =$ the total recombination rate of the nth electron, and

$P(n,t) =$ the probability that n electrons are present in the nanocrystal at time t.

The boundary condition that is applied during the program condition (from Eq. 2.47) is as follows:

$$P(n,0) = 0, \forall \geq 1$$
$$P(0,0) = 1 \tag{2.48}$$

Hence, in a nanocrystal the average number of electrons as a function of time is given by

$$\langle n(t) \rangle = \sum_{n=0}^{\infty} nP(n,t). \tag{2.49}$$

Subsequently, the boundary condition for the erase operation is

$$P(k,0) = 1$$
$$P(n,0) = 0, \forall n < k' \tag{2.50}$$

where $k =$ the number of electrons inside the nanocrystal initially.

The value considered is such that it equals the initially programmed threshold.

2.7   Programming Time Model [60]

In the model Singaraju and Venkat [60] explain the effect of the discrete energy levels accounting for charging of nanocrystal-based nonvolatile memory. In the model proposed the tunneling of electrons only in the allowed energy levels was considered. In the nanocrystal the number of electrons stored was considered to be more than one. Also considered was the charging energy of the nanocrystal due to stored charge.

Since the nanocrystal used in the device were considered to be spherical and within the range of 3–12 nm, behaviors like quantum confinement became dominant. Hence, in the conduction band of the nanocrystals discrete energy levels were observed. The same was solved by Schrödinger's equation in a spherical coordinate given by [60]

$$\frac{\hbar^2}{2m_{Si}}\left(\frac{\partial^2}{\partial r^2}+\frac{2\partial}{r\partial_r}+\frac{1\Lambda^2}{r^r}\right)\psi(r)+V(r)\psi(r)=E_r\psi(r), \qquad (2.51)$$

where $\Delta^2 = \dfrac{1}{\sin^2\theta}\dfrac{\partial^2}{\partial\varphi^2}+\dfrac{1}{\sin\theta}\dfrac{\partial}{\partial\theta}\sin\theta\dfrac{\partial}{\partial\theta}$

for a spherical finite barrier and

$\hbar$ = $h/2\pi$ (reduced Plank's constant),

m_{Si} = the effective mass of the electron in silicon,

$V(r)$ = the electrostatic potential energy,

$\psi(r)$ = the wave function, and

E_r = energy, with index r indicating carrier confinement along the radical direction.

The wave function is spherically symmetrical because the nanocrystals have been considered to be spherical. The diameter of the nanocrystal influences the energy level, as reported by Singaraju and Venkat in Fig. 2.12 [60]. Observations reveal that the energy levels in the nanocrystals increase exponentially with a decrease in their diameters, with smaller diameters having energy levels with a larger spacing and vice versa.

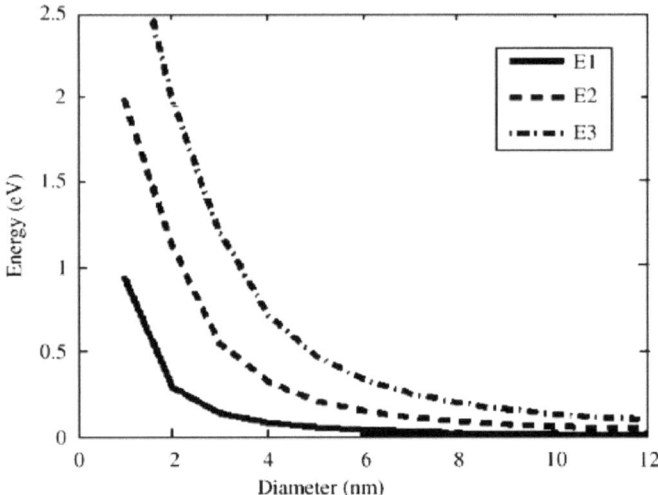

Figure 2.12 Energy values for the first, second, and third quantum levels as a function of the diameter of the Si nanocrystals. Reprinted from Ref. [60], Copyright (2008), with permission from Elsevier.

Any gate bias induces charging to take place due to the direct tunneling of electrons from the substrate into the nanocrystal embedded in the gate dielectric and discharging due to back tunneling in the reverse direction. The model provided by Busseret et al. [61] deduces the direct tunneling (J_{Fi}) current comprising i electrons, given by

$$J_{Fi} = \sum_{i=0}^{n} \frac{4\pi m_{Si} q kT}{h^3} \int_{E_{1\min}}^{E_{1\max}} T(E_1 F_i) \ln(1 + e^{(E_f(F_i - E_1)/kT)}) dE , \quad (2.52)$$

where

q = the charge of an electron,

k = Boltzmann's constant,

F_i = the electric field at the nanocrystal with i electrons,

E_f = the position of the Fermi level in the substrate, and

$T(E_1,F_i)$ = the transparency factor, given by

$$T(E_1,F_i) = \exp\left[-b\frac{\phi_b^{3/2} - (\phi_b^{3/2} - F_i d_b)^{3/2}}{F_i}\right].$$

$$b = \frac{8\pi\sqrt{2m_{ox}}}{3hq} \qquad (2.53)$$

In Eq. 2.53, m_{ox} = the effective mass of the electron in the oxide and

ϕ_b = the barrier energy between the gate/substrate interface.

The discharging current for the electrons tunneling back into the substrate, J_{ri+1}, is given by [62]

$$J_{r_i+1} = \sum_{i=0}^{n} \frac{AF_{i+1}^2}{(1-(\phi+F_{i+1}d_{fb}/\phi_s)^{1/2})^2} \exp\left[-b\frac{(\phi_b^{3/2}-F_{i+1}d_{fb})^{3/2}}{F_{i+1}}\right]. \quad (2.54)$$

$$A = \frac{q^3 m_{Si}}{8\pi m_{ox}\phi_b}$$

Figure 2.13 shows the simulation result for the effect of the threshold voltage on time for different gate voltages as reported by Singaraju and Venkat [60].

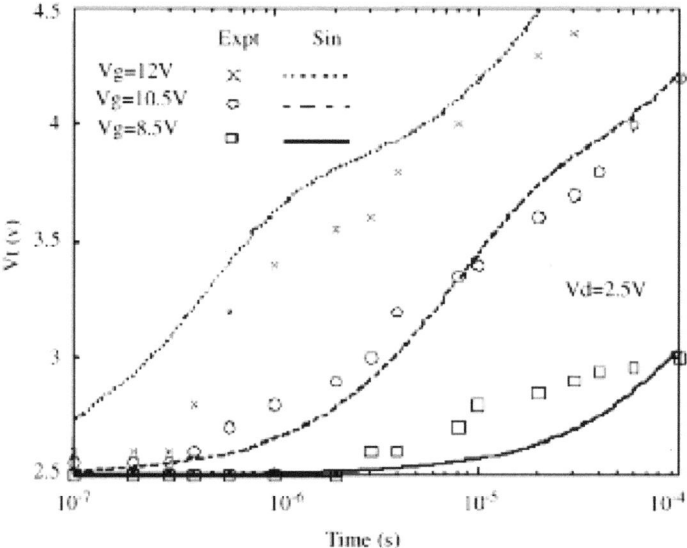

Figure 2.13 Simulation and experimental results for shift in V_T versus programming time for various gate voltages at drain voltage V_d = 2.5 V. Reprinted from Ref. [60], Copyright (2008), with permission from Elsevier.

The model's simulation followed various observations in the nanocrystals, like the formation of discrete energy levels. Nanocrystals are capable of storing more than one electron as the electrons are held in the potential traps.

2.8 Growth of Metal (Au) Nanocrystals in High-κ Dielectrics [63]

With the advantage of using metal nanocrystals in nonvolatile memory applications, there has been extensive study on the various types of metals being adopted for use as nanocrystals. The effect of size has been explained earlier. Among the various metals being used, Au has been adopted for various reasons, as shown later in the model proposed by Chan et al. [63]. The large work function and small energy perturbation give Au an advantage over some of the other noble metals [3, 64–66]. Pulsed-laser deposition was adopted for the formation of nanocrystals, which provided the chemical stability. Also, high-κ dielectrics were extensively used, as they made provision for a thinner oxide layer and improved retention time without giving into the nonvolatile nature of the device. Among various high-κ materials HfAlO was considered by Chan et al. due to its good thermal stability and the programming and retention modes having unique band symmetry [67, 68]. The application of high-κ dielectrics was also demonstrated by Schaijk et al. [93]. Chan et al. have also reported a three-layer floating-gate memory structure with HfAlO/Au (Au-nanocrystal-embedded layer)/HfAlO-based floating. The model considered the FN tunneling mechanism for program and erase operations. The model has shown very good results [69]. The same could be deployed for storing multilevel charges too [70]. Different types of gas deposition were used for the fabrication of the structure. The preliminary oxides deposited were cleaned using a wet-chemical process where the p-type silicon wafer was dipped into a hydrofluoric acid (HF) solution. Pulsed-laser deposition was carried out at 550°C under 2.0 Pa of oxygen partial pressure to deposit a 10 nm (approx.) thick HfAlO tunnel oxide layer. The deposition of Au nanocrystals as a very thin layer could be carried out by various methods (as reported by Chan et al. [63] in different samples of Fig. 2.14). Sample A depicts the deposition of the thin nanocrystal layer by applying 2.0 Pa of oxygen partial pressure. Sample B depicts deposition by relatively high vacuum (1×10^{-10} Pa). Sample C depicts the use of 5.9 Pa of Argon partial pressure. Au nanocrystals are formed rather than a Au film due to a quenching effect and a lower substrate temperature. This is used so as to give better information about the influence of the ambient gas on the nanocrystal size. Further, a 20-nm-thick (HfAlO) tunnel layer was deposited using the above-mentioned mechanism. Thermal annealing was carried out at 850°C for 30 min. in N_2 gas.

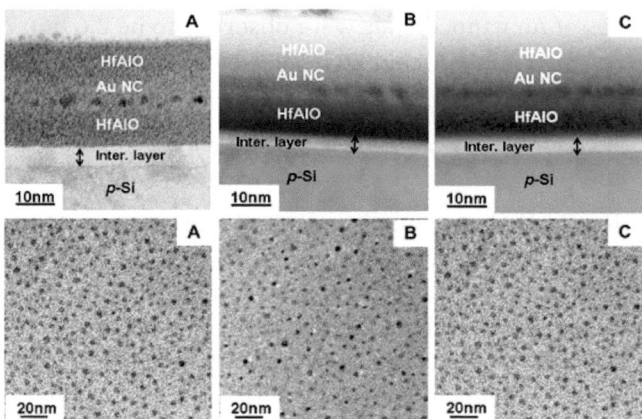

Figure 2.14 Cross-sectional and plan-view HRTEM images of the three-layer floating-gate memory structure. Reprinted from Ref. [63], Copyright (2008), with permission from Elsevier.

Figure 2.14 shows the three layers of the floating-gate structure of Au nanocrystals embedded into the HfAlO layer as reported by Chan et al. [63]. The tunnel and control oxide layers formed were thinner than 3 nm. Annealing enabled the formation of Au nanocrystals. Interface thicknesses of 4.8 nm, 3.2 nm, and 3.8 nm were observed for the three samples, sample A, sample B, and sample C, respectively. This was a result of oxygen diffusion and reaction with the silicon substrate. Further, studies established that due to the interfacial layers no significant degradation of the electrical properties was caused. Au nanocrystals observed were found to be self-organized and distributed uniformly, as depicted in the plan-view transmission emission microscopy (TEM) images (Fig. 2.14, reported by Chan et al. [63]). The densities of the nanocrystals were found to be about 2×10^{12} cm^{-2}, 1.4×10^{12} cm^{-2}, and 1.6×10^{12} cm^{-2} in sample A, sample B, and sample C, respectively. From the TEM images it was observed that Au nanocrystals of the lowest density existed when fabricated at 1×10^{-4} Pa. It was due to the plasma cloud being less concentrated and the presence of fewer particles that collide with the Au atoms during transportation. It was also observed that the effect on the density was not significant, which was caused due to the gaseous ambient where the density difference of the three layers of the Au nanocrystals was not apparent. From the TEM plan-view images, the mean sizes of Au nanocrystals were observed to be 2.6 nm, 2.7 nm, and 3.8 nm for sample A, sample B, and sample C, respectively.

2.9 Retention Characteristics Model [71]

The trade-off to balance the program and erase time and the charge retention time gets very important as a thinner gate dielectric favors a faster program and erase time and a thicker gate dielectric endorses a longer retention time. The same is proposed in the model by Guan et al. [71]. A model for nonvolatile floating-gate nanocrystal-based memory was demonstrated by Hasaneen et al. [95]. Hence, the conventional nonvolatile memory and silicon-based dielectrics were proving to reach their scaling limits. Thus, the idea of using different materials, especially high-κ materials, as gate dielectrics was worked upon. Also, geometric engineering of the shape was adopted, like deepening the depth of the potential well. This created a small asymmetrical barrier for a faster program and erase time and at the same time there was a deeper well that further facilitated longer retention time due to a larger barrier for the charge to escape. Engineering such a mechanism was difficult in the case of Si-, Ge-, or semiconductor-based nonvolatile memory but was easily possible in the case of metal-based memory. The same was introduced by Liu et al. [3] where fabrication was simpler due the self-assembly formation of metals. The depth of the potential well too could be adjusted as different metal work functions are available.

The Coulomb blockade effect [72] and the quantum confinement effect get prominent with the size of the nanocrystals getting smaller (a few nanometers). The electrostatic potential of the nanocrystal is increased due to the Coulomb blockade effect, while the quantum confinement effect shifts the energy band edge upward in the nanocrystal. This results in the lowering of the energy band offset between the nanocrystal and the dielectric [73]. A separate energy band is formed due to the quantum confinement effect in metallic nanocrystals, as reported by Kubo theory [74]. The average energy spacing can be determined by

$$\delta = 4E_F/(3N),\qquad(2.55)$$

where

$E_F =$ the Fermi potential of the bulk metal and

$N =$ the total number of conducting electrons in the nanocrystal.

The charge density and energy (E) relation is given by

$$E(n) = \frac{\hbar^2}{2m^*}(3\pi^2 n)^{2/3}.\qquad(2.56)$$

Ni and Au nanocrystals were considered, and their respective parametric values were taken ($E_{FNi} = 11.7$ eV; $E_{FAu} = 5.53$ eV; and $m^* = m_0$, where m_0 is the free electron mass, 9.1×10^{-31} kg). Thus the relation between the charge density and spacing were obtained as follows:

$$\delta_{Ni} = \frac{0.1639}{d_{Ni}^3}\qquad(2.57)$$

and

$$\delta_{Au} = \frac{0.2388}{d_{Au}^3}.\qquad(2.58)$$

Unlike the tunneling back of the electrons due to the positive bias during programming this model assumed the electrons to be stored in the Fermi level of the metal nanocrystal. The threshold voltage due to the stored charge was given by [45]

$$\Delta V_{th} = \frac{Q}{\varepsilon_{tun}}\left(t_{con} + \frac{1}{2}\frac{\varepsilon_{tun}}{\varepsilon_{nc}}d_{nc}\right),\qquad(2.59)$$

where

$Q =$ the density of charge in the nanocrystal,

$\varepsilon_{tun} =$ permittivity of the tunnel dielectric,

$t_{tun} =$ the thickness of the tunnel dielectric,

$t_{con} =$ the thickness of the control dielectric,

$\varepsilon_{nc} =$ the permittivity of the nanocrystal, and

$d =$ the diameter of the nanocrystal.

This model considered only the discharging current induced by the tunneling of electrons into the substrate from the nanocrystal. Due to the limited number of charges retained during the zero bias of the control gate, only direct tunneling existed. Hence, a one-dimensional model based on vertical discharge through direct tunneling was proposed, as depicted in Fig. 2.15, by Guan et al. [71]. The direct tunneling of electrons into the substrate from the nanocrystal dictates the density of the discharge current [75],

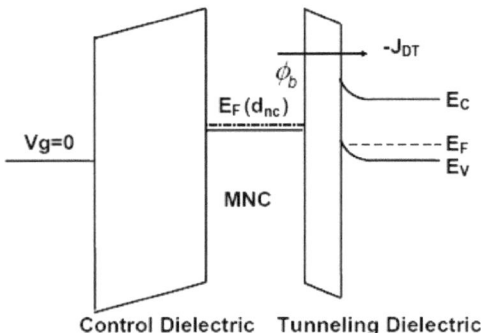

Figure 2.15 Schematic energy band diagram of metal nanocrystal memory during retention, demonstrating the reduced effective barrier offset caused by the quantum confinement effect. Reprinted from Ref. [71], Copyright (2007), with permission from Elsevier.

$$J_{DT} = AE_{tun}^2 \exp\left\{-\frac{B}{E_{tun}}\left[\phi_b^{3/2} - (\phi_b - qV_{tun})^{3/2}\right]\right\},\qquad (2.60)$$

where

$$A = q^3 m_0 / (16\pi^2 \hbar m_{tun}^* \phi_b),$$

$$B = 4\sqrt{2m_{tun}^*} / (3\hbar q),$$

q = the charge of a single electron,

m_0 = the mass of a free electron,

$\hbar$ = the reduced Plank's constant,

m_{tun}^* = the effective mass of an electron in a tunnel dielectric,

ϕ_b = the barrier height from the nanocrystal to the substrate, and

V_{tun} = the voltage drop across the tunnel dielectric.

Also, $\phi_b = \phi_{b_0} - \phi_{b_{up}}$ and

$$V_{tun} = E_{tun} \cdot t_{tun},$$

where

ϕ_{b_0} = barrier offset for a bulk material,

$\phi_{b_{up}}$ = energy band shift considering the quantum confinement, and

E_{tun} = the electric field in the tunnel dielectric, given by

$$E_{tun} = \frac{Q}{\varepsilon_{tun} R}, \tag{2.61}$$

where R = the nanocrystal coverage ratio.

The direct tunneling current density (J_{DT}) and the charge density (Q) are related as follows:

$$\frac{dQ}{dt} = -J_{DT}(t). \tag{2.62}$$

Thus from the above equations the charge retention characteristics for a nanocrystal-embedded nonvolatile memory were derived. The electric field was seen to be the strongest in the tunneling dielectric. The charge loss got slower due to the decrease in the charge density (Q) and the electric field as electrons tunneled out of the nanocrystals.

2.10 Tunneling Characteristics of Metal-Nanocrystal- and Semiconductor-Nanocrystal-Based Gate Dielectrics [76]

For a MOS structure whose gate oxide is embedded with nanocrystals, a tunneling current density study is being discussed in this section.

The nanocrystals that are embedded in the gate oxide are metallic and semiconductor in nature. As the scaling limits for the traditional floating-gate nonvolatile memory is reaching its limit, nanocrystal-based nonvolatile memory has been introduced [77]. Initially, MOS devices embedded with semiconductor nanocrystals like Si or Ge were in use for memory devices [45, 78]. Tiwari et al. [79] introduced nonvolatile memory with Si nanocrystals. The programming efficiency of the memory device depends largely on the thickness of the oxide. A model for the study of the tunneling characteristics in a MOS structure whose gate dielectric is embedded with metal or semiconductor nanocrystals is being discussed here.

With the scaling of the tunneling oxide, the program voltages of nanocrystal-based memory devices can be reduced to a level below 10 V. But these devices face difficulties due to high leakage current. The problem of leakage current is addressed by the use of metal nanocrystals as had been proposed [3, 64]. Presently in this field attention is shifting toward metallic nanocrystals as charge storage nodes. The larger work function of metallic nanocrystals makes them better than semiconductor nanocrystals in reducing the leakage current in MOS structures. Due to the high density of the energy states metal nanocrystals have proved to increase program efficiency. The work function of many metal nanocrystals (Ag, Ni, Pt, Au, Al, etc.) is higher than that of silicon [28]. And due to the higher electron barrier height the possibility of trapped electrons in metal nanocrystals tunneling back to the substrate is less. The charging and discharging of the nanocrystals embedded in the gate oxide govern the operation of the devices used for memory applications. The electron tunneling takes place through the very thin tunnel oxide layer between the nanocrystal layer and the Si substrate, inducing charging and discharging in the nanocrystals [80]. The theoretical modeling of the charging of the nanostructure materials (mainly metallic nanocrystals) embedded within the gate oxides of MOS nonvolatile memory devices is discussed here. Leakage current accounts for significant charge loss in memory devices with metal nanocrystals and SiO_2 as the gate dielectric is ultrathin. The replacement of SiO_2 with a SiO_2 high-κ dielectric (e.g., HfO_2) stacks can solve this problem to a significant extent. Previously, pure HfO_2- and SiO_2-HfO_2-stacked tunnel oxide multilayer gate dielectrics to solve the leakage current problem and improve charge retention in such MOS nonvolatile memory have been reported [81, 82]. Therefore a composite multilayer structure of stacked SiO_2-HfO_2 and HfO_2 partly embedded with metal nanocrystals is expected to exhibit better leakage characteristics. The metallic nanocrystals get contained in a thin layer of gate dielectric. The gate dielectric was divided into four layers and made up of high-κ dielectric HfO_2. The bottom layer was made up of a very thin, pure SiO_2 tunnel oxide layer in contact with the silicon substrate. The next layer, also a part of the tunnel oxide, was a pure HfO$_2$ region, and the third layer was the oxide layer embedded with Ag nanocrystals. The topmost layer of the dielectric composite stack was a pure HfO$_2$ control oxide layer, which is thicker than the tunnel oxide. The proposed model was validated with the fabricated devices, as reported by Ryu et al., Schnippering et al., and Ryu [83–85]. A MOS structure with a composite gate dielectric is depicted in Fig. 2.16.

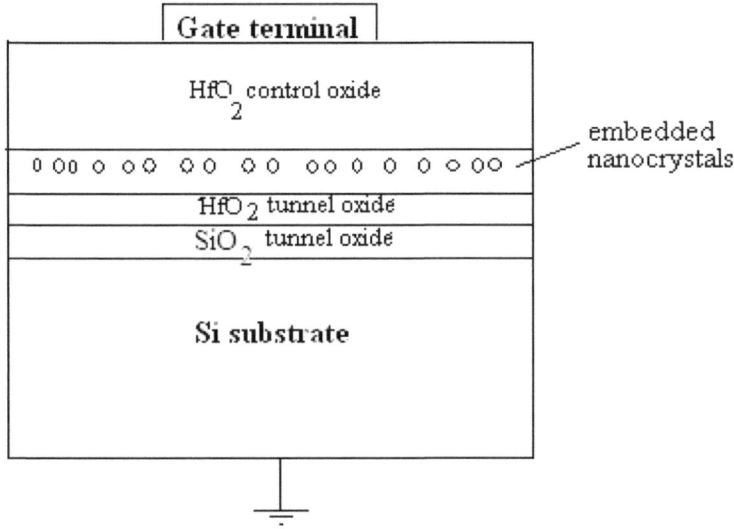

Figure 2.16 Schematic structure of the nanocrystal-embedded gate dielectric MOS structure [76].

In this model mainly Ag nanocrystals were considered as embedded elements since they have a reasonably high work function of 5.53 eV [84] and because of the availability of the experimental results [85]. The case of Si nanocrystals embedded in the same gate dielectric HfO$_2$ was used to compare the tunneling currents through the semiconductor and the metallic nanocrystals. Also Ag nanocrystals as well as Si nanocrystals embedded in SiO$_2$ were considered to compare the tunneling characteristics of both metallic and semiconductor nanocrystals in the conventional gate oxide (SiO$_2$). The theoretical tunneling currents with the nanocrystallites silicon and Ni embedded in HfO$_2$ were compared with the experimental results reported by Lee and Kwong [86].

The increasing gate current (I_G) through the ultrathin gate oxides is a key issue limiting MOSFET scaling. Understanding I_G is vital for different parameters that control the leakage current. The effective barrier height (ϕ_{eff}) of the oxide is a crucial parameter that affects I_G, where the channel is embedded with nanocrystals that encounter the tunneling electrons. The leakage was explained by the very important quantum mechanical tunneling phenomenon in MOS structure. To calculate the tunneling current at the interface of the silicon substrate and the composite gate dielectric WKB approximation, a conventional method was used. Direct tunneling dominates at a low applied gate voltage whereas the FN tunneling mechanism dominates at high applied gate voltages, which is basically field-assisted tunneling.

Generally the Maxwell Garnett theory [87] is applicable for the spherical nanocrystal embedded in the dielectric. The shape of the embedded nanocrystal in the dielectric is important to determine the effective dielectric constant. In the case of Ag and Si nanocrystals, the inclusions were considered to be spherical in shape. The model expresses the relationship between the dielectric constant of the embedded nanocrystals, the gate oxide, and the volume fraction of the nanocrystals.

The dielectric constant of Ag nanocrystals embedded with the high-κ dielectric HfO_2 can be written as

$$\varepsilon_{nc\text{-}HfO_2} = \frac{\varepsilon_{ox}\left[2v\left(\varepsilon_{nc\text{-}Ag} - \varepsilon_{ox}\right) + \left(\varepsilon_{nc\text{-}Ag} + 2\varepsilon_{ox}\right)\right]}{\varepsilon_{nc\text{-}Ag} + 2\varepsilon_{ox} - v\left(\varepsilon_{nc\text{-}Ag} - \varepsilon_{ox}\right)}, \quad (2.63)$$

where

v = the volume fraction of the Ag nanocrystallites,

$\varepsilon_{nc\text{-}Ag}$ = the dielectric constant of the Ag nanocrystals, and

ε_{ox} = the dielectric constant of pure HfO_2.

The diameter of Ag nanocrystals used was 5 nm and its static dielectric constant was −11 using the curve fitting method obtained from Yeh et al. [88]. The dielectric constants of metallic nanocrystals were determined at various microwave frequencies from the resonant frequency and the quality factor. In this MOS model the composite gate dielectric has a four-layered gate dielectric.

In this model the barrier height ϕ_b was determined as the difference between the conduction band edge of the composite gate dielectric and the conduction band edge of the silicon substrate. Thus, the barrier height ϕ_b is given by [81],

$$\phi_b = \phi_0 - \frac{(E_{g_{eff}} - E_{Si})}{2}, \qquad (2.64)$$

where

ϕ_0 = the intrinsic Si-SiO$_2$ barrier height (~3.15 eV),

$E_{g_{eff}}$ = the effective bandgap of the composite gate dielectric (which is a combination of pure SiO$_2$-HfO$_2$ tunnel oxide, embedded with Ag nanocrystals, and pure HfO$_2$ control oxide), and

$E_{g_{Si}}$ = the bandgap of Si.

In one work, Sengupta et al. [82] proposed a model for evaluating the electrical parameters of such a multilayer stack considering a unit cross-sectional stack and the relative thickness of each constituent layer. Following this method, the effective bandgap of the composite gate dielectric is $E_{g_{eff}}$ and is formulated as

$$E_{g_{eff}} = E_{g_{SiO2}} \times \frac{d_{SiO2}}{d} + E_{g_{nc\text{-}ox}} \times \frac{d_{embed}}{d} + E_{g_{HfO2}} \times \frac{d_{HfO2}}{d}, \qquad (2.65)$$

where

$E_{g_{SiO2}}$ = the bandgap of only pure SiO$_2$, which is the first layer of the composite gate dielectric,

$E_{g_{HfO2}}$ = the bandgap of pure HfO$_2$, and

$E_{g_{nc\text{-}ox}}$ = the bandgap of HfO$_2$ embedded with nanocrystals, which is the second layer of the composite gate dielectric.

The bandgaps of pure HfO$_2$ and the silver-nanocrystal-embedded HfO$_2$ differ slightly. The bandgap of the embedded dielectric was formulated by considering the analogy of the formation of an alloy of nanocrystals (Si nanocrystals as semiconductor nanocrystals or Ag nanocrystals as metallic nanocrystals) and high-κ dielectric HfO$_2$. According to the virtual crystal approximation, $E_{g_{nc\text{-}ox}}$ is given by

$$E_{g_{nc\text{-}ox}} = E_{g_{ox}}(1-v) + E_{g_{nc\text{-}Ag}}v, \qquad (2.66)$$

where

$E_{g_{ox}}$ = the bandgap of the gate dielectric HfO$_2$, which is 5.7 eV [89], and

$E_{g_{\text{nc-Ag}}}$ = the energy bandgap of Ag nanocrystallites.

As Ag is a metal, the energy bandgap of bulk Ag is zero. The work function of 4.30–4.81 eV has been reported for bulk Ag. And the most frequently used value is 4.5 eV. However, the work function of fabricated metal nanocrystal differs from the ideal case due to the different surface condition. The measured work function of Ag nanocrystals increased from 5.29 eV to 5.53 eV as the particle size decreased to 3.5 nm from 4 nm [84]. Hence the bandgap of Ag nanocrystals of 5 nm would be 0.24 eV.

2.10.1 Fowler–Nordheim Tunneling

With the application of the gate voltage and neglecting the image lowering force, the effective field is lowered to $1/\varepsilon_{\text{embd}}$ and the field near the silicon surface is lowered by $1/\varepsilon_{\text{Si}}$ times that in vacuum. Using the linear potential approximation near the silicon surface,

$$\varepsilon_{\text{Si}} F_{\text{Si}} = F_{\text{embd}} \varepsilon_{\text{embd}} = F, \tag{2.67}$$

where

$\qquad F_{\text{Si}}$ = the surface field of Si,

$\quad F_{\text{embd}}$ = the field within the composite gate dielectric,

$\qquad\quad F$ = the field in vacuum,

$\quad \varepsilon_{\text{embd}}$ = the dielectric constant of the composite gate dielectric, and

$\qquad \varepsilon_{\text{Si}}$ = the dielectric constant of silicon.

The tunneling current is estimated by [47]

$$J_{\text{FN}} = q N_{\text{I}} V_0 D(E_0), \tag{2.68}$$

where

$\quad J_{\text{FN}}$ = the tunneling current density and

$D(E_0)$ = the tunneling probability of the electrons in the lowest (0th) sub-band of the energy levels.

The potential barrier through which the electrons tunnel needs to be known for determining the tunneling probability of the electrons. The leakage current originates with the application of a biasing field in the different conditions, and the fundamental transmission process is tunneling.

Following the WKB approximation, the tunneling probability is written as

$$D(E_0) = \exp\left(\frac{-2}{\hbar} \int_0^a \{2m(V(x)-E_0)\}^{1/2}\, dx\right) \quad \text{when } qF_{embd}d < (\Phi_b - E_0).$$

(2.69)

When $\qquad (\Phi_b - E_0) < qF_{embd}d < (\Phi - E_0),$

$$D(E_0) = \left\{\sin^2\theta_2 \cosh^2(\theta_3 - \theta_1) + \cos^2\theta_2 \cosh^2\left[\theta_3 + \theta_1 + \ln(4)\right]\right\}^{-1}, \quad (2.70)$$

where $\qquad\qquad \hbar\theta_i = \int_{x_{i-1}}^{x_i} \left\{2m^*\left[V(x)-E_0\right]\right\}^{1/2}\, dx.$

For $\qquad\qquad\qquad qF_{embd}d > (\Phi - E_0),$

$$D(E_0) = \exp\left(-\frac{4\sqrt{2m_{eff}}}{3q\hbar F_{embd}}(\phi_b - E_0)^{3/2}\right), \qquad (2.71)$$

where m_{eff} = the equivalent effective mass of the composite dielectric with pure SiO_2-HfO_2 tunnel oxide and HfO_2 embedded with either silicon nanocrystals or silver nanocrystals.

2.10.2 Direct Tunneling

Direct tunneling dominates the gate current at low voltages. The barrier height and the effective mass affect the tunneling equation and also depend on the oxide thickness. Charge gets accumulated near the semiconductor surface with a reduction in the thickness of the oxide layer. According to the Simmons model [57], the tunneling current density through a thin barrier in the direct tunneling regime $\left(V < \dfrac{\phi_b - E_0}{q}\right)$ is written as

$$J_D = \left(\frac{e}{4\pi^2\hbar d^2}\right)\left\{\left(\phi_b - E_0 - \frac{eV}{2}\right)\exp\left[-\frac{2(2m_{eff})^{\frac{1}{2}}}{\hbar}\alpha\left(\phi_b - E_0 - \frac{eV}{2}\right)^{\frac{1}{2}} d\right]\right\}.$$

$$-\left(\phi_b - E_0 + \frac{eV}{2}\right)\exp\left[-\frac{2(2m_{\text{eff}})^{\frac{1}{2}}}{\hbar}\alpha\left(\phi_b - E_0 + \frac{eV}{2}\right)^{\frac{1}{2}}d\right] \quad (2.72)$$

The current density for very low gate applied voltages is written as

$$J_D = \frac{\left[2m_{\text{eff}}(\phi_b - E_0)\right]^{\frac{1}{2}}\alpha q^2 V}{\hbar^2 d}\exp\left[-\frac{2\alpha\sqrt{2m_{\text{eff}}(\phi_b - E_0)}}{\hbar}d\right], \quad (2.73)$$

where

m_{eff} = the electron effective mass,

d = the barrier width,

ϕ_b = the barrier height for the modified dielectric due to the embedded nanocrystals,

V = the applied gate bias, and

α = a unitless variable for effective mass in a modified rectangular barrier model.

The direct tunneling current density can be adjusted by the two parameters ϕ_b and α. Figure 2.17 shows the variation in the field emission current, which is the so-called FN tunneling gate current, as a function of the gate voltages as reported by Sharma et al. [76].

In Fig. 2.17, two different nanocrystals, Si nanocrystals and Ag nanocrystals, were considered to be embedded in the device structure as reported by Sharma et al. [76]. It was assumed that the diameters of both nanocrystals (Si nanocrystals and Ag nanocrystals) are 5 nm. To match the model with the fabricated device reported by Ryu et al. [83], it was assumed that the thickness of the tunnel oxide is 4.5 nm, the floating-gate region—which is HfO_2 embedded with the nanocrystals—is 5-nm-thick, and the thickness of the control oxide is 30 nm. The charge carriers tunnel through the nanocrystals when the gate bias voltage is applied across the composite dielectric. It was seen from Fig. 2.17 that FN tunneling started at lower (about 0.5 volts less) applied gate voltages in the metallic (Ag) nanocrystals embedded in oxide than in the semiconductor (Si) nanocrystals embedded in oxide. The FN onset voltage for the Ag-nanocrystal-embedded gate dielectric was about 11.35 V, whereas that for the

Si-nanocrystal-embedded gate dielectric was about 11.85 V. It may be due to the fact that the value of the dielectric constant of the gate oxide embedded with metallic nanocrystals is higher than that of the gate oxide embedded with semiconductor nanocrystals. Moreover, the triangular barrier height for FN tunneling is lower in the case of the Ag-nanocrystal-embedded gate dielectric than in the case of the Si-nanocrystal-embedded gate dielectric. So the FN tunneling current increases, and tunneling onset voltage decreases. The lower onset voltage of FN tunneling in the metallic-nanocrystal-embedded oxide demonstrates a lower programming voltage, which is used for writing in nonvolatile memory devices.

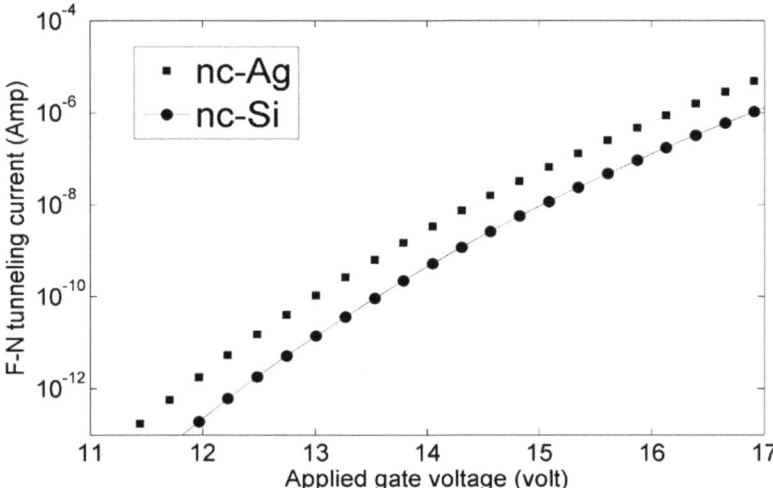

Figure 2.17 Comparison of theoretical FN tunneling currents of Ag-nanocrystal- and Si-nanocrystal-embedded gate dielectrics [76].

Figures 2.18 and 2.19 are the plots of direct tunneling current at very low applied gate voltages as reported by Sharma et al. [76]. Figure 2.18 illustrates direct tunneling at positive applied gate voltages, whereas Fig. 2.19 illustrates the same at negative applied gate voltages. In this work, metal nanocrystals were used together with the SiO_2-HfO_2 tunneling barrier in order to propose an alternative approach for resolving leakage current problems. The barrier height for electron injection from the substrate is identical for both Si and Ag nanocrystals, while it differs for the electrons stored in the nanocrystals due to the work function difference between the metal and silicon nanocrystals.

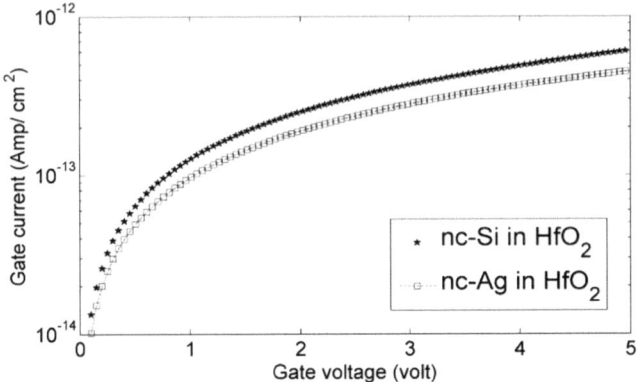

Figure 2.18 Theoretical direct tunneling current of metal-nanocrystal- and semiconductor-nanocrystal-embedded HfO$_2$ gate dielectrics at positive gate voltages [76].

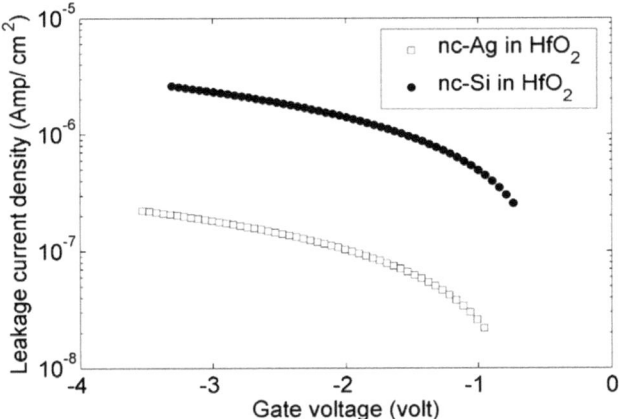

Figure 2.19 Theoretical direct tunneling gate leakage current at a very low negative gate bias [76].

Lower leakage current can be realized by introducing metal nanocrystals like Ag, which have higher work functions than semiconductors (Si). Figure 2.19 shows the calculated leakage or direct tunneling currents of a 5-nm-thick tunneling HfO$_2$ film, a 5-nm-thick nanocrystal-embedded oxide layer, and a 15-nm-thick control dielectric for both Ag and Si nanocrystals as reported by Sharma et al. [76]. The overall thickness of the gate dielectric is nearly 25 nm. The theoretical leakage current of Si nanocrystals embedded in HfO_2 was compared with the experimental results of Lee and Kwong [86], and for this reason the thickness of the three layers was taken according to Lee's model for exact verification. The leakage current was calculated by direct tunneling mechanisms, depending on the nanocrystals' potential and the electron barrier height. The charges are trapped in the nanocrystals, and the nanocrystals break the continuous tunneling path from the Si substrate to the gate electrode by storing charges through it (nanocrystals). In both positive and negative gate voltages the leakage current is lowered more in the case of Ag nanocrystals than in the case of Si nanocrystals. This proved that the charge storage capacity was better in the metal nanocrystals than in the semiconductor nanocrystals.

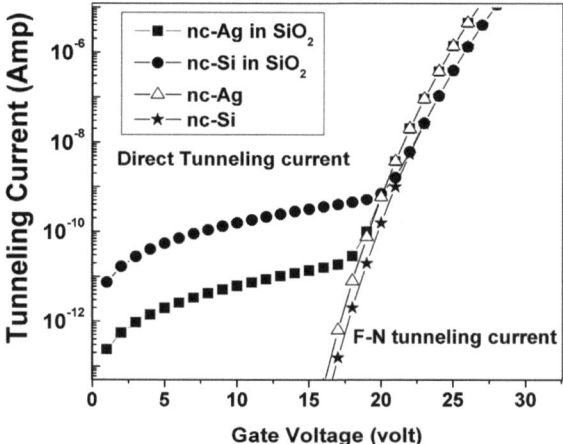

Figure 2.20 Theoretical direct and FN tunneling currents of Ag and Si nanocrystals embedded in a SiO_2 gate dielectric [76].

Figure 2.20 shows the gate current versus the gate voltage from two different sets of composite gate dielectrics as reported by Sharma et al. [76]. One is the semiconductor nanocrystals (Si) embedded in SiO_2 and another one is the metallic nanocrystals (Ag) embedded in the same gate oxide (SiO_2). The gate tunneling current in this figure is the combination of both direct and FN tunneling currents. The assumptions of surface potential and equivalent height [90, 91] and dependence on size [88] were also observed and relatively deduced.

Figure 2.21 is the plot of the FN tunneling current for the gate dielectric HfO_2 embedded with two different metallic nanocrystals (Ag and Ni) separately as reported by Sharma et al. [76, 92]. The figure shows that the FN tunneling onset voltage is less for Ag nanocrystals than for Ni nanocrystals embedded in HfO_2. This is because both the work function and the dielectric constant of Ag nanocrystals are higher than those of Ni nanocrystals. As a result, Ag nanocrystals show better efficiency in the application of nonvolatile memory devices than Ni nanocrystals.

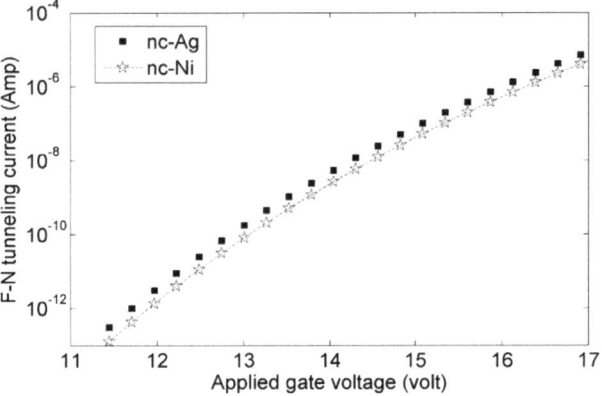

Figure 2.21 Comparison between theoretical FN currents of Ag and Ni nanocrystals [76].

2.11 Conclusion

This chapter presents the progressive development in nanocrystal-based nonvolatile memory. Various references have been cited to showcase the progress made through the models adopted and the results based on simulations and comparative experimental data. The chapter has dealt with the initial nitride-trap-based memory, where the basic floating-gate approach has been described briefly, and then the idea of nanocrystal-embedded memory depicting better scaling capability than their predecessors. The tunneling current characteristic has been summarized for silicon-nanocrystal-based nonvolatile memory. Their programming time characteristics and a model are also highlighted. Because of the limitations of semiconductor (silicon in this chapter) nanocrystals in terms of scaling and tunneling characteristics, metal-nanocrystal-based memory is a better option for nonvolatile memory. The use of high-κ dielectrics has further enhanced the scaling capabilities and tunneling and retention characteristics. Various models have been discussed in brief. Finally, a comparative study between semiconductor-nanocrystal-based and metal-nanocrystal-based memory identified the various advantages provided by the latest, most widely accepted models of metal nanocrystals based on high-κ dielectric memory. Also, stack-based models have found larger acceptance for development. Lately, there have been studies [96–98] on the use of low-dimensional materials for the application in nonvolatile memory, showing great prospect for such materials to be used in future memory devices.

References

1. Casperson, J.D. (2004). Design and characterization of layered tunnel barriers for nonvolatile memory applications (Doctoral dissertation, California Institute of Technology).

2. Chang, T.C., Jian, F.Y., Chen, S.C., and Tsai, Y.T. (2011). *Mater. Today*, **14**(12), 608–615.

3. Liu, Z., Lee, C., Narayanan, V., Pei, G., and Kan, E.C. (2002). *IEEE Trans. Electron Devices*, **49**(9), 1606–1613.

4. De Blauwe, J. (2002). *IEEE Trans. Nanotechnol.*, **99**(1), 72–77.

5. Muralidhar, R., Steimle, R.F., Sadd, M., Rao, R., Swift, C.T., Prinz, E.J., et al. (2004). *International Conference on Integrated Circuit Design and Technology (IEEE Cat. No.04EX866)*, 31–35.

6. Chindalore, G., Yater, J., Gasquet, H., Suhail, M., Kang, S.T., Hong, C.M., et al. (2008). *Symposium on VLSI Technology,* 136–137.

7. Molas, G., Masoero, L., Della Marca, V., Gay, G., and De Salvo, B. (2014). *Advances in Non-volatile Memory and Storage Technology*, 120–157.

8. Lee, C., Gorur-Seetharam, A., and Kan, E.C. (2003). *IEEE International Electron Devices Meeting*, 22.6.1–22.6.4.

9. Tseng, J.Y., Cheng, C.W., Wang, S.Y., Wu, T.B., Hsieh, K.Y., and Liu, R. (2004). *Appl. Phys. Lett.*, **85**(13), 2595–2597.

10. Takata, M., Kondoh, S., Sakaguchi, T., Choi, H., Shim, J.C., Kurino, H., and Koyanagi, M. (2003). *IEEE International Electron Devices Meeting*, 22.5.1–22.5.4.

11. Lin, Y.H., Chien, C.H., Lin, C.T., Chang, C.Y., and Lei, T.F. (2005). *IEEE Electron Device Lett.*, **26**(3), 154–156.

12. Choi, S., Kim, S.S., Chang, M., Hwang, H., Jeon, S., and Kim, C. (2005). *Appl. Phys. Lett.*, **86**(12), 123110.

13. Liu, Y., Chen, T.P., Zhao, P., Zhang, S., Fung, S., and Fu, Y.Q. (2005). *Appl. Phys. Lett.*, **87**(3), 033112.

14. Tan, Z., Samanta, S.K., Yoo, W.J., and Lee, S. (2005). *Appl. Phys. Lett.*, **86**(1), 013107.

15. Lee, J.J., Harada, Y., Pyun, J.W., and Kwong, D.L. (2005). *Appl. Phys. Lett.*, **86**(10), 103505.

16. Lee, J.J., Bai, W., and Kwong, D.L. (2005). *43rd Annual Proceedings IEEE International Reliability Physics Symposium*, 668–669.

17. Yeh, P.H., Yu, C.H., Chen, L.J., Wu, H.H., Liu, P.T., and Chang, T.C. (2005). *Appl. Phys. Lett.*, **87**(19), 193504.

18. Yeh, P.H., Wu, H.H., Yu, C.H., Chen, L.J., Liu, P.T., Hsu, C.H., and Chang, T.C. (2005). *J. Vac. Sci. Technol. A*, **23**(4), 851–855.

19. Kim, J.H., Jin, J.Y., Jung, J.H., Lee, I., Kim, T.W., Lim, S.K., et al. (2005). *Appl. Phys. Lett.*, **86**(3), 032904.

20. Paul, S., Pearson, C., Molloy, A., Cousins, M.A., Green, M., Kolliopoulou, S., et al. (2003). *Nano Lett.*, **3**(4), 533–536.

21. Chandra, A., and Clemens, B.M. (2005). *Appl. Phys. Lett.*, **87**(25), 253113.

22. Liu, Z., Lee, C., Narayanan, V., Pei, G., and Kan, E.C. (2003). *IEEE Electron Device Lett.*, **24**(5), 345–347.

23. Kouvatsos, D.N., Ioannou-Sougleridis, V., and Nassiopoulou, A.G. (2003). *Appl. Phys. Lett.*, **82**(3), 397–399.

24. Yang, H.G., Shi, Y., Pu, L., Gu, S.L., Shen, B., Han, P., et al. (2003). *Microelectron. J.*, **34**(1), 71–75.

25. Kapetanakis, E., Normand, P., Tsoukalas, D., Beltsios, K., Stoemenos, J., Zhang, S., and Van den Berg, J. (2000). *Appl. Phys. Lett.*, **77**(21), 3450–3452.

26. Yang, F.M., Chang, T.C., Liu, P.T., Yeh, P.H., Yu, Y.C., Lin, J.Y., et al. (2007). *Appl. Phys. Lett.*, **90**(13), 132102.

27. Chang, T.C., Yan, S.T., Liu, P.T., Chen, C.W., Lin, S.H., and Sze, S.M. (2004). *Electrochem. Solid-State Lett.*, **7**(1), G17–G19.

28. Yeh, P.H., Chen, L.J., Liu, P.T., Wang, D.Y., and Chang, T.C. (2007). *Electrochim. Acta*, **52**(8), 2920–2926.

29. Hu, C.W., Chang, T.C., Tu, C.H., Huang, Y.H., Lin, C.C., Chen, M.C., et al. (2010). *Electrochem. Solid-State Lett.*, **13**(3), H49–H51.

30. Chen, S.C., Chang, T.C., Hsieh, C.M., Li, H.W., Sze, S.M., Nien, W.P., et al. (2010). *Thin Solid Films*, **519**(5), 1677–1680.

31. Lin, C.C., Chang, T.C., Tu, C.H., Chen, W.R., Hu, C.W., Sze, S.M., et al. (2008). *Appl. Phys. Lett.*, **93**(22), 222101.

32. Yang, S.M., Chien, C.H., Huang, J.J., and Lei, T.F. (2007). *Jpn. J. Appl. Phys.*, **46**(6R), 3291.

33. De Salvo, B., Ghibaudo, G., Pananakakis, G., Masson, P., Baron, T., Buffet, N., et al. (2001). *IEEE Trans. Electron Devices*, **48**(8), 1789–1799.

34. Frohman-Bentchkowsky, D., and Lenzlinger, M. (1969). *J. Appl. Phys.*, **40**(8), 3307–3319.

35. Groeseneken, G., Maes, H.E., Van Houdt, J., and Witters, J.S. (1998). *Nonvolatile Semiconductor Memory Technology*, 1–88.

36. Schuegraf, K.F., King, C.C., and Hu, C. (1992). *Symposium on VLSI Technology Digest of Technical Papers,* 18–19.

37. Weinberg, Z.A. (1977). *Solid-State Electron.*, **20**(1), 11–18.

38. Yano, K., Ishii, T., Hashimoto, T., Kobayashi, T., Murai, F., and Seki, K. (1995). *Appl. Phys. Lett.*, **67**(6), 828–830.

39. De Salvo, B., Ghibaudo, G., Pananakakis, G., Guillaumot, B., and Baron, T. (2000). *Superlattices Microstruct.*, **28**(5–6), 339–344.

40. Bhattacharyya, A. (1984). *Solid-State Electron.*, **27**(10), 899–906.

41. Heiman, F.P., and Warfield, G. (1965). *IEEE Trans. Electron Devices*, **12**(4), 167–178.

42. Schulz, M. (1993). *J. Appl. Phys.*, **74**(4), 2649–2657.

43. Steimle, R.F., Muralidhar, R., Rao, R., Sadd, M., Swift, C.T., Yater, J., et al. (2007). *Microelectron. Reliab.*, **47**(4), 585–592.

44. Yater, J.A., Kirichenko, T., Prinz, E.J., Sadd, M., Steimle, R., Swift, C.T., and Chang, K.M. (2006). *21st IEEE Non-Volatile Semiconductor Memory Workshop,* 60–61.

45. Tiwari, S., Rana, F., Hanafi, H., Hartstein, A., Crabbé, E.F., and Chan, K. (1996). *Appl. Phys. Lett.*, **68**(10), 1377–1379.

46. Swift, C.T., Hoefler, A., Kirichenko, T., Muralidhar, R., Prinz, E., Rao, R., et al. (2006). *21st IEEE Non-Volatile Semiconductor Memory Workshop*, 56–57.

47. Chakraborty, G., Chattopadhyay, S., Sarkar, C.K., and Pramanik, C. (2007). *J. Appl. Phys.*, **101**(2), 024315.

48. Chen, T.P., Tse, M.S., and Zeng, X. (2001). *Appl. Phys. Lett.*, **78**(4), 492–494.

49. Huang, S., Banerjee, S., Tung, R.T., and Oda, S. (2003). *J. Appl. Phys.*, **93**(1), 576–581.

50. De Salvo, B., Ghibaudo, G., Luthereau, P., Baron, T., Guillaumot, B., and Reimbold, G. (2001). *Solid-State Electron.*, **45**(8), 1513–1519.

51. Cassan, E., Galdin, S., Dollfus, P., and Hesto, P. (1999). *J. Appl. Phys.*, **86**(7), 3804–3811.

52. Quan, W.Y., Kim, D.M., and Cho, M.K. (2002). *J. Appl. Phys.*, **92**(7), 3724–3729.

53. Weinberg, Z.A. (1982). *J. Appl. Phys.*, **53**(7), 5052–5056.

54. Chen, J., Lee, T., Su, J., Wang, W., and Reed, M.A. (2004). Molecular electronic devices, in *Encyclopedia of Nanoscience and Nanotechnology*. American Scientific, Valencia, California, Vol. 5, 633–662.

55. Sun, C.Q., Sun, X.W., Tay, B.K., Lau, S.P., Huang, H.T., and Li, S. (2001). *J. Phys. D: Appl. Phys.*, **34**(15), 2359.

56. Berashevich, J.A., Danilyuk, A.L., Kholod, A.N., and Borisenko, V.E. (2003). *Mater. Sci. Eng. B*, **101**(1), 111–118.

57. Simmons, J.G. (1963). *J. Appl. Phys.*, **34**(6), 1793–1803.

58. Campera, A., and Iannaccone, G. (2005). *Solid-State Electron.*, **49**(11), 1745–1753.

59. Iannaccone, G., Crupi, F., Neri, B., and Lombardo, S. (2003). *IEEE Trans. Electron Devices*, **50**(5), 1363–1369.

60. Singaraju, P., and Venkat, R. (2008). *Physica E*, **40**(9), 2851–2858.

61. Busseret, C., Ferraton, S., Montès, L., and Zimmermann, J. (2006). *IEEE Trans. Electron Devices*, **53**(1), 14–22.

62. Depas, M., Vermeire, B., Mertens, P.W., Van Meirhaeghe, R.L., and Heyns, M.M. (1995). *Solid-State Electron.*, **38**(8), 1465–1471.

63. Chan, K.C., Lee, P.F., and Dai, J.Y. (2008). *Microelectron. Eng.*, **85**(12), 2385–2387.

64. Liu, Z., Lee, C., Narayanan, V., Pei, G., and Kan, E.C. (2002). *IEEE Trans. Electron Devices*, **49**(9), 1614–1622.

65. Lee, D.U., Lee, M.S., Kim, J.H., Kim, E.K., Koo, H.M., Cho, W.J., and Kim, W.M. (2007). *Appl. Phys. Lett.*, **90**(9), 093514.

66. Wang, C.C., Chiou, Y.K., Chang, C.H., Tseng, J.Y., Wu, L.J., Chen, C.Y., and Wu, T.B. (2007). *J. Phys. D: Appl. Phys.*, **40**(6), 1673.

67. Yang, J.Y., Kim, J.H., Choi, W.J., Do, Y.H., Kim, C.O., and Hong, J.P. (2006). *J. Appl. Phys.*, **100**, 066102

68. Lee, P.F., Dai, J.Y., Wong, K.H., Chan, H.L., and Choy, C.L. (2003). *J. Appl. Phys.*, **93**(6), 3665–3667.

69. Chan, K.C., Lee, P.F., and Dai, J.Y. (2008). *Appl. Phys. Lett.*, **92**(22), 223105.

70. Lu, T.Z., Alexe, M., Scholz, R., Talelaev, V., and Zacharias, M. (2005). *Appl. Phys. Lett.*, **87**(20), 202110.

71. Guan, W., Long, S., Liu, M., Liu, Q., Hu, Y., Li, Z., and Jia, R. (2007). *Solid-State Electron.*, **51**(5), 806–811.

72. Likharev, K.K. (1999). *Proc. IEEE*, **87**(4), 606–632.

73. She, M., and King, T.J. (2003). *IEEE Trans. Electron Devices*, **50**(9), 1934–1940.

74. Ryogo, K. (1962). *J. Phys. Soc. Jpn.*, **17**, 975–986.

75. Schuegraf, K.F., and Hu, C. (1993). *31st Annual Proceedings Reliability Physics 1993*, 7–12.

76. Sharma, B., Bhattacharyya, G., Sengupta, A., Rahaman, H., and Sarkar, C.K. (2014). *International Conference on Advanced Materials and Energy Technology (ICAMET)*.

77. Ng, C.Y., Chen, T.P., Yang, M., Yang, J.B., Ding, L., Li, C.M., et al. (2006). *IEEE Trans. Electron Devices*, **53**(4), 663–667.

78. Duguay, S., Burignat, S., Kern, P., Grob, J.J., Souifi, A., and Slaoui, A. (2007). *Semicond. Sci. Technol.*, **22**(8), 837.

79. Tiwari, S., Rana, F., Chan, K., Hanafi, H., Chan, W., and Buchanan, D. (1995). *IEEE International Electron Devices Meeting* 521–524.

80. Chakraborty, G., and Sarkar, C.K. (2008). *J. Appl. Phys.*, **104**(3), 034313.

81. Sengupta, A., Sarkar, C.K., and Requejo, F.G. (2011). *J. Phys. D: Appl. Phys.*, **44**(40), 405101.

82. Sengupta, A., and Sarkar, C.K. (2014). *J. Nanotechnol.*, **11**(12), 1073–1080.

83. Ryu, S.W., Mo, C.B., Hong, S.H., and Choi, Y.K. (2008). *IEEE Trans. Nanotechnol.*, **7**(2), 145–150.

84. Schnippering, M., Carrara, M., Foelske, A., Kötz, R., and Fermín, D.J. (2007). *Phys. Chem. Chem. Phys.*, **9**(6), 725–730.

85. Ryu, S.W. (private communication with Sengupta, A.).

86. Lee, J.J., and Kwong, D.L. (2005). *IEEE Trans. Electron Devices*, **52**(4), 507–511.

87. Garnett, J.M. (1906). *Philos. Trans. R. Soc. London Ser. A*, 237–288.

88. Yeh, Y.S., Lue, J.T., and Zheng, Z.R. (2005). *IEEE Trans. Microwave Theory Tech.*, **53**(5), 1756–1760.

89. Gritsenko, V., Gritsenko, D., Shaimeev, S., Aliev, V., Nasyrov, K., Erenburg, S., et al. (2005). *Microelectron. Eng.*, **81**(2), 524–529.

90. Mondal, I., and Dutta, A.K. (2008). *IEEE Trans. Electron Devices*, **55**(7), 1682–1692.

91. Ghosh, B. (2008). *Appl. Surf. Sci.*, **254**(15), 4908–4911.

92. Nandy, S., Mallick, S., Ghosh, P.K., Das, G.C., Mukherjee, S., Mitra, M.K., and Chattopadhyay, K.K. (2008). *J. Alloys Compd.*, **453**(1), 1–6.

93. Van Schaijk, R., Slotboom, M., Van Duuren, M., Dormans, D., Akil, N., Beurze, R., et al. (2005). *Solid-State Electron.*, **49**(11), 1849–1856.

94. Thean, A., and Leburton, J.P. (2001). 3 *IEEE Electron Device Lett.*, **22**(3), 148–150.

95. Hasaneen, E.S., Heller, E., Bansal, R., Huang, W., and Jain, F. (2004). *Solid-State Electron.*, **48**(10), 2055–2059.

96. Bertolazzi, S., Krasnozhon, D., and Kis, A. (2013). *ACS Nano*, **7**(4), 3246–3252.

97. Zhang, E., Wang, W., Zhang, C., Jin, Y., Zhu, G., Sun, Q., et al. (2014). *ACS Nano*, **9**(1), 612–619.

98. Sharma, B., Mukhopadhyay, A., Sengupta, A., Rahaman, H., and Sarkar, C.K. (2016). *J. Comput. Electron.*, **15**(1), 129–137.

Chapter 3

Charge Trapping and High-κ Nanocrystal Flash Memory

Meng Chuan Lee* and Hin Yong Wong*

Faculty of Engineering, Multimedia University, Persiaran Multimedia,
63100 Cyberjaya, Selangor, Malaysia
hywong@mmu.edu.my

One of the most critical challenges that floating-gate and nitride-based charge storage nonvolatile memory (CS-NVM) devices face is the conflicting requirements of the bottom oxide layer due to the incessant technology scaling. This conflicting requirement emerges because of the need to reduce the overall critical dimensions of the CS-NVM cell, while maintaining (if not enhancing) the charge retention performance. The overall critical dimensions of CS-NVM are consistently miniaturized to fit more transistors in a chip, as predicted by Moore's law. By 2015, CS-NVM was predicted to be scaled down to 16 nm. With this aggressive technology scaling trend, the device fundamental limit of CS-NVM is fast approaching because of the scale down of the critical dimensions of the CS-NVM

---

*Both authors have contributed equally to this work.

Nanocrystals in Nonvolatile Memory
Edited by Writam Banerjee
Copyright © 2018 Pan Stanford Publishing Pte. Ltd.
ISBN 978-981-4774-73-4 (Hardcover), 978-1-351-20327-2 (eBook)
www.panstanford.com

cell, especially in the tunnel oxide layer. This has also caused the reconsideration of keeping the practical limit to balance the economic gain and the gross investment required to resolve technology issues arising because of aggressive scaling. To improve the scalability of CS-NVM devices, another promising candidate of CS-NVM that is being heavily researched is nanocrystal-based CS-NVM, which will be elucidated comprehensively in this chapter. Along the technology scaling trend, these variants of CS-NVM devices have significant advantages in terms of compatibility with the standard CMOS fabrication process (and thus cost effective) and scaling capacity, while they maintain high-quality and high-reliability performance. In this review chapter, the evolution of charge-trapping nanocrystal CS-NVM is elucidated thoroughly. Furthermore, the enhancement in the reliability performance of nanocrystal CS-NVM devices in endurance and data retention is discussed comprehensively in a later section. In view of reliability, the key challenges for nanocrystal CS-NVM are also explained, with an emphasis on the charge loss mechanism, which critically impacts data retention. Moreover, in a later section, the innovation of tunnel barrier engineering with variants of metal nanocrystals and its major potential advantages are discussed comprehensively. This chapter will serve as a good reference to understand the key advantages, reliability challenges, and technical mitigations of nanocrystal CS-NVM and the future innovation trend in nanocrystal CS-NVM to overcome the critical challenges due to technology scaling in adherence to Moore's law.

3.1 Introduction to Charge Storage Nonvolatile Memory

Sequential actions of a typical algorithm that include receiving a stimulus from the environment via sensors, interpreting/decoding the stimulus, and then responding accordingly on the basis of the built-in artificial intelligence and smart algorithm are critical for almost all information processing systems [1]. Without memory elements, these sequential actions will not be possible to enable the systems to recall any items of interest during any operation/computations to respond to any raw stimulus. Up until the 1980s, the semiconductor industry relied heavily on volatile memory, such as dynamic random access memory (DRAM) and static random access memory (SRAM), for temporary coding or data storage operations. However, the main disadvantage of implementing volatile memory in the system is that the data will be lost once the volatile memory devices are powered off. The volatile characteristic of random access memory (RAM) has created a demand for charge storage nonvolatile memory (CS-NVM), which has led to the emergence of read-only memory (ROM) [2, 3]. Variants of ROM devices, for example, programmable read-only memory (PROM), erasable programmable read-only memory (EPROM), and electrical erasable and programmable read-only memory (EEPROM), have provided temporary relief to the market demand for CS-NVM to replace RAM [4, 5]. Nevertheless, even for various ROM devices, the voltages required for program and erase (P/E) operations are high. This creates a strong opportunity to aggressively lower the voltage required to perform P/E operations. Furthermore, this has created a strong demand in the electronic industry to have a new variant of EEPROM that (i) enables us to store digital data reliably, immune to the external environment, (ii) enables us to electrically modify the stored data using low voltages and the least amount of time possible, and (iii) is compatible with the standard complementary-metal-oxide-semiconductor (CMOS) fabrication process. These requirements can be met by inventing a memory transistor that is able to electrically modify the stored data by changing the threshold voltage (V_T) to two distinct levels, high and low. The switch between low and high states of the transistor's threshold voltage corresponds to the binary values of 1 and 0, which can be translated to the erase state and the program state, respectively, of the CS-NVM. The change of the logical states of CS-NVM strongly depends on the quantity of charges stored in the designed charge storage layer. The number of electrons stored in the charge storage layer directly impacts the read current flow in the channel of the CS-NVM transistor. Thus, V_T is defined as the gate voltage that produces a predetermined read current (I_{read}) of the CS-NVM transistor. Figure 3.1 shows the typical erase state and program state V_T distributions of a CS-NVM device. Figure 3.1 also shows the definition of a memory window (MW) in a typical V_T distribution of a CS-NVM device; a MW is defined as the delta voltage between the highest points of the erase state V_T distribution and the program state V_T distribution. For P/E operations, several charge injection mechanisms were researched to inject different species of charges into the designed charge storage layer in CS-NVM, for example, Fowler–Nordheim (FN) tunneling, channel hot electron injection (CHEI), and hot hole injection (HHI) [14, 15]. These charge injection mechanisms have different charge injection efficiencies and charge injection profiles for electron and hole distributions.

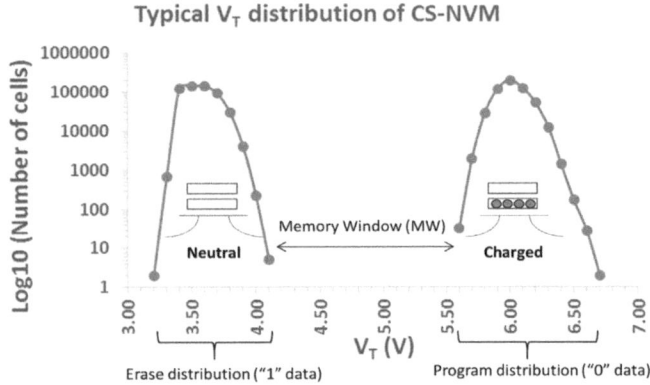

Figure 3.1 Representations of the logical states of 1 and 0, corresponding to erase and program states in typical V_T distributions of a CS-NVM device.

The critical turning point in the development of CS-NVM came when flash memory device was invented by Dr. Fujio Masuoka [1, 2, 6–8]. The flash memory market soared from $310 million to $20.5 billion market in 2005 and close to $23 billion worldwide in 2007 [1]. The flash memory market rapidly soared especially when embedded computing devices and the recent Internet of things (IoT) started to bloom. In this new era of IoT, the application of flash memory to embedded computing devices (including system on chip, SoC) is imperative to ensure the interconnection of these devices with the Internet infrastructure functions properly. This has generated an insatiable demand for a larger storage space at a lower cost per bit, a demand which continues to increase since the digital world of data has approximately doubled every year and had expanded to 2.72 zettabytes (10^{21} bytes) by 2012 [1, 7, 16]. This trend of digital data expansion hit the mark of 8 zettabytes by 2015 [90]. Thus, it is important that the trend of innovating advanced flash memory devices to achieve a higher memory density continues even as the cost per bit is driven down. This trend is aggressively driven by the well-known Moore's law cost reduction through incessant technology scaling [2, 9–15]. Moore's law predicted that the number of transistors that fit into an integrated circuit (IC) chip will double approximately every 18 months [5, 6]. And as per the technology scaling trend predicted by Moore's law, the memory density of advanced flash memory devices has increased by leaps and bounds. The outlook for flash memory devices is getting brighter in various IT and electronic applications since flash memory devices are flexible enough to cater to both code and data storage depending on the architecture of the device [17].

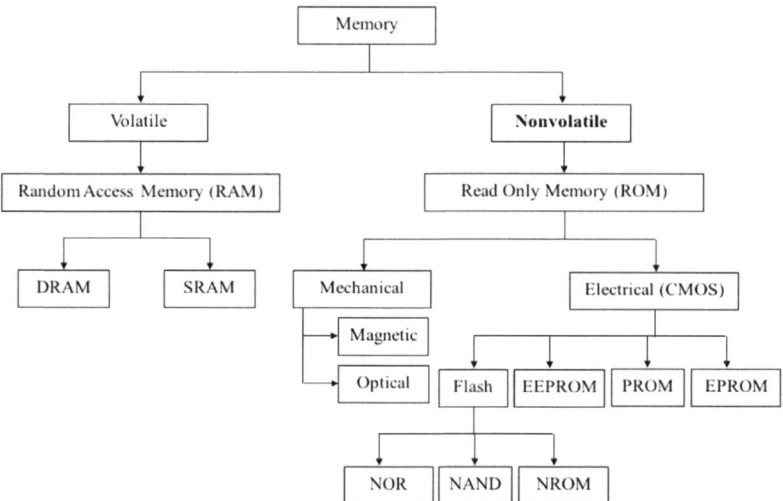

Figure 3.2   Categories of semiconductor memory available in the market.

Figure 3.2 shows the categories of semiconductor memory available in the market. Flash memory is a nonvolatile electrical read-only memory based on CMOS technology. There are three main types of flash memory architectures: NOR, NAND, and nonvolatile read-only memory (NROM). As shown in Fig. 3.3a, NOR architecture is addressable at the bit level since the bit lines are connected to each memory cell. Furthermore, it is optimized for code storage that requires fast random access, which is critical for systems that require a boot up within seconds, and it is ultrareliable and doesn't allow any bad blocks in the flash memory array. As compared to NOR architecture, NROM architecture's bit lines can be interchangeable between source and drain, depending on the program, erase, and read algorithms [4]. NROM is flexible for use in both data and code applications since it allows bit accessibility and is relatively fast in program, erase, and read operations [16]. On the other hand, as shown in Fig. 3.3b, NAND architecture has fewer bit lines and, thus, it is accessed in blocks or pages. As compared to NOR architecture, NAND architecture is optimized for mass data storage because a high packing density of flash memory cells can be achieved at a low cost [17].

(a) NOR architecture

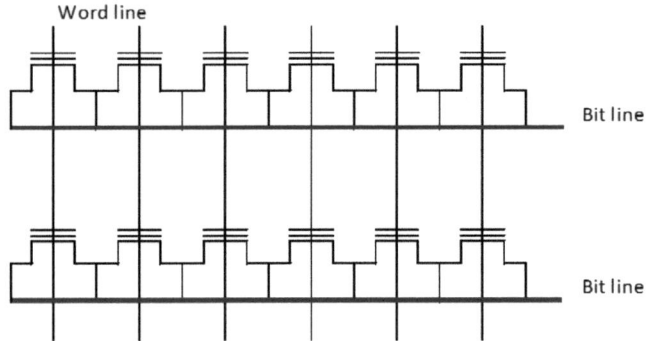

(b) NAND architecture

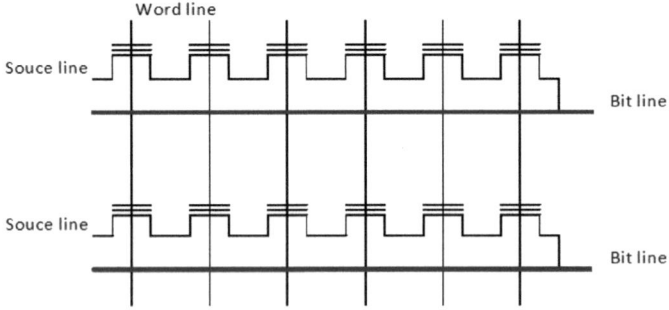

Figure 3.3   (a) NOR architecture and (b) NAND architecture.

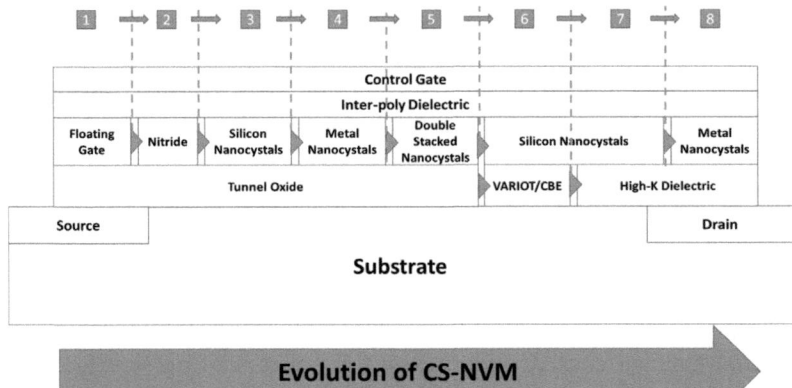

Figure 3.4 Evolution of CS-NVM devices started from floating-gate- and nitride-based CS-NVM and moved to possible variants of CS-NVM that combine nanocrystals and high-κ dielectric constant (high-κ) materials: (1) floating-gate CS-NVM, (2) nitride-based CS-NVM, (3) silicon-nanocrystal-based CS-NVM, (4) metal-nanocrystal-based CS-NVM, (5) double-stack-nanocrystal-based CS-NVM, (6) silicon-nanocrystal-based CS-NVM with a variable oxide thickness (VARIOT)/ crested barrier engineering (CBE) tunnel dielectric, (7) silicon-nanocrystal-based CS-NVM with a high-κ tunnel dielectric, and (8) metal-nanocrystal-based CS-NVM with a high-κ tunnel dielectric.

Figure 3.4 summarizes the evolution of CS-NVM devices, which started from the conventional floating-gate (FG) CS-NVM devices and includes the possible variants of CS-NVM devices that combine the implementation of nanocrystals and tunnel barrier engineering (TBE) methods such as variable oxide thickness (VARIOT)/CBE/ high-κ tunnel dielectric. As shown in Fig. 3.4, there are several variants of CS-NVM devices available in the market that differ predominantly by the charge storage media, that is, continuous versus discrete charge traps [2–6]. Examples of these CS-NVM devices are FG CS-NVM, nitride-based CS-NVM, and nanocrystal-based CS-NVM. With different designed charge storage layers, these CS-NVMs exhibit different behaviors in terms of endurance and charge retention performance/mechanisms. With respect to various charge storage media, the key focus is to engineer a charge storage layer that will retain the injected charges reliably, regardless of the temporal fluctuations of the usage environment [2–4, 7–17]. These injected charges that are stored in the charge storage layer regulate the internal electric field and exert proper control on the current flow through the channel of the memory cell. As shown in Fig. 3.4a, the charge storage layer for FG CS-NVM is designed as the conductive polysilicon sandwiched between top interpoly dielectric and bottom silicon oxide layers [2–5]. The top and bottom dielectric layers act as a barrier to retain the injected charges in the conductive polysilicon layer, which is also known as FG. Since the FG is conductive, the injected charges are stored in a continuous state. Therefore, from the reliability point of view, the most important concern for FG-based CS-NVM is the finite probability of a huge tunnel leakage current induced by the clusters of defects existing in the tunnel oxide layer attributed to the damages incurred after an extensive P/E cycling performed on the CS-NVM device [25, 33]. This anomalous leakage current in the tunnel oxide layer induced by oxide defects determines the anomalous failure bits (also known as anomalous tail bits) that jeopardize the data storage integrity of the FG-based CS-NVM [7–17]. Furthermore, this phenomenon of the oxide-defect-assisted leakage current is also known as stress-induced leakage current (SILC), which is a well-known Achilles' heel of FG-based CS-NVMs [7, 8, 10–12, 14–17]. Moreover, SILC is one of the key limiters that prevent the physical scaling of a tunnel oxide layer beyond 8 nm [4–7]. For FG-based CS-NVMs, the storage media, which is a doped polysilicon, is located above the channel and isolated from the channel by a dielectric usually composed of SiO_2. Because of this structure, the FG is able to exert influence on the channel form lying beneath the bottom oxide layer [7–13]. Therefore, the current in the channel can be turned off by having sufficient electrons stored in the FG, which leads to the first logical state, 0, also known as the programmed state [9–17]. On the other hand, the current in the channel can flow from drain to source if a sufficient number of electrons is removed from the FG, which leads to the second logical state, 1, also known as the erased state [14–17]. Thus, the qualities of dielectric layers surrounding the FG guarantee the nonvolatility feature of the device while the thickness of the dielectric layers allows the injection and removal of charges through P/E operations through the application of electrical pulses with optimized V_{gate}, V_{drain}, V_{source}, pulse width, and number-of-pulses settings [16, 17]. The silicon oxide layer located between the FG and the channel is known as tunnel oxide because charge carriers can be tunneled through the tunnel oxide layer and stored in the FG through the application of a charge injection mechanism, for example, the FN tunneling mechanism. The control gate (CG) and the FG are separated by a silicon oxide– silicon nitride–silicon oxide a triple layer, also known as an oxide-nitride-oxide (ONO) stack, to improve the quality of the tunnel oxide layer. Figure 3.5 shows the schematic energy band diagram of the two logical states of a FG flash memory cell. The FG can function as a potential well to store charge carriers, and the ONO stack (generally known as interpoly dielectric layers) and the tunnel oxide act as potential barriers that prevent the charge carriers in the FG from leaking out to the substrate [15, 16]. Thus maintaining the charge carriers in the FG can help modulate the cell's V_T to have two clearly distinguished logical states. On the basis of this fundamental concept, it can be said that the information retained in a flash memory device is essentially analog as the cell's V_T can be modulated between two distinct logical states [4].

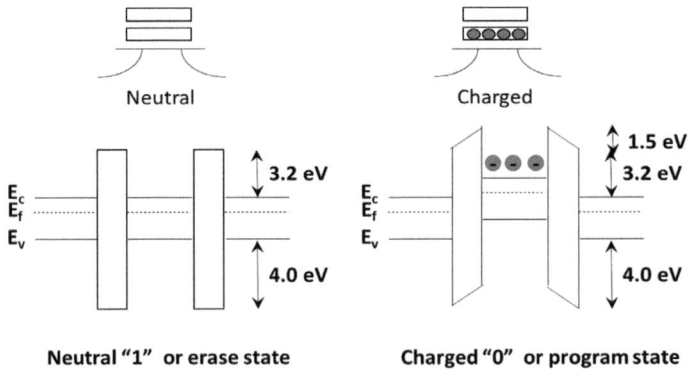

Figure 3.5 Schematic energy band diagram that refers to two logical states of a floating-gate flash memory cell [1].

On the other hand, as shown in Fig. 3.4b, the charge storage layer for nitride-based CS-NVM is engineered as the nonconductive silicon nitride layer between top and bottom oxide layers [17–21]. The injected charges are trapped in the nitride layer that is cladded between top and bottom oxide layers. The inherent intrinsic defects of the silicon nitride layer are able to trap the injected charges. This effectively suppresses the possibility of the trapped charges in the nitride storage layer leaking out to the substrate through point defects generated in the bottom oxide layer. Compared to FG-based CS-NVM, the fabrication process for nitride-based CS-NVM yielded fewer steps and this CS-NVM was compatible with the standard CMOS fabrication process. Hence, the fabrication cost of nitride-based CS-NVM is relatively lower than that of FG-based CS-NVM. Amorphous silicon nitride (Si_3N_4) has the distinguished ability to localize and retain charged particles, that is, electrons and holes, in existing traps with a long lifetime, approximately 10 years at 450 K [18]. Therefore, amorphous silicon nitride is widely used as the storage medium of nitride flash memory cells such as silicon-oxide-nitride-oxide-silicon (SONOS) and MirrorBit technologies. Figure 3.4b shows the cell structure of NROM (a variant of MirrorBit technology) is simpler, with less fabrication complexity, and the cell size is generally smaller than a FG cell [1–5, 7–17, 22]. Silicon nitride has been widely employed as the storage medium of nonvolatile memory (NVM) devices, for example, NROM and SONOS devices, and as a dielectric layer in the ONO stack between FG and CG to reduce the discharge of electrons from the FG [19]. However, the nature of electrons and holes trapped in the nitride layer remains unclear [18]. There are three main models that describe the charge-trapping effect in silicon nitride [18, 19]. The first model suggests the dangling Si bonds form the charge capturing centers [18–20]. According to this model, the $N_3Si\cdot$ pair, that is, the paramagnetic neutral defects in Si_3N_4, could form two diamagnetic defects that consist of positively charged N_3Si^+ and negatively charged N_3Si^- and through this process, a hole and an electron can be trapped in each defect, as shown in Eq. 3.1.

$$N_3Si\cdot + \cdot SiN_3 \rightarrow N_3Si^+ + :{}^-SiN_3 \qquad (3.1)$$

The second model suggests the presence of amphoteric charged silicon atoms that act as traps for electrons and holes [19, 20]. The amphoteric traps in silicon nitride can be in a neutral state D^0, a negatively charged state D^-, and a positively charged state D^+ [18]. All the traps in Si_3N_4 are in the D^0 state since Si_3N_4 shows a neutral behavior. On the basis of this model, the emission (capture) of an electron (hole) from a D^- (D^0) trap will bring it back to the D^0 (D^+) state and vice versa [21]. For a programmed state, the charge retention characteristics are mainly determined by the transition from the D^- state to the D^0 state [21]. On the other hand for an erased state, the charge retention characteristics are mainly determined by the transitions from the D^+ state to the D^0 state [21]. In this model, the amphoteric charged silicon atoms capture two charge carriers at the same time (a hole and an electron) [19, 20]. The third model suggests that the Si–Si bond can act as a trap in silicon nitride to capture holes and electrons [19, 20]. According to this model, the interaction between two ≡Si· defects will result in the creation of a diamagnetic positively charged (≡Si) and a negatively charged (≡Si:) defect, which process is determined by Eq. 3.2 [19, 20]. In this case, the negatively charged defect is a hole trap while the positively charged defect is an electron trap in Si_3N_4 [18]. The simulation result from a previous work also shows that neutral diamagnetic Si–Si bonds can trap electrons and holes.

$$≡Si· + ·Si ≡ \rightarrow Si + :Si ≡ \tag{3.2}$$

Compared to FG-based CS-NVM devices, the charge transport and charge retention of nanocrystal-based CS-NVM strongly depends on the quantum confinement effect and the Coulomb blockade effect [22–67]. During programming operation for a typical silicon-nanocrystal-based CS-NVM, electrons were injected and trapped in the nanocrystals through a tunnel oxide layer. After the electrons were trapped, the nanocrystal's potential energy increased by an electrostatic charging energy of $q^2/2C_{tt}$ [22, 23, 26–29]. C_{tt} is defined as the capacitance of the nanocrystal, which significantly depends on the size of the nanocrystal, the thickness of the tunnel oxide, and the thickness of the control oxide layer [26–29]. Rao et al. [28, 29] have reported that the change in V_T (ΔV_T) of silicon-nanocrystal-based CS-NVM can be expressed as Eq. 3.3. In this expression, n is the nanocrystal number density, p is the average number of electrons stored per nanocrystal, q is the electronic charge, ε represents the medium permittivity, and $t_{control}$ and t_{nc} represent the thickness of the dielectric layer and the nanocrystal layer, respectively [28, 29].

$$\Delta V_T = \frac{npq}{\varepsilon_{ox}}\left(t_{control} + \frac{\varepsilon_{ox}t_{nc}}{2\varepsilon_{Si}}\right) \tag{3.3}$$

The injected electrons trapped in nanocrystals will hence charge up the nanocrystals and the internal electric field across the tunnel oxide layer will be lowered accordingly. The reduction of the internal electric field across the tunnel oxide layer suppresses the tunneling current density during the program operation [22–29, 35–37]. Therefore, as reported by She and King [27], the potential energy of the charged nanocrystal will impede the other electrons from tunneling through the tunnel oxide layer and injecting into the charged nanocrystals. This impedance effect is also known as the Coulomb blockade effect. Moreover, the Coulomb blockade effect will effectively prevent the tunneling of electrons at a low internal electric field resulting from a low gate voltage (similar to the gate voltage applied during read operation) [22–29, 35–37]. This increases the immunity of nanocrystal-based CS-NVM to the disturb mechanism that impacts FG-based CS-NVM devices [33–61]. With a sufficient number of electrons trapped in nanocrystals, the Coulomb blockade effect improves the charge retention performance of nanocrystal-based CS-NVM [35–37]. Despite the key advantages of the Coulomb blockade effect, this effect will negatively impact the programming speed and voltage [35–38]. Increasing the size of nanocrystals can counter this negative impact and achieve faster programming speed and larger tunneling current during the programming operation. Compared to smaller nanocrystals, larger nanocrystals favor charging of electrons under a smaller Coulomb blockade effect/ quantum confinement effect since quantum confinement energy is approximately inversely proportional to the diameter of nanocrystals [27–29, 35–37]. Furthermore, the quantum confinement energy becomes significant because the dimensions of nanocrystals are in the nanometer range [27]. This causes the conduction band of the nanocrystal to shift upward while the conduction band offset between the nanocrystal and the surrounding control oxide layer lowers [27].

In brief, a larger nanocrystal is preferable for faster programming speed since the increment in the diameter of the nanocrystal results in larger tunneling current, which improves the programming speed. With respect to the charge retention effect, a smaller nanocrystal is preferable for improved reliability performance, attributed to the enhanced Coulomb blockade effect. Evidently, there is a conflicting requirement with respect to the size of the nanocrystals in terms of achieving higher programming speed versus better charge retention performance. Experimental results show the optimum size for nanocrystals is 5 nm, as reported by Min She et. al [27]. Thus, during the design of nanocrystal-based CS-NVM devices, it is necessary to determine the Coulomb blockade effect and the quantum confinement effect with respect to the typical 10-year data retention requirement in order to determine the size of nanocrystals, the density of nanocrystals in the control dielectric layer, and the thickness of the tunnel dielectric layer [33–66]. On the basis of technology computer-aided design (TCAD) simulations performed on nanocrystals, Gasperin et al. [39] have reported that the dimension, quantity, size, and positions of nanocrystals can impact the charge localizations of nanocrystal memory cells, which in turn affect the program window in the subthreshold as well as the linear region.

In summary, the critical potential advantages that discrete charge trap memory (i.e., nanocrystal and nitride-based CS-NVM) offer are:

- Suppression of anomalous failure bits induced by SILC driven by oxide defects clusters

- Scaling of the tunnel oxide layer beyond 8 nm without impacting the data retention performance

- Enhanced immunity of discrete charge traps against bulk oxide defects and P/E-cycling-induced charge-trapping interface states

- Suppression of the drain turn-on phenomenon due to the absence of drain-to-FG coupling ratio

- Innovation of multibits per cell by taking advantage of the trapping of multiple charges at different physical locations of the cells

- Compatibility of the fabrication process of nanocrystal and nitride-based CS-NVM with the standard CMOS fabrication process, which renders lower cost of device fabrication

3.2 Evolution of Nanocrystal-Based CS-NVM

Discrete silicon-nanocrystal-based CS-NVM was proposed by Tiwari et al. in 1995 [22, 23] as a promising alternative to standard FG-based CS-NVM [27, 33, 35–37] to satisfy the conflicting requirements of tunnel oxide resulting from the incessant technology scaling as per Moore's law. To improve the speed of P/E and reduce the operating voltage without aggravating the drain turn-on phenomenon, thinner tunnel oxide of FG-based CS-NVM is desirable to allow fast and efficient transfer of charges into and out of the FG. At the same time, tunnel oxide isolation between the FG and the silicon substrate has to be sufficient to meet the data retention criterion of 10 years, typical for industrial applications. Thus with discrete nanocrystal NVM as an alternative, the scaling of tunnel dielectric is feasible for achieving a lower operating voltage, faster program, erase, and read speed, and a desirable charge retention time, which also prolongs the longevity of Moore's law in the semiconductor CS-NVM road map. For P/E operations of nanocrystal-based CS-NVM devices, a uniform charge tunneling mechanism through the tunnel oxide layer was found to be most suitable to transport charges into and out of nanocrystals. This is because the uniform charge tunneling mechanism results in uniform charging/discharging of the nanocrystals across the control oxide layer. Examples of uniform charge tunneling mechanisms are direct tunneling and FN tunneling. Molas et al. [46] have shown that programming voltages for nanocrystal-based CS-NVM can be further optimized with the implementation of high-κ materials in the control dielectric stack [46–48, 61, 64, 65]. This implementation also helps to reduce the risk of the drain turn-on phenomenon that plagues FG-based CS-NVM devices, attributed to the effect of tunnel oxide layer scaling. Incremental step programming pulse (ISPP) programming algorithm [30, 38, 53] was reported to be able to enhance the programming performance and the silicon nanocrystal memory performance. On the other hand, lower P/E voltages and time can be achieved by substituting the tunnel oxide layer with an engineered tunnel barrier that is made of multilayer dielectric stacks [44, 58, 65] and TBE references without altering the 10-year data retention performance expected of nanocrystal-based CS-NVM devices.

Figure 3.6a shows the simplified schematic cross section of silicon nanocrystal NVM. Band diagrams during charge injection, retention, and removal are shown in Figures 3.6b, 3.6c, and 3.6d, respectively [22, 23, 26, 81]. Fabrication process techniques of a nanocrystal-based CS-NVM devise have been the subject of many research studies in published literatures. Various fabrication techniques have been developed to realize silicon nanocrystals in nanocrystal-based CS-NVM, including electron beam lithography, chemical vapor deposition (CVD), and excessive silicon precipitation. As thoroughly reported by Chang et al. [24], the most common techniques to form nanocrystals as quantum storage dots are self-assembly, precipitation, and chemical reaction. Among these three techniques, precipitation and chemical reaction are found to be more robust in controlling the size and density of nanocrystals [24, 25]. Nanocrystal-based CS-NVM has a distinct advantage in that the density and distribution of the intrinsic traps can be controlled by the density and physical locations of the nanocrystals through the adjustment of fabrication process conditions [22–26]. Silicon nanocrystal NVM replaces the conductive polysilicon FG charge storage layer of standard FG CS-NVM with discrete and mutually isolated charge storage nodes in the silicon nanocrystals distributed in the control oxide layer [22–29]. Each nanocrystal, or "dot," stores several electrons in the control oxide layer and collectively, these charges will modulate the channel conduction of each memory cell. The deep electron storage traps of each typical silicon nanocrystal (~3 eV) enhance the charge retention within the nanocrystal as compared to the nitride traps (1–2 eV) utilized in nitride-based CS-NVM [28, 29]. Because of the nature of the distributed discrete charge storage, nanocrystal-based CS-NVM exhibits excellent inherent immunity toward defect-assisted charge leakage (SILC) through defects in tunnel oxide, a problem that critically limits the scaling down of tunnel oxide to a value below 8 nm for standard FG CS-NVM [1–5, 6–17]. Thus, the tunnel oxide of a silicon nanocrystal can be further scaled down to a value below 8 nm keeping in view the trade-off between operating voltage, speed, and charge retention time. Furthermore, because of the discrete charge storage, multiple-bit-per-cell architecture was achieved on nanocrystal-based CS-NVM devices [40].

Despite many salient advantages of silicon-nanocrystal-based CS-NVM, there are critical technological challenges that prove to be key obstacles for it to transition from research mode to high-volume manufacturing mode. One of the critical challenges for silicon-nanocrystal-based CS-NVM is that these nanocrystals are not fabricated through the patterning (or lithography) process but through the nucleation process, excess silicon precipitation techniques, the aerosol deposition technique, or the direct growth technique [35–42]. This indicates that the fabrication process variation of nanocrystals, the variation in dimensions of each nanocrystal, and the distance between nanocrystals will critically impact the charge retention performance of nanocrystal-based CS-NVM devices [48–53]. Furthermore, as compared to nitride-based CS-NVM devices, the fabrication process of nanocrystal-based CS-NVM is more complex since the size and density of nanocrystals have to be preserved throughout the entire process. This reduces the thermal budget and room for process optimization of the fabrication process of nanocrystal-based CS-NVM. Hence, the fabrication process of nanocrystal-based CS-NVM proves to be a stern challenge for the high-volume manufacturing of nanocrystal-based CS-NVM, similar to nitride- and FG-based CS-NVM.

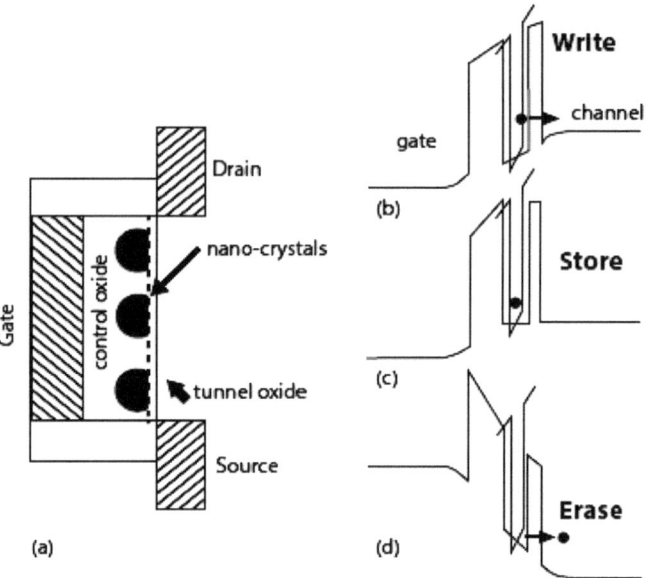

Figure 3.6 (a) Schematic cross section of silicon nanocrystal NVM, (b) band diagram during charge injection through program operation, (c) band diagram during charge retention, and (d) band diagram during charge removal through erase operation [22, 23, 26, 81].

On the other hand, there is concern regarding the leakage of charge retained in nanocrystals through the surrounding oxide layer if the tunnel oxide layer is consistently reduced and the density of nanocrystals increases accordingly. Molas et al. have shown that a tunnel oxide layer thickness of more than 4.2 nm is required to satisfy the general expected data retention performance of 10 years

[46, 60, 63, 64]. Furthermore, with silicon nanocrystals ~4 nm in size, the experimental result shows that ~0.3 V residual MW was observed on a tunnel oxide layer of ~7 nm at the extrapolated 10-year data retention performance through an accelerated reliability test [45]. On the other hand, Ng et al. [45] reported that ~70% charge loss (CL) was observed on a tunnel oxide layer of ~4 nm through an accelerated reliability test of 10-year data retention performance. This finding corroborates the finding reported by Molas et al. [46] that a tunnel oxide layer of more than 4 nm is crucial to fulfill the typical 10-year data retention performance, which is expected for embedded flash memory solutions [45, 46, 61, 64, 65]. Furthermore, to alleviate the lateral charge migration, the nanocrystals have to be substantially electrically isolated from each other with respect to the lateral charge tunneling transport. Experimental and simulation results have shown that the typical separation of each nanocrystal has to be greater than ~4 nm [28, 29]. As compared to FG-based CS-NVM devices, nanocrystal-based CS-NVM devices have shown superior tolerance to CL attributed to heavy ion irradiations bombarded on the memory array [66–72]. Furthermore, as reported by Cester et al. [66], at least a triple-ion hit on a memory cell is needed to induce a significant V_T shift of the memory cell. This finding clearly demonstrates the superior immunity of nanocrystal-based CS-NVM since substantial CL occurs in FG-based CS-NVM cells when hit by a single high-linear-energy-transfer (LET) ion [66–72]. Moreover, the V_T distribution shift was only observed in program cells of nanocrystal-based CS-NVM and not in erase cells. This observation can be explained since the nanocrystals are almost neutral in the erase state and thus a single high-LET ion hit on the memory cell did not result in any significant V_T shift [66–72]. Heavy ion irradiation on nanocrystal-based CS-NVM devices will not result in permanent damage in the data retention behavior. Any immediate V_T distribution shift of nanocrystal-based CS-NVM induced by CL due to heavy ion bombardment can be remedied by performing an P/E operation on the memory cell. Cester et al. [66] reported the programmed tail cells (induced by partial CL) completely disappear after the memory cells are reprogrammed, which indicates the CL induced by heavy ions irradiation is recoverable [66]. Furthermore, Cester et al. [66] also reported that the tail cells did not appear anymore after 20 days of room temperature storage for the irradiated memory devices, which indicates that no permanent damage was observed in the data retention performance [66–72]. The salient advantages and key disadvantages of nanocrystal-based CS-NVM devices are summarized in Table 3.1 and Table 3.2.

Table 3.1 Key advantages of nanocrystal-based CS-NVM

	References
It has superior inherent immunity to SILC through defects in tunnel oxide.	[22–27]
The tunnel oxide thickness can be further scaled down to a value below 8 nm.	[27–29, 35–37, 45, 46, 61, 64, 65]
It involves low operating voltage, fast write and erase speeds, and an enhanced MW.	[22, 23, 27–29]
Multiple bit storage significantly enhances memory density.	[33, 61, 64, 65]
FG interference is eliminated, especially for ultradense NAND flash memory	[33]
Existing Si material technology can be leveraged, and it is a CMOS-compatible process.	[35–45]
A simpler device structure yields fewer process steps, which further reduces fabrication costs.	[22, 23, 27–29]
There is superior tolerance to radiation effects as compared to standard FG flash memory.	[66–72]
Nitrided silicon nanocrystal NVM shows a larger MW and a faster program speed.	[33, 61, 64, 65, 87]

Table 3.2 Key disadvantages of nanocrystal-based CS-NVM

	References
• Nanocrystals are not fabricated through the lithography process; thus variability in the fabrication process is critical for memory properties of nanocrystal NVM.	[24, 33, 61, 64, 65]
• There is concern regarding charge leakage through the surrounding oxide if thinning of the surrounding oxide continues and the density of nanocrystals increases.	[37–41]

	References
• Optimal nanocrystal memory performance requires formation of nanocrystals of optimal size and density and their preservation during subsequent processing steps.	[22–30, 35, 36, 46]
• There is concern regarding the impact of charge localization of nanocrystals on the operating window.	[30, 38]

3.3 Reliability Challenges of Nanocrystal-Based CS-NVM

After P/E cycling and baking, V_T instability in the form of V_T distribution broadening and shifting remains one of the major reliability challenges for the future development of CS-NVM devices, such as FG-based CS-NVM, nitride-based CS-NVM, and nanocrystal-based CS-NVM [77–87]. V_T instability mechanisms of CS-NVM are detrimental to the fundamental data retention performance and may compromise the integrity of the stored data and result in undesirable field usage failures, especially for mission-critical or public safety applications. V_T instability mechanisms can be categorized into cell-level V_T instability and array-level V_T instability. Array-level V_T instability mechanisms include cell-to-cell interference and disturb phenomenon (e.g., read disturb, erase disturb, and program disturb mechanisms) that impact part of or the entire memory array of FG- and nitride-based CS-NVM devices [73–76]. These mechanisms result in an inadvertent change in the stored data of neighboring cells while program, erase, and read operations are performed on other memory cells on the same memory array. Because of the inadvertent change in the stored data, the failures induced by array-level V_T instability mechanisms are recoverable failures. The recovery involves erasing the data and programming the impacted memory cells with the expected data. These disturb mechanisms are devastating to FG- and nitride-based CS-NVM devices and deteriorate when the technology scaling trend continues [73–76]. As compared to these devices, nanocrystal-based CS-NVM devices have better immunity toward array-level V_T instability mechanisms. This enhanced immunity is attributed to the inherent Coulomb blockade effect in nanocrystal-based CS-NVM devices that impedes the tunneling of electrons at a low gate voltage.

Cell-level V_T instability mechanisms focus on the intricate interactions between the charges stored in the charge storage layer and its surrounding [77–87]. Moreover, this interaction induces variations in the read current flowing in the channel of the memory transistor. The variations in the read current will, hence, result in the V_T instability of the impacted memory cells, which directly impacts the accuracy of the data readout. Therefore, cell-level V_T instability mechanisms are a genuine reliability concern for CS-NVM devices. Numerous published literatures have reported program state V_T distribution broadening and shifting induced by CL mechanisms observed on CS-NVM devices [77–87]. The CL mechanism is one of the major cell-level V_T instability mechanisms of CS-NVM devices. CL mechanisms typically cause program state V_T distribution to broaden and shift from a high V_T level to a low V_T level, which possibly causes critical read failures due to the deteriorating read margin of the affected memory cell [77–87].

Figure 3.7 shows the typical charge leakage paths for FG-, nitride-, and nanocrystal-based CS-NVM devices. For a FG-based CS-NVM cell, extensive P/E cycling induces point defects in the form of bulk oxide traps and interface traps at the tunnel oxide layer. These point defects in the tunnel oxide induce the electrons from the conductive FG to tunnel out through a Frenkel–Poole (FP) mechanism at a much higher rate [77, 78, 81]. Therefore, the thickness of the tunnel oxide must be greater than 8 nm to prevent the tunneling of charges stored in the FG to the substrate through the percolation mechanism [1–4, 77, 78, 81]. The percolation mechanism is explained as shown in Fig. 3.8. In Fig. 3.8a, if the point defect is located within the length that an electron at the interface between the FG and the tunnel oxide layer can potentially percolate through (as circled in red), then this point defect can potentially induce FP tunneling, which will unavoidably cause CL. In Fig. 3.8b, if the point defect is located far from said length and an electron cannot tunnel through (as circled in red), then even with the presence of multiple point defects in the tunnel oxide layer, these point defects will not induce defect-assisted tunneling that results in CL [78, 81]. This process of charge carrier leak out from FG to substrate is also known as trap-assisted tunneling (TAT) or SILC.

Since the charge is stored in the conductive polysilicon of the FG, the TAT defect is able to drain all stored charge of the FG and hence cause data retention failure (due to a V_T shift).

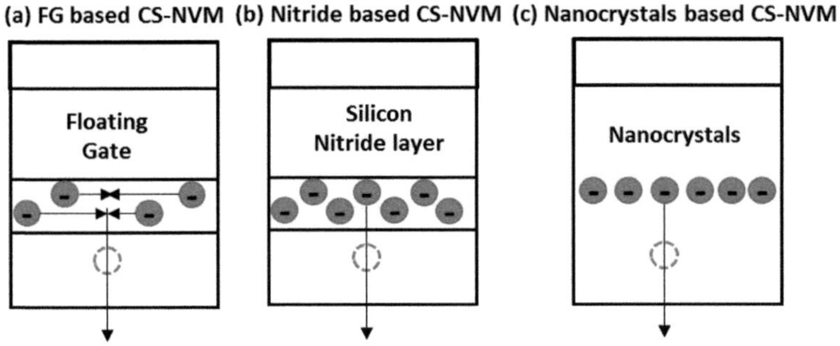

(a) FG based CS-NVM (b) Nitride based CS-NVM (c) Nanocrystals based CS-NVM

Figure 3.7   CL mechanisms for various CS-NVM devices: (a) FG-based CS-NVM, (b) nitride-based CS-NVM, and (c) nanocrystal-based CS-NVM.

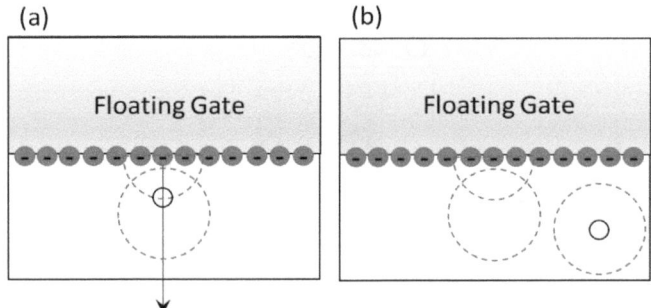

Figure 3.8   (a) A point defect located near the electron at the interface and (b) a point defect located far from the electron at the interface [1–4, 77, 78, 81].

For nitride-based CS-NVM devices, CL was reported to be attributed to FP emission of trapped charges in the nitride storage layer and then their tunneling out through P/E-cycling-induced point defects in the tunnel oxide layer [1–4, 77, 78, 81]. This tunneling mechanism is shown in Fig. 3.8b, which is similar to the SILC mechanism that impacts FG-based CS-NVM devices. Because of the discrete charge trap nature of the nitride storage layer in nitride-based CS-NVM, the point defects in the tunnel oxide layer will not drain out all charges stored in the nitride charge storage layer. The point defect will only drain out the charges trapped in the discrete trap sites that are in close proximity to the point defect in the tunnel oxide layer. The point defects in the tunnel oxide layer are generated because of extensive P/E cycling whereby the P/E operations are performed in a high internal electric field. At an elevated storage temperature, stored charges are detrapped and undergo TAT through bulk oxide traps, which leads to a V_T distribution shift. The program state V_T shift was reported to be a function of the negative gate bias used in P/E operations, storage temperature, and P/E cycle count [77, 78, 81]. In summary, as shown in Fig. 3.9, a CL-induced program state V_T shift is attributed to the thermally activated FP emission of trapped charges from the nitride storage layer and followed by TAT through the tunnel oxide layer of nitride-based CT NVM. This thermally activated mechanism was characterized and found to have activation energy of approximately 1.1 eV to 1.2 eV [77, 78, 81].

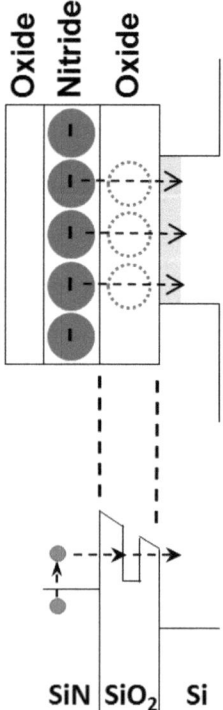

Figure 3.9 Vertical CL path through TAT of detrapped charges via bulk oxide traps [77, 78, 81].

During P/E cycling, charge injection mechanisms utilized by CS-NVM devices are typically performed in a high-electric-field environment. This is to ensure the charges in the channel have substantial energy to overcome the tunnel oxide barrier and are successfully injected into the charge storage layer of CS-NVM devices that have tunnel oxide layers. Therefore, charge injection mechanisms during P/E cycling will only stop when the cell's V_T level has reached the predefined program or erase verify V_T level. Hence, extensive P/E cycling will damage the tunnel oxide by generating charge-trapping interface states and point defects in the tunnel oxide layer. These interface states and point defects are able to trap charges during P/E operations. During the read or verify operation of CS-NVM devices, the sensing operation cannot determine the location of trapped charges but will constantly measure the cell's V_T level on the basis of current flow in the channel of the memory transistor. Therefore, as shown in Fig. 3.10, the annealing of interface states (together with the trapped charges) at an elevated temperature was reported to cause apparently fewer number of charges to store in the memory transistor. Hence, this mechanism lowers the V_T level, as was observed through subthreshold slope measurements as reported in Refs. [1–4, 77, 78, 81].

Compared to nitride-based CS-NVM, the charges are stored in the inherent discrete trap sites (deep-level traps) of the nanocrystals (e.g., silicon or metal) surrounded by the control dielectric layer [22–29, 33, 35, 36, 46, 60, 63, 64]. Therefore, as shown in Fig. 3.11, there are four potential CL mechanisms for the stored charges in nanocrystals to leak out to the substrate layer:

- A vertical leakage path through the intrinsic direct tunneling mechanism [30, 38]
- A vertical leakage path through the extrinsic defect-assisted tunneling mechanism, similar to the SILC mechanism, which severely impacts FG-based CS-NVM [30, 38, 44–48]
- Lateral redistribution of trap charges within the nanocrystal layer [30, 38, 44–48]
- Annealing of P/E-cycling-induced charge-trapping interface states in the tunnel oxide layer [77, 78, 81]

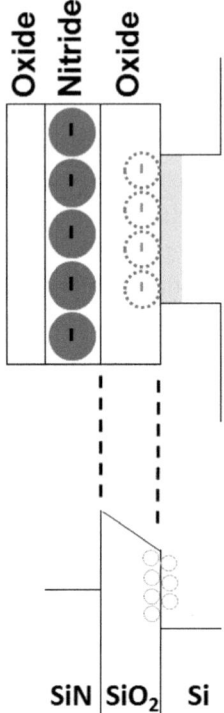

Figure 3.10 Annealing of P/E-cycling-induced interface states induces the cell's V_t level to reduce to a lower voltage level [77, 78, 81].

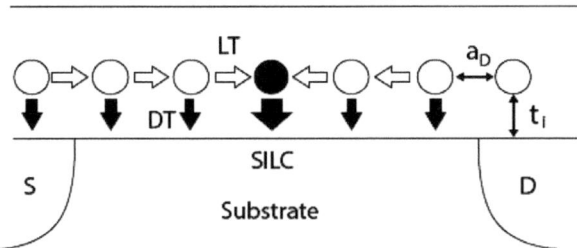

Figure 3.11 Schematic diagram of the vertical leakage path, or SILC, through direct tunneling (DT) and lateral tunneling (LT). a_D is the lateral spacing between each nanocrystal. t_1 is the tunnel oxide thickness [30, 38].

On the basis of comprehensive modeling work done by Compagnoni et al., the retention time was modeled and calculated as a function of nanocrystal spacing [30, 38]. Direct tunneling emerges as the dominant discharge mechanism for the large distance between nanocrystals and the primary limiting mechanism of the retention performance for nanocrystal-based CS-NVM devices [30, 35–46]. On the other hand, lateral tunneling of trap charges in nanocrystals dominates at smaller distances between nanocrystals [30, 35–46]. Therefore, larger spacing between nanocrystals is able to effectively suppress lateral tunneling and improves data retention performance [30, 38]. Compagnoni et al. reported that the minimum distance between nanocrystals of 3.7 nm and the minimum tunnel oxide thickness (t_1) of 4.2 nm were required to satisfy the typical 10-year data retention requirement for CS-NVM devices [30, 38]. However, in the context of increasing the nanocrystals' density, reducing the distance between nanocrystals significantly enhances lateral tunneling over direct tunneling and as such reduces overall data retention time. On the basis of the characterization result, Compagnoni et al. reported that the combination of t_1 = 6 nm and the minimum distance between nanocrystals a_D = 3.7 nm (with doping density N_D = 1.2 × 10^{12} cm^{-2}) is sufficient to meet the expectation of 10-year data retention lifetime even with the presence of bulk oxide defects or interface traps in the tunnel oxide layer that could lead to SILC leakage [30, 38].

As shown in Fig. 3.11, the thickness of the tunnel oxide layer for nanocrystal-based CS-NVM is critical for modulating the CL mechanism and substantially impacts the data retention performance of nanocrystal-based CS-NVM devices. This finding corroborates the characterization result reported by Ng et al. [45]. The CL rate of a thicker tunnel oxide layer, at 7 nm, is reported to be 0.05 V/decade at 25°C and 0.07 V/decade at 85°C. For a thinner tunnel oxide layer, at 3 nm, the CL rate is estimated to be ~0.10 V/ decade at both 25°C and 85°C [45]. This finding indicates that the CL mechanism for a thicker tunnel oxide layer is a thermal accelerated process that elevates the charge decay rate at a higher temperature [30, 38]. This temperature-dependent CL process is hypothesized as the combination of the CL effect of the charge detrapping process from bulk oxide traps to the silicon substrate, the annihilation of trap charges at charge-trapping interface states, and the tunneling of charges from nanocrystals to the silicon substrate [30, 38, 45]. For a thinner oxide layer, the temperature-independent behavior of the CL process indicates the stored charges trapped in nanocrystals could easily escape to the silicon substrate through the direct tunneling process [38, 45]. In summary, for nanocrystal-based CS-NVM devices, the MW is determined by several critical factors, such as the number of electrons stored in each nanocrystal, the size of the nanocrystal, the thickness of the control/tunnel dielectric layers that isolate the nanocrystals from the control gate (CG) and the substrate, the gate capacitive coupling ratios, and the area density of nanocrystals [30, 38, 45–65].

3.4 Technical Mitigations

To further enhance the MW or ΔV_T of nanocrystal-based CS-NVM devices, several viable technical solutions are there that revolve around the engineering work done to increase the number of charges stored in nanocrystals, the density of the nanocrystals in the control dielectric layer, and the scaling/engineering of the tunnel dielectric layer while not aggravating CL performance. In this section, several critical and viable technical mitigations will be elucidated in terms of innovation, background, and improvement involving the endurance/ retention performance of nanocrystal-based CS-NVM devices. These methods are metal-nanocrystal-based CS-NVM and TBE, where TBE involves the implementation of nitridation on tunnel oxide, the engineering work done on modulating the tunnel oxide structure (i.e., VARIOT and CBE), and the substitution of alternative high-κ materials to form high-κ tunnel dielectrics.

Several published literatures have discussed the findings obtained through the implementation of metal nanocrystals that inherently possess larger work function and larger density of states [34, 42, 43, 48, 50, 54–57, 62, 65]. Furthermore, metal nanocrystals have a lower probability of interdot charge tunneling, which enhances the charge retention performance of nanocrystal-based CS-NVM devices [48]. Nanocrystals with higher work functions have demonstrated minimal lateral charge diffusion among the nanocrystals in the control dielectric layer, which is preferable in scaled memory devices with expected prolonged charge retention duration [48]. Metal nanocrystals provide higher confinement barriers to enhance the charge retention performance and smaller confinement barriers for charge injection during P/E operations in nanocrystal-based CS-NVM devices [34]. Compared to silicon nanocrystals, metal nanocrystals were found to be superior as asymmetry in the potential barrier between the substrate and nanocrystals enhances the charge injection during P/E operations and charge retention as well [56]. In several published literatures, germanium (Ge) nanocrystals have demonstrated better charge retention performance compared to silicon nanocrystals [50, 54–57, 62, 65]. Furthermore, Kim et al. reported that the use of silicon germanium nanocrystals coupled with a HfO_2 tunnel dielectric has resulted in improved endurance performance as compared to the silicon-nanocrystal-based CS-NVM devices [34]. (i) Metal nanocrystals offer larger density of states around the Fermi level, (ii) there is absence of band splitting, and (iii) work function engineering is attributed to the wide array of choices of metals [34, 42, 43, 48, 50, 54–57, 62, 65]. On the other hand, the primary concern in the implementation of metal nanocrystals is the thermal stability of metals at elevated temperatures, especially during the fabrication process. Another key concern for metal nanocrystal-based CS-NVM devices is the compatibility of the fabrication process for metal nanocrystals with the conventional metal-oxide-semiconductor field-effect transistor (MOSFET) fabrication process. Kao et al. [56] have reported that the formation of Ge nanocrystals is much more challenging than the formation of Si nanocrystals because of the lower evaporation temperature of Ge, which complicates the capability of high-volume manufacturing of metal nanocrystal-based CS-NVM devices. Furthermore, with metal nanocrystals, the other key concern is the contamination risk of metals onto the front-end tools and memory gate stacks, which may lead to a critical deterioration in the performance of the fabrication tools and the reliability performance of the memory devices [60, 63, 64]. To alleviate the incompatibility issue, one of the proposed mitigation approaches is to choose and implement MOSFET-compatible metals, such as tungsten (W) and TiN through atomic layer deposition, which enhances the uniformity of the nanocrystals in the control dielectric layer [42, 43, 54–57, 62, 65, 89]. Moreover, SiN is best coupled with metal nanocrystals since SiN acts as a diffusion barrier for metal nanocrystals such as TiN and at the same time acts as an additional charge-trapping layer [60, 63, 64]. An approach other than implementing metal nanocrystals, reported in published literature, is to perform nitridation on silicon nanocrystals, which results in a larger MW, faster P/E speed, and enhanced retention performance [52, 88]. Figure 3.12 shows a typical cell structure of nitrided silicon-nanocrystal-based CS-NVM. The enhancement of endurance and retention performance of nitrided silicon-nanocrystal-based CS-NVM is attributed to the nitrogen passivation of the traps on the surface of silicon nanocrystals, which substantially suppresses the CL from the nanocrystals to the surroundings because of electrostatic repulsion among the charged silicon nanocrystals [52, 88].

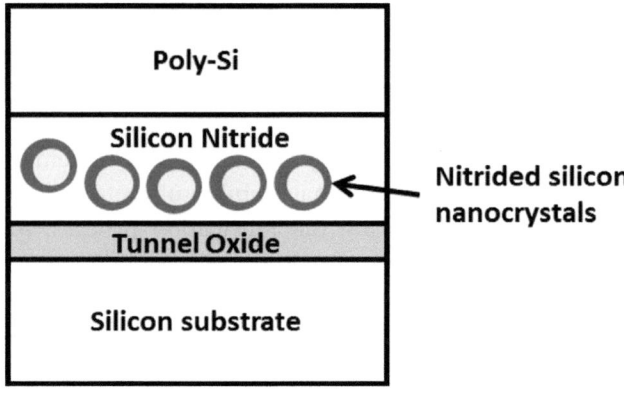

Figure 3.12 Nitridation was performed on silicon nanocrystals that reside in the silicon nitride layer to enhance its reliability performance.

The reliability performance of nanocrystal-based CS-NVM devices is enhanced by increasing the density of nanocrystals in the control dielectric layer through the implementation of stacked nanocrystals layers. The main purpose of implementing double-stack nanocrystal layers is to increase the number of charge-trapping sites and MW of the silicon nanocrystal CS-NVM [63, 64]. Molas et al. reported that the MW of double-stack nanocrystals was significantly improved with an increase in ΔV_T by 50% at $V_{gate} = \pm17$ V with 2 ms pulses while the charge-trapping layer remained discrete [60, 63, 64]. This is because the erase state V_T distribution of double-stack nanocrystals is significantly lower than the single-layer nanocrystals, which further opens up the MW upon completion of the P/E operation. This finding indicates that the erase state V_T of single-layer nanocrystals gets saturated at the native state V_T after erase while the erase state V_T of double-stack nanocrystals gets saturated well below the native state V_T (similar to the overerase condition) [60, 63, 64]. Native state V_T is the V_T level of a memory cell without any history of P/E performed. Thus for double-stack nanocrystals, the lower erase state V_T is attributed to the valence band electron tunneling model whereby the valence band electrons tunnel from the top silicon nanocrystal layer to the bottom silicon nanocrystal layer despite the thin oxide layer between both nanocrystal layers. Hence because of the electrons' tunneling effect, holes are left in the top silicon nanocrystal layer valence band, which indirectly positively charge the memory cell [63, 64]. Therefore, the enhancement of the MW for double-stack-silicon-nanocrystal-based CS-NVM can be attributed to (i) larger channel area surface coverage and (ii) valence band electron tunneling from the top silicon nanocrystal layer, leaving holes in the top silicon nanocrystal layer during erase. Moreover, with double-stack nanocrystals, the erase time was found to be faster than that for single-layer nanocrystals. The conventional FG-based CS-NVM devices that typically implement SiO_2 as the tunnel oxide layer have met crucial trade-offs between high P/E speed and improved data retention performance [89]. To further enhance the endurance and retention performance of nanocrystal-based CS-NVM devices, TBE is one of the attractive and viable technical mitigations. The goal of the TBE method is to modulate the electrical properties of the tunnel dielectric to achieve better tunnel dielectric layer scalability while enhancing endurance and retention performance. TBE can be implemented through several techniques: (i) implementing nitridation in the top and bottom oxide layers [97–102], (ii) implementing CBE and VARIOT [91–96], and (iii) replacing SiO_2 with an alternative high-κ material as the tunnel oxide layer to strengthen the immunity toward vertical charge [90, 91, 103, 104].

Nitridation is done by incorporating nitrogen in the control and tunnel oxide layers of the CS-NVM device structure, which can be implemented through thermal nitridation and chemical/physical nitridation process techniques to enhance the reliability performance of CS-NVM devices [97, 98]. As shown in Fig. 3.13, nitridation can be accomplished through bottom nitridation of the tunnel oxide layer and top nitridation (TN) of the tunnel oxide layer. Bottom nitridation is implemented by introducing nitrogen in the

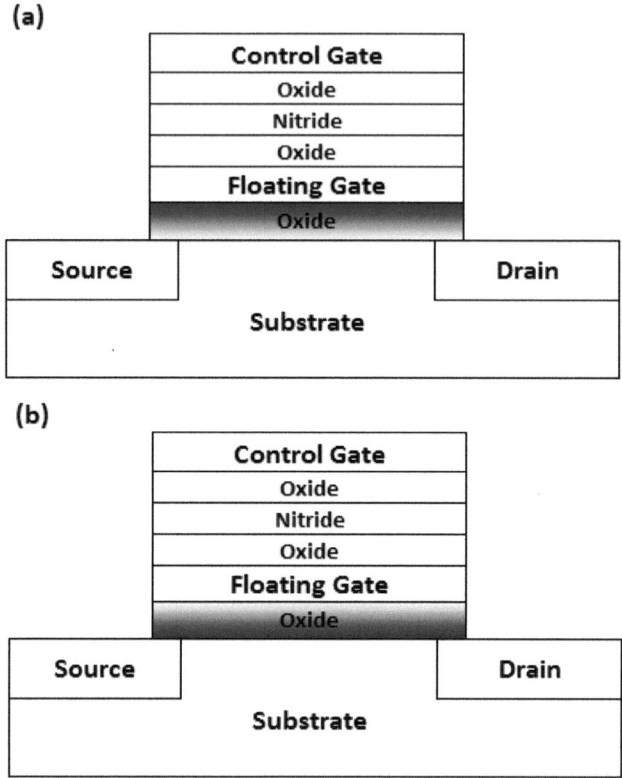

Figure 3.13 Nitridation done on CS-NVM devices, that is (a) top nitridation and (b) bottom nitridation.

tunnel oxide close to the channel of the CS-NVM device. This will strengthen the endurance performance of the CS-NVM device since the introduction of nitrogen in the tunnel oxide layer will selectively substitute the dangling bond of Si–O with the relatively stronger Si–N bond. A wider MW was achieved with the nitrided oxide layer as compared to the conventional silicon oxide layer after extensive P/E cycling was performed on both structures [99]. Kim et al. reported that the improvement in endurance and MW performance is attributed to the increment in the nitrogen content in the tunnel oxide layer [100]. Nevertheless, the drawback of bottom nitridation is that the excess Si–N bonds in the tunnel oxide layer increase the chances of a defect-related breakdown that may induce a vertical CL path from the charge storage layer to the silicon substrate.

Furthermore, bottom oxide nitridation was reported to yield large quick electron detrapping (QED) and random telegraph noise (RTN) [100]. Both QED and RTN were reported to induce MW closure and hence further impacted the data retention performance of nitrided NVM devices [99–101]. On the other hand, TN of the tunnel oxide layer can be achieved by incorporating nitrogen into the top of the tunnel oxide layer, which is located below the FG, through rapid thermal nitridation with ammonia anneal and decoupled plasma nitridation [70, 77]. The nitrided tunnel oxide layer, also known as the oxynitride layer, is located at the interface of the FG and the tunnel oxide layer. This oxynitride layer takes on the role of a diffusion barrier to prevent the penetration of contaminants (e.g., boron or carbon) from the doped polysilicon of the FG [98–100]. With TN, the detrimental V_T shift of programmed cells on CS-NVM devices had reduced after 10,000 P/E cycles as compared to non-nitrided CS-NVM devices [100–102]. The suppression of the V_T shift of programmed cells was attributed to the high gate coupling frequency and fast programming speed exhibited on the nitrided device as compared to the conventional non-nitrided devices [101]. However, a trade-off between the endurance/retention performance of nitrided CS-NVM devices and the nitrogen profile in the bottom oxide was reported. Figure 3.14a shows the relationship between the normalized ΔV_T attributed to a different TN profile and the P/E cycle count. A higher nitrogen concentration implemented in the TN profile has resulted in a larger ΔV_T. This finding indicates that a higher nitrogen concentration induces more charges to be trapped because of the deep-level energy traps resulting from the incorporation of nitrogen in the tunnel oxide layer. Figure 3.14b demonstrates the normalized ΔV_T attributed to a different TN profile after 10,000 P/E cycling and 32 h of bake at 85°C. A higher concentration of nitrogen (i.e., TN-C) has apparently demonstrated a larger ΔV_T as compared to a lower nitrogen concentration (i.e., TN-A and TN-B). Kim et al. [100] have attributed the endurance and retention behavior of various TN to the incorporation of nitrogen, which increases the deep-level energy traps, resulting in more charges being trapped, especially during P/E cycling. With appropriate nitrogen concentration, the reduction of normalized ΔV_T is similar to that in non-nitrided CS-NVM devices. However, with the maximum nitrogen concentration, TN-C has exacerbated the normalized ΔV_T. This finding has led to the explanation that a larger reduction of normalized ΔV_T is attributed to the reduction of the tunnel oxide barrier height and (from 3.07 eV to 2.84 eV) and the increase of the local electric field in the tunnel oxide layer (from 0.74 to 0.77 eV/nm). The modulation of the tunnel oxide barrier height and the electric field increases the probability of stored charges leaking out from FG to substrate, exacerbating the normalized ΔV_T shift.

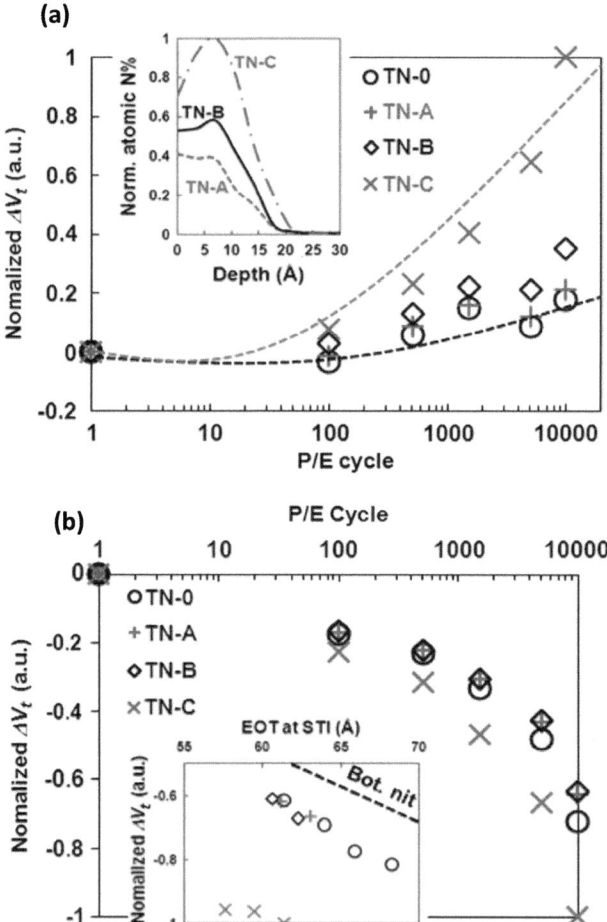

Figure 3.14 (a) Normalized ΔV_T is plotted as a function of the P/E cycle for each top nitridation (TN) profile and (b) normalized ΔV_T after bake at each P/E cycle for each TN profile. N% is normalized by the maximum concentration of TN-C. TN-0 in this figure indicates no TN [80].

Other than nitridation, the other method to implement TBE concept is to alter the structure of the tunnel barrier; this concept is known as the engineered tunnel barrier. The goal of an engineered tunnel barrier was to prolong the scalability of the tunnel dielectric and enhance the reliability of CS-NVM devices. Furthermore, the implementation of the engineered tunnel barrier achieves the same equivalent oxide thickness (EOT) with a thicker physical tunnel dielectric stack. There are two main approaches for implementing multilayered engineered tunnel barriers on CS-NVM devices, CBE and VARIOT. This concept of utilizing a multilayered engineered tunnel dielectric stack as replacement for a tunnel oxide layer in CS-NVM applications was first introduced by Govoreanu et al. in 1998 [91,93]. Figure 3.15 exhibits various conduction band edge diagrams of different tunnel dielectric stacks. Figure 3.15a shows the typical uniform tunnel barrier, and the dashed line indicates the barrier height lowering, which is attributed to the applied voltage. On the other hand, Fig. 3.15b shows the conduction band edge diagram of a crested asymmetric tunnel barrier. As shown in Fig. 3.15b, the potential barrier height is the maximum at the middle and gradually decreases toward both ends of the conducting electrodes. On the basis of research reported by Govoreanu et al. [93], the combinational dielectric stack of silicon nitride–aluminum nitride–silicon nitride exhibited superior performance as compared to the conventional single-layer silicon oxide tunnel barrier in terms of programming speed and data retention performance. Recent research reported by Park et al. has implied that the improvement in the endurance/ retention performance of the engineered tunnel barrier is primarily attributed to the suppressed leakage current induced by direct/FN tunneling since the physical oxide thickness has increased for an engineered tunnel barrier with respect to the conventional tunnel oxide layer [95]. As shown in Fig. 3.15a, the transparency of uniform tunnel barrier changes more slowly as compared to that of the crested tunnel barrier since the highest part of the uniform tunnel barrier closest to the electrode is weakly affected by the applied voltage. However, for the crested tunnel barrier, the highest part in the middle can be pulled quickly; thus, barrier lowering can be achieved faster for the crested tunnel barrier as compared to the uniform tunnel barrier. With this faster barrier lowering, the crested tunnel barrier enables faster carrier injection from the electrodes. Hence, the current can be modulated between high and low fields. The critical drawback of the CBE concept is the challenges related to its fabrication process—it is very challenging to obtain good interface quality between a high-κ and a silicon channel.

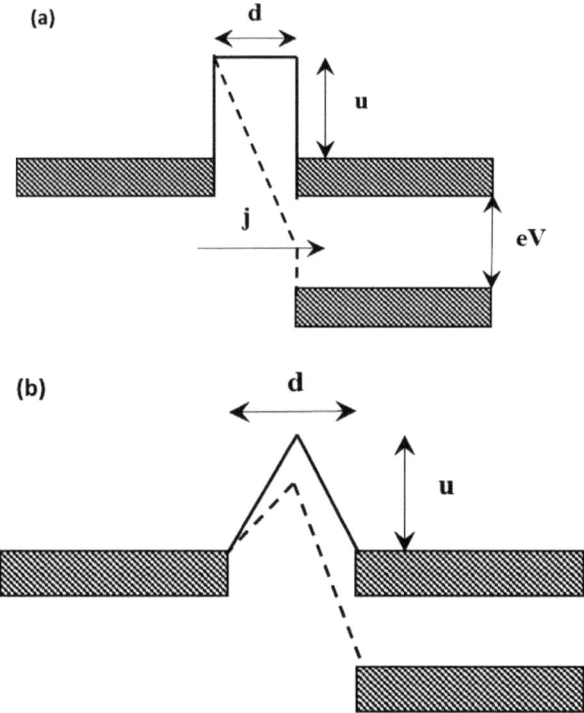

Figure 3.15 Conduction band edge diagrams for different tunnel dielectric barriers: (a) conventional tunnel barrier and (b) crested symmetric barrier. The u and d represent the conduction band of a high-κ layer, the conduction band offset of the tunnel oxide layer, the physical thickness of the high-κ layer, and the physical thickness of the tunnel oxide layer, respectively. The dashed lines indicate the barrier height lowering induced by the modulated internal electric field attributed to the applied voltage.

The other key method to achieve TBE is variable oxide thickness, also known as VARIOT. Govoreanu et al. [93] were first to propose VARIOT to achieve enhancement in endurance and retention performance of CS-NVM devices. As shown in Fig. 3.16, the implementation of the novel VARIOT concept involves a two-layer dielectric stack with a combination of low-κ/high-κ (asymmetric) or a three-layer dielectric stack with a combination of low-κ/high-κ/low-κ (symmetric) to regulate the tunnel barrier height [92]. As compared to the crested tunnel barrier, the bandgap and the dielectric constant of VARIOT-type dielectric stacks are totally opposite. For an asymmetric barrier, the first dielectric layer has a lower dielectric constant and hence a higher barrier height as compared to the second dielectric layer, which has a higher dielectric constant. With the applied bias voltage, the combination of a low-κ/high-κ dielectric stack results in enhanced sensitivity of the tunnel current through it. This increased sensitivity is attributed to a significant variation in the redistribution of the electric field across the low-κ/high-κ stack in which the electric field is higher at the low-k dielectric layer. On applying the bias voltage, at a certain point, direct or FN tunneling will occur only through the thin low-k dielectric layer of the low-κ/high-κ stack. This phenomenon observed is the barrier thinning of the stack. Compared to the conventional tunnel oxide layer, the VARIOT stack has a more enhanced field sensitivity, which is critical for improvement in the programming and retention performance [92, 93]. Furthermore, with enhanced field sensitivity, VARIOT enables higher programming speed and better retention performance as compared to the EOT low-k tunnel oxide layer. Recent research has reported that CS-NVM devices that have VARIOT-type barriers have superior endurance/retention performance as compared to CS-NVM devices with CBE-type barriers [94]. The endurance/retention performance degradation of a CBE-type barrier is attributed to the increased trap density at the interface of the tunneling barrier and the silicon channel on the basis of the interface trap density measurement made. This finding implies that the quality of the interface between the tunnel dielectric barrier and the silicon channel is imperative to the retention/endurance performance of CS-NVM devices that employ the engineered tunnel barrier solution. On the other hand, research has reported that the combinational approaches of tunnel oxide nitridation and bandgap engineering have exhibited excellent endurance performance for NAND flash memory [95, 96].

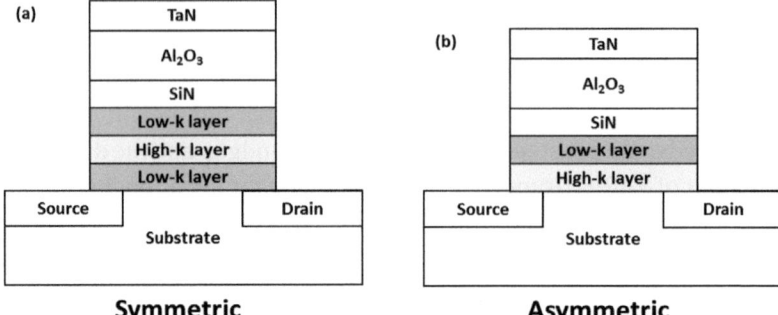

Figure 3.16 Two implementations of engineered tunnel barriers for nitride-based CS-NVM devices: (a) symmetric and (b) asymmetric.

Other than nitridation and engineered tunnel barrier (i.e., CBE and VARIOT) concepts, the implementation of an alternative high-κ material as a tunnel dielectric stack is another attractive and viable method to further enhance the endurance/retention performance of CS-NVM devices [103, 104]. High-κ materials are promising candidates to replace SiO_2 as the tunnel dielectric layer and the control dielectric layer since they are able to achieve a smaller equivalent oxide thickness with a lower tunnel barrier. Recent research has also shown that one of the key drivers to enable planar FG flash memory device to downscale beyond 1X nm is the implementation of a high-κ material as the intergate dielectric (IGD) between the FG and the CG [105, 106]. For high-κ materials to be implemented in CS-NVM applications, there are several criteria that the materials are required to meet. On the basis of the research done by Verma [92], here are the critical requirements of high-κ material with respect to silicon oxide materials:

- High permittivity to ensure prolong technology scaling capability beyond one technology node. High permittivity of a high-κ material will hence result in a higher physical thickness with the same EOT. This is important to enhance the data retention performance.

- Suitable characteristics of conduction and valence bands to achieve enhanced endurance/retention performance. This is because the balance characteristics of conduction and valence bands ensure that optimum endurance and retention performance can be achieved.

- Minimum bulk traps to suppress TAT-induced leakage from the charge storage layer. This characteristic is important to alleviate the risk of CL-induced retention loss, which may impact the data integrity in CS-NVM applications.

- Compatibility to existing technology and fabrication process in view of manufacturability and cost while ensuring the quality of CS-NVM devices can be sustained. The targeted high-κ material should be able to integrate with silicon.

- Optimum effective mass of the high-κ material to obtain a balanced endurance and retention performance. This is because a lower effective mass of a high-κ material prompts faster endurance speed and a higher effective mass of a high-κ material results in a better retention performance.

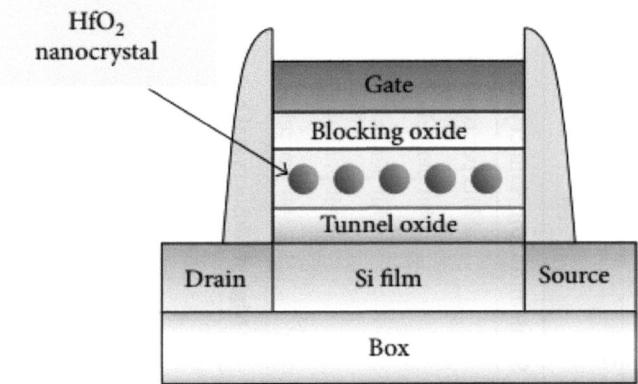

Figure 3.17 Novel HfO$_2$-based nanocrystal NVM proposed to achieve better P/E characteristics and retention performance than that achieved by FG-based NVM devices [80].

As shown in Fig. 3.17, the use of HfO$_2$ as the control dielectric layer was reported to enhance P/E characteristics and charge retention performance [40, 41]. Therefore, the idea of utilizing high-κ materials as control dielectric and tunnel dielectric layers has garnered much attention among researchers. Table 3.3 summarizes the research findings of reliability performance and endurance performance with respect to the combinations for various types of nanocrystals, control dielectric layers, and tunnel dielectric layers. Despite the

Table 3.3 Reliability performance and endurance performance of various types of nanocrystal-based CS-NVM reported in numerous published literatures

Structure	Nanocrystal	Number of bits per cell	Initial MW	Endurance	Ref
		Semiconductor Nanocrystal based NVM			
HfO_2/SiO_2	HfO_2	2 bits per cell	2.25 V	106	[40]
ONO/SiO_2	Si dots	1 bit per cell	0.5 V to 2.2 V		[22, 23, 40]
SiO_2/SiO_2	SiGe	1 bit per cell	0.5 to 2.25 V	104	[34]
SiO_2/SiO_2	Silicon (4 nm)	1 bit per cell	1 V	105	[45]
Silicon Nitride/SiO_2	Nitrided Si	1 bit per cell	1.5 V to 1.8 V	N/A	[88]
		High-κ Nanocrystal based NVM			
TAO/Al_2O_3	$TiAl_2O_5$	1 bit per cell	2.3 V	105	[89]
Al_2O_3/Al_2O_3	Ti-Al-O	1 bit per cell	2.5 V to 2.8 V	105	[53]
SiO_2/SiO_2	Gd_2O_3	1 bit per cell	1.25 to 1.75 V	104	[42]
SiO_2/SiO_2	Iridium Silicide	1 bit per cell	2.15 V	104	[62]
Al_2O_3/HfO_2	Platinum (Pt)	1 bit per cell	6.6 V	N/A	[50]
SiO_2/Al_2O_3	Ruthenium Oxide (RuOx)	1 bit per cell	5.2 V	N/A	[65]

enhancements in the endurance and retention performance of nanocrystal-based CS-NVM, the common challenge of implementing metal nanocrystals, high-κ tunnel dielectrics, and double-stack nanocrystals is the compatibility of its fabrication process with the standard CMOS fabrication process. As such, exhaustive studies or research is still required to ensure these innovative enhancements of nanocrystal-based CS-NVM devices are suitable for high-volume manufacturing.

3.5 Summary

CS-NVM devices are still the key enabler to the meet the demand for fast and reliable embedded storage space for smart appliances and applications especially in the upcoming Internet of things (IOT) area. To further enhance the cost per bit ratio and fulfill the insatiable demand for larger storage space, the aggressive effort to consistently scale down the dimensions of the memory cells has to continue along the trend set by Moore's law. Nevertheless, alternative CS-NVM devices are heavily explored and researched as existing CS-NVM devices, such as FG- and nitride-based CS-NVM devices, have met the key scaling limiter in their tunnel oxide layers. In other words, technology scaling alone will no longer be sufficient to achieve the technological and economical advantage of Moore's law. The scaling has to be complemented with alternative technical mitigations. One of these mitigations is nanocrystal-based CS-NVM devices. In this review chapter, the background of CS-NVM devices has been elucidated in detail, including the history and evolution of CS-NVM devices from their birth till the present. Next, the evolution of nanocrystal-based CS-NVM was discussed, with emphasis on its salient advantages and key disadvantages with respect to the existing CS-NVM devices. Critical reliability challenges of nanocrystal-based CS-NVM devices reported by numerous published literatures were reviewed to further the understanding of their endurance, retention, and immunity toward irradiation performance. The charge transport mechanism of CL-induced V_T decay of nanocrystal-based CS-NVM devices was discussed. Last but not least, several key technical mitigations to enhance the retention and endurance performance of nanocrystal-based CS-NVM devices were illustrated to explain the key research findings of existing mitigation methods and their challenges. In conclusion, we can say that nanocrystal-based CS-NVM devices are one of the most promising candidates for prolonging the dominance of charge-based NVM devices in embedded electronic solutions, with significant research effort being made to enable high-volume manufacturing of these devices.

References

1. Lai, S. (2008). *IEEE International Electron Devices Meeting*, 1–6.

2. Moore, G.E. (1998). *Proc. IEEE*, **86**(1), 82–85.

3. Kim, K., and Jeong, G. (2009). *IEEE International Electron Devices Meeting*, 27–30.

4. Kim, K. (2008). *International Symposium on VLSI Technology, Systems and Applications (VLSI-TSA)*, 5–9.

5. Pellizzer, F., and Bez, R. (2010). *IEEE International Conference on Nanotechnology Joint Symposium with Nano Korea*, 21–24.

6. Kim, K. (2010). *IEEE International Memory Workshop*, 1–2.

7. Kim, K. (2012). *Proceedings of the European Solid-State Device Research Conference (ESSDERC)*, 1–6.

8. Kim, K., and Choi, J. (2006). *IEEE Non-volatile Semiconductor Memory Workshop*, 9–11.

9. Wellekens, D., and Van Houdt, J. (2008). *IEEE International Conference on Integrated Circuit Design and Technology and Tutorial*, 189–194.

10. Kim, K. (2006). *International Conference on Solid-State and Integrated Circuit Technology*, 685–688.

11. Koh, Y. (2009). *IEEE International Memory Workshop (IMW)*, 1–3.

12. Kim, K., and Koh, G.H. (2004). *24th International Conference on Microelectronics*, 377–384.

13. Prall, K. (2007). *IEEE Non-volatile Semiconductor Memory Workshop*, 5–10.

14. Bez, R., Camerlenghi, E., Modelli, A., and Visconti, A. (2003). *Proc. IEEE*, **91**(4), 489–502.

15. Pavan, P., Bez, R., Olivo, P., and Zanoni, E. (1997). *Proc. IEEE*, **85**(8), 1248–1271.

16. Spinelli, A.S., and Compagnoni, C.M. (2011). *IRPS 2011 Tutorial*, Topic 121.

17. Brewer, J.E., and Mansur, G. (2008). *Nonvolatile Memory Technologies with Emphasis on Flash: A Comprehensive Guide to Understanding and Using NVM Devices*. Wiley Interscience, Hoboken, New Jersey.

18. Baklanov, M., Green, M., and Maex, K. (2007). *Dielectric Films for Advanced Microelectronics*. John Wiley & Sons, Chichester.

19. Gritsenko, V.A., Nekrashevich, S.S., Vasilev, V.V., and Shaposhnikov, A.V. (2009). *Microelectron. Eng.*, **86**(7), 1866–1869.

20. Gritsenko, V.A., Novikov, Y.N., Morokov, Y.N., and Wong, H. (1998). *Microelectron. Reliab.*, **38**, 1457–1464.

21. Arreghini, A., Akil, N., Driussi, F., Esseni, D., Selmi, L., and Duuren, M.J.V. (2008). *Solid-State Electron.*, **52**, 1460–1466.

22. Tiwari, S., Rana, F., Hanafi, H., Hartstein, A., Crabbe, E.F., and Chan, K. (1996). *Appl. Phys. Lett.*, **68**(10), 1377–1379.

23. Tiwari, S., Rana, F., Chan, K., Hanafi, H., Chan, W., and Buchanan, D. (1995). *IEEE International Electron Devices Meeting*, 521–524.

24. Chang, T.C., Jian, F.Y., Chen, S.C., and Tsai, Y.T. (2011). *Mater. Today*, **14**(12), 608–615.

25. De Blauwe, J. (2002). *IEEE Trans. Nanotechnol.*, **1**(1), 72–77.

26. Hanafi, H.I., Tiwari, S., and Khan, I. (1996). *IEEE Trans. Electron Devices*, **43**(9), 1553–1558.

27. She, M., and King, T. (2003). *IEEE Trans. Electron Devices*, **50**(9), 1934–1940.

28. Rao, R.A., Gasquet, H.P., Steimle, R.F., Rinkenberger, G., Straub, S., Muralidhar, R., Anderson, S.G.H., Yater, J.A., Ledezma, J.C., Hamilton, J., and Acred, B. (2005). *Solid-State Electron.*, **49**(11), 1722–1727.

29. Rao, R.A., Steimle, R.F., Sadd, M., Swift, C.T., Hradsky, B., Straub, S., Merchant, T., Stoker, M., Anderson, S.G.H., Rossow, M., and Yater, J. (2004). *Solid-State Electron.*, **48**(9), 1463–1473.

30. Compagnoni, C.M., Ielmini, D., Spinelli, A.S., Lacaita, A.L., Previtali, C., and Gerardi, C. (2003). *IEEE International Reliability Physics Symposium Proceedings*, 506–512.

31. Chang, K.M. (2006). *IEEE International Conference on Solid-State and Integrated Circuit Technology*, 725–728.

32. Lee, C., Kwon, J.H., Lee, J.S., Kim, Y.M., Choi, Y., Shin, H., Lee, J., and Sohn, B.H. (2007). *Appl. Phys. Lett.*, **91**(15), 153506.

33. De Salvo, B., Gerardi, C., Lombardo, S., Baron, T., Perniola, L., Mariolle, D., Mur, P., Toffoli, A., Gely, M., Semeria, M.N., and Deleonibus, S. (2003). *IEEE International Electron Devices Meeting*, 26–1.

34. Kim, D.W., Kim, T., and Banerjee, S.K. (2003). *IEEE Trans. Electron Devices*, **50**(9), 1823–1829.

35. Steimle, R.F., Rao, R., Sadd, M., Swift, C., Hradsky, B., Straub, S., Merchant, T., Stoker, M., Parikh, C., Anderson, S., and Rossow, M. (2004). *4th IEEE Conference on Nanotechnology*, 290–292.

36. Steimle, R.F., Muralidhar, R., Rao, R., Sadd, M., Swift, C.T., Yater, J., Hradsky, B., Straub, S., Gasquet, H., Vishnubhotla, L., and Prinz, E.J. (2007). *Microelectron. Reliab.*, **47**(4), 585–592.

37. De Salvo, B., Gerardi, C., Van Schaijk, R., Lombardo, S., Corso, D., Plantamura, C., Serafino, S., Ammendola, G., Van Duuren, M., Goarin, P., and Mei, W.Y. (2004). *IEEE Trans. Device Mater. Reliab.*, **4**(3), 377–389.

38. Compagnoni, C.M., Ielmini, D., Spinelli, A.S., Lacaita, A.L., Gerardi, C., Perniola, L., De Salvo, B., and Lombardo, S. (2003). *IEEE International Electron Devices Meeting*, 22.4.1–22.4.4.

39. Gasperin, A., Amat, E., Porti, M., Martín-Martínez, J., Nafría, M., Aymerich, X., and Paccagnella, A. (2009). *IEEE Trans. Electron Devices*, **56**(10), 2319–2326.

40. Lin, Y.H., and Chien, C.H. (2012). *IEEE Trans. Nanotechnol.*, **11**(2), 412–417.

41. Cheng, P.H., Huang, S.H., and Wu, F.M. (2012). *IEEE Trans. Nanotechnol.*, **11**(1), 164–171.

42. Wang, J.C., Lin, C.T., Huang, P.W., Chang, L.C., and Lai, C.S. (2013). *IEEE International Symposium on VLSI Technology, Systems, and Applications (VLSI-TSA)*, 1–2.

43. Lee, C., Kwon, J.H., Lee, J.S., Kim, Y.M., Choi, Y., Shin, H., Lee, J., and Sohn, B.H. (2007). *Appl. Phys. Lett.*, **91**(15), 153506.

44. Baik, S.J., Choi, S., Chung, U.I., and Moon, J.T. (2003). *IEEE International Electron Devices Meeting*, 22.3.1–22.3.4.

45. Ng, C.Y., Chen, T.P., Sreeduth, D., Chen, Q., Ding, L., and Du, A. (2006). *Thin Solid Films*, **504**(1), 25–27.

46. Molas, G., De Salvo, B., Mariolle, D., Ghibaudo, G., Toffoli, A., Buffet, N., and Deleonibus, S. (2003). *Solid-State Electron.*, **47**(10), 1645–1649.

47. Dimitrakis, P., Kapetanakis, E., Tsoukalas, D., Skarlatos, D., Bonafos, C., Asssayag, G.B., Claverie, A., Perego, M., Fanciulli, M., Soncini, V., and Sotgiu, R. (2004). *Solid-State Electron.*, **48**(9), 1511–1517.

48. Lwin, Z.Z., Pey, K.L., Liu, C., Liu, Q., Zhang, Q., Chen, Y.N., Singh, P.K., and Mahapatra, S., 2011. *Appl. Phys. Lett.*, **99**(22), 222102.

49. Gerardi, C., Lombardo, S., Ammendola, G., Costa, G., Ancarani, V., Mello, D., Giuffrida, S., and Plantamura, M.C. (2007). *Microelectron. Reliab.*, **47**(4), 593–597.

50. Chen, Y.N., Pey, K.L., Goh, K.E.J., Lwin, Z.Z., Singh, P.K., and Mahapatra, S. (2011). *18th IEEE International Symposium on the Physical and Failure Analysis of Integrated Circuits (IPFA)*, 1–4.

51. Gay, G., Molas, G., Bocquet, M., Jalaguier, E., Gély, M., Masarotto, L., Colonna, J.P., Grampeix, H., Martin, F., Brianceau, P., and Vidal, V. (2009). *IMW'09 IEEE International Memory Workshop*, 1–4.

52. Yu, J., Ma, Z., Wang, Y., Ren, S., Fang, Z., Huang, X., Chen, K., Wu, G., Zhang, Y., and Wang, L. (2014). *12th IEEE International Conference on Solid-State and Integrated Circuit Technology (ICSICT)*, 1–3.

53. Cao, W., Kang, J., Bertolazzi, S., Kis, A., and Banerjee, K. (2014). *IEEE Trans. Electron Devices*, **61**(10), 3456–3464.

54. Pavel, A., Khan, M., Kirawanich, P., and Islam, N. (2007). *IEEE International Semiconductor Device Research Symposium*, 1–2.

55. Lee, C., Hou, T.H., and Kan, E.C.C. (2005). *IEEE Trans. Electron Devices*, **52**(12), 2697–2702.

56. Kao, C.H., Lai, C.S., Huang, C.S., and Fan, K.M. (2008). *Appl. Surf. Sci.*, **255**(5), 2512–2516.

57. Wang, J.C., Chen, C.H., and Lin, C.T. (2015). *Microelectron. Eng.*, **138**, 52–56.

58. Vatajelu, E.I., Aziza, H., and Zambelli, C. (2014). *9th International Design & Test Symposium (IDT)*, 61–66.

59. Gerardi, C., Lombardo, S., Ammendola, G., Costa, G., Ancarani, V., Mello, D., Giuffrida, S., and Plantamura, M.C. (2007). *Microelectron. Reliab.*, **47**(4), 593–597.

60. Molas, G., De Salvo, B., Mariolle, D., Ghibaudo, G., Toffoli, A., Buffet, N., and Deleonibus, S. (2003). *Solid-State Electron.*, **47**(10), 1645–1649.

61. Compagnoni, C.M., Ielmini, D., Spinelli, A.S., Lacaita, A.L., and Gerardi, C. (2004). *Solid-State Electron.*, **48**(9), 1497–1502.

62. Wang, T.T.J., Hung, S.W., Chuang, P.K., and Kuo, C.T. (2010). *Thin Solid Films*, **518**(24), 7287–7290.

63. Molas, G., Bocquet, M., Vianello, E., Perniola, L., Grampeix, H., Colonna, J.P., Masarotto, L., Martin, F., Brianceau, P., Gély, M., and Bongiorno, C. (2009). *Microelectron. Eng.*, **86**(7), 1796–1803.

64. Molas, G., Masoero, L., Della Marca, V., Gay, G., and De Salvo, B. (2014). *Advances in Non-volatile Memory and Storage Technology*, 120–157.

65. Das, A., Maikap, S., Lin, C.H., Tzeng, P.J., Tien, T.C., Wang, T.Y., Chang, L.B., Yang, J.R., and Tsai, M.J. (2010). *Microelectron. Eng.*, **87**(10), 1821–1827.

66. Cester, A., Wrachien, N., Gasperin, A., Paccagnella, A., Portoghese, R., and Gerardi, C. (2007). *IEEE Trans. Nucl. Sci.*, **54**(6), 2196–2203.

67. Gerardin, S., Bagatin, M., Paccagnella, A., Visconti, A., and Greco, E. (2011). *IEEE Trans. Nucl. Sci.*, **58**(3), 827–833.

68. Cellere, G., Pellat, P., Chimenton, A., Wyss, J., Modelli, A., Larcher, L., and Paccagnella, A. (2001). *IEEE Trans. Nucl. Sci.*, **48**(6), 2222–2228.

69. Cellere, G., Paccagnella, A., Larcher, L., Chimenton, A., Wyss, J., Candelori, A., and Modelli, A. (2002). *IEEE Trans. Nucl. Sci.*, **49**(6), 3051–3058.

70. Larcher, L., Cellere, G., Paccagnella, A., Chimenton, A., Candelori, A., and Modelli, A. (2003). *IEEE Trans. Nucl. Sci.*, **50**(6), 2176–2183.

71. Cellere, G., Paccagnella, A., Visconti, A., Bonanomi, M., and Candelori, A. (2004). *IEEE Trans. Nucl. Sci.*, **51**(6), 3304–3311.

72. Cellere, G., Larcher, L., Paccagnella, A., Visconti, A., and Bonanomi, M. (2005). *IEEE Trans. Nucl. Sci.*, **52**(6), 2144–2152.

73. Cellere, G., Paccagnella, A., Visconti, A., and Bonanomi, M. (2006). *IEEE Trans. Nucl. Sci.*, **53**(6), 3291–3297.

74. Cai, Y., Luo, Y., Ghose, S., Haratsch, E.F., Mai, K., and Mutlu, O. (2015). *45th Annual IEEE/IFIP International Conference on Dependable Systems and Networks*, 438–449.

75. Fang, H.K., Chang-Liao, K.S., Chen, C.Y., Chen, P.H., Li, D.Y., Huang, C.P., Shen, C.H., and Shieh, J.M. (2015). *Microelectron. Eng.*, **147**, 5–9.

76. Wang, W., Xie, T., Khoueir, A., and Kim, Y. (2015). *IEEE 31st Symposium on Mass Storage Systems and Technologies (MSST)*, 1–8.

77. Chang, Y.M., Li, Y.C., Chang, Y.H., Kuo, T.W., Hsieh, C.C., and Li, H.P. (2015). *Proceedings of the IEEE/ACM International Conference on Computer-Aided Design*, 479–486.

78. Janai, M., and Lee, M.C. (2012). *IEEE Trans. Electron Devices*, **59**(3), 596–601.

79. Lee, M.C., and Wong, H.Y. (2013). *IEEE Trans. Electron Devices*, **60**(10), 3256–3264.

80. Lee, M.C., and Wong, H.Y. (2013). *J. Nanomater.*, **2013**, Article ID 650457, 8 p.

81. Lee, M.C., and Wong, H.Y. (2013). *J. Nanomater.*, **2013**, Article ID 195325, 17 p.

82. Lee, M.C., and Wong, H.Y. (2014). *J. Nanosci. Nanotechnol.*, **14**(7), 4799–4812.

83. Lee, M.C., and Wong, H.Y. (2014). *J. Nanosci. Nanotechnol.*, **14**(2), 1508–1520.

84. Lee, M.C., Wong, H.Y., and Lee, L. (2014). *Solid-State Electron.*, **99**, 78–83.

85. Lee, M.C., and Wong, H.Y. (2014). *IEEE Electron Device Lett.*, **35**(9), 918–920.

86. Lee, M.C., Wong, H.Y., and Lee, L. (2014). *Microelectron. Reliab.*, **54**(11), 2392–2395.

87. Lee, M.C., and Wong, H.Y. (2015). *Microelectron. Reliab.*, **55**(2), 337–341.

88. Lee, M.C., and Wong, H.Y. (2016). *J. Nanosci. Nanotechnol.*, **16**, 663–669.

89. Qian, X.Y., Chen, K.J., Wang, Y.F., Jiang, X.F., Ma, Z.Y., Fang, Z.H., Xu, J., and Huang, X.F. (2012). *J. Non-Cryst. Solids*, **358**(17), 2344–2347.

90. Zhou, Y., Yin, J., Xu, H., Xia, Y., Liu, Z., Li, A., Gong, Y., Pu, L., Yan, F., and Shi, Y. (2010). *Appl. Phys. Lett.*, **97**(14), 3504.

91. Banerjee, W. (2012). Nonvolatile memories using nanoscale IrOx metal nanocrystals (Doctoral dissertation).

92. Verma, S. (2010). Tunnel barrier engineering for flash memory technology (Doctoral dissertation).

93. Govoreanu, B., Blomme, P., Rosmeulen, M., Van Houdt, J., and De Meyer, K. (2003). *IEEE Electron Device Lett.*, **24**(2), 99–101.

94. Jung, J., and Cho, W.-J. (2008). *J. Semicond. Technol. Sci.*, **8**(1), 32–39.

95. Park, G.H., Cho, W.J. (2010). *Appl. Phys. Lett.*, **96**, 043503.

96. Lue, H., Wang, S., Lai, E., Shih, Y., Lai, S., Yang, L., Chen, K., Ku, J., Hsieh, K., Liu, R., and Lu, C. (2005). *IEEE International Electron Devices Meeting*, 547–550.

97. Wang, S., Lue, H., Hsu, T., Du, P., Lai, S., Hsiao, Y., Hong, S., Wu, M., Hsu, F., Lian, N., Lu, C., Hsieh, J., Yang, L., Yang, T., Chen, K., and Hsieh, K. (2010). *IEEE International Reliability Physics Symposium*, 951–955.

98. Ghidini, G. (2012). *Microelectron. Reliab.*, **52**(9–10), 5–11.

99. Ganguly, U., Guarini, T., Wellekens, D., and Date, L. (2010). *IEEE Electron Device Lett.*, **31**(2), 2009–2011.

100. Kim, T., Sarpatwari, K., Koka, S., and Wang, H. (2013). *IEEE Electron Device Lett.*, **34**(3), 396–398.

101. Ho, C.Y., Lien, C., Sakamoto, Y., Yang, R.J., Fijita, H., Liu, C.H., Lin Y.M., Pittikoun, S., and Aritome, S. (2008). *IEEE Electron Device Lett.*, **29**(11), 1199–1202.

102. Kim, T., Koka, S., Surthi, S., and Zhuang, K. (2013). *IEEE Electron Device Lett.*, **34**(3), 405–407.

103. Molas, G., Bocquet, M., Vianello, E., Perniola, L., Grampeix, H., Colonna, J.P., Masarotto, L., Martin, F., Brianceau, P., Gély, M., Bongiorno, C., Lombardo, S., Pananakakis, G., Ghibaudo, G., and De Salvo, B. (2009). *Microelectron. Eng.*, **86**(7–9), 1796–1803.

104. Larcher, L., and Padovani, A. (2010). *Microelectron. Reliab.*, **50**(9–11), 1251–1258.

105. Tang, B., Zhang, W.D., Degraeve, R., Breuil, L., Bloome, P., Zhang, J.F., Ji, Z., Zahid, Mohammed, Toledano-luque, M., Van den Bosch, G., and Van Houdt, J. (2014). *IEEE Trans. Electron Devices*, **61**(5), 1299–1306.

106. Breuil, L., Lisoni, J.G., Bloome, P., Van den Bosch, G., and Van Houdt, J. (2014). *IEEE Electron Device Lett.*, **35**(1), 45–47.

Chapter 4

Silicon Nanocrystal Flash Memory

Lili Zhao,[a,c,*] **Tiezheng Lv,**[b,*] **and Guofeng Fan**[c,*]

[a]*Chemistry and Chemical Engineering, Harbin Institute of Technology, Harbin 150001, China*
[b]*Materials Science and Engineering, Hunan University, Changsha 410082, Hunan, China*
[c]*Soft-Impact China (Harbin), Harbin 150001, China*
zhaolili@hit.edu.cn

Nanocrystal (NC) flash memory, going by its competitive performance, has a very promising role in future electronic devices. So knowing how to measure and characterize NC flash memory is an important aspect for developing it for commercial devices. This chapter reviews the progress in the electrical characterizations of silicon (Si) NCs as materials and also in floating gate memory devices. This chapter is mainly divided into four parts. The first will discuss the introduction of Si NCs in flash memory, including structure development of a Si NC floating gate and electrical characteristics of Si NCs in flash memory. In the structure development of a Si NC floating gate, we will introduce the Si NC floating-gate story and preparation of Si NCs for flash memory. In the second part, we will discuss the Si NC

---

*All authors have contributed equally to this work.

Nanocrystals in Nonvolatile Memory
Edited by Writam Banerjee
Copyright © 2018 Pan Stanford Publishing Pte. Ltd.
ISBN 978-981-4774-73-4 (Hardcover), 978-1-351-20327-2 (eBook)
www.panstanford.com

trap center studied by deep-level transient spectroscopy (DLTS). The third one will introduce engineering for improved Si NC flash memory. Special care is taken to discuss in detail DLTS, which is the method used for measuring Si NCs in flash memory and the characterization of the promising role Si NC flash memory. Also, the chapter identifies the challenges Si-NC-based floating-gate memory devices face in the future.

4.1 Introduction

During the past decades, memory has been rapidly developed at higher integration density and lower cost. Memory devices, which are an integral part of PCs, mobile phones, digital cameras, smart media, networks, automotive systems, global positioning systems, etc., must be reliable and nanosized, and at this point, nanocrystals (NCs) in flash memory, because of special advantages, are promising candidates for playing a major role in future electronic devices. Using NC floating-gate flash memory can solve major scaling issues in order to meet the growing demand for increased information storage density, faster data transfer speed, and low-voltage operation. These issues originate mainly from the scalability limitation of the tunnel dielectric (usually made of silicon dioxide) and the floating-gate interference effect between adjacent cells.

To obtain NC floating-gate flash memory, researchers have made many attempts with several nonvolatile memory (NVM) approaches; different concepts or technologies have been used to address flash downscaling limitations. Many methods are reported for preparing Si NCs, such as chemical vapor deposition (CVD), ion implantation of silicon, and preparation of silicon-rich silica. CVD methods are the usual technology, including plasma-enhanced chemical vapor deposition (PECVD) [10], high-pressure vapor annealing (HPVA) [11], and low-pressure chemical vapor deposition (LPCVD) [12]. Carrada and coworkers synthesized ordered Si NCs by ultralow-energy ion implantation in 7-nm-thick SiO_2 and subsequent annealing [13]. In this chapter, we introduce two methods usually used for preparing Si NCs.

Once NC floating-gate flash memory is obtained, people also need to confirm that their performance meets the need. However, conventional routes don't provide good data on the performance of Si NCs in flash memory. So we will discuss a new method, deep-level transient spectroscopy (DLTS), for measuring the charge emission from interface states of Si NC. For a Si-NC-embedded metal-oxide semiconductor (MOS) structure, if the charge emission from interface states of the Si NC is detectable, it is worthwhile to use DLTS for characterization. The DLTS signal, which is the difference of capacitances at two different times after a filling pulse, can also be used for improving the Si NC flash memory.

However, conventional Si-based memory is still the leader in the current market. Si NC flash memory is a field of intensive research, and the generated results can be beneficial not only for the broader scientific community with interests in NC-based optoelectronics or photovoltaics applications but also for accelerated commercialization.

4.2 Si NCs in Flash Memory

4.2.1 Structure Development of a Si NC Floating Gate

In the past 30 years, microelectronics have rapidly developed in terms of higher integration density and lower cost. Research on microelectronics has focused on the miniaturization of devices with the reliability constraint on the device, and at this point, memory devices are very important. Memory chips are used in PCs, mobile phones, digital cameras, smart media, networks, automotive systems, global positioning systems, etc. Chip sizes are shrinking, and memory content and applications are increasing [1]. Semiconductor memory can be divided into two main types, both based on complementary metal-oxide semiconductor (CMOS) technology: volatile memory and NVM.

Volatile memory is fast but loses its contents when power is removed. NVM is slower and used to store data even if it is nonpowered. Because of this, NVMs are widely used in mission-critical applications, such as data centers. Most NVMs used today can be categorized into four types: flash memory, ferroelectric random access memory (FeRAM), magnetic random access memory

(MRAM), and phase change memory. Flash memory is needed in many fields of microelectronics to store data.

And flash memory has developed very fast. In 1967, Kahng and Sze first suggested the idea of a polysilicon floating-gate device at Bell Labs [2], which is attractive because of its advantages and has led to the floating-gate age now. Though flash memory based on floating gates dominates the market for NVMs, there are scaling issues that must meet the growing demand for increased information storage density, faster data transfer speed, and low-voltage operation. These issues originate mainly from the scalability limitation of the tunnel dielectric (usually made of silicon dioxide) and the floating-gate interference effect between adjacent cells [1]. Recently, researchers have developed NC floating-gate technology, which is different from the classical stacked-gate NVM devices. The NC chargestorage process based on floating gates offers several advantages to solve various limitations of earlier devices, such as (i) simplicity, keeping the cost of fabrication low (no process complications); (ii) better retention (resulting from Coulomb blockade and quantum confinement effects [5]), attaining thinner layers of tunnel oxides and lower operating voltages; (iii) optimized anti-punch-through performance (due to the absence of drain-to-floating-gate coupling, thereby reducing drain-induced punch-through), offering higher drain voltages for readout, shorter channel lengths, and consequently, a smaller cell area; and (iv) excellent resistance to stress-induced leakage current (SILC) and defects in terms of the distributed nature of the charge storage in the NC layer. Because of these four advantages, more and more researchers are moving their focus to NC chargestorage devices.

4.2.1.1   Si NC floating-gate story

After looking at the advantages mentioned above, let us look at the recent NVM approaches using different concepts or technologies to address flash downscaling limitations. There is considerable focus on replacing the conventional floating gate (poly-Si layer) with laterally isolated floating gate in the form of nanoparticles. This floating gate could realize the so-called nanocrystal memory (NCM), which hold the potential for operating at lower voltages and higher speeds compared to the conventional NVMs. Basically, an NCM cell includes a metal-oxide semiconductor field-effect transistor (MOSFET) with monodisperse nanometer-scale crystals embedded within the gate dielectric and associates the finite-size effects of NCs and the benefits (robustness and fault tolerance) of a stored charge distribution.

In the past few years, researchers have made significant advances in NC fabrication (see, for example, Refs. [2, 3]) and prototype NCM-based products used for low-power microcontroller applications have recently been demonstrated [4]. Nevertheless, NCM technologies developed till now still face issues related to producing high-density size-homogeneous NCs with a uniform distribution that cannot avoid any effect in device performance and can't deal with size-dependence effects [2, 5–7]. While more researchers need to address these limitations and any other related to dielectric materials and the device design [5–7], the NCM approach still requesting device results with Si, SiGe NCs inset in HfO$_2$, or HfAlO gate dielectrics has been demonstrated [20–22]. The long retention times for NVM applications are still questionable. In addition, these studies have not succeeded in showing that the device performances clearly leads to an increase in the conduction band offset. In such dielectric materials, the deep traps are often at the origin of electron trapping because of a lack of control of the interfacial layer with silicon. So, even if there are a lot of studies in the literature on these systems, there is still some room for their optimization.

Now we must introduce the insulated floating-gate flash memory, especially with an NC floating gate, which can improve the scaling limitations to meet the growing need for storing more and more data into a smaller and smaller chip. This kind of memory has its own special performance compared with the conventional ones, as follows:

It has two gates instead of just one. One gate is the control gate, as in metal-oxide semiconductor (MOS) transistors; the other is a floating gate that is insulated all around by an oxide layer, as shown in Fig. 4.1. The floating gate is isolated by its insulating oxide layer. Thus any carriers placed on it get trapped there and store the information.

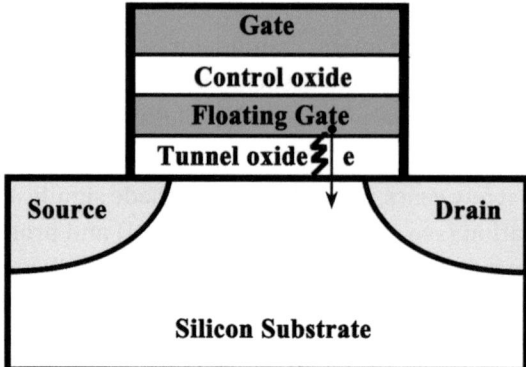

Figure 4.1 Structure of a typical memory with a floating gate. Other than a traditional MOSFET, two gates are designed, a floating gate and a control gate, in which is sandwiched the trapping layer.

A widely exploited NVM uses a continuous layer of polycrystalline silicon or Si_3N_4 as a trap layer for carriers [2, 3]. A polycrystalline silicon floating gate or a Si_3N_4 charge trap layer is sandwiched between a tunnel oxide and an interpoly oxide as a control oxide to form a charge storage layer. The concept of NC memory was introduced in 1996 [4–6], when the formation of NC-based charge trap flash memory devices was reported by Tiwari and coworkers. Figure 4.2 shows schematically a proposal to distribute stored charge among isolated silicon NCs [7]. In this memory configuration, any defects that form in the oxide cause only a localized disturbance and minimal shift in the threshold voltage (V_{th}). As a result, a lower operating voltage and high write/erase speeds for highly scalable peripheral devices are enabled through the use of NC memory [8].

Because of the advantages above, recently, we focused on Si floating gate memory with SiO_2 layers and high-dielectric-constant (high-k) dielectric stacks. In the different routes explored recently for preparing NCs in the gate oxide of MOS devices, two major techniques have been applied, LPCVD (see, for example, Refs. [5, 6] and references therein) and ion beam synthesis (IBS) [23, 24]. The second technique has been paid attention because of its manufacturing advantages and full compatibility with standard MOS technology. In the recent decade, researchers have extensively used ultralow-energy ion beam synthesis (ULE-IBS) to synthesize single planes of Si NCs.

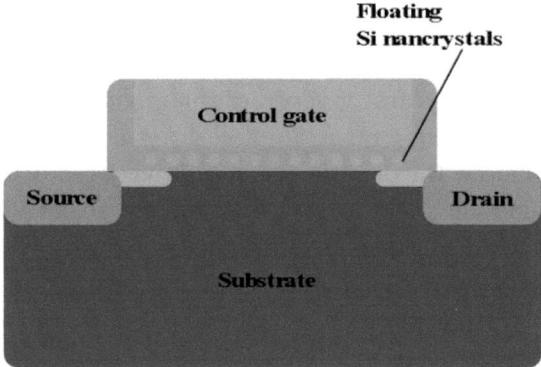

Figure 4.2   Scheme of Si NC memory.

We paid more attention to fabricating the main materials with controlled 2D arrays of NCs in dielectric layers by ULE-IBS. The focus was on the better tuning of the depth positioning and density of 2D arrays of NCs. We stress that Si NCs, within the so-called classical gate oxides (SiO_2), could be successfully attained. In addition, many researchers give the same single layer of Si NCs in an oxide-nitride-oxide (ONO) dielectric stack, combining the charge-trapping advantages of Si_3N_4 materials and Si NCs by ULE-IBS. Meanwhile, researchers also report the synthesis of Si NCs directly within HfO_2 layers to be limited by oxidation phenomena. Including both charge storage (e.g., silicon-oxide-nitride-oxide-silicon [SONOS]) and noncharge storage (e.g., FeRAM, MRAM, and phase-change random access memory [PCRAM]) devices, NVM has been selected [9] as a midterm solution for flash.

NC memory with Si NCs floating in SiO_2 has been introduced and extensively studied [10–13]. It has very good data retention because of the higher valence-band offset that restricts the tunneling of holes during the retention test. Until recently, metal arrays sandwiched in SiO_2 have been proposed for obtaining a higher conduction-band offset [17–19]. Even if it has attractive results, this latter approach always has limitations in terms of the thermal budget after array fabrication, which may make oxide and metal react with the silicon substrate and, thereby, influence the device performance.

For both these systems (Si or metallic NCs involved in SiO_2 layers), charge transfer from the substrate to the NC floating gate, which is in a SiO_2 layer 3–5 nm thick (thickness required for ensuring data storage for over 10 years), is very tardy and charging times typically above the microsecond range are necessary to obtain a functional memory window. On the other hand, the voltages of the erasing and programming stand at very high values (10–14 V) for many applications (mobile phones, MP3 players, laptop, etc.), needing low-voltage progress. Compared to SiO_2 tunnel barriers, when we use NCM layers in dielectric engineering, we can achieve lower programming voltages and improved data retention. Because of these advances, flash memory is still paid attention by the market. In the future, the development of flash memory must be in the areas of speed, write cycle, and size. In addition, the production cost should be relatively low. However, due to the immature manufacturing technology, this method has not been formally commercialized. Here are a few ways to make this kind of flash memory.

4.2.1.2 Preparation of Si NCs for flash memory

Many methods are reported for preparing Si NCs, such as CVD, ion implantation of silicon, and preparation of silicon-rich silica. Along with CVD methods other methods such as PECVD [10], HPVA [11], and LPCVD [12] can also be used. Carrada and coworkers synthesized ordered Si NCs by ultralow-energy ion implantation in 7-nm-thick SiO_2 and subsequent annealing [13]. We will now introduce two methods usually used for preparing Si NCs.

First, we will introduce the method of preparing NC silicon particles by the pyrolysis of diluted silane at 950°C. Initially, NC silicon particles are prepared by homogeneous gas-phase nucleation and grown by vapor deposition and coagulation. Next, the product is quenched with an ultrahigh-purity nitrogen flow to reduce the coagulation. Many factors, like silane concentration, furnace temperature, and silane residence time, influence the generation of NCs with a spherical, single-crystalline structure. We can also do many experiments to optimize these factors to get better NCs (Fig. 4.2, inset) with well-controlled diameters (Fig. 4.3) as small as 3 nm. Then, 1.5–2 nm high-quality thermal oxide shells are grown at 1000°C over the particles. These shells are insulating and reduce lateral crystal-to-crystal conduction in the NC layer. The step of oxidation has an additional advantage of improving the particle size distribution through a decrease in the oxidation rate with a smaller NC size [6]. Finally, a layer of NCs with a particle density as high as 10^{13}/cm^3 is thermophoretically deposited on an 8″ wafer by a temperature gradient of 200°C.

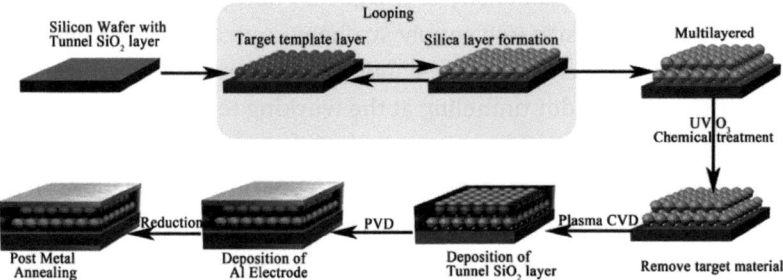

Figure 4.3 Schematic diagram of the typical process adopted to fabricate NC MOS-based flash memory. First, the silicon substrate is ready for depositing the target template layer (a special layer used for getting the NC layer easily due to the stronger combining ability between substrate and target template). Then a silica layer is deposited on it. By looping these cycles, multilayered structures are fabricated. After getting the NC layers, the target material that is not needed is removed by UV, oxidation, or even some type of chemical treatment. By plasma CVD, the NC layer used for the floating gate is embedded in a control gate, and aluminum electrodes are deposited on the substrate [17].

Moreover, Si NCs can be synthesized by a plasma technique at a low temperature of 430°C and it is suitable for fabricating a flash memory on a bulkSi wafer. The problem is how to fabricate Si NCs on a low-temperature polycrystalline silicon (LTPS) substrate. And the quality of the tunneling oxide is the key factor in flash memory using the LTPS substrate. However, the process temperature can't be over 600°C. The PECVD was used for tunneling oxide. On the other hand, it is widely known that HPVA has a high degree of influence on the optimizing electrical properties of low-temperature polycrystalline silicon thin-film transistor (LTPS-TFT), such as mobility and on-current. Therefore, HPVA has an excellent performance in improving the tunneling oxide as well as the memory properties.

In conclusion, we can say that all the mentioned methods of manufacturing Si NCs are used to prepare Si NC floating gates of flash memory devices directly. With this method, we can make Si NCs in the floating gate more uniform and more stable so that we can get more reliable memory devices.

The properties of flash memory will be influenced by the size, size distribution, and density of Si NCs for charge storage devices. Si NCs with smaller size will supply stronger control of the number of carriers; there must be a large enough space for the NCs to prevent significant dot-to-dot tunneling at the working temperature [9].

As mentioned earlier, the ordered Si NCs were fabricated by Carrada et al. with the method of ultralow-energy ion implantation in 7-nm-thick SiO_2 and subsequent annealing [13]. Different oxidizing annealing ways can achieve Si NC population characteristics (size and density). Seo et al. developed Er- and Tm-codoped silicon-rich silicon oxide by spatial separation [14]. Adjustment of the size and size distribution in a single layer and multilayers was demonstrated [15, 16]. Lu et al. [16] demonstrated Si-NC-based metal-oxide-semiconductor structures for memory by depositing SiO_2/SiO_x/SiO_2 multilayer stacks. The various stoichiometries resulted in a varied area density of the NCs after high-temperature annealing of the multilayer stacks. In 2013, a 3D-multilayered nanoparticle composition was fabricated by the biological target layer-by-layer method and controlled in a MOS-based device structure as a floating gate of a flash memory device [17]. The schematic diagram is shown in Fig. 4.3. The 3D-integrated multilayered nanoparticle architecture is attained successfully, resulting in improving the charge storage capacity compared to that of a conventional structure device in flash memory devices.

4.2.2 Electrical Characteristics of Si Nanocrystal in Flash Memory

Memory devices are the fastest and easiest way to evaluate basic charge storage characteristics such as programming/erasing (P/E) voltage and speed, P/E cycling endurance, and charge retention. This is usually done through high-frequency capacitance–voltage (C–V) measurements and monitoring of the flat-band voltage [18]. Lu et al. investigated the characteristic of multilevel charge storage for our metal-oxide-semiconductor structures containing different layers of Si NCs, as shown in Fig. 4.4 [16]. In the figure, the left part, which is sweep-up, represents the discharge stages, and the right, which is sweep-down, represents the charge stages. Changing several programming voltages and a fixed erasing voltage, we can easily get the relationship between the flat-band voltage shift and the programming voltage. In the case of MOSFETs [19], measurement of the transfer characteristics (I_{DS}–V_{GS}) and monitoring of the extracted threshold voltage V_{th} are the usual methods to realize flash memory. Figure 4.5 shows the drain voltage (V_d) and drain current (I_d) characteristics of the fabricated nano-floating-gate memory (NFGM) device measured as a function of the gate voltage [20]. As the diagram indicates, NFGM has the general characteristics of a MOSFET. So to investigate the NFGM operation of electrical characteristics for the memory application, we can measure the erasing state and the programming state.

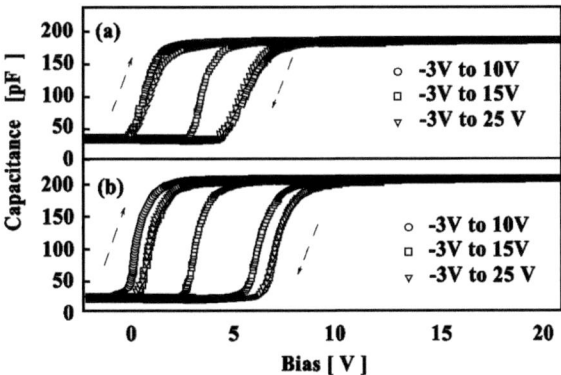

Figure 4.4 Characteristics of high-frequency C–V of a sample (a) with two layers and (b) with three layers. The widths of hysteresis gradually widen with increasing programming bias.

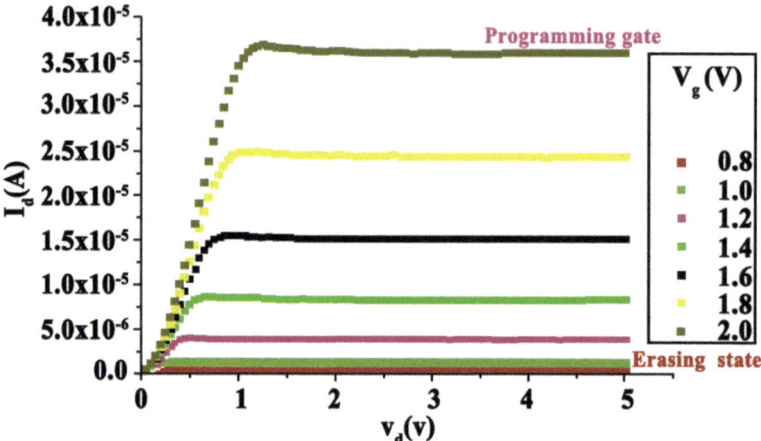

Figure 4.5 Characteristics of V_d and I_d the fabricated device, measured as a function of the gate voltage.

4.3 Si Nanocrystal Trap Center Studied by Deep-Level Transient Spectroscopy

In contrast to the steady-state high-frequency $C-V$ methods described so far, transient capacitance spectroscopy gives information by measuring how the non-steady-state high-frequency capacitance changes with time. DLTS, invented by Lang [21], is now widely used to detect traps of so-called deep levels in the bandgap. Initially, the method utilized measurements of transient capacitance following the pulsed bias in a p-n junction or a Schottky barrier diode to monitor changes in the charge state of defect centers. Schulz and Johnson [22] extended the application of DLTS to study charge emission from the interface states in MOS structures. Therefore, for a Si-NC-embedded MOS structure, if charge emission from the interface states of Si NC is detectable, it is worthwhile to use DLTS for characterization. The DLTS signal is the difference of capacitances at two different times after a filling pulse. It shows peaks for different trap levels in the sample at the respective temperatures T. If traps are filled by a filling pulse and the reverse bias is switched on again, the sample is in thermal nonequilibrium and relaxes into equilibrium by detrapping the charges. This relaxation process is related with capacitance transient. Its time constant is governed by the thermal emission rate $e_{n,p}$, which depends on the trap energy E_t and the temperature T:

$$e_{n,p} = \frac{1}{\tau_e} = \frac{N_{n,p}\sigma_{n,p}\mathbf{v}_{n,p}}{g}\exp\left(-\frac{E_t}{kT}\right). \tag{4.1}$$

Here N is the effective density of states in the conduction band or valence band, σ is the capture cross section of the trap, $\mathbf{v}$ is the carrier velocity, and g is the degeneracy factor. We can see here that the emission rate is exponentially dependent on $1/T$, so the thermal energy (or activation energy E_t) determining its slope and the capture cross section σ determining its intercept are the main features of the curve.

$$\text{DLTS} = a[C(t_2)-C(t_2)]^{\exp\left(-\frac{E_t}{kT}\right)}. \tag{4.2}$$

The DLTS signal is scaled in units of capacitance (usually pF). In principle, a DLTS measurement starts at a low temperature. The signal is recorded, and the temperature is ramped up for measurement. During the raising of temperature, according to the dependence of the thermal emission rate, as long as the temperature is too low for significant thermal emission until t_2, the difference in Eq. 4.2 is zero. If T is so high that the thermal emission is already over at t_1, the difference in Eq. 4.2 is also zero. Only if the emission time constant (or its inverse, the emission rate for convenience) of one level falls into the so-called rate window, given by the definition of t_1 and t_2, a DLTS peak appears. We obtain the following condition for the DLTS peak to appear:

$$e_{n,p} = \frac{\ln\left(\dfrac{t_2}{t_1}\right)}{(t_2-t_1)}. \tag{4.3}$$

The rate window is the inverse of τ_e, according to the formula of the relaxation time constant τ_e, and has the unit of s^{-1}. On the basis of Eq. 4.1, for each trap, the trap energy and the capture cross section are determined by its slope and intercept, respectively, and may be determined from an Arrhenius plot $\ln(e_{n,p}) \approx 1/T$.

Few researchers have used DLTS to observe the Si NC charging process [23], however. Because of the drawback of their fabrication methods, only limited information about Si nanodots was obtained and was usually hidden by artificial parasitic traps. We present our DLTS measurements of the Si NC sample based on the SiO/SiO_2 superlattice. Several results are explained and are consistent with previous reports [16]. These results might help us to understand the trap process of the Si nano floating gates and optimize the future operation of NVM devices based on Si NCs. The charge-pumping method is a kind of non-steady-state measurement, but it is widely used to evaluate the interface states in MOS transistors, which have a more complex four-terminal structure, so we don't concentrate on this method.

Nonsymmetrical sandwich structure samples were prepared as usual on highly doped n-type (100) silicon substrates (0.05–0.1 Ω cm). First, a 4 nm SiO_2 film was deposited as a tunnel oxide; then a 4 nm layer of SiO was deposited by SiO powder evaporation. On top of this structure, an additional SiO_2 layer was evaporated as the upper control oxide with a layer thickness of 24 nm. To form the Si NCs, thermal annealing was performed in a quartz tube furnace under N_2 ambient (1100°C, 0.5 h) for phase separation and crystallization. For comparison, a control sample, that is, a pure SiO_2 film with the same total dioxide thickness as the sample with Si NCs was prepared on the same type of substrate using the same deposition conditions. A schematic sample structure is shown in Fig. 4.6a; atomic-resolution plan-view TEM images were investigated using a JEOL JEM4010 at 400 KV. Because the electron beam is vertically transmitted through the sample, the texture of the (111) lattice plane is easier to visualize for a diamond cubic crystal of Si. Several Si NCs are marked in Fig. 4.6b. Zooming into this in-plane image, only NCs having the right orientation to the incident electron beam can be seen. We roughly estimate the interplanar spacing (d spacing) to be about 3.14 Å and the size of Si dots is about 5–7 nm. These interplanar spacing values are identical to the bulk values from literature and demonstrate that dots are unstrained by the surrounding SiO_2. This evidence could help us to understand the mechanical property of the Si dots in the matrix. Strain could influence the carrier mobility; hence, the message that the Si NCs are obviously not under strain is an important property.

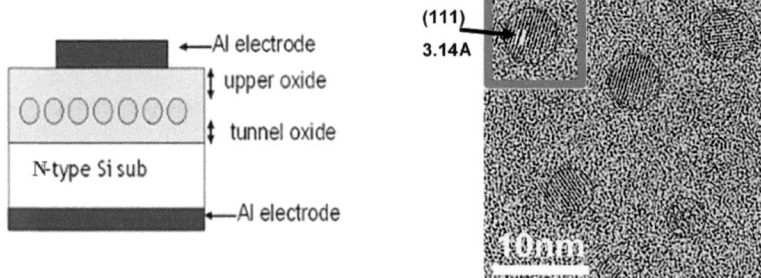

Figure 4.6 (a) A schematic model of a sample structure and (b) a plan-view high-resolution TEM image of a sample; this image was taken by JEOL JEM400. The average size of a Si NC is about 5–7 nm.

The electrical measurements were performed at 100 kHz using an HP4194A impedance analyzer at room temperature. For more information about the traps of the samples, DLTS was performed in the 100–310 K temperature range with variable pulse biases and rate windows. The heating rate was 0.4 K/min. The cryostat containing the sample was attached to the capacitance meter with a preamplifier, and the amplified transient capacitance. The capacitance of the sample was measured under depletion region conditions. The rate windows were chosen as 34.7, 17.3, and 8.7 s^{-1} for the measurements.

Figure 4.7 presents typical C–V curves of the sample. The loop of the forward and reverse sweep C–V characteristics indicates the electron charging and discharging process of Si NCs embedded in SiO_2. The details of the charging behavior of Si NC were discussed previously. A DLTS study gives more detailed information, in addition to C–V characterization, as shown in Fig. 4.7. The process of measurement is as follows: electron injection into dots appears as a positive reverse bias on the gate, whereby, near the interface of Si/SiO_2, the potential slope is so steep that electrons are populated and confined by the oxide and the steep potential. The electron energy perpendicular to the interface is now quantized into discrete states; this phenomenon is called surface quantization and it helps the electron injection [24]. Electron escape (emission) is from the interface or quantum levels of the dots through repeat bias filling pulses. The emission of electrons is recorded by measuring the capacitance transient. Figure 4.8 shows the DLTS signal of the control SiO_2 sample in Fig. 4.8a and a sample with Si NC in Fig. 4.8b.

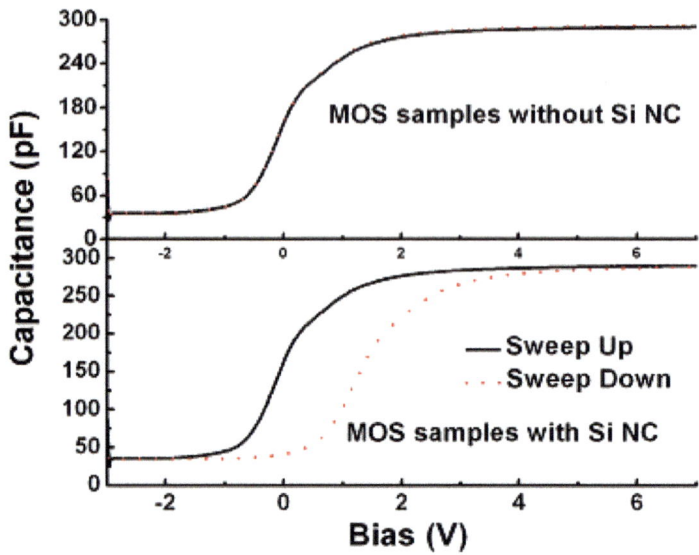

Figure 4.7 High-frequency *C–V* curves of the control sample and the sample with Si NCs.

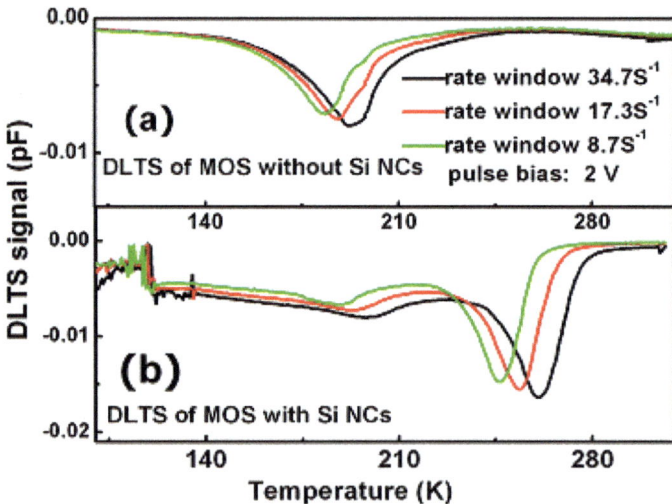

Figure 4.8 DLTS spectra of the control sample (a) and the sample with the Si NC layer (b) dependent on the rate window. The samples have the same thickness, but (a) is the sample without Si NCs (b) is the sample with Si NCs.

Negative peaks mean that the electrons (majority carriers) are in the space charge region. In Fig. 4.8a, only one peak, around 190 K, is obtained in the spectra, but the spectra in Fig. 4.8b present two peaks around, 190 K and 250 K. The signal noise in its left part is due to problems associated with the metal contact at low temperatures. For the simple MOS control sample, the signal consists of the interface charge states between the oxide and the Si substrate, and for the MOS sample with Si NC, apart from the above peak, a second and dominant peak appears. It should be related to the charge within the Si NCs. Furthermore, to distinguish whether the charges are stored in the quantum-confined conduction band or in deep levels of the bandgap in the Si NCs, we tuned the rate window of DLTS to obtain the thermal activation energy and capture cross sections to explain this peak. In Fig. 4.8, we measure the DLTS spectra under three rate windows, 34.7 s^{-1}, 17.3 s^{-1}, and 8.7 s^{-1}, by choosing certain t_2 and t_1, and the corresponding peak shifts are shown.

Yamasaki theoretically analyzed and measured the bulk traps and interface states in Si MOS [25] diodes and demonstrated the distinction between interface charge and bulk traps. Because the emission rate of bulk traps at a temperature is constant regardless of the pulse bias, the peak temperature and shape of $\Delta(C)$ should not change with the pulse bias. We keep the same rate window and a constant reverse bias 1 V but variable pulse biases 0.5 V, 1 V, and 2 V on the samples. The obtained spectra are presented in Fig. 4.9. The same peak temperature, at around 250 K, and shape convince us that this trap is probably a bulk trap, unrelated to the interface states. Without applying a pulse bias, the MOS sample is in the accumulation region and electrons are injected from the substrate and stored in the Si NCs. It is well known that these stored electrons alter the electrostatic potential and cause the positive flat-band shift. Therefore, the sample actually is in the inversion region under the pulse bias. Electron emission from Si NCs occurs and contributes to transient capacitance. Similar phenomena were observed by Souifi et al. [23] in their CVD samples. However, the results from Souifi show other noisy peaks in the DLTS spectra, which means there exist trap sites that are different from the Si quantum dots and interface in their oxide, such as hydrogen-related traps in CVD process or defects. Their sample had only a 2 nm tunnel oxide, whereas our sample had a 4 nm tunnel oxide. We believe it is because of the high density ($>10^{12}$ cm^{-2}) of Si NCs in our sample, since the tunnel probability is proportional to the total electron emission number. Normally, because of the Si–O bond breaking in the phase separation process, the sample used here might contain some dissociating dangling bond in the matrix at a deep level, but sufficient heat treatment nucleates them for the crystallization of Si dots, so their amount appears to be below the detection limit of DLTS.

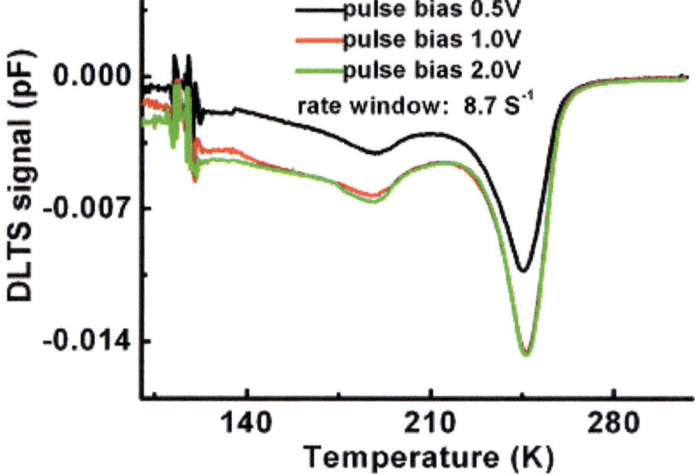

Figure 4.9 DLTS spectra of a MOS sample containing Si NCs dependent on the fitting pulse bias; as the pulse bias rises, the transient signal rises too until saturation.

In the case of pulse biases 1 V and 2 V, the DLTS signals at the peak, around 250 K, are nearly the same because the stored electrons in Si NCs are fully released under these pulse biases and contribute to the same transient capacitances. From the shape of the DLTS peaks, we consider these deep traps as consistent with the expected discrete behavior of the NC. We observe the shift of DLTS peak related to Si NCs for different rate windows. This emission rate shift as a function of temperature is presented in the Arrhenius plot in Fig. 4.10. Each DLTS peak has its own corresponding temperature T, so the Arrhenius plot is used to determine the parameters of the trap center in Fig. 4.10. We can find the value of carrier velocity $v = 10^7$ cm/s and $N_c = 8.4 \times 10^{18}$ cm^{-3} from the literature and use them in Eq. 4.1. For measuring E_f and σ, we fitted the Arrhenius plot in Fig. 4.10 according to Eq. 4.1. The calculated activation energy E_t of the main peak is about 0.56 eV. Using this value and the above value of m*, $\mathbf{v}_{n,p}$, and Nc in Eq. 4.1, the capture cross section is found to be about $(1-7) \times 10^{-13}\,cm^2$.

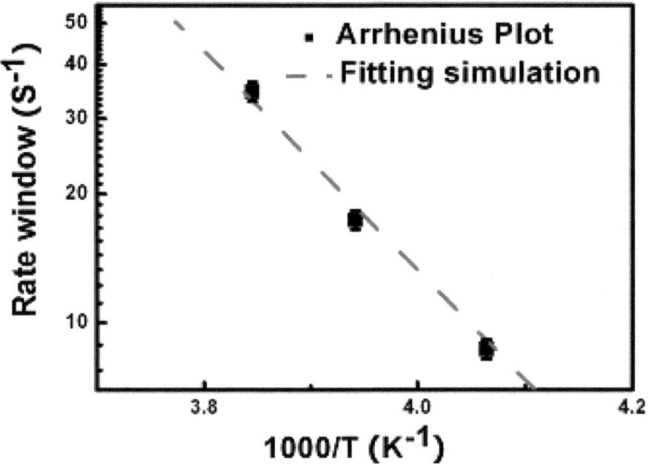

Figure 4.10 The Arrhenius plot of a MOS sample containing Si NC, together with a parameter simulation.

These estimations are much larger than the counterparts of trap impurities associated with the deep levels of silicon bandgap but are adjacent to the values of Si NCs in the SiO_2 matrix. For example, the obtained potential barrier of Si NCs by a SiO_2 insulator is 1.6 eV in the previous *I–V* characterization [26]. Biteen [27] and Kovalev [28] showed that the large cross section of silicon NCs is at the $10^{-13}\,cm^2$ level and much larger than the cross section of the usual deep-level trap centers. Silva confirmed the large capture cross section of Si NCs [29] in a dielectrics matrix as well. In contrast, interfacial traps or single dangling bands are mainly located at the gap center with a capture cross section of $10^{-15}\,cm^2$ to $10^{-17}\,cm^2$ [30]. The capture cross section is an important parameter for memory operation; a reduced capture cross section and a Coulomb blockade will result in a lower programming speed and saturation V_{th}. The larger capture cross sections of Si NCs not only benefit NVM operation but also play a key role as a sensitizer in the Er-doped silicon dioxide containing the Si NC system already discussed in Chapter 1.

Using DLTS, we were able to observe thermal emission from Si NC at around 250 K for certain rate windows. To produce the DLTS curve we have estimated some intrinsic parameters, such as the capture cross section σ at the 10^{-13} cm^2 level and activation energy E_t of about 1.6 eV. These results could be a clue that the trapping mechanism in MOS systems containing Si NCs is related to the quantum levels of the Si quantum dots at around 300 K. This trapping mechanism is also suitable for other types of quantum dots embedded in the dielectric matrix as a system. DLTS supplied information on other electrical characterizations, information other than that obtained through C–V measurement.

4.4 Engineering for Improved Si Nanocrystal Flash Memory

Flash memory cells have caused a dramatic increase in memory capacity and reduction in cost per bit, while in principle one wishes to scale down the tunnel oxide at the same rate as that for the gate oxide of complementary CMOS transistors [31]. To overcome this scaling limit, it appears that any candidate tunnel dielectric must exhibit low leakage current and SILC. The thickness of the tunnel oxide was chosen on the basis of models developed to predict the long-term charge retention behavior of silicon NCs [32]. The expected charge retention of the silicon NC memory device as a function of tunnel oxide thickness is shown in Fig. 4.11. The cases are calculated for the range of the likely effective mass in the oxide. The oxide barrier height is 3.15 eV.

There have been several technical breakthroughs in fabricating dynamic random access memory (DRAM) capacitors during the past decades, including Si/ONO/Si, where ONO is a SiO$_2$/Si$_3$N$_4$/SiO$_2$-stacked dielectric. The individual storage node device, making use of traps in Si$_3$N$_4$, was proposed by Wegener et al. in 1967 [3]. Sougleridis et al. demonstrated ONO structures using a silicon nitride layer with an embedded 2D array of Si NCs, as shown in Fig. 4.12 [33]. Electrical investigations of the produced oxide/Si NCs nitride/oxide stacks by means of Al gate capacitors indicated reduction of the onset voltage for electron and hole injection, reduction in electron storage ability, and establishment of hole storage [33]. There is particular interest in the fabrication of NCs into high-*k* capacitor dielectric matrices instead of SiO_2 for achieving NC memory with low programming voltages and improved data retention. In 2015, Yim et al. analyzed some novel high-dielectric-constant (high-κ) materials for next-generation electronic devices screened by automated ab initio calculations [34]. The promising device results using Si NCs embedded in HfO_2 or Al_2O_3 gate dielectrics have been presented in Refs. [18, 35–37].

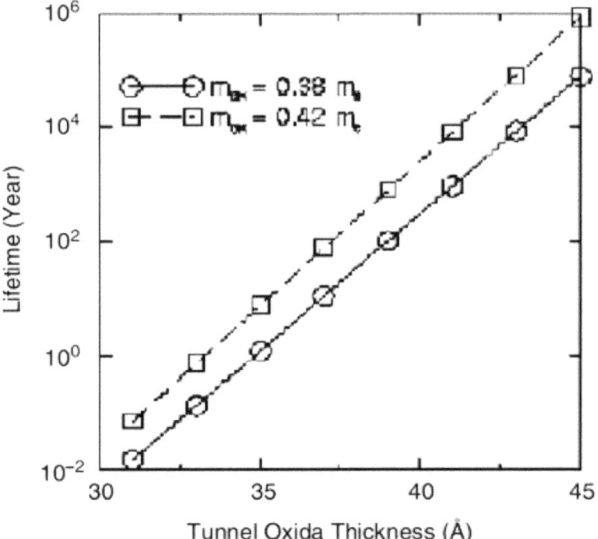

Figure 4.11 Lifetime of charge decay in a nanocrystal, calculated as a function of tunnel oxide thickness, assuming one electron stored per nanocrystal in the conduction band, a nanocrystal density of $9 \times 10^{11}\ cm^{-2}$, and an equivalent control oxide thickness of 7 nm.

Resistive random access memory (RRAM) and ferroelectric random access memory/ferroelectric field-effect transistor (FeRAM/FeFET) are types of flash memory that are considered promising candidates for next-generation NVM device. Jiang et al. prepared a high-performance and ultralow-power RRAM based on an $Al/a\text{-}SiN_x{:}H/p^+\text{-}Si$ structure, which can be achieved by tuning the Si dangling bond conduction paths [38]. Guo and coworkers demonstrated ferroelectric tunnel junctions ($Pt/BaTiO_3/$ $La_{0.67}Sr_{0.33}MnO_3$) epitaxially grown on silicon substrates [39]. The results prove that the silicon-based ferroelectric tunnel junction is a very promising candidate for application in future NVMs. In addition, Si nanowire field-effect transistors (FETs) have been prepared. Nanowire-based ferroelectric complementary-metal-oxide-semiconductor (NW FeCMOS) NVM devices were successfully fabricated by utilizing single n- and p-type Si nanowire ferroelectric-gate field-effect transistors (NW FeFETs) as individual memory cells [40]. Si NW FeCMOS memory devices exhibit a direct readout voltage and ultralow power consumption. The high-performance top-gated nanowire molecular flash memory fabricated with redox-active molecules was proposed by Zhu and coworkers [41]. Ren et al. designed a nonplanar flash memory architecture with ultrahigh-density ($\sim 1.5 \times 10^{12}$ cm^{-2}) NiSi NCs as the floating gate, which was demonstrated using a triangular-shaped Si nanowire array as the memory transistor channel [42]. Recently, graphene flash memory was reported by Hong et al. [43]. Han et al. adopted functionalized reduced graphene oxide (RGO) as the charge-trapping layer in ambipolar flash memory, and a dramatic transition of charging behavior from unipolar trapping of electrons to ambipolar trapping and eventually to unipolar trapping of holes was achieved [44].

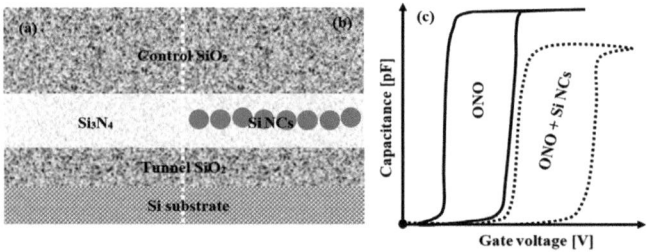

Figure 4.12 Schematic of the cross-sectional images images of the high-dose sample (1.5×10^{16} Si$^+$/cm^2) taken (a) around zero loss and (b) at 17 eV, showing the location of the Si NC band in the nitride layer, and (c) bidirectional high-frequency (1 MHz) $C-V$ characteristics of MONOS capacitors with and without Si NCs embedded in the silicon nitride layer [33].

In summary, the conventional Si-based memory still is the leader in the current market. However, there is a lot of research and development happening on flash memory. Si NC flash memory is a field of intensive research, and the generated results can be beneficial not only for the broader scientific community with interest in NC-based optoelectronics or photovoltaics applications but also for accelerated commercialization.

References

1. International technology roadmap for semiconductors (2015). http://www.itrs2.net/itrs-reports.html.

2. Kahng, D., and Sze, S.M. (1967). *Bell. Sys. Tech. J.*, **46**, 1288–1295.

3. Wegener, H.A.R., Lincoln, A.J., Pao, H.C., O'Connell, M.R., Oleksiak, R.E., and Lawrence, H. (1967). The variable threshold transistor, a new electrically-alterable, non-destructive read-only storage device. *IEEE International Electron Devices Meeting, 70.*

4. Tiwari, S., Rana, F., Hanafi, H., Hartstein, A., Crabbe, E.F., and Chan, K. (1996). *Appl. Phys. Lett.*, **68**, 1377–1379.

5. Tiwari, S., Rana, F., Chan, K., Shi, L., and Hanafi, H. (1996). *Appl. Phys. Lett.*, **69**, 1232–1234.

6. Hanafi, H.I., Tiwari, S., and Khan, I. (1996). *IEEE Trans. Electron Devices*, **43**, 1553–1558.

7. Muralidhar, R., Sadd, M.A., and White Jr, B.E. (2009). *Silicon Nanocrystal Nonvolatile Memories*, Springer.

8. Yater, J.A. (2013). *Phys. Status Solidi*, **8**, 1505–1511.

9. Lu, T.Z., Alexe, M., Scholz, R., Talalaev, V., Zhang, R.J., and Zacharias, M. (2006). *J. Appl. Phys.*, **100**, 014310.

10. Ichikawa, K., Uraoka, Y., Yano, H., et al. (2007). *Jpn. J. Appl. Phys.*, **46**, 661–663.

11. Sameshima, T., Satoh, M., Sakamoto, K., Ozaki, K., and Saitoh, K. (1998). *Jpn. J. Appl. Phys.*, **37**, 4254.

12. Della Marca, V., Amouroux, J., Molas, G., Postel-Pellerin, J., Lalande, F., Boivin, P., and Ogier, J.L. (2013). *ECS Trans.*, **53**, 129–139.

13. Carrada, M., Wellner, A., and Paillard, V. (2005). *Appl. Phys. Lett.*, **87**, 251911.

14. Seo, S.Y., and Shin, J.H. (2004). *Appl. Phys. Lett.*, **85**, 4151.

15. Lu, T.Z., Alexe, M., Scholz, R., Talalaev, V., Zhang, R.J., and Zacharias, M. (2006). *Appl. Phys. Lett.*, **100**, 014310.

16. Lu, T.Z., Alexe, M., Scholz, R., Talelaev, V., and Zacharias, M. (2005). *Appl. Phys. Lett.*, **87**, 202110.

17. Sano, K.I., Miura, A., Yoshii, S., Okuda, M., Fukuta, M., Uraoka, Y., Fuyuki, T., Yamashita, I., and Shiba, K. (2013). *Langmuir*, **29**, 12483–12489.

18. Bonafos, C., Carrada, M., Benassayag, G., Schamm-Chardon, S., Groenen, J., Paillard, V., Pecassou, B., Claverie, A., Dimitrakis, P., Kapetanakis, E., Ioannou-Sougleridis, V., Normand, P., Sahu, B., and Slaoui, A. (2012). *Mater. Sci. Semicond. Process.*, **15**, 615–626.

19. Dimitrakis, P., Kapetanakis, E., Tsoukalas, D., Skarlatos, D., Bonafos, C., Ben Asssayag, G., Claverie, A., Perego, M., Fanciulli, M., and Soncini, V. (2004). *Solid-State Electron.*, **48**, 1511.

20. Yang, J.S., Kim, S., Kim, Y.T., Cho, W.J., and Park, J.H. (2008). *Microelectron. J.*, **39**, 1553–1555.

21. Lang, D.V. (1974). *J. Appl. Phys.*, **45**, 3023.

22. Schulz, M., and Johnson, N.M. (1978). *Solid. State. Commun.*, **25**, 481.

23. Souifi, A., Brounkov, P., Bernardini, S., Busserret, C., Militaru, L., Guillot, G., and Baron, T. (2003). *Mater. Sci. Eng. B*, **102**, 99.

24. Oda, S., and Ferry, D.K. (2005). *Silicon Nanoelectronics*, Taylor & Francis, New York.

25. Yamasaki, K., Yoshida, M., and Sugano, T. (1979). *Jpn. J. Appl. Phys.*, **18**, 113.

26. Lv, T.Z., Shen, J., Mereu, B., Alexe, M., Scholz, R., Talalaev, V., and Zacharias, M. (2005). *Appl. Phys. A*, **80**, 1631.

27. Biteen, J.S., Pacifici, D., Lewis, N.S., and Atwater, H.A. (2005). *Nano Lett.*, **5**, 1768.

28. Kovalev, D., Diener, J., Heckler, H., Polisski, G., Künzner, N., and Koch, F. (2000). *Phys. Rev. B*, **61**, 4485.

29. Silva, H., Kim, M.K., Avci, U., Kumar, A., and Tiwari, S. (2004). *MRS Bull.*, **3**, 845.

30. Militaru, L., and Souifi, A. (2003). *Appl. Phys. Lett.*, **83**, 2456.

31. Ma, T.P., Brewer, J.E., and Gill, M. (2008). *Nonvolatile Memory Technologies with Emphasis on Flash: A Comprehensive Guide to Understanding and Using Flash Memory Devices*, Wiley-IEEE.

32. Suzuki, M., Kusunoki, T., Shinada, H., and Yaguchi, T. (1995). *J. Vac. Sci. Technol. B*, **13**, 2201.

33. Sougleridis, V.I., Vamvakas, V.E., Dimitrakis, P., Normand, P., Bonafos, C., et al. (2007). *Nanotechnology*, **18**, 215204.

34. Yim, K., Yong, Y., Lee, J., Lee, K., Nahm, H.H., Yoo, J., Lee, C., Hwang, C.S., and Han, S. (2015). *NPG Asia Mater.*, **7**, e190.

35. Wang, M., Bi, C., Li, L., Long, S., Liu, Q., Lv, H., Lu, N., Sun, P., and Liu, M. (2014). *Nat. Commun.*, **5**, 4598.

36. Schmidt, B., Grambole, D., and Herrmann, F. (2002). *Methods Phys. Res. Sect. B*, **191**, 482.

37. Lu, J.X., Gong, C.J., Ou, X., Lu, W., Yin, J., Xu, B., Xia, Y.D., Liu, Z.G., and Li, A.D. (2014). *AIP Adv.*, **4**, 117110.

38. Jiang, X.F., Ma, Z.Y., Xu, J., Chen, K.J., Xu, L., Li, W., Huang, X.F., and Feng, D. (2015). *Sci. Rep.*, **5**, 15762.

39. Guo, R., Wang, Z., Zeng, S.W., Han, K., Huang, L.S., Schlom, D.G., Venkatesan, T., Ariando, Chen, J.S. (2015). *Sci. Rep.*, **5**, 12576.

40. Van, N.H., Lee, J.H., Whangb, D., and Kang, D.J. (2015). *Nanoscale*, **7**, 11660–11666.

41. Zhu, H., Pookpanratana, S.J., Bonevich, J.E., Natoli, S.N., Hacker, C.A., Ren, T., Suehle, J.S., Richter, C.A., and Li, Q.L. (2015). *ACS Appl. Mater. Interfaces*, **7**, 27306–27313.

42. Ren, J.J., Li, B., Zheng, J.G., Olmedo, M., Zhou, H.M., Shi, Y., and Liu, J.L. (2012). *IEEE Electron Device Lett.*, **33**, 1390–1392.

43. Hong, A.J., Song, E.B., Yu, H.S., Allen, M.J., Kim, J., Fowler, J.D., Wassei, J.K., Park, Y., Wang, Y., Zou, J., Kaner, R.B., Weiller, B.H., and Wang, K.L. (2011). *ACS Nano*, **5**, 7812–7817.

44. Han, S.T., Zhou, Y., Sonar, P., Wei, H.X., Zhou, L., Yan, Y., Lee, C.S., and Roy, V.A.L. (2015). *ACS Appl. Mater. Interfaces*, **7**, 1699–1708.

Chapter 5

Synthesis, Characterization, and Memory Application of Germanium Nanocrystals in Dielectric Matrices

Wee Kiong Choi[a] and Writam Banerjee[b,c,d]

[a]*National University of Singapore, Singapore*
[b]*Key Laboratory of Microelectronics Devices & Integrated Technology,*
Institute of Microelectronics, Chinese Academy of Sciences, No. 3,
BeiTuCheng West Road, ChaoYang District, Beijing 100029, China
[c]*University of Chinese Academy of Sciences, Beijing 100049, China*
[d]*Jiangsu National Synergetic Innovation Center for Advanced*
Materials (SICAM), Nanjing 210009, China
elechoi@nus.edu.sg

As zero-dimensional systems, nanocrystals (NCs) are very attractive candidates for and will play a major role in future electronic devices. This chapter reviews in depth the progress in the growth and physical and electrical characterizations of germanium (Ge) NCs as materials and also their application in floating-gate memory devices. This chapter is mainly divided into three parts. The first part will discuss the synthesis method of Ge NCs and the dependence on concentration, annealing conditions, influence of dielectric material,

Nanocrystals in Nonvolatile Memory
Edited by Writam Banerjee
Copyright © 2018 Pan Stanford Publishing Pte. Ltd.
ISBN 978-981-4774-73-4 (Hardcover), 978-1-351-20327-2 (eBook)
www.panstanford.com

etc. In the second part, we will discuss the physical characteristics, such as photoluminescence and electroluminescence, and the stress effect on Ge NCs. The third part will introduce the size-dependent, density-dependent electrical properties of Ge-NC-based floating-gate memory devices. Special care is taken to discuss in detail the fabrication process of memory devices embedding Ge NCs of different sizes, densities, annealing treatments, and dielectric materials and their characterization as transistors. The chapter outlines the challenges faced by Ge-NC-based floating-gate memory devices, with the scope of future development.

5.1 Introduction

The possibility of integrating group IV elements with silicon (Si)-based technology for optoelectronics has spurred much investigation in the field of low-dimensional systems [1–75]. Germanium (Ge), as a group IV element, is structurally similar to Si [76–125]. The energy difference between the indirect gap (0.66 eV) and the direct gap (0.8 eV) is smaller in Ge (0.14 eV) than in Si (1.1 eV) [126–147]. As Ge also has a larger permittivity and smaller effective masses for electrons and holes than Si, the Bohr radius of the excitons in Ge (24.3 nm) is larger than that of Si (4.9 nm) [32]. The quantum size effect is thus predicted to be stronger in Ge than in Si [8]. These facts had suggested that it is easier to modify the electronic structure around the bandgap of Ge. Hence there has been an intense interest in Ge nanocrystals (NCs) embedded in a silicon oxide (SiO_2) matrix as such a structure has potential applications in optoelectronics [126, 127] and electronics [40, 72, 75, 132]. The interest in optoelectronic applications stems from the observation of visible photoluminescence (PL) or electroluminescence (EL) exhibited by Ge NCs in a SiO_2 matrix [65, 141]. For electronic applications, Tiwari et al. [132] have proposed a Si NC memory device that can be programmed at hundreds of nanoseconds using low voltages for direct tunneling and storage of electrons in the Si NCs.

Several methods have been used in the synthesis of Ge NCs in an oxide matrix, for example, cosputtering [21, 34], chemical vapor deposition (CVD) [39, 118], ion implantation [98], pulsed laser deposition [35, 36], e-beam irradiation [28], reduction of silicon germanium oxides [69, 112], and the use of anodic alumina membrane (AAM) [15]. We will briefly summarize the results of Ge NCs fabricated by these methods in this section. However, by far, the most common method in the synthesis of Ge NCs is by annealing cosputtered Ge plus silicon oxide.

Ion implantation is a versatile technique for forming Ge NCs in the near-surface region of a substrate. For example, Mestanza [96] synthesized Ge NCs in a SiO_2 matrix by implanting Ge74+ ions at room temperature using 250 keV energy, at doses of 0.5–4 × 10^{16} cm^{-2}. The samples were subsequently annealed at 1000°C for 1 h in a forming gas ambient. Similar implantation conditions were also adopted by Yang [140] and Barba [4]. However, the disadvantage with the synthesis of Ge NCs by implantation is that the NCs usually have a relatively wide size distribution. Multienergy ion implantation and thermal annealing produce a uniform depth-concentration profile of Ge NCs over the SiO_2 film with a narrow size distribution. NCs are usually formed by implanting Ge at doses of 3 × 10^{16} cm^{-2} or higher using 100 keV and annealing at temperatures varying from 700°C to 1000°C. Oxidation of Ge due to the diffusion of oxidizing species from the annealing ambient at high temperatures plays an important role in samples with a projected range closer to the surface.

E-beam irradiation has been used to synthesize Ge NCs in a SiO_2 matrix [28]. The samples consisted of a trilayer structure with a radio frequency (rf)-cosputtered Ge-plus-SiO_2 layer sandwiched between two SiO_2 layers. It was concluded that the thermal effect of e-beam radiation was only able to initiate Ge clustering as it does not impart enough energy for long-range diffusion for NC formation. The NC formation can only be achieved with a substrate heated to 250°C under e-beam irradiation.

Baron et al. [5] were the first to report the synthesis of Ge NCs by CVD. The NCs were synthesized by a two-step method: first, a SiO_2 surface was functionalized with Si nuclei by CVD using SiH_4; this was followed by the growth of Ge NCs by the introduction of GeH_4 in the second step. By carefully controlling the number of Si nuclei, the density of the Ge NCs can be varied between 10^9 to 10^{12} cm^{-2}. The most recent work reported Ge NCs prepared using the plasma-enhanced chemical vapor deposition (PECVD) technique [31, 147]. For instance, in the investigation of light harvesting properties of Ge NCs in SiO_2 and Si_3N_4 matrices by Cosentino et al. [31], the Ge NCs were prepared by annealing thin films of SiGeO and SiGeN alloys. The thin films were deposited by PECVD on fused silica quartz or Si substrates kept at 250°C. Different Ge concentrations were obtained by varying the flux of GeH_4 while keeping constant the fluxes of SiH_4 and N_2O (or NH_3) gases for the growth of SiGeO (or SiGeN) films. The as-deposited samples were annealed at a temperature range of 600°C–850°C in a N_2 atmosphere to induce the phase separation of Ge in SiGeO and SiGeN alloys and the precipitation of excess Ge into NCs.

Paine et al. [112] were the first to report the synthesis of Ge NCs via the reduction of $Si_{0.6}Ge_{0.4}$ films. $Si_{0.6}Ge_{0.4}$ films of 400 nm were grown on Si by CVD and were oxidized in stream at 25 MPa at 475°C. The samples were then annealed in forming gas (N_2/H_2:80/20) at 750°C, 800°C, and 850°C for times ranging between 1 min and 120 min Kan et al. [69] prepared amorphous $Si_{0.54}Ge_{0.46}$ films (~10 nm) by rf sputtering at room temperature on a 5 nm thermally grown SiO_2 layer. The conversion from amorphous to polycrystalline $Si_{0.54}Ge_{0.46}$ films was achieved by annealing the sample at 800°C in N_2 for 6 h. The films were then wet-oxidized at 600°C for 60 min to form mixed oxides. The structures were lastly capped with a 50 nm layer of a rf-sputtered silicon oxide film. In synthesizing Ge NCs, the samples underwent rapid thermal annealing (RTA) in a pure N_2 ambient. More recently, Rodríguez et al. [116] prepared amorphous $Si_{0.7}Ge_{0.3}$ layers by low-pressure CVD at 425°C with Si_2H_6 and GeH_4 as precursor gases. The films were annealed at 725°C in vacuum for 72 h to completely crystallize the amorphous films. Dry and wet oxidation were performed on the crystallized films at 850°C and 650°C, respectively. It should be noted that, however, Ge NC formation via reduction of mixed oxides from SiGe films is kinetically complex and, therefore, makes the prediction and control of NC size distribution difficult.

Finally, Ge NCs have been synthesized by evaporating Ge through an AAM mask [94] on a Si substrate. The size of the NCs was controlled by varying the aspect ratio (the height of the mask to the pore diameter) of the mask from 4:1 to 1:1, and the NC diameter can vary from 50 nm to 100 nm. However, with this method of fabrication, some NCs were missing due the imperfection of the mask. The problem of missing NCs can be reduced by using a thinner mask or a mask with a smaller pore aspect ratio.

5.2 Synthesis of Ge Nanocrystals

5.2.1 Ge Atoms for Nanocrystal Growth

It is well recognized that an as-prepared rf-cosputtered film of Ge and SiO_2 contains elemental Ge and oxides of Si and Ge [23, 33]. In such a Si-Ge-O ternary system, the NC formation process can be viewed as:

1. GeO_2 (or GeO_x) reduction leading to the creation of elemental Ge atoms
2. Diffusion of liberated Ge in the oxide matrix
3. Nucleation due to Ge-Ge collisions
4. Growth by diffusing Ge atoms bonding to existing Ge nuclei
5. Coarsening of NCs due to Ostwald ripening

The decomposition of GeO_2 and GeO_x to Ge by Si is given by the following equations:

$$GeO_2 \rightarrow Ge + O_2 \tag{5.1}$$

$$Si + GeO_x \rightarrow Ge + SiO_x \tag{5.2}$$

$$Si + GeO_2 \rightarrow Ge + SiO_2 \tag{5.3}$$

Figure 5.1 shows an X-ray photoelectron spectroscopy (XPS) analysis of as-prepared and rapidly thermally annealed cosputtered Ge and SiO_2 samples. The samples were cosputtered in argon at an ambient temperature with a target that consisted of a 4-inch SiO_2 disk attached with pieces of Ge ~99.999% pure (10 mm × 10 mm × 0.3 mm), with the Ar pressure and rf power fixed at 3 × 10^{-3} mbar and 100 W, respectively. Figure 5.1a shows the compositions of elemental Ge, GeO_2, and the suboxides as a function of sample etch depth *(x)* for the as-sputtered sample. The Ge concentration stays at

10% ≈ 11% throughout the film. The concentration of GeO_2 reduces from 32% to 23%, and the concentration of Ge suboxides increases from 55% to 66% as x increases from 150 Å to 2500 Å. The high proportions of GeO_2 and suboxides in the as-sputtered samples was reported by Maeda [91] and Maeda et al. [92]. A film annealed at 800°C (see Fig. 5.1b) shows an increase in the concentration of Ge from 13% to 29% as x increases from 150 Å to 3000 Å. The concentration of GeO_2 is slightly higher than that of the as-sputtered sample (i.e., 39% and 30% at x = 150 Å and 3000 Å, respectively) but also reduces as x increases. Note that the suboxide's concentration, however, has reduced significantly (41%–47%) compared to the as-sputtered samples. The reduction in suboxides corresponds well to the increase in the Ge concentration in the film. It is also possible that some of the suboxides may have formed GeO_2. Note that this annealing temperature is where we observed uniform NC growth (as shown by the transmission electron microscopy [TEM] image in Fig. 5.2). This means that the Ge for NC growth most likely comes from dissociation of suboxides. Figure 5.1c shows a very pronounced increase (8%–85%) in Ge concentration (1000 Å ≈ 2750 Å) for a film annealed at 1000°C. There is also a reduction in the concentrations of GeO_2 and the suboxides, the latter to a larger extent. This results in big NCs with multiple twin structures. Again, the results of Fig. 5.1c show that the dissociation of suboxides and GeO_2 supplies the Ge for NC growth. The high percentage of suboxides at x = 150–1000 Å may be due to the reduction of Ge.

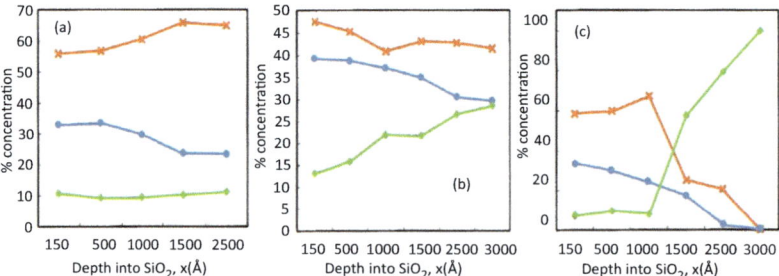

Figure 5.1 Composition of elemental Ge (green lines), GeO_2 (sky-blue lines), and Ge suboxides (orange lines) as a function of depth of a SiO_2 film for (a) an as-sputtered film, (b) the sample annealed at 800°C, and (c) the sample annealed at 1000°C. Note that x = 0 refers to the sample surface. Reprinted from Ref. [23], with the permission of AIP Publishing.

5.2.2 Effect of Ge Concentration and Annealing Temperature

It is obvious that the amount of Ge in the cosputtered Ge and SiO_2 samples will be intimately linked to the formation and properties of Ge NCs. It is, however, surprising to note that very little has been reported on this topic. Figure 5.2a shows a schematic illustrated image of Ge NCs about 30 nm from the SiO_2/Si interface, prepared by annealing oxidized cosputtered SiGe films at 900°C in N_2 ambient for 2 h [33]. The NCs are spherical and were well dispersed in the SiO_2 matrix, with a diameter in the range 4–7 nm. Figures 5.2b and 5.2c show the cross-sectional TEM results of two sets of samples prepared for the investigation of the effect of Ge concentration on the formation of NCs in a silicon oxide matrix. The concentrations of germanium, germanium oxide, and germanium suboxides (Ge, GeO_2, and GeO_x) were determined by XPS to be 20%, 10%, and 70% for sample A and 75%, 5%, and 20% for sample B. Figure 5.2b shows that sample A, which has been annealed at 800°C for 60 min, consists of a region (i), of approximately 50 nm from the oxide surface that is void of NCs, followed by a region (ii), of nanoclusters/NCs of bigger and varying sizes, and a region (iii), which is again void of NCs. As sample A contains a high concentration of GeO_x, most of the Ge that is responsible for the formation of nanoclusters/NCs should come primarily from the reduction of GeO_x by the Si atoms originally present in the silicon oxide matrix and those diffused from the substrate.

The absence of NCs in region (i) was attributed to the relatively low melting point of Ge, which preferentially aids Ge evaporation near the surface into the vapor phase during the annealing process. The reduction of GeO_x should result in evenly populated NCs in regions (ii) and (iii), with the number of NCs increasing as one approaches the Si/SiO_2 interface. However, as Ge and Si are completely miscible, there is also a tendency for Ge, once reduced, to diffuse toward the Si substrate [27]. Figure 5.2b suggests that Ge atoms residing within approximately 105 nm from the Si/SiO_2 interface would diffuse into the Si substrate, which results in region (iii) being void of NCs.

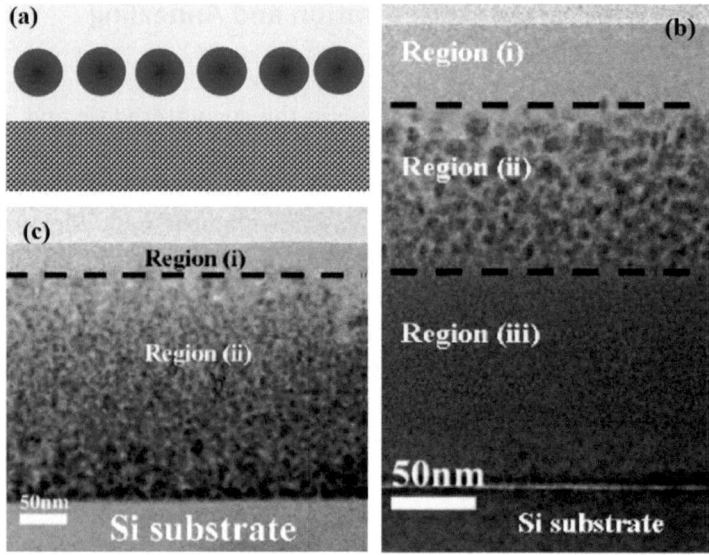

Figure 5.2 (a) Schematic illustration of Ge NCs embedded in a SiO$_2$ matrix. Das et al. [33] have reported Ge NCs approximately 30 nm from the Si/SiO$_2$ interface. (b and c) Cross-sectional TEM micrographs of cosputtered Ge-plus-SiO$_2$ samples that were furnace-annealed at 800°C in Ar ambient for 60 min for sample A and sample B, respectively [16].

For sample B, annealed at 800°C for 60 min (see Fig. 5.2b,c), two regions exist in the oxide, with region (i) void of NCs and region (ii) densely populated with nanoclusters. As the Ge concentration of sample B is significantly higher than that of sample A (75% compared with 20%), the contribution of the Si reduction reaction is less important in the formation of nanoclusters in sample B. The abundance of Ge atoms means that nucleation can bypass the reduction process of Ge oxides by Si to supply the elemental Ge atoms, and it can occur at the same time throughout the oxide and results in the rather uniform distribution of Ge nanoclusters in region (ii). It is also more favorable for the Ge to form numerous and small NCs initially as they are easier to nucleate because of the high Ge concentration. The high Ge concentration leads to an increase in the Ge supersaturation and hence increases the driving force for nucleation and decreases the critical size for nuclei formation. The nucleation rate would also increase. Once formed, the Ge NCs attain a dynamically stable configuration and it requires a driving force to lead to the dissolution of these NCs, which is not sufficiently provided by the relatively low annealing temperature of 800°C. When the nucleation rate is faster than the rate of diffusion of Ge into the Si substrate the denuded region mentioned previously, in the sample with a low Ge concentration, would not form. This accounts for the absence of the denuded zone in region (ii) and also the presence of the densely populated nanoclusters.

Raman spectroscopy was also used to examine the effect of Ge concentration on NC formation. The samples with different Ge concentrations underwent RTA at 500–1000°C for 300 s. For samples that underwent RTA and were sputtered with two pieces of Ge attached to the SiO_2 target, no NC formation could be detected by Raman spectroscopy. For films sputtered with four pieces of Ge, only the sample annealed at 1000°C showed the Ge peak in the Raman spectrum. Figure 5.3 shows typical Raman spectra for rapidly thermally annealed samples sputtered with six pieces of Ge.

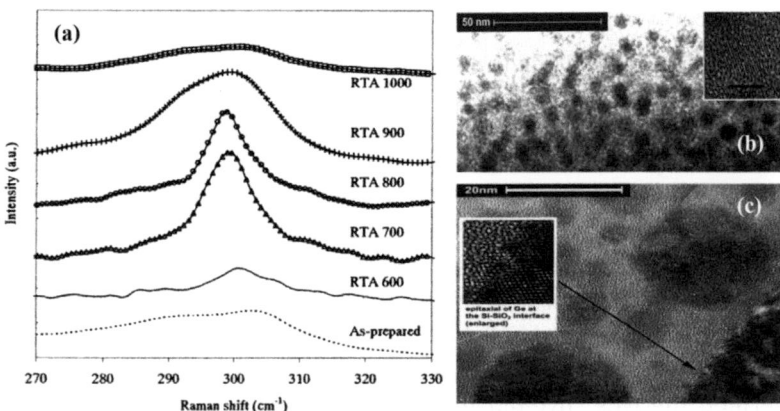

Figure 5.3 (a) Raman spectra of as-sputtered samples rapidly thermally annealed at 600°C, 700°C, 800°C, 900°C, and 1000°C. Reprinted from Ref. [56], Copyright (2017), with permission from Elsevier. (b) TEM image of a sample rapidly thermally annealed at 800°C for 300 s. The inset shows a magnified picture of a Ge NC. (c) TEM image of a sample rapidly thermally annealed at 1000°C for 300 s. Note the multiple twin structure Ge NCs near the Si/SiO_2 interface.

The concentrations of Ge, GeO_2, and GeO_x of the as-sputtered sample were determined by XPS to be 10%, 30%, and 60%. The as-sputtered sample shows a broad Raman band at 303.4 cm^{-1}, very similar to that of amorphous Ge. The appearance of a sharp peak at 299 cm^{-1} for a film annealed at 700°C indicates the initiation of a crystallization process. The sample annealed at 800°C for 300 s shows a sharp Raman peak with asymmetrical broadening over the low-frequency portion of the spectrum. We attributed the broadening of the Raman peak to phonon confinement [42, 56, 67, 101, 115]. It is interesting to note that for the sample annealed at 1000°C, the Raman spectrum is very similar to that of the as-sputtered film.

Figure 5.4a shows a TEM picture of a sample rapidly thermally annealed at 800°C for 300 s. The dark patches are Ge NCs of ~6 nm in diameter in the SiO$_2$ matrix. The NCs are almost spherical and are well dispersed in the matrix. The inset shows the formation of good NCs, as indicated by clear lattice fringes. The crystal lattice fringe separation of 0.33 nm corresponds to the {111} planes of Ge with a diamond structure. This observation is similar to that reported by Fujii et al. [44] and Maeda [91]. Figure 5.4b shows a TEM picture of a sample after RTA at 900°C for 300 s. It shows Ge NCs with diameters (d) of 20 nm ≈ 28 nm near the Si/SiO$_2$ interface, and d reduces further from the interface. The size variation of the NCs is also greater as compared to the one annealed at 800°C, indicating that coarsening has taken place or is in the process of occurring. This is due to the fact that at 900°C and 1000°C, Ge atoms are able to overcome kinetic limitations and enhance the nucleation and growth of the NCs. The large NC contains pockets of amorphous regions and disordered boundaries between single crystals. The defects could be due to the rapid growth rate and high Ge concentration at the Si/SiO$_2$ interface. Similar twin structures have been reported by Oku et al. [109], who termed the NCs "fivefold multiply twinned Ge nanoparticles." We have also observed that at the Si/SiO$_2$ interface, there are regions of crystalline material with different crystal orientation from that of the Si substrate and with a lattice constant of 0.552 nm. This is different from the bulk Si lattice constant of 0.5431 nm, and this may be due to Ge diffusion into the Si substrate. The epitaxial regions near the Si/SiO$_2$ interface have lattice fringes of 0.201 nm separation, which corresponds to the lattice fringe separation of the Ge {2 2 0} plane.

Heinig et al. [54] and Fukuda et al. [45] have also reported such observation of Ge diffusion into the Si/SiO$_2$ interface for samples annealed at elevated temperatures. Fukuda et al. pointed out that the diffusion of Ge in SiO_2 and nucleation of Ge depend on the annealing temperature. They suggested that when the sample is annealed at 800°C, which is below the melting temperature of Ge, only Ge nucleation occurs. When the sample is annealed at 1000°C, Ge diffuses into the Si/SiO_2 interface and precipitates at the interface. Heinig et al. [54] argued that since the concentration of Ge dissolved in SiO_2 is lower than the solubility at the Si/SiO_2 interface and higher at the bulk of the oxide, the concentration gradient can lead to diffusion flux and result in an accumulation of Ge at the interface. The Ge at the interface can either grow epitaxially on top of the Si substrate or diffuse into the substrate if the annealing temperature is sufficiently high.

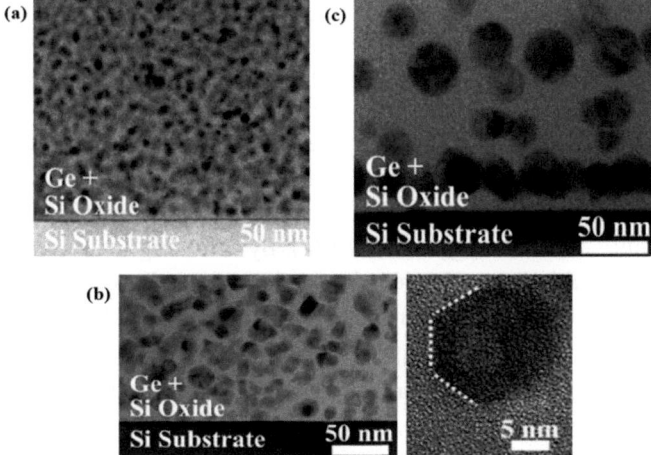

Figure 5.4 Ge-plus-SiO_2 samples furnace-annealed at (a) 800°C, (b) 900°C, and (c) 1000°C for 15 min. The inset is the HRTEM micrograph of a Ge NC from the corresponding sample.

The cross-sectional TEM micrographs of samples furnace-annealed at 800°C, 900°C, and 1000°C for 15 min are shown in Figs. 5.4a, 5.4b, and 5.4c, respectively. In comparison to the samples annealed by RTA, the furnace-annealed samples generally exhibit larger NCs at the same annealing temperatures. The size variation of the NCs is attributed to the much longer annealing duration of the furnace annealing, which assists the diffusion of Ge atoms and therefore the growth of the NCs. Note from the inset of Fig. 5.4b that the NCs synthesized at 900°C were well formed, showing facets that are bound by crystal planes. This implies that it is possible to attain the equilibrium-interface-energy-minimizing configuration in these conditions. Moreover, one can observe the line-up of the Ge NCs near the Si/SiO$_2$ interface in Fig. 5.4c. This is probably due to the huge increase of Ge diffusivity at 1000°C, which allows the diffusion of Ge atoms toward the Si substrate.

5.2.3 Effect of Annealing Ambient

It has been demonstrated that [112] the presence of H$_2$ in the annealing ambient (i.e., in forming gas) could act as a reducing agent for GeO$_2$ in a reduction reaction given by

$$GeO_2 + 2H_2 \rightarrow Ge + 2H_2O. \tag{5.4}$$

Annealing samples in such ambient can cause the incorporation of hydroxyl groups (–OH) into the oxide matrix [6]. The –OH acts as a network modifier in the system as its presence opens up the oxide structure, consequently enhancing the diffusivity of Ge. H$_2$ is also important in assisting the nucleation of the Ge NCs due to its high value of diffusivity in silica [82]. By diffusing through the SiO$_2$ matrix rapidly, H$_2$ can hasten the nucleation and growth processes by reducing Ge oxide to increase the supply of Ge in the matrix.

Chew et al. [17] examined the effect of the ambient on the formation of Ge NCs in a SiO$_2$ matrix on cosputtered samples (with concentrations of Ge, GeO$_2$, and GeO$_x$ at 7%, 46%, and 47%, respectively) in N$_2$ and forming gas ambient. The Raman results showed that annealing in N$_2$ ambient up to 1000°C resulted in no Ge NC formation. However, when the sample was annealed in forming gas at 900°C, Ge NCs could be obtained.

Figure 5.5 shows the cross-sectional TEM images of a sample annealed in forming gas at 800°C, 900°C, and 1000°C for 15 min at 800°C (i.e., Fig. 5.5a), numerous small Ge NCs can be seen to be distributed throughout the entire bulk of the film that give rise to a weak Raman peak. At 900°C (see Fig. 5.5b), the NCs become larger in the bulk of the film. There also exists a region void of NCs between the substrate and the band of NCs in the bulk of the film. From the various steps in the formation of Ge NCs in a SiO$_2$ matrix outlined in Section 5.2.1, the Ge oxides and suboxides could be reduced to elemental Ge by Si at a temperature above 800°C and the main source of Si for the reduction of GeO_2 was Si atoms diffused from the Si substrate. The voided region at the surface of the film annealed at 800°C (i.e., Fig. 5.5a) can be explained by the out-diffusion of Ge due to the low solubility of Ge in SiO_2 [16] or by the reoxidation of Ge by the small concentration of oxidants present in the annealing gas to form GeO_2 [54]. The voided region between the substrate and the band of NCs observed for samples annealed at 900°C and 1000°C can be attributed to the diffusion of Ge into the Si substrate due to the complete miscibility between Ge and Si. The significant increase in the diffusivity of Ge at a temperature of 900°C or higher is most likely due to the fact that such temperatures are very close to the melting point of bulk Ge and enable the Ge atoms to overcome kinetic limitations and diffuse into the Si substrate [27].

Figure 5.5 Cross-sectional TEM micrograph of a cosputtered Ge-plus-SiO_2 sample annealed at (a) 800°C, (b) 900°C, and (c) 1000°C in forming gas (10% H_2 + 90% N_2) for 15 min.

5.2.4 Effect of an Oxide Barrier Layer

In the previous sections, we have shown that due to the complete miscibility between Ge atoms and a Si substrate and the total insolubility of Ge in a SiO_2 matrix, very different distributions of Ge

NCs can be found annealed at different temperatures and ambient. In this section, we investigate the effectiveness of an oxide layer between the cosputtered Ge-plus-SiO$_2$ layer and the Si substrate in a Ge NC formation. This is very important to the fabrication of the Ge-NC-based trilayer memory devices.

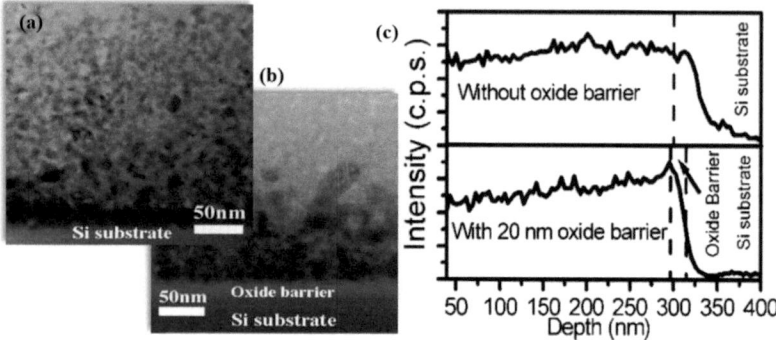

Figure 5.6 Cross-sectional TEM micrographs of a cosputtered Ge-plus-SiO$_2$ sample that was furnace-annealed at 800°C in N$_2$ ambient for 60 min (a) without a barrier and (b) with a thermal oxide barrier (20 nm) inserted between the Si substrate and the cosputtered layer. (c) SIMS depth profiles of Ge of a cosputtered Ge-plus-SiO$_2$ sample that was furnace-annealed at 800°C without and with a thermal oxide barrier (20 nm) inserted between the Si substrate and the cosputtered layer.

Figure 5.6a shows a TEM micrograph of a cosputtered sample with concentrations of Ge, GeO$_2$, and GeO$_x$ at 75%, 5%, and 20%, respectively, annealed at 800°C for 60 min in N$_2$. Figure 5.6b shows the TEM micrograph of a sample prepared in exactly the same conditions as the sample shown in Fig. 5.6a but with a thermal oxide layer (20 nm) inserted between the Si substrate and the cosputtered layer. For the sample shown in Fig. 5.6a, the nanoclusters at the Si/SiO$_2$ interface become larger due to coarsening and a significant Ge diffusion into the Si (indicated by a stratified layer at the Si surface). This is in agreement with the secondary ion mass spectroscopy (SIMS) results shown in Fig. 5.6c, which indicates a significant amount of Ge diffusion into the Si substrate for the sample without the oxide barrier. For the sample shown in Fig. 5.6b, there exists a band of extremely dense nanoclusters near the Si/SiO$_2$

interface. This is possibly due to the confinement of Ge within the cosputtered layer by the oxide barrier, which results in a higher Ge supersaturation in contrast to the sample without the barrier [99, 107]. The effectiveness of the oxide barrier in terms of Ge diffusion is clearly shown in Fig. 5.6c for the sample with the barrier. It shows a decrease in Ge concentration to almost zero from the oxide barrier into the Si surface in contrast to the sample without the barrier, which has a diffusion tail into the Si substrate. Finally, out-diffusion of Ge is again observed with a decreasing Ge content toward the surface of the film.

5.2.5   Influence of Dielectric Matrices

Apart from silicon oxide, Ge NCs were synthesized in other oxide matrices, such as hafnium oxide (HfO_2) [35], aluminum oxide (Al_2O_3) [36], and hafnium aluminum oxide (HfAlO) film [19, 136]. Das et al. [35] prepared a dielectric stack consisting of Ge NCs sandwiched between tunneling and capping layers of HfO_2. The tunneling layer (~4 nm) was deposited on Si by rf sputtering from a 3-inch HfO_2 (99.999% pure) target at 50 W rf power in Ar + O_2 ambient. The intermediate layer was then deposited by cosputtering of HfO_2 and Ge for 30 min in the same condition followed by the deposition of a final cap layer for 40 min. The trilayer structure was then subjected to thermal annealing in nitrogen ambient at 800°C and 900°C for 30 min. TEM (Figs. 5.7a and 5.7b) shows a cross-sectional micrograph of sample annealed at 800°C and 900°C, respectively, for 30 min in a N_2 atmosphere. The dark patches are Ge NCs 4–6 nm in diameter in the HfO_2 matrix [35].

The NCs are almost spherical and are well dispersed in the HfO_2 matrix. Note that there is some clustering of neighboring NCs in the annealed sample. Ge NCs embedded in an Al_2O_3 matrix after the sample had been annealed at 850°C and 950°C have also been reported [36]. Again, the NCs are well dispersed in the host matrix. The inset shows the lattice fringes with an interplanar separation of 0.33 nm, which corresponds to the {111} plane of a Ge NC with a diamond structure.

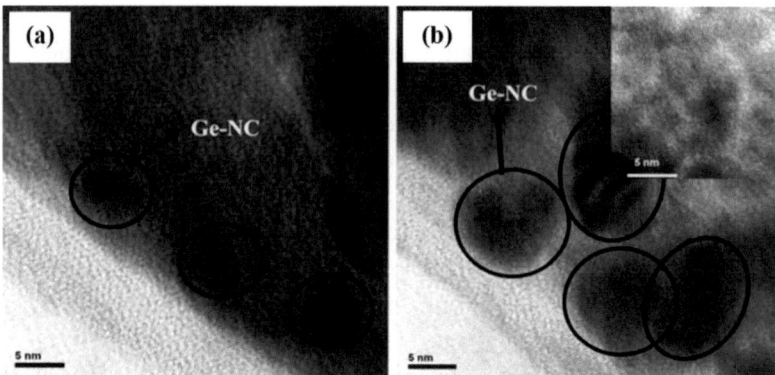

Figure 5.7 Cross-sectional TEM micrograph of Ge NCs embedded in HfO$_2$ annealed at (a) 800°C and (b) 900°C. From Ref. [35]. With permission from Springer.

Chew et al. [19] synthesized Ge NCs in a HfAlO matrix by cosputtering a Ge-plus-HfAlO target in Ar at an ambient temperature. The target was a 3-inch HfAlO (99.95% pure, HfO$_2$:Al$_2$O$_3$ = 1:1) disc with Ge pieces attached. The Ge content in the cosputtered samples was estimated to be 10.5 atomic percent. The Ar pressure and the rf power for the sputtering were fixed at 3×10^{-3} mbar and 100 W for all the samples. The thickness of the cosputtered films was approximately 360 nm. The formation of Ge NCs in the matrix was achieved by furnace-annealing the cosputtered films in N$_2$ from 700°C to 1000°C for min. The Raman spectra of the annealed samples showed a peak at ~299.6 cm^{-1} for the sample annealed at 800°C, indicating the presence of Ge NCs in the HfAlO matrix. Wang et al. [136] also suggested that the phase structure of Hf + GeO$_2$ changes to Ge + HfO$_2$ at a temperature of about 700°C. However, Chew et al. [19] noted that at higher annealing temperatures (i.e., 900°C and 1000°C), no Ge peak can be detected. The cross-sectional TEM image of the sample annealed at 800°C (Fig. 5.8a) shows the presence of Ge NCs ~4 nm in diameter (Fig. 5.8c) uniformly distributed throughout the matrix.

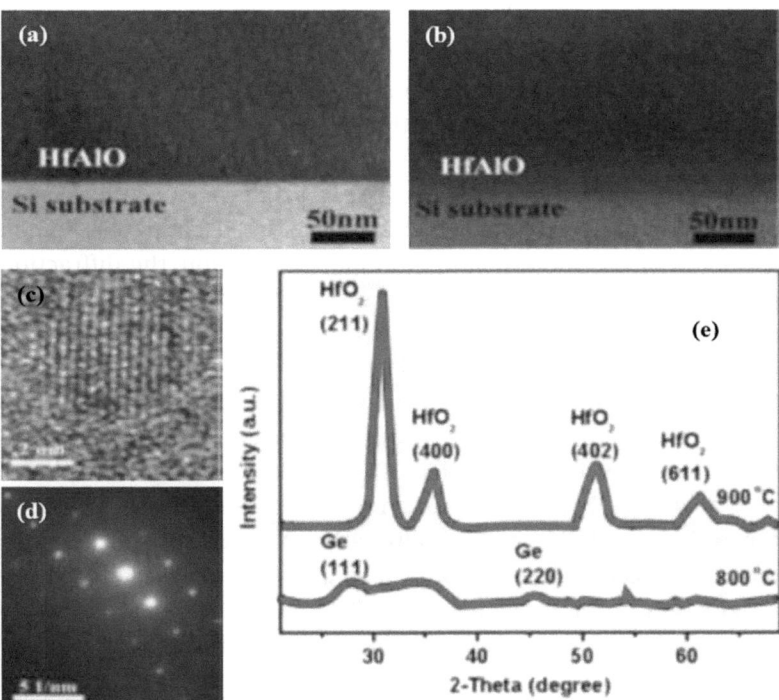

Figure 5.8 Cross-sectional TEM micrograph of Ge-plus-HfAlO samples annealed at (a) 800°C and (b) 900°C in nitrogen for 15 min. (c) The HRTEM of an NC formed at 800°C and (d) the diffraction pattern of the sample annealed at 900°C, showing crystallization of the matrix. (e) XRD patterns of Ge-plus-HfAlO samples annealed at 800°C and 900°C in nitrogen for 15 min.

The size of the NCs is smaller and the density lower in the HfAlO matrix than what is usually found in the Si oxide matrix with a similar Ge concentration. The higher nucleation and growth rate observed in the Si oxide matrix as compared to the HfAlO matrix can be explained as follows. At 800°C, the HfAlO matrix will be closer to its crystallization temperature of 900°C and there will be a reduction in defects in the HfAlO matrix as the matrix atoms start to develop order. When the defect density in the matrix is reduced, nucleation and growth will become more difficult for the Ge atoms, resulting in a smaller size and lower density of NCs in the HfAlO film. On the other hand, it has been suggested that thermal processing of the Ge-plus-HfO system can lead to the formation of hafnium germinate

(HfGeO$_x$) [41]. The formation of this hafnium germinate phase during annealing of our Ge-plus-HfAlO samples could reduce the Ge atoms available for nucleation and growth by binding them to the matrix atoms, which results in the smaller size and lower density of NCs in the HfAlO matrix. The cross-sectional TEM image of the sample annealed at 900°C (Fig. 5.8b) reveals that there is an absence of Ge NCs in the HfAlO film, which agrees with the featureless Raman spectrum of the corresponding sample. In addition, the diffraction pattern of this sample (Fig. 5.8d) shows that this HfAlO film had crystallized. The XRD spectrum of the sample shown in Fig. 5.8e clearly confirms that there is indeed no Ge reflection peak and that the matrix has indeed crystallized with the occurrence of the HfO$_2$ (211), (400), (402), and (611) reflection peaks [61]. Note that the HfAlO matrix when annealed at 800°C remains mainly amorphous, with a weak HfO$_2$ halo, at ~35° and Ge (111) and (220) reflections can be detected [62]. The crystallized HfAlO structure could enhance the out-diffusion of Ge atoms through grain boundaries and lead to a reduction in the Ge concentration in the HfAlO film. This would reduce the Ge supersaturation in the matrix and raise the barriers to nucleation.

5.3 Characterizations of Ge Nanocrystals

5.3.1 Photoluminescence Properties

Germanium, due to its larger effective Bohr radius, is predicted to exhibit strong confinement for NC size around 10–15 nm [60]. This has prompted extensive study of the PL properties of Ge NCs embedded in the SiO$_2$ matrix at the visible and near-infrared range [22, 34, 98, 117, 124, 125, 141]. The light emission mechanism from Ge NCs has been attributed either to radiative recombination via quantum confinement states or to defects at the NC/matrix interface and in the matrix itself. We have carried out a comprehensive investigation of the PL properties of Ge NCs in a silicon oxide matrix. Figure 5.9 shows the room temperature PL spectra of samples that underwent RTA at different temperatures. Note that the as-deposited Ge-plus-SiO$_2$ sample exhibits a weak PL peak P1 at ~2.2 eV and a rather broad peak P2 at 1.8 eV. The luminescence intensity of

P1 increases as the annealing temperature increases and shows a maximum at 750°C whereas that of P2 shows a maximum at 800°C. Further increase in the annealing temperature results in a decrease in the PL intensity. The PL peak at 3.1 eV (P3) shows a reduction in intensity as the annealing temperature increases from 800°C to 1000°C.

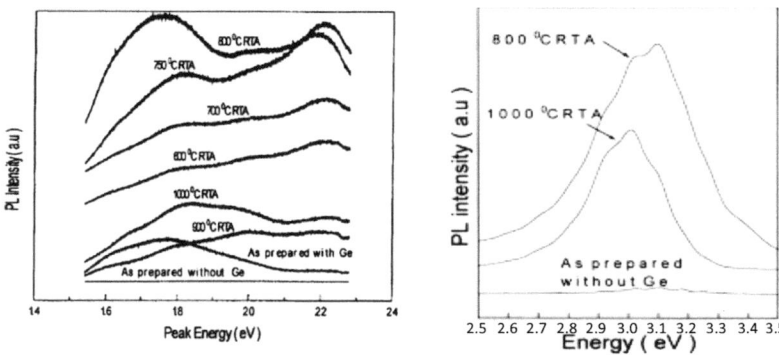

Figure 5.9 Room temperature PL spectra of samples after RTA at different temperatures (a.u., arbitrary units). The curves have been displaced vertically intentionally for clarity.

As the annealing temperature increases from 600°C to 800°C, the P1 peak decreases from 2.21 to 2.18 eV. Maeda [91], Paine et al. [111], Okamoto and Kanemitsu [108], Wu et al. [139], and Das et al. [37] have also reported a similar decrease in the peak energy as the crystal size increases from 2 to 16 nm. Figure 5.10 plots the theoretical value of the optical bandgap of the Ge NC versus the NC diameter estimated from the effective mass approximation model and experimental results from several research groups. Note that while Takeoka et al. [125] reported a reduction of the optical bandgap of the Ge NC as the NC diameter increases, other researchers [22, 37] reported minimal change in the optical bandgap of the NC as the diameter increases. Maeda [91], Paine et al. [112], and Wu et al. [139] have also reported a similar decrease in the optical bandgap (not shown in Fig. 5.10) from approximately 2.35 to 2.2 eV as the NC size increased from 2 to 16 nm. This is in contrast to the prediction of the quantum confinement model reported by Takagahara and Takeda [124] and has been attributed to a defect-related mechanism. The existence of the P2 peak has been reported less in the literature than the existence of the P1 peak.

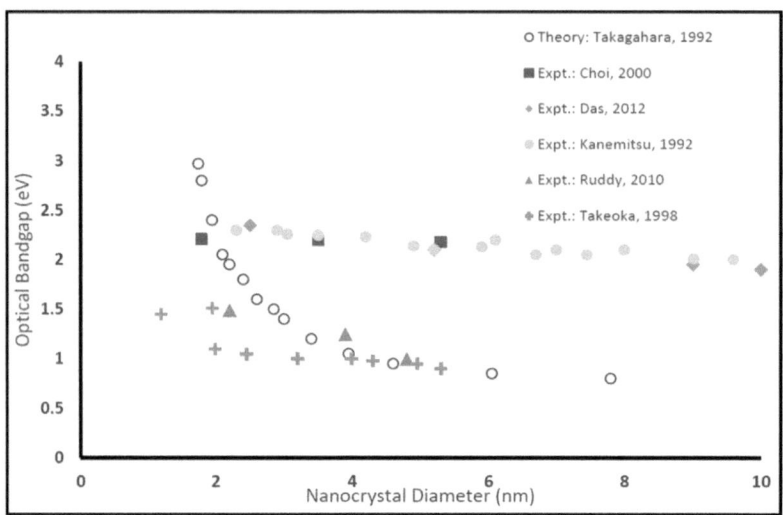

Figure 5.10 Theoretical and experimental results on the variation of optical bandgap as a function of diameter for Ge NCs.

Wang et al. [135] showed the existence of the P1 and P2 peaks in their cosputtered samples with a 550°C and 580°C furnace anneal. They attributed P1 to the local environments of Ge NCs in annealed samples. No explanation, however, was offered for the P2 peak. Min et al. [98] have shown clearly the PL peak at around 1.8 eV for Ge NC samples made by precipitation from a supersaturated solid solution of Ge in Ge-implanted SiO_2. They suggested that the 1.8 eV peak was not due to the radiative recombination of quantum-confined excitons in Ge NCs but due to radiative defect centers in the SO_2 matrix. We have subjected the annealed samples to a further anneal at 400°C for 15 min in forming gas (10% H_2 + N_2). The PL intensity of the forming-gas annealed sample was much reduced compared with the sample without the forming-gas anneal. This seems to suggest that the PL peaks were associated with defects in SiO_2. This means that as long as we are working on Ge NCs embedded in a silicon oxide matrix, there is no way to avoid such oxide-related defects. This places a serious challenge to the realization of high-performance luminescence devices from Ge NCs.

5.3.2 Electroluminescence Properties

The effect of the Ge NC size on the electroluminescence (EL) spectra of a metal-insulator-semiconductor (MIS) structure had been investigated by Zhang et al. [142] and Shen et al. [121]. The EL peak at 590 nm from the MIS structure with Ge NCs embedded in SiO_2 was observed to be size independent (see Fig. 5.11a) and was attributed due to defect-mediated emission. EL characteristics of Ge NCs embedded in Al_2O_3 have also been examined [13, 143]. Three EL peaks were observed, located at 586–629 cm^{-1}, 691–736 cm^{-1}, and 824–878 cm^{-1}, and they were attributed to defect-related recombination at the NC/Al_2O_3 interfaces, the electron-hole recombination in Ge NCs in Al_2O_3, and oxygen-related defects in GeO_2. Kan et al. [71] reported EL spectra of Ge NCs in the near-infrared region. Figure 5.11b shows the room temperature EL spectrum of an MIS structure, Al-capped SiO_2 (20 nm)-rapid thermal oxide (RTO, 5 nm)-p-type Si, with 10 nm diameter Ge NCs embedded within the silicon oxide. Under accumulation, for a constant current density bias of 8.3–13.8 A/cm^2, two distinct EL peaks at 1200 nm and 1350 nm are observed. The peak at 1200 nm has been determined as due to Si interband recombination. The spectra have been normalized with reference to this 1200 nm peak at 13.8 A/cm^2. The wavelength or energy positions of both peaks remain constant for the different current densities applied (i.e., different applied voltages). The inset shows the EL spectrum of a similar structure, Al-capped SiO_2 (20 nm)-Ge NCs (5 nm)-RTO (5 nm)-p-type Si, with Ge NCs of 5 nm diameter. A similar peak at 1350 nm was observed. The peak at 1350 nm was related to the presence of Ge NCs.

Figure 5.12 shows a series of EL spectra obtained from MIS structures under different oxidations and annealing conditions. For comparison, the spectra have again been normalized with reference to the 1200 nm peak. The current density across the structures was set at 13.8 A/cm^2. Figure 5.12a is the EL spectrum of the control sample (pure SiO_2).

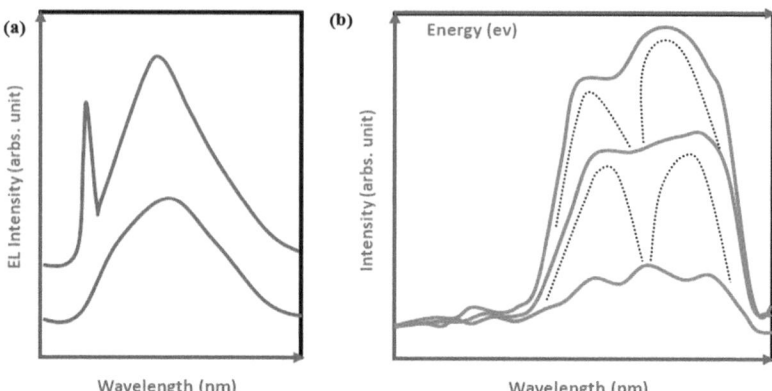

Figure 5.11 (a) Schematic illustration of typical electroluminescent spectra of the MIS under high applied voltages obtained by Zhang et al. [142]. (b) Schematic illustration of room-temperature EL spectrum (solid line) of an MIS structure with 10 nm Ge NCs embedded in between the 20-nm-thick capped SiO$_2$ and 5-nm-thick RTO layers formed by annealing in pure N$_2$. The dashed lines show the deconvolution of the measured EL spectrum into two peaks at different energies [71].

Figure 5.12b shows that the EL spectrum of a 5-nm-thick film consists entirely of a mixed oxide (Si$_x$Ge$_y$O$_z$) and the 1350 nm EL peak is not observed. Figure 5.12c corresponds to the EL spectrum of a 5-nm-thick Si$_{0.54}$Ge$_{0.46}$ film that has been partially oxidized for 4 min and was not subjected to annealing, so Ge NCs were not formed and only elemental Ge was present. A broad and weak peak at 1350 nm is observed with a full width at half maximum (FWHM) of 160 nm. Therefore, the results from Figs. 5.12b and 5.12c suggest that the existence of the 1350 nm EL peak is due to the presence of elemental Ge, and not a result of defects, in the oxide matrix. This is supported by the unchanging peak energy position when the applied bias increases. At higher bias values, the excess energy that the injected carriers gain from the oxide electric field will cause a blue shift in the peak energy position if recombination was to happen at luminescent defect centers in the oxide. The structure with the EL spectrum in Fig. 5.12d was annealed in pure N$_2$, which results in the formation of Ge NCs. The peak intensity at 1350 nm is enhanced by a factor of 1.5 and the FWHM is reduced by a factor of 1.6 when compared to the spectrum of the unannealed structure in Fig. 5.12c. Consequently, the 1350 nm peak (due to the presence

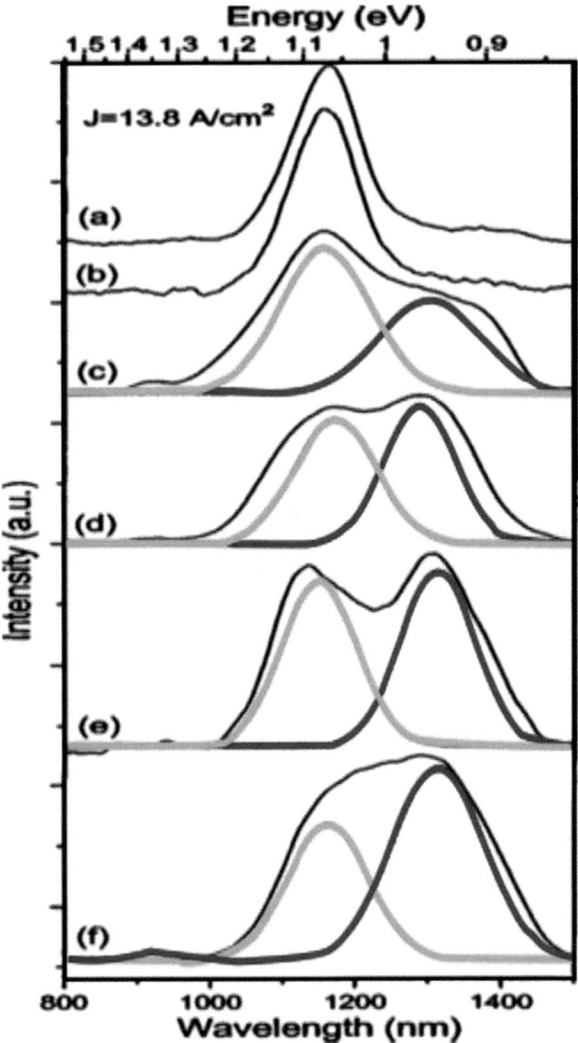

Figure 5.12 A series of electroluminescence spectra, acquired under an accumulation mode constant current density bias of 13.8 A/cm^2. (a) Control sample (pure SiO$_2$) and (b) a fully oxidized 5-nm-thick Si$_{0.54}$Ge$_{0.46}$ film with subsequent annealing in pure N$_2$. (c) A partially oxidized 5-nm-thick Si$_{0.54}$Ge$_{0.46}$ film without subsequent annealing. (d) A partially oxidized 5-nm-thick Si$_{0.54}$Ge$_{0.46}$ film with subsequent annealing in pure N$_2$. (e) A partially oxidized 5-nm-thick Si0.54Ge0.46 film with subsequent annealing in forming gas and (f) the same as sample (d) but acquired under an inversion mode bias.

of elemental Ge) is enhanced in the presence of better crystallinity Ge NCs that provide improved electronic band structures for greater radiative recombination. Figure 5.12e shows the EL spectrum of a structure similar to that of Fig. 5.12d but which has been annealed in forming gas instead of pure N_2. The peaks in Fig. 5.12e are better defined (resolved) as compared to that in Fig. 5.12d. The presence of H_2 during annealing in forming gas effectively reduces GeO_2 to elemental Ge and passivates dangling bonds at the interface of the NCs. The elimination of GeO_2 should not be the reason for the better-resolved 1350 nm peak as it has been shown in Fig. 5.12b that GeO_2 does not contribute to this peak. The only plausible reason for the better-resolved 1350 nm peak is the passivation of dangling bonds by H_2, thus removing any traps around the NC that could generate additional energy levels in the bandgap of Ge NCs. These trap levels are more likely to be distributed rather than discrete energy levels as they cause a spread in the 1350 nm peak in Fig. 5.12d, rather than a peak shift as suggested by Niquet et al. [106]. To further establish radiative recombination of confined carriers in Ge NC, Fig. 5.12f shows the EL spectrum under an inversion mode bias for the same device as that of Fig. 5.12d. Since the current density is the same for both accumulation and inversion biases, the carriers injected through the oxide and into the NC should be equivalent. Hence, the peak intensity at 1350 nm remains significant for both biases. However, the peak at 1160 nm originates from the Si interband recombination and the peak intensity is related to the amount of carriers at the Si surface. A lower intensity peak is, therefore, observed during inversion as the concentration of inversion carriers is smaller than that of accumulation carriers under the same constant current density bias.

5.3.3 Stress in Ge Nanocrystals Embedded in Dielectrics

Liu et al. [88] have recently shown that for NCs embedded in a dielectric, the distribution of the stress and strain field plays an important role in deciding the physical and thermodynamic properties of the NCs. Hence, Liu et al. [88] have suggested a way of fabricating the potential high-performance photodetector by varying the Ge bandgap via tensile strain. However, there has been a lack of a systematic study on the growth of Ge NCs in a silicon oxide matrix in relation to the stress state of these NCs. There were a few reports that indicated the Ge NCs were under compressive stresses on the order of a few gigapascals [119, 138] in the oxide matrix when synthesized under certain conditions. The stress arises mainly from the inability of the matrix to accommodate the growing NCs and/or the volumetric expansion of Ge of about 6% when it solidifies from the liquid state [7, 138].

We have carried out a systematic study on the stress experienced by Ge NCs embedded in SiO_2 and HfAlO matrices. The samples were prepared by cosputtering a Ge-plus-SiO_2 [30] or Ge-plus-HfAlO [114, 145] target in Ar at room temperature. The target was a 3-inch SiO_2 disk or HfAlO disk with pieces of Ge (99.999% pure, 5 mm × 10 mm × 0.3 mm) attached.

The sputtering pressure and power were fixed at 3×10^{-3} mbar and 100 W, respectively. The thickness of the sputtered film was approximately 300 nm. The silicon nitride (SiN) capping stressor with a thickness of ~200 nm was grown by the PECVD technique at 280°C with SiH_4 and NH_3 as source gases. The synthesis of Ge NCs was carried out either by furnace-annealing in N_2 ambient from 800°C to 1000°C for 15 min or by RTA in N_2 ambient from 800°C to 1000°C for 60 s. We followed the procedure outlined by Sharp et al. [119] and dipped the samples in a 1:1 $HF:H_2O$ solution to selectively etch away the oxide to obtain freestanding NCs (see Fig. 5.13). After the selective etching, the piling up of the freestanding NCs on the Si substrate was observed by a cross-section TEM image as shown in Fig. 5.13d. These freestanding NCs were stable as the surface oxidation of the NCs was self-limiting (see Fig. 5.13e). Another round of Raman experiments was performed on these samples to obtain the Raman peak position of the freestanding NCs.

Cosputtered Ge-plus-SiO_2 samples were prepared and underwent RTA at 800°C, 900°C, and 1000°C for 60 s for the investigation of the stress experienced by NCs. For the sample annealed at 800°C, numerous small Ge NCs were uniformly distributed in the entire bulk of the film. When the annealing temperature was increased to 900°C and 1000°C, the NCs grew in size and generally adopted a spherical shape. The size variation of the NCs was also greater as compared to the NCs in the sample annealed at 800°C, indicating that coarsening has taken place. The NCs prepared by annealing at 1000°C were found to contain a twinning structure. For samples annealed with conventional furnace at 800°C, 900°C, and 1000°C for 15 min, the furnace-annealed samples generally exhibit larger NCs at the same annealing temperatures.

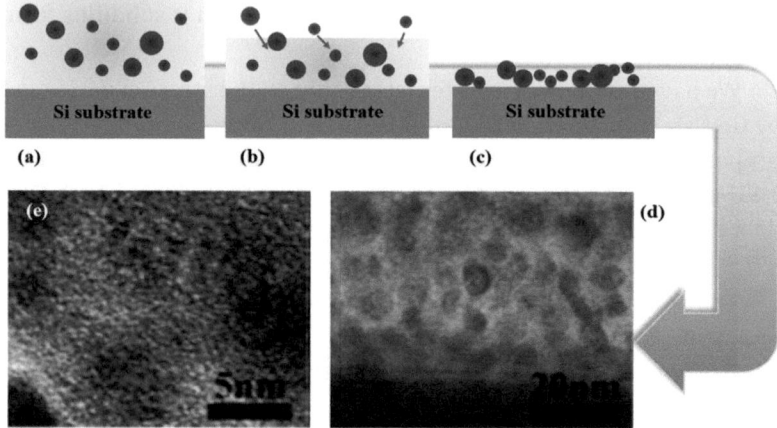

Figure 5.13 (a–c) The schematic of the chemical etching process to obtain freestanding Ge NCs, (d) the cross-sectional TEM micrograph of freestanding Ge NCs, and (e) the high-resolution TEM micrograph of the NCs.

Figure 5.14 summarizes the compressive stress experienced by the Ge NCs. For samples that have undergone RTA, the compressive stress increases gradually from 0.3 GPa to 1.2 GPa as the annealing temperature increases from 800°C to 1000°C. For the furnace-annealed samples, the compressive stress decreases from 0.4 GPa to 0.16 GPa when the annealing temperature increases from 800°C to 900°C, followed by a sharp increase to 1.3 GPa at 1000°C. It is reasonable to expect that for the rapidly thermally annealed samples, the low compressive stress experienced by the NCs when annealed at 800°C is because of the small size of the NCs. The size of the NCs increases when the annealing temperature reaches 900°C and 1000°C; therefore, it becomes more difficult for the silicon oxide matrix to accommodate those NCs, resulting in a build-up and increase in stress. On the other hand, the furnace annealing at 800°C allows the growth of the NCs because of its longer annealing time. This accounts for a relatively higher value obtained at 800°C. For NCs annealed at 900°C, the formation of facets means that it is energetically favorable for the NCs to grow along planes that exert the least pressure on the SiO_2 matrix as it enables them to minimize their strain energy and thus to minimize stress for the NCs. At 1000°C, the Ge atoms would become molten and would lose their atomic ordering. This results in a significant increase in the diffusivity of the Ge atoms, and the NCs will form very rapidly, giving rise to a large compressive stress exerted on the NCs. This large compressive stress will then cause the NCs to adopt a spherical shape to minimize the surface-to-volume ratio of the NCs and thus to minimize the strain energy of the NCs. Under such a large compressive stress, NCs are observed to be defective with multiple twinning.

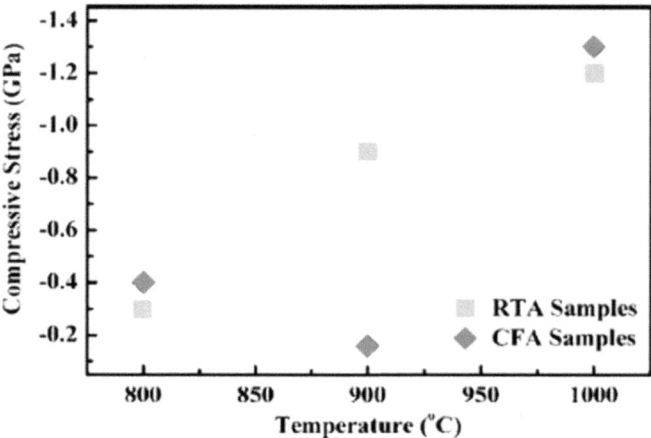

Figure 5.14 Comparison of the stress experienced between Ge NCs in samples that underwent RTA and Ge NCs in furnace-annealed samples.

We have also evaluated the stress or strain experienced by the Ge NCs in different dielectric matrices (i.e., Si oxide and HfAlO). Note that at 900°C, the HfAlO matrix is likely to recrystallize and aids the out-diffusion of Ge atom, so there is no formation of Ge NCs [84]. Therefore, only the stress development on the samples that have undergone RTA at 800°C will be investigated. Figures 5.15a and 5.15b show the cross-sectional TEM micrographs of samples employing Si oxide and HfAlO as the host materials with RTA at 800°C. The Ge NCs in the HfAlO matrix were much less than those in Si oxide matrix. The slower nucleation and growth rate of the NCs in the HfAlO matrix implies that the enthalpy of mixing between the HfAlO and Ge phase is more negative than the Si oxide and Ge phase.

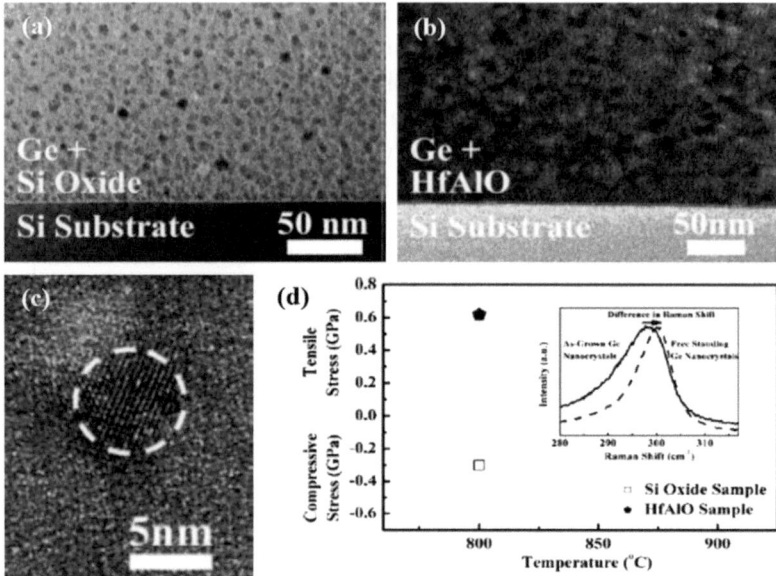

Figure 5.15 Cross-sectional TEM micrographs of (a) Ge-plus-Si oxide and (b) Ge-plus-HfAlO samples that have undergone RTA at 800°C for 60 s. (c) The HRTEM micrograph of a single Ge NC from the Ge-plus-HfAlO sample. (d) Comparison between the stress experienced by Ge NCs in the Si oxide matrix and HfAlO matrix samples that have undergone RTA at 800°C for 60 s; the inset shows the typical Raman spectra of as-grown and etched Ge-plus-HfAlO samples rapidly thermally annealed at 800°C for 60 s.

From thermodynamic calculations and study on the Ge-plus-Si oxide system, it has been concluded that Ge is almost insoluble in Si oxide [11, 111]. However, thermal processing of the Ge-plus-HfO system can lead to the formation of hafnium germinate (HfGeO$_x$) [41]. Therefore, the formation of the hafnium germinate phase during annealing of the Ge-plus-HfAlO samples could reduce the Ge atoms available for nucleation and growth by binding them to the matrix atoms. There is a typical Raman shift from 297 cm^{-1} (for the annealed Ge-plus-HfAlO sample) to 300 cm^{-1} (corresponding sample with freestanding NCs) observed from the inset of Fig. 5.15d. This suggests that the NCs embedded inside the HfAlO matrix experience a tensile strain of about 0.62 GPa. However, for the Si oxide matrix sample, the Ge NCs experience a compressive stress of about 0.3 GPa. The tensile strain experienced by the HfAlO film upon cooling because of the thermal mismatch between the Si substrate and the HfAlO film can be approximated as

$$\sigma_{\text{thermal}} = \left[\left(\frac{Y}{1-\gamma} \right)_{\text{film}} (\alpha_{\text{sub}} - \alpha_{\text{film}})(T - T_{\text{d}}) \right], \tag{5.5}$$

where Y and γ are Young's modules and Poisson's ratio of the film, respectively, which were estimated from the mechanical data of HfO$_2$ and Al$_2$O$_3$ to be 357 GPa and 0.25 [38]; α_{sub} = 3.1 × 10^{-6} K^{-1} and α_{film} = 7.7 × 10^{-6} K^{-1} are the thermal expansion coefficients of the Si substrate and the HfAlO film, respectively, [131]; and T and T_{d} are the process temperature and the environmental temperature, respectively.

The pressure acting on the NCs should be different from the pressure applied from the outside because of the difference in stiffness between the matrix and the NC material. Therefore, the pressure exerted on the NC, P_{NC}, could be estimated by

$$P_{\text{NC}} = \frac{9B(1-\gamma)}{2Y + 3(1+\gamma)B} P_{\text{film}}, \tag{5.6}$$

where B is the bulk modulus of the NC material (i.e., B = 75 GPa for Ge) and P_{film} is the hydrostatic pressure experienced by the matrix [53].

From Eqs. 5.5 and 5.6, we estimated that the Ge NCs in the HfAlO matrix were under a tensile strain of around 0.86 GPa when annealed at 800°C, which is fairly close to the experimental value of 0.62 GPa. Therefore, among the tuning methods of stress on Ge NCs, we are only able to change the stress state of the NCs from compressive to tensile by introducing the dielectric matrix to that of HfAlO. With the other two methods, we can only engineer the amount of compressive stress experienced by the NCs in the silicon oxide matrix.

We also carried out the synthesis of Ge NCs embedded in a silicon oxide film deposited on the surface of the silicon Si wafer etched with V or U grooves [145]. We found that the substrate geometry has a significant influence on the distribution of the Ge NCs in the SiO$_2$ matrix. The variation in the distribution of the NCs in the silicon oxide matrix may be due to defective oxide regions caused by a sputter-deposited silicon oxide film on the V or U grooves or may also be related to the different strain fields in the oxide layer.

5.4 Ge Nanocrystal-Based Floating-Gate Memory Devices

5.4.1 Fabrication of Ge Nanocrystal Memory Structures

The NC-based memory devices were basically following the floating-gate memory device invented by Sze and Kahng [68], as shown in Fig. 5.16a. The top gate acts as the control gate, and the bottom (floating) gate is used for charge storage. Once charged, the storage node can be discharged with an external electric field. Hence by injecting ("write" operation) charges into or removing ("erase" operation) charges from the floating gate, information can be stored digitally. The tunnel oxide acts as a blockage that prevents charges from leaking into the silicon substrate. It has to be very thin in order to achieve faster write and erase times. Because of this requirement, such devices suffer from charge leakage through the tunnel oxide. To address this problem, metal-oxide-silicon (MOS) memory cells with NCs as charge-storage sites were proposed by Tiwari et al. [132] (see Fig. 5.16b).

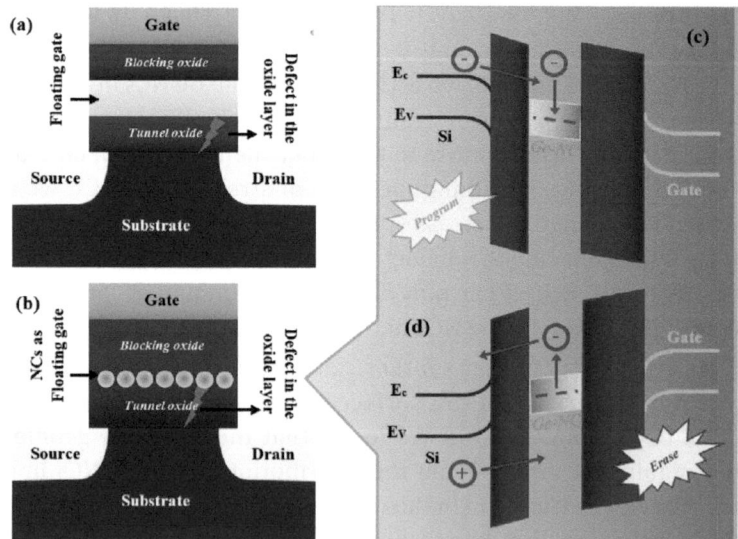

Figure 5.16 Schematic diagrams of (a) a conventional floating-gate memory device and (b) an NC floating-gate MOS memory device. (c, d) Schematic band diagrams of the write and erase processes in an NC memory device.

The advantages of using NCs as a storage node are the improved performance of nonvolatility and charge retention due to the Coulomb blockade effects and the reduction in charge leakage from isolated NCs through the weak spots in the tunnel oxide. The band diagrams illustrating the write and erase operations in an NC memory device are shown in Figs. 5.16c and 5.16d. During the write process, a positive gate voltage is applied to the control gate to inject electrons from the inversion layer into the floating gate containing isolated NCs. With a reverse bias, electrons will tunnel back from the NCs into the channel or the accumulation-layer holes to tunnel into the NCs. Direct tunneling, Fowler–Nordheim tunneling, and channel hot electron injection have been suggested as the tunneling mechanisms.

Memory effect of NC and associated device [40, 74, 110, 122] has been reported since the paper by Tiwari et al. [132] on Si NC-based memory devices. A substantial effort was devoted to the control of the size and distribution of silicon NCs in silicon oxide and the tunnel oxide [2, 52, 81, 122, 133] and the selection of the dielectric layer [60, 93] in order to optimize the device structure to function as nonvolatile memory. To date, NC-based floating-gate memory devices are used in the 4 MB NOR flash memory [97], the 128 kB NOR memory of Freescale Semiconductor [147], and the dual-layer NC floating-gate memory devices of CEA-Leti [47].

We reported memory effects in Ge NCs embedded in a trilayer structure [22]. The memory device consisted of a tunnel silicon oxide layer, a layer of silicon oxide embedded with Ge NCs, and a SiO_2-capping oxide layer. We believe this is the first report of memory effect of Ge NCs embedded in a silicon oxide matrix. The tunnel oxide layer was a thin (~5 nm) SiO_2 layer grown on a p-type silicon substrate in dry oxygen ambient using rapid thermal oxidation at 1000°C. A Ge-plus-SiO_2 layer of 20 nm thickness was then deposited by a rf cosputtering technique. A third layer, of pure SiO_2, (~50 nm) was then deposited by rf sputtering. The trilayer structure was then RTA in Ar ambient at 1000°C for 300 s. The RTA ramp-up and ramp-down rates were fixed at 30°C/s. Figure 5.17a shows the capacitance-versus-voltage (*C–V*) characteristics of a trilayer structure device (device A). The device exhibits counterclockwise hysteresis (~6 V). Figure 5.17a also shows the *C–V* curve of another trilayer structure device (device B), with the middle layer consisting of 20-nm-thick pure sputtered oxide. Counterclockwise hysteresis was also observed but was of a smaller width (0.73 V).

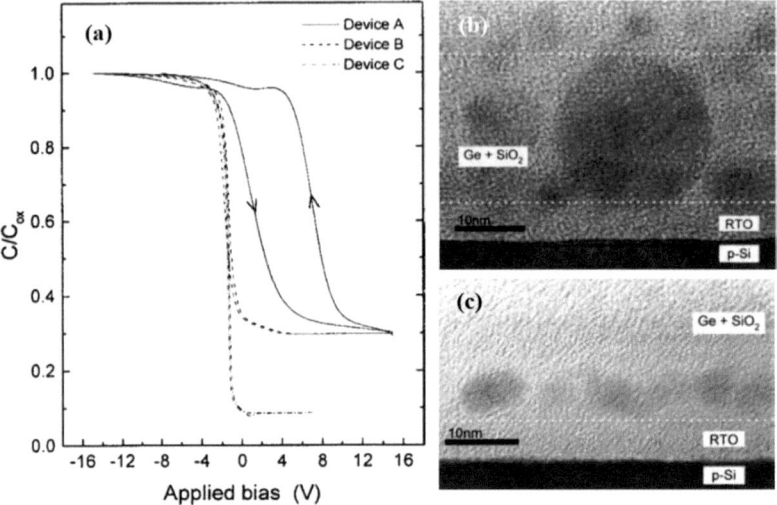

Figure 5.17   (a) Capacitance-versus-voltage characteristics of trilayer-structure devices: device A, RTO SiO_2 (5 nm)-Ge plus SiO_2 (20 nm)-sputtered SiO_2 (50 nm); device B, RTO SiO_2 (5 nm)-sputtered SiO_2 (20 nm)-sputtered SiO_2 (50 nm); and two-layered structure device C, RTO SiO_2 (5 nm)-Ge plus SiO_2 (20 nm). (b) TEM micrograph of device A, which has undergone RTA at 1000°C for 300 s. (c) TEM micrograph of device C, which has undergone RTA at 1000°C for 300 s.

Note that a trilayer structure similar to device B, but with no RTA step, showed hysteresis of 1.09 V. This means that the RTA process improved the sputtered oxide's quality and reduced the trapped charge density in device B from 3.62×10^{11} (as prepared) to 1.98×10^{11} cm^{-2} (after RTA). The pronounced hysteresis exhibited by device A must be due to charge storage in the Ge NCs located in the middle layer. Shi et al. [122] attributed the memory effects of their devices to charge stored in deep traps in Si NCs. We will examine the charge storage mechanism of our devices in detail in the following sections. Note that device A also shows a significant positive shift (~4 V) and a *C–V* curve with a gentler slope compared to device B. Since the hysteresis width is approximately 6 V, this means that device A has better charge storage capability than device B.

Figure 5.17b shows a TEM micrograph of device A. It can be seen from this micrograph that the middle layer consists of Ge NCs of different sizes. Ge NCs of diameter (d) ~20 nm formed near the RTO SiO$_2$-sputtered Ge-plus-SiO$_2$ interface and smaller Ge NCs, with d of ~6 nm, further away from the interface. There seem to be more Ge NCs near the RTO SiO$_2$-sputtered Ge-plus-SiO$_2$ interface than near the sputtered Ge-plus-SiO$_2$-pure sputtered oxide interface. The center region of the middle layer contains far fewer Ge NCs. Figure 5.17c shows a TEM micrograph of a two-layered device (device C) that consists of a 5 nm RTO SiO$_2$ layer and a 20 nm Ge-plus-SiO$_2$ layer. The device underwent RTA at 1000°C for 300 s. It can be seen from Fig. 5.17c that Ge NCs are located only at the RTO SiO$_2$-sputtered Ge-plus-SiO$_2$ interface. Because this device was fabricated without an oxide cap layer, it is reasonable to expect significant out-diffusion of Ge to occur during RTA at 1000°C. The *C–V* characteristic of device C in Fig. 5.17a exhibits small hysteresis of <0.5 V. Since the Ge NCs are far fewer in number in device C, it is reasonable to expect the charge storage capacity of this device to be more reduced compared to that of device A.

5.4.2 Control of Nanocrystal Size

Chang et al. [14] have pointed out that NC memory properties are influenced by NC size, shape, and configuration. According to the International Technology Roadmap for Semiconductors [63], memory cell size is likely to reduce to 1000 nm^2 by 2020. This means that only about 10 NCs can be contained in every memory cell. On such a small scale, it becomes increasingly challenging to synthesize suitable materials of a uniform size and shape and assemble them into a well-ordered NC matrix [90].

In this section, we will show that the Ge NC size can be controlled by varying the middle-layer thickness [58, 129] of a trilayer structure. Figure 5.18a shows a high-resolution TEM micrograph of a sample with a middle-layer thickness of 6 nm (device A). It shows that most of the NCs are confined within the middle layer. Note that 80% of the NCs are ellipsoidal in shape and that the average diameter of the NCs in the horizontal *x–y* plane is 8 nm, with a standard deviation of 2.1 nm (see Fig. 5.18c).

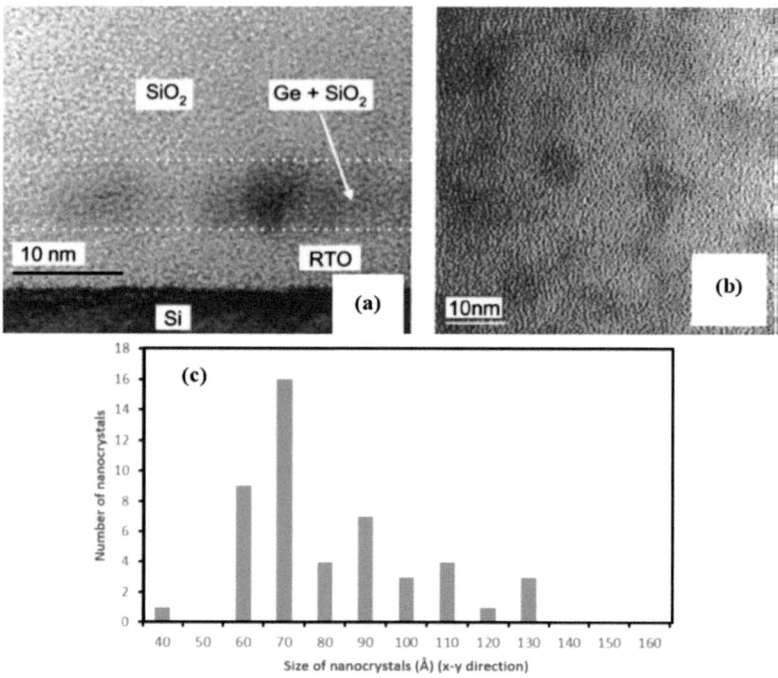

Figure 5.18 (a) Cross-sectional micrograph, (b) planar TEM micrograph, and (c) histogram of Ge NC size distribution of a trilayer structure consisting of 5 nm of rapid thermal oxide, 6 nm of cosputtered Ge-plus-SiO_2 middle layer, and 50 nm (capping oxide) of pure sputtered SiO_2 (device A).

The diameter in the z direction (i.e., direction perpendicular to the surface of the device) is ~6 nm. This is very different from our TEM results of a structure with a thicker middle layer (20 nm), which exhibited NCs of varying sizes distributed in the middle layer. Figure 5.18a also suggests clustering of neighboring NCs in device A. This is further verified from the planar (x–y plane) TEM picture shown in Fig. 5.18b.

Figure 5.19 shows the TEM micrographs of a structure with a middle-layer thickness of 3 nm (device B). The average size of the NCs in the x–y plane is 3.3 nm, with a standard deviation of 1.3 nm (see Fig. 5.19c). The maximum size of the NCs here is ~3 nm in the z direction. We found that 70% of the NCs in device B were spherical and the rest were slightly ellipsoidal. The planar TEM micrograph of device B in Fig. 5.19b shows NCs with d varying from 2 to 4 nm but mostly well separated from one another. Note that there are a few bigger clusters in Fig. 5.19b. The densities of the NCs were estimated from Fig. 5.18b and 5.19b to be 5.73×10^{11} and 1.63×10^{12} cm^{-2}, for devices A and B, respectively.

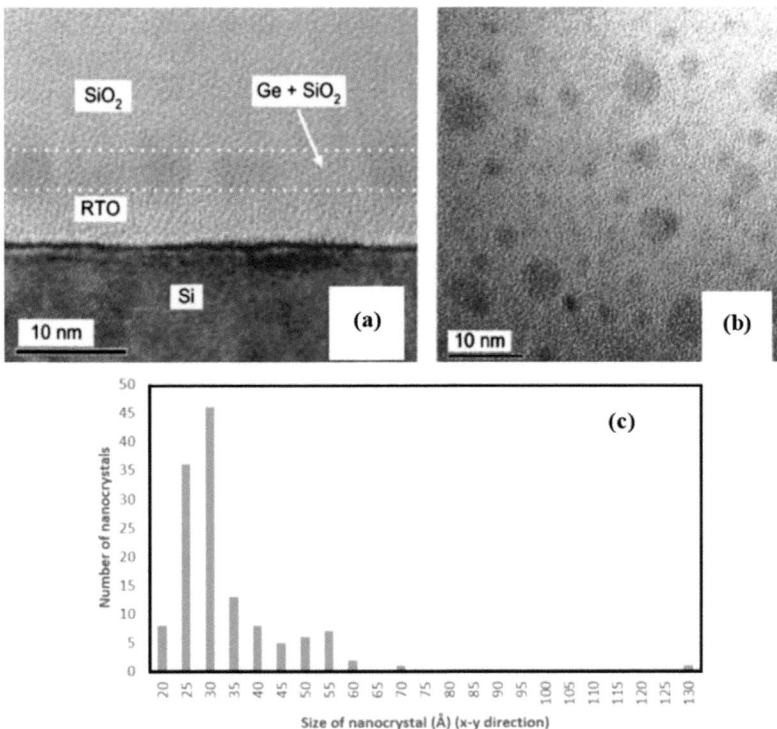

Figure 5.19 (a) Cross-sectional micrograph, (b) planar TEM micrograph, and (c) histogram of Ge NC size distribution of a trilayer structure consisting of 5 nm of rapid thermal oxide, 3 nm of cosputtered Ge-plus-SiO$_2$ middle layer, and 50 nm (capping oxide) of pure sputtered SiO$_2$ (device B).

Figure 5.20 shows typical high-frequency C–V characteristics of devices A, B, C (middle-layer thickness of 20 nm), and D (an MIS structure with 20 nm of pure SiO$_2$ as the middle layer). Note that in all the devices, the thicknesses of the RTO and the capping oxide layer were fixed at 5 nm and 50 nm, respectively. It is clear from these four devices that with Ge NCs in the structure, a significant counterclockwise hysteresis and shift of the C–V curves are observed. There is also a pronounced change in the slope of the C–V curves of devices A, B, and C from D. Note that the slope of the *C–V* curves of devices A, B, and C is affected by the charge stored in the structure. This makes it difficult to differentiate the influence of the interface traps, normally found in an MIS structure, in a device exhibiting charge storage. The voltage shift appears to be rightward (toward positive gate voltage) as the middle-layer thickness increases.

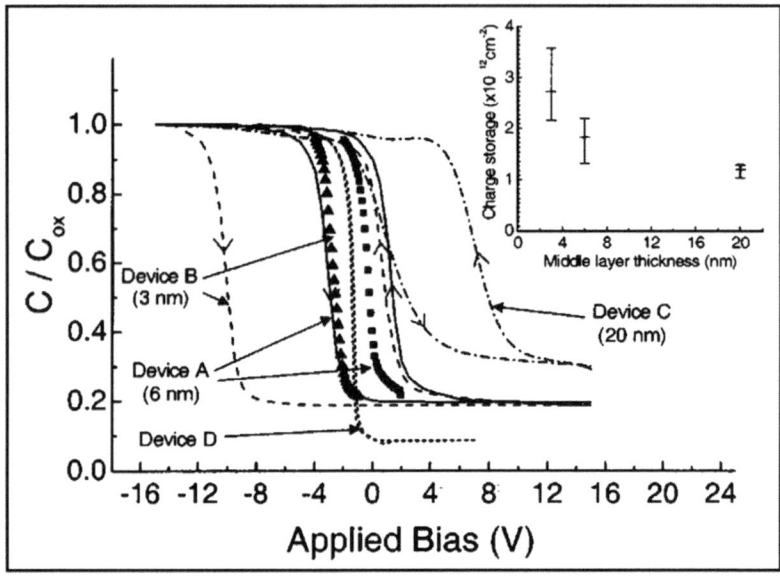

Figure 5.20 Capacitance-versus-voltage characteristics of devices with 6 nm (device A), 3 nm (device B), and 20 nm (device C) middle-layer thickness. The RTO and capping layers were fixed at 5 nm and 50 nm, respectively. Device D is an MIS structure with 20 nm of pure SiO_2 as the middle layer. The inset shows the estimated densities of the stored charge for devices A, B, and C. Note that the quasi-neutral *C–V* curves for devices A and B (symbols ■ and ▲) were obtained by restricting the gate bias to a very narrow range to minimize charging up of the Ge NCs.

Ahn et al. [1] have suggested that in a system that contained Si–O–Si and Si–O–Ge bonds, the Ge–O bond is weaker and can be easily broken, leaving a Si–O dangling bond structure. The dangling bond can capture an electron and become negatively charged. Our XPS results on cosputtered Ge-plus-SiO_2 samples showed that there was a substantial reduction in the amount of GeO_x bonds and a corresponding increase in the amount of elemental Ge after the samples underwent RTA at 1000°C. Therefore, it is possible that in the devices that have undergone RTA, the number of Si–O dangling bonds increases and these dangling bonds subsequently capture electrons during the *C–V* measurements. The trapped electrons will then result in a rightward shift in the measured *C–V* curves. As the middle-layer thickness decreases, the number of Si–O dangling bonds reduces, resulting in less rightward shift in the *C–V* curves. The number of trapped electrons (i.e., the rightward shift of the *C–V* curve) would therefore be proportional to the middle-layer thickness. This is in agreement with the results shown in Fig. 5.20.

The inset in Fig. 5.20 shows the estimated densities of the stored charge for devices A, B, and C. The inset shows that the amount of charge stored in the device increases with a reduction in the middle-layer thickness. The stored charge (which includes both electrons and holes) for devices A and B was estimated from the area of the hysteresis loop in Fig. 5.20 to be 1.33×10^{12} and 3.63×10^{12} cm^{-2}, respectively. Busseret et al. [9] have pointed out the difficulty in deciding from *C–V* measurements whether the charge was stored in the NCs or at the interfaces between the NCs and the oxide matrix. We will examine these two possibilities as follows. We have estimated earlier the densities of the NCs for devices A and B to be 5.73×10^{11} and 1.6×10^{12} cm^{-2}, respectively. A simple calculation using surface areas of the ellipsoidal and spherical NCs (based on the NC size distributions shown in Figs. 5.18c and 5.19c) indicated that the total surface area of the NCs in device A is about 1.6 times that of the NCs in device B. As the charge storage capability of device A is lower than that of device B by about three times (from results in Fig. 5.20 explained above), it suggests that the charge trapping at the interfaces of the NCs with the oxide matrix is less likely to be the dominant charge storage mechanism.

The quasi-neutral *C–V* curves in Fig. 5.20 for devices A and B were obtained by restricting the gate bias to a very narrow range to minimize charging up of the Ge NCs. Note that the quasi-neutral *C–V* curve is not exactly at the middle of the hysteresis loop for both devices. Also, the flat-band voltage obtained from the quasi-neutral *C–V* curve is close to zero for device A but negative for device B. Note that the control structure that does not contain any Ge (i.e., device D) shows a negative flat-band voltage, indicating the presence of positive fixed charges. For the structures containing the Ge layer, there is an additional negative charge component due to the Si–O dangling bonds trapping electrons, as mentioned above. The position of the quasi-neutral C–V curve is therefore dependent on the overall effect of the positive fixed charge and the negative charge component. For device A (6-nm-thick Ge layer), there seems to be almost complete compensation of these two charge components, resulting in a flat-band voltage close to zero. For device B (3-nm-thick Ge layer), we expect the positive fixed charge component to be slightly more dominant than the negative charge component (proportional to the thickness of the Ge layer) because of the thinner Ge layer, which leads to a negative flat-band voltage. However, there seems to be a discrepancy when we compare the quasi-neutral C–V curve of device B and the C–V curve of the control device D in Fig. 5.20. We would expect the flat-band voltage of device B to be less negative than that of the control device D since there is some compensation of the positive fixed charge by negative charges in device B. However, this is not the case in Fig. 5.20. This discrepancy is explained possibly by the fact that a small but non-negligible amount of hole trapping/storage still occurs, since the charge storage capability of device B is higher than device A by about three times.

If the quasi-neutral C–V curve is assumed to be in the middle of the hysteresis loop for each device, then the amount of hole storage or electron storage in Fig. 5.20 will be half of the area of the hysteresis loop. This works out to 6.53×10^{11} and 1.83×10^{12} cm^{-2} for devices A and B, respectively, from the values stated above. As previously mentioned, the densities of the NCs for devices A and B are 5.73×10^{11} and 1.63×10^{12} cm^{-2}, respectively, each value corresponding closely to the number of holes or electrons stored in the respective device. This suggests that one electron or one hole is stored per NC for both devices A and B. For this reason, we propose that the charge storage in our samples is more likely to occur within the NCs, rather than at the interfaces between the NCs and the oxide matrix.

5.4.3 Retention Properties

A major problem with NC memory is charge leakage. The advantage of NC-based memory devices is that high-density NCs can store more charge in a memory device and mitigate the influence of fluctuations between individual devices because there are more NCs in each memory cell. However, in high-density NC systems, the charges stored in NCs leak more easily though the surrounding oxide. If the surrounding oxide cannot effectively prohibit the charge transport between NCs, NC memory will lose the advantage of having a discrete charge storage node, that is, preventing the loss of total stored charge through a leakage path in the tunnel oxide. We have examined the effect of Ge concentration and the tunnel oxide layer thickness on the retention characteristics of Ge-NC-based memory devices [55, 57]. MIS structures with tunnel oxide layers of 2.5 to 5 nm and varying Ge concentrations were fabricated. The capping sputtered oxide layer was 46–64 nm thick. Figure 5.21a shows that a number of NCs are in contact with the Si substrate for the structure sputtered with a pure Ge middle layer and a 5-nm-thick tunnel oxide layer (device P5). There are also NCs that have penetrated into the oxide layer but are not in contact with the Si substrate. Note that the planar TEM image of Fig. 5.21b shows even smaller (and more numerous) NCs, which were difficult to identify from the cross-sectional TEM image of Fig. 5.21a. The density of the NCs was estimated to be about 9.33 $\times 10^{11}$ cm^{-2} from several planar TEM images (similar to the image in Fig. 5.21b) of device P5. We have estimated the densities of the NCs of the cosputtered Ge-plus-SiO$_2$ systems (with the tunnel and capping oxide layer thicknesses fixed at 5 nm and 50 nm, respectively) to be about 5.73 $\times 10^{11}$ cm^{-2} and 1.63 $\times 10^{12}$ cm^{-2} for devices with middle layer thicknesses of 6 nm and 3 nm, respectively. Therefore, we were not able to obtain a higher density of NCs with a middle-layer structure of pure Ge.

It should be noted that with a 100% Ge concentration in the middle layer and a thin tunnel oxide layer, significant diffusion of Ge through the RTO layer is possible, and Ge atoms can also coalesce to form larger NCs in contact with the Si substrate (see Fig. 5.21a). As a significant number of Ge atoms were used up in the formation of the large NCs and some would have diffused out through the RTO layer, this could have resulted in a lower density of NCs formed. Note in

Fig. 5.21b that the smaller NCs are generally separated from each other by a few nanometers and are not in contact with the Si substrate. This is important as charge loss, either through tunneling between adjacent NCs or tunneling into the Si substrate, can be minimized by such a separation.

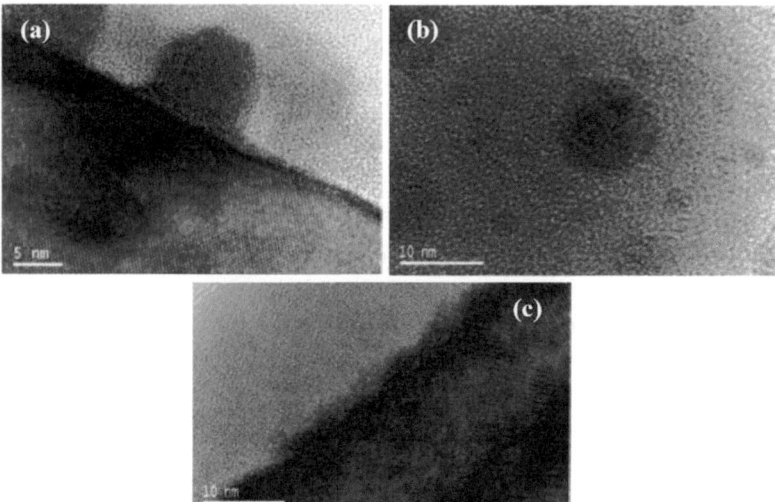

Figure 5.21   (a) Cross-sectional and (b) planar HRTEM images of device P5 and (c) cross-sectional HRTEM image of device P2–5. The devices have a 4-nm-thick pure sputtered Ge middle layer and 46- and 65-nm-thick capping oxide layers for devices P2–5 and P5, respectively. The tunnel oxide thicknesses of devices P5 and P2–5 are 5 nm and 2.5 nm, respectively. Note that both devices P5 and P2–5 have been subjected to RTA at 1000°C for 300 s.

Figure 5.21c shows the TEM image of a device with a pure Ge middle layer and a 2.5-nm-thick tunnel oxide layer that has undergone RTA at 1000°C for 300 s (device P2–5). The original pure Ge layer has disappeared after RTA, and we could not locate any NCs in the device. The tunnel oxide/Si substrate interface also appears to be rather uneven. We have shown that for devices with a cosputtered Ge-plus-SiO$_2$ middle layer, the tunnel oxide layer and the capping oxide layer have been effective in confining the NCs to the middle layer. In device P2–5, due to the steep Ge concentration gradient between the middle layer and the Si substrate and the thin tunnel oxide barrier for diffusion, it is likely that most of the Ge may have diffused into the Si substrate during RTA at 1000°C [54]. As a consequence, no (or very few) NCs were formed in the MIS structure. The uneven RTO/Si interface may be due to the pronounced diffusion of Ge from the middle layer to the Si substrate.

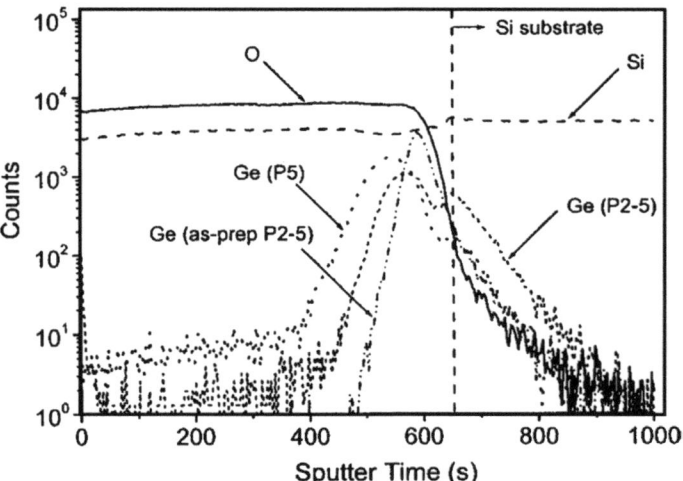

Figure 5.22 SIMS results of as-prepared (as-prep P2–5) and annealed (P2–5 and P5) devices. Due to the difference in the capping oxide and tunnel oxide thickness of devices P5 and P2–5, we have adjusted the SIMS results by shifting the SIMS profiles horizontally using the Si profile as the reference (i.e., matching the Si profiles of devices P5 and P2–5 at the tunnel oxide/silicon interface, as indicated by the vertical dashed line).

Figure 5.22 shows the SIMS results (obtained using a 2 keV Cs + source) of as-prepared and annealed devices P5 and P2–5. The SIMS results of device P2–5 show that Ge from the middle layer diffuses into the capping layer and the tunnel oxide–Si substrate region. Compared to the Ge profile of the as-prepared device, there appears to be a pronounced diffusion of Ge into the tunnel oxide–Si substrate region. This is in agreement with our discussion earlier. If we compare the Ge profiles of the annealed devices P5 and P2–5, it is obvious that the thicker tunnel oxide layer has reduced the diffusion of Ge to the RTO-Si region. As a consequence, a more pronounced out-diffusion of Ge with a higher Ge concentration into the capping layer is observed in device P5 compared to device P2–5. This may account for the formation and location of NCs observed in device

P5 and the general absence of NCs in device P2–5, as shown in Figs. 5.21b and 5.21c.

Figure 5.23a shows the forward and reverse sweep *C–V* characteristics of devices P5 and P2–5 measured with the bias held at –15 and +15 V for 240 s before each *C–V* sweep commences. We found that the device P5 structure exhibited a counterclockwise *C–V* hysteresis, with an average hysteresis width of ~12 V.

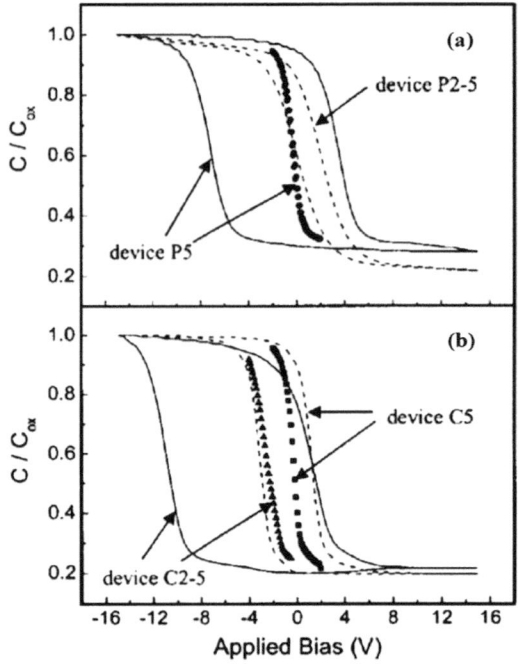

Figure 5.23 High-frequency *C–V* characteristics of (a) devices P5 and P2–5 and (b) devices C5 and C2–5. Note that the quasi-neutral *C–V* curves for the respective devices (symbols ●, ▲, and ■) were obtained by restricting the gate bias to a very narrow range to minimize charging up of the Ge NCs.

The charge storage density was estimated to be 2.33×10^{12} cm^{-2}. However, only 50% of the device P2–5 structure exhibited a *C–V* characteristic, as shown in Fig. 5.23a; the other 50% exhibited no proper *C–V* curves. For those P2–5 devices that exhibited proper *C–V* characteristics, a significantly smaller hysteresis of ~2.4 V was observed. One would expect the hysteresis of device P2–5 to be larger than that of device P5 as a larger number of charge carriers are able to tunnel across the thinner tunnel oxide layer to be stored in the NCs. However, as no (or very few) NCs were formed in device P2–5 (see Fig. 5.21c), it is reasonable to expect a much smaller hysteresis in such devices. Figure 5.23a also shows the quasi-neutral *C–V* curve, obtained by restricting the gate bias to a very narrow range to minimize the charging up of NCs, for device P5. The *C–V* curve of device P2–5 shifts toward a more positive gate voltage as compared to the quasi-neutral *C–V* curve of device P5. This may be due to Ge penetration through the tunnel oxide layer and into the Si substrate. Figure 5.23b shows the *C–V* characteristics of devices with a 6-nm-thick cosputtered middle layer and tunnel oxide layer thicknesses of 2.5 nm (device C2–5) and 5 nm (device C5). Both *C–V* curves show counterclockwise hysteresis, and the charges stored for devices C2–5 and C5 were estimated to be 3.73×10^{12} cm^{-2} and 1.33×10^{12} cm^{-2}, respectively. This is reasonable as a larger number of charge carriers from the Si substrate are therefore able to tunnel across the thinner oxide layer, resulting in an increase in the charge stored in device C2–5 as compared to that stored in device C5. Note that the flat-band voltage of the uncharged device C5 is close to zero (obtained from the quasi-neutral curve in Fig. 5.23b) but is negative for the uncharged device C2–5.

Note that in Fig. 5.23b, there exists a gentler slope of the *C–V* curve for device C2–5 as compared to that for device C5. We have pointed out previously that it is difficult to separate the influence of the charging/discharging processes and the interface traps on the slope of the *C–V* curves of memory devices. We have carried out a series of *C–V* measurements on device C2–5 at different sweep rates, ranging from 0.1 V/s to 10 V/s, and found that a higher sweep rate resulted in a steeper *C–V* curve as the NCs have lesser time to charge and discharge. This suggests that the gentler *C–V* slope in device C2–5 is likely a result of the smaller RTO thickness affecting the ease of the charging and discharging processes occurring concurrently during the *C–V* sweep. This also means that even though device C2–5 has a better charge storage capability (i.e., larger hysteresis), it is likely to have a poorer charge retention capability than device C5. Returning to Fig. 5.23a, one can also observe a gentler slope in the *C–V* curve of device P5 as compared to its quasi-neutral curve, which is not surprising as the quasi-neutral curve represents the uncharged condition of the device. Device P2–5 also exhibits a gentler C–V slope as compared to device P5. Since device P2–5 exhibits negligible charge storage, it is likely that this gentler C–V slope is attributed to a poorer Si/tunnel oxide interface, possibly a result of the Ge penetration causing an uneven interface, as shown previously in Fig. 5.21c.

5.4.4 High-κ Dielectrics

The charge retention properties of Ge-NC-based memory devices could be improved by using a high-dielectric-constant (high-κ) material with a large barrier height, such as Al_2O_3 or HfO_2. The leakage current of HfO_2 is several orders of magnitude smaller than that of SiO_2 for the same equivalent oxide thickness (EOT) [83], thus ensuring superior data retention performance. HfO_2 can also provide a higher tunneling current than SiO_2 during the write operation due to a lower electron barrier height (1.2 eV) than SiO_2 (3.1 eV) [146]. In this section, we present results of using another type of silicon nitride/hafnium dioxide (Si_3N_4/HfO_2) stack structure as the tunnel dielectric layer to minimize the Ge penetration effect. The leakage current of the Si_3N_4/HfO_2 stack film is also several orders of magnitude smaller than that of SiO_2 with a similar EOT, resulting in better charge retention performance for the high-κ trilayer memory structure as compared to the trilayer memory device with a SiO_2 tunnel oxide layer.

The trilayer insulator structure was either a 1.9 nm EOT high-κ tunnel dielectric (device HK1-9) or a 2.5-nm-thick tunnel oxide (device RTO2-5). The high-κ tunnel dielectric stack, consisting of a 1-nm-thick Si_3N_4 layer in contact with Si and a 5-nm-thick HfO_2 layer, was grown on (100) n-type Si substrates. This high-dielectric-constant (high-k) stack, grown by metal-organic CVD, has an EOT of 1.9 nm and a measured gate leakage current density of 1 mA/cm^2 at 1 V accumulation bias. The 4-nm-thick Ge middle layer was then formed by sputter deposition. This was followed by a capping silicon oxide layer of 50 nm thickness. The trilayer structure was then rapidly thermally annealed at 1000°C for 300 s in Ar before forming the gate electrode of the MIS device.

Figure 5.24 shows the cross-sectional TEM image of device RTO2-5, where the formation of Ge NCs is not successful due to the penetration and diffusion of Ge through the 2.5-nm-thick SiO_2 layer during the 1000°C RTA step. The original pure Ge layer has disappeared after the RTA step and no NCs could be detected in the device. The tunnel oxide/Si interface also appears to be rather uneven. We have suggested earlier (see Fig. 5.21c) that it was likely that most of the Ge may have diffused into the Si substrate (through the 2.5-nm-thick RTO) during RTA at 1000°C. Consequently, no (or very few) NCs were formed in the MIS structure. The uneven tunnel oxide/Si interface may be due to the pronounced diffusion of Ge from the middle layer into the Si substrate. Figure 5.24b shows the TEM image of a well-formed Ge NC in device HK1-9.

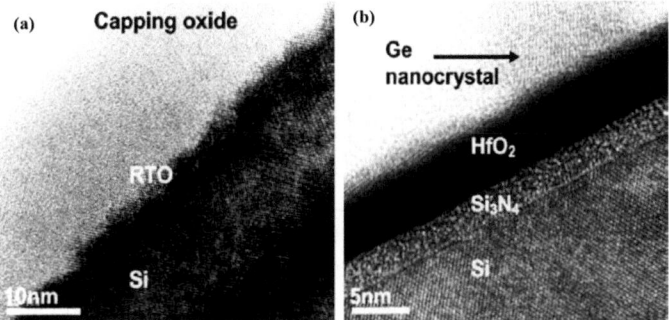

Figure 5.24 Cross-sectional TEM image of (a) device RTO2-5 and (b) device HK1-9. Note the absence of Ge NCs and the uneven tunnel oxide/Si interface caused by the Ge penetration in device RTO2-5.

The high-k tunnel dielectric stack is effective in preventing Ge penetration due to its greater physical thickness, although it has a much smaller EOT than device RTO2-5. The charge storage capability of devices with SiO_2 and high-κ tunnel dielectric layers is compared in Fig. 5.25. As the device with a 2.5-nm-thick SiO_2 layer (device RTO2-5) does not exhibit significant charge storage capability because of the Ge penetration effect, comparison of charge storage and retention of the high-κ device HK1-9 is thus performed with a 5-nm-thick layer device (device RTO5). The physical thickness of the tunnel dielectric layer of device RTO5 is comparable to that of device HK1-9, and device RTO5 also shows good charge storage characteristics (see Fig. 5.25). For device RTO5, when the gate voltage is swept from 25 V to 15 V, the area enclosed by the hysteresis loop is 1.80×10^{-7} C cm^{-2}

and the corresponding NC density was estimated to be 1.12×10^{12} cm^{-2}. Note that a similar amount of charge storage could be obtained from device HK1-9 over a smaller gate voltage sweep range of –3.4 V to +3.4 V. This is due to the smaller EOT of device HK1-9 and the lower barrier height (as compared to SiO_2) of the high-κ layer, which allows easier tunneling of charge carriers from the substrate into the NCs.

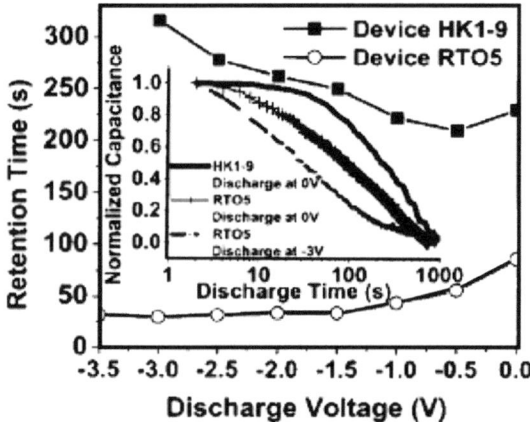

Figure 5.25 Comparison of retention time versus discharge bias of devices RTO5 and HK1-9. The inset shows representative normalized *C–t* curves during discharge of devices RTO5 and HK1-9 at a constant discharge voltage of either 0 or –3 V as indicated.

For evaluating the charge retention performance, the devices were first fully charged up at a fixed positive gate voltage until saturation in the gate capacitance was observed, followed by switching abruptly to a constant negative discharging voltage. We defined the retention time as the time for the capacitance to decrease to 50% of its initial value. Note that the total amount of charge storage for device HK1-9 over the gate voltage sweep range of –3.4 V to +3.4 V is similar to that for device RTO5 over the gate voltage sweep range of –5 V to +5 V. Hence, in the charge retention time studies, device HK1-9 is first charged at +3 V and then discharged at voltages ranging from –3 V to 0 V. For device RTO5, the sample is charged at +5 V, followed by discharging at voltages ranging from –5 V to 0 V. It is seen from Fig. 5.25 that device HK1-9 shows better charge retention performance, with generally longer retention times, than device RTO5.

5.4.5 Characterization Ge-Nanocrystal-Based Transistors

One of the major issues in NC-based memory is the poor long-term charge retention performance, as compared to the 10-year charge retention duration requirement that is typically achievable by floating-gate nonvolatile memory. While it is important to achieve a short write time by using a thinner tunnel barrier, it would, however, inevitably compromise the charge retention capability of the device. Likharev [86, 87] proposed the use of a tunnel barrier, which is able to increase the electric field sensitivity across the tunnel barrier, such that higher and lower tunneling efficiencies can be obtained during the write and retention processes, respectively. Another method to improve on charge retention is to replace the tunnel barrier with a high-κ material, as demonstrated in the last section. Another interesting problem is regarding the role of traps at the tunnel oxide/silicon interface in charge storage in NC-based memory. Shi et al. [122] demonstrated that the NC memory device lost its memory effect after it was annealed in hydrogen ambient. Since traps in NCs may play a dominant role during the discharging process, the trap energy level is critical in determining the charge retention capability of the NC memory device. We have investigated the role of traps in Ge NC memory in charge storage and retention through theoretical modeling and experimental measurements. The model consists of a self-consistent quantum mechanical numerical method for multilayered dielectric stack structures with a layered or crested potential profile tunnel dielectric [79]. This model was used in our investigation of traps in Ge NCs [80].

The NC memory device can be subjected to various operating bias conditions such as write, erase, read, and off-state retention operations. Figure 5.26a shows the basic structure of the Ge-NC-based transistor memory device.

Figure 5.26b shows the one-dimensional (through the gate stack) conduction band diagram of the NC memory device during the program or write operation. During programing, the control gate is biased at a high positive voltage. This would induce a positive electric field across the tunnel barrier, which would attract electrons to tunnel from the Si substrate toward the NCs. The electrons that have successfully tunneled through the tunnel barrier would subsequently either tunnel through the thick cap oxide (cap SiO_2)

or be captured by the traps that reside in the Ge NC bandgap. Due to the thick cap oxide used, it is more likely that the majority of the tunneled electrons are captured by the traps. Figure 5.26c shows the conduction band diagram of the Ge NC memory device during the erase process. The control gate is biased at a high negative voltage, which would induce a negative electric field across the tunnel barrier. The electrons in the traps would respond to this negative electric field either by thermal detrapping out of the traps to the NC conduction band followed by direct tunneling to the conduction band of the Si substrate or by direct tunneling to the available states in the Si bandgap. Another possible process during discharging is the direct tunneling of holes from the Si substrate to the NCs.

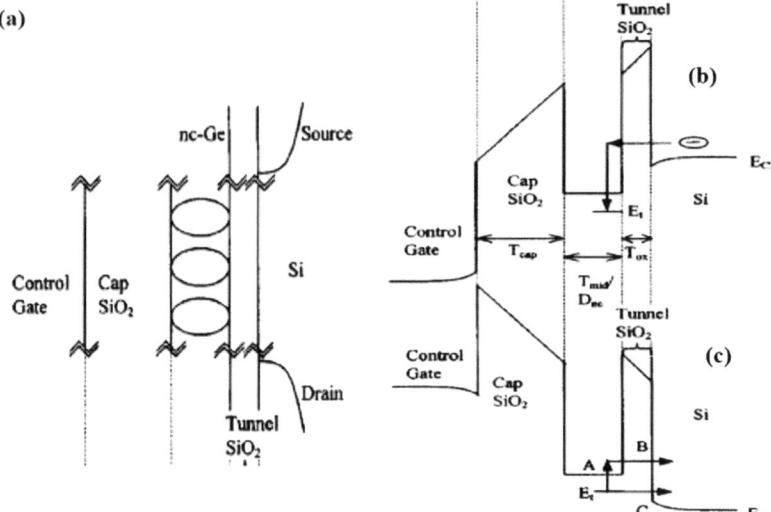

Figure 5.26 (a) Schematic of Ge NC memory device structure. (b) Electron tunneling and capture by a trap in an NC during the write process. (c) Thermal detrapping of an electron and subsequent electron tunneling into the Si substrate during the erase process.

It has been proposed that conduction band storage may not be the dominant mechanism for charge storage in NC memory [122], primarily due to the unusually long retention time observed in many reported experiments [49, 73]. Due to the charging effect of the trapped electrons in the NCs, the conduction band edge of the NC is higher than that of the Si substrate. Thus, the trapped electrons in the conduction band should easily tunnel out of the NCs to the Si substrate. King [76] reported that a temperature-dependent effect was observed in the charge retention time. This observation supports the existence of trap levels in the NC bandgap; the thermal detrapping of electrons from the NC traps is responsible for the temperature dependency of charge retention. Shi et al. [122] have performed hydrogen annealing for NC memory devices and found that the memory or charge storage effect disappeared after the annealing process. This suggests that hydrogen passivated the traps in the NCs and thus decreased the available trap sites for electron storage.

We propose a discharge mechanism, shown schematically Fig. 5.26c, in which during the charge retention process, the electron at a trap site would first detrap thermally to the conduction band of the NC as it is energetically more favorable to do so, as compared to direct tunneling to available states in the Si substrate bandgap. This is followed by the direct tunneling of the detrapped electron in the conduction band of the NC to the conduction band of the Si substrate. The discharge model is similar to that of She and King [120]. However, a temperature-dependent Shockley–Read–Hall for thermally detrapped electrons [95] is used in the discharge model in this work. The thermal equilibrium emission constant of electrons can be derived as

$$e_n(E_t) = AT^2 \exp\left(-\frac{qE_t}{kT}\right),$$ (5.7)

where k is Boltzmann's constant, q is the electronic charge, T is the temperature, E_t is the trap energy level measured with respect to the conduction band edge of the NC, and A is the temperature-independent constant.

The discharge model is obtained by first letting $P(t)$ be the probability of an electron escaping from the NC per unit time. The change in charge retained in the NC can be expressed as

$$\frac{dQ(t)}{dt} = -P(t)Q(t),$$ (5.8)

Therefore, the charge retained in the NC at time t_1 can be derived as

$$Q(t_1) = Q(0)\exp\left[-\int_{t=0}^{t_1} P(t)dt\right]. \tag{5.9}$$

$P(t)$ can be expressed as

$$P(t) = \int_{E \geq E_c}^{1} f_{imp}\rho(E)T_m(E)e_n(E_t)\beta dE, \tag{5.10}$$

where E_c is the NC conduction band edge, $\rho(E)$ is the density of states at energy E, f_{imp} is the Weinberg impact frequency and is given as E/h [137], $T_m(E)$ is the transmission coefficient across the tunnel barrier, and β is a fitting parameter, which is calibrated using experimental measurements for a particular process/technology. Note that the trap energy level E_t in Eq. 5.7 was extracted by fitting to the experimental data.

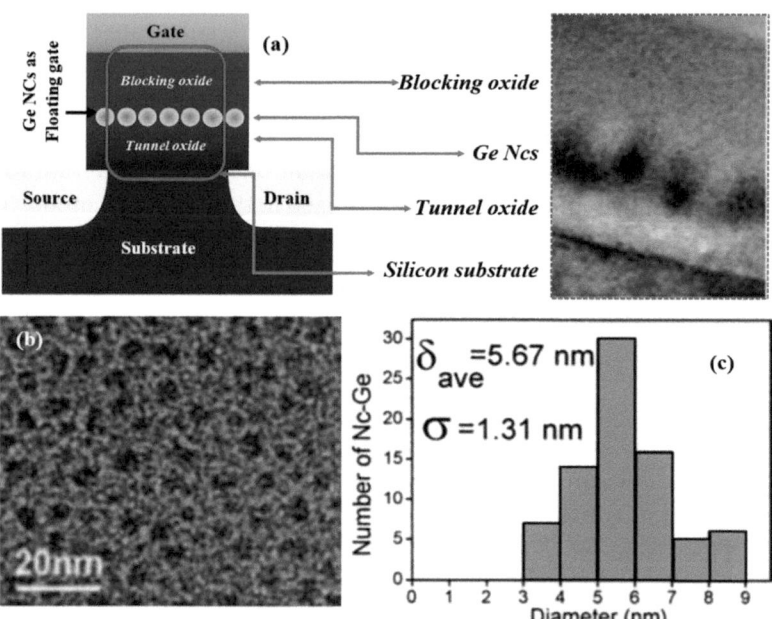

Figure 5.27 (a) Schematic cross-sectional structure of fabricated device and HRTEM image of the SiO_2-Ge-NC-SiO_2 transistor memory structure. (b) Planar TEM image of self-assembled Ge NCs embedded in SiO_2. The histogram plot shows the size distribution of the NCs.

Ge NC memory was fabricated on p-type (100) Si wafers using ring transistor test structures with 10 μm gate length. The dielectric stack consisted of a 5-nm-thick (Ge plus SiO_2) layer sandwiched between a 5-nm-thick tunnel oxide and a control or cap oxide of 40 nm thickness (see Fig. 5.27a). Figure 5.27b shows the planar TEM image of the synthesized NCs. The mean diameter of the NC was estimated to be 5.67 ± 1.31 nm, with an area density of ~9 × 10^{11} cm^{-2}. A poly-Si gate electrode was deposited at 600°C using low-pressure CVD. The NCs that remained after the gate etch step outside the channel region was removed by a subsequent 10% HF etch. After the source/drain and gate n+ implantation, RTA was performed at 950°C for 30 s in N_2. Aluminum metal was deposited before a forming gas sintering process at 450°C for 5 min.

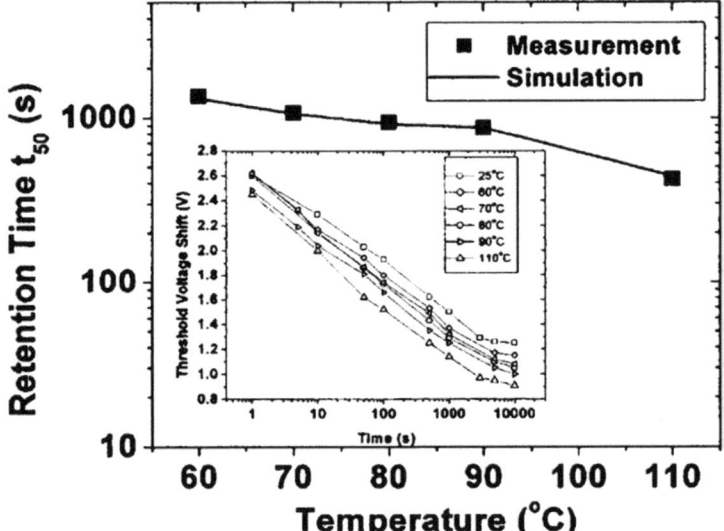

Figure 5.28 Plot of retention time (t_{50}) versus temperature. The solid squares represent the experimental measurements (taken from the plot in the inset) while the bold line represents the simulated result using E_t = 0.16 eV and β = 13.65. The inset shows the variation in threshold voltage (V_{th}) with discharge time of the Ge NC memory device at different temperatures, after a charging or write operation at 15 V for 20 s.

Figure 5.28 shows the retention time versus temperature for our devices. The inset shows the threshold voltage (V_{th}) shift with time.

The V_{th} shift corresponds to the change in charge retained in the NC. The measurements were performed after applying a write voltage of 15 V for 20 s. The electron loss rate was estimated to be about 0.31 V/decade at 25°C, 0.35 V/decade at 80°C, and 0.45 V/decade at 110°C. We define the retention time t_{50} as the time taken for 50% of the total stored charges in the NCs to leak away. The t_{50} values were obtained from the inset in Fig. 5.28 and plotted versus temperature in Fig. 5.28. The t_{50} plot shows a temperature dependence that agrees with the results by King [76] and indicates the presence of thermal detrapping of electrons from traps.

In Fig. 5.26c, the first discharging mechanism is portrayed by paths A and B. This charge decaying mechanism involves the thermal excitation of trapped electrons from the occupied trap state to the conduction band of the NCs, followed by band-to-band tunneling into the Si substrate. The second decaying mechanism, depicted by path C, which has a smaller temperature dependence, is trap-to-trap tunneling of electrons from the occupied trap state to the Si/SiO$_2$ interface states. By performing transient drain current (I_{DS}) measurement at different temperatures, the discharging time constant (τD) can be extracted [3]. τD is defined as the time corresponding to 90% of the I_{DS} at steady state when the drain current gradually returns to its steady-state value when the electrons are detrapped during the discharge phase following the initial charging of the NCs. Using the temperature dependence of the discharging time constant, the two different discharging mechanisms were resolved in the experimental measurements. Baik et al. [3] have derived an expression for the expected value of drain current during the discharge phase as

$$I_{DS(transient)} = q\left[n_1\left\{1-\exp(-\lambda_1 t)\right\}+n_2\left\{1-\exp(-\lambda_2 t)\right\}\right], \quad (5.11)$$

where $n_1 + n_2 = n$ and n is the total number of Ge NC charging sites. λ_1 and λ_2 represent the discharging rate constants for the two different discharging mechanisms.

Figure 5.29a shows the temperature-dependent I_{DS} transients with both read (control gate) voltage (V_R) and drain voltage (V_D) held at 5 V, after the NC transistor device has been charged with a write bias at 15 V for 20 s. The I_{DS} transient at each temperature exhibits an S-shaped monotonically increasing curve due to the discharging of stored electrons in the NC. The increase in I_{DS} for higher temperatures could be explained by the larger carrier mobility and the higher intrinsic carrier concentration of the substrate. The increased intrinsic carrier concentration changes the Fermi level of the substrate. This causes the voltage drop at the substrate during inversion to reduce, which would then enhance the vertical electric field across the gate stack.

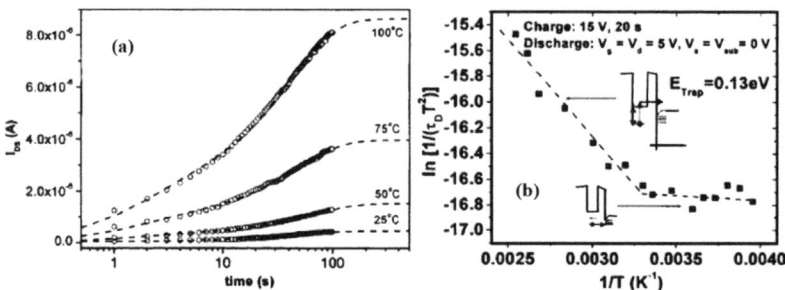

Figure 5.29 (a) Drain current (I_{DS}) transient at a read voltage of 5 V after a write operation at 15 V for 20 s. The symbols represent the measured data, and the lines are fitted data from simulation. (b) Inverse of the discharging time constant ($1/\tau D$) divided by the square of the temperature (T) plotted against the inverse of T.

Each I_{DS} transient was fitted with a first-order double exponential decaying equation (Eq. 5.3). Using Eq. 5.7, the rate of emission (R_e) of electrons from traps in the NCs comes to be proportional to the number of occupied electron traps [51, 123],

$$R_e = e_n N_t f_t ,$$ (5.12)

where e_n is the emission constant, N_t is the total number of electron traps (cm^{-3}), and f_t is the trap occupation probability, which is given by the Fermi–Dirac distribution function. Since the rate of tunneling has little dependency on the temperature, an expression for the rate of discharge R_D can be expressed as

$$R_D \propto e_n = T^2 \exp\left(\frac{E_t}{kT}\right).$$ (5.13)

Figure 5.29b shows the $\ln[1/(\tau D T^2)]$ versus $1/T$ plot whereby the trap energy E_t (~0.13 eV) is extracted from the gradient of the fitted straight line at high temperatures. At lower temperatures, the small slope of the second fitted straight line shows the insignificance of temperature dependence. This region is probably due to the dominance of direct tunneling of electrons from deep-level traps to the interface states at the Si/SiO_2 interface.

By tracking the threshold voltage as a function of time at specified temperatures during discharge of a charged Ge NC transistor device (see the inset in Fig. 5.28), the charge retention times were obtained for different temperatures. The trap energy level was obtained theoretically by fitting the simulated charge retention time versus temperature to the experimental results. The t_{50} values were fitted by using β and E_t as fitting parameters. Charge retention simulation was performed, and the simulated results were found to fit the experimental data with β of 13.65 and E_t of 0.16 eV [80], results of which are shown in Fig. 5.28. The value of E_t obtained from the theoretical simulation is close to the experimentally extracted trap energy level of 0.13 eV from transient drain current measurements.

The origin of traps in Ge has been investigated by many researchers. Haesslein et al. [50] identified the monovacancy and self-interstitial in Ge using perturbed angular correlation spectroscopy (PACS) and deep-level transient spectroscopy (DLTS). They found a donor level lying 0.04 ± 0.02 eV below the conduction band edge. Our experimentally and theoretically extracted trap energy level lies near the predicted trap energy due to self-interstitial defects in Ge. This suggests that self-interstitial defects in the Ge NCs may possibly be responsible for the charge storage in our devices. The interface of bulk Ge and SiO_2 has been reported to have surface defect densities on the order of 10^{12} cm^{-2} [100]. Although the defect density is low on an NC (one defect per Ge NC 5 nm in diameter), the highly strained surface of each NC induces large surface stress, which increases its defect density, and it could be the likely source and location of self-interstitial defects.

Recently, Peibst et al. [113] have investigated the electronic properties of the interface between Ge NCs and a SiO_2 matrix by analyzing the frequency-dependent response of charge stored in the NCs to an alternating external field. The cluster charge with respect to the position of the quasi-Fermi level in the NC was extracted from the experiments and calculated by density functional theory. Note that the model considered only charge storage in the quantized NC states and ignored trap states at the NC/SiO$_2$ interface. A fairly high NC/SiO$_2$ interface state density of 2.8×10^{13} cm^{-2} eV^{-1} was obtained. This high density of states posed a serious problem for the application of the Ge-NC-SiO$_2$ system in optoelectronics and in photovoltaics. Also due to the efficient nonradiative processes mediated by these traps, the EL from the Ge NCs, and the photocurrent extractable from the NCs can be expected to be only marginal. In another piece of work by Peibst et al. [114], electron and hole charging and discharging processes for Ge NCs embedded in the gate oxide of long channel MOS transistors were examined. The results showed fundamental contradictions to theoretical predictions in that the observed retention times for holes were much shorter than the predicted values. This was attributed to the tunneling of electrons from an inversion layer in the Si substrate into the NCs, which compensated the positive charge stored in the NCs.

5.5 Summary

In conclusion, we can say that the origin of the charge storage property of Ge NCs still remains elusive. We have not been able to locate many publications in this area in the last five years except that from Peibst et al. [113, 114]. This is an area that more effort can be devoted to. However, given the challenges faced by NC-based memory devices outlined in the previous paragraph, the theoretical modeling of Ge NC memory devices may be hampered by a lack of experimental data for verification.

References

1. Ahn, C.G., Kang, H.S., Kwon, Y.K., and Kang, B. (1998). *Jpn. J. Appl. Phys.*, **37**, 1316.

2. Ammendola, G., Vulpio, M., Bileci, M., Nastsi, N., Geradi, C., Crupi, G., Crupi, I., Nicotra, G., and Lombardo, S. (2002). *J. Vac. Sci. Technol. B*, **20**, 2075.

3. Baik, S.J., Choi, S., Chung, U-I., and Moon, J.T. (2004). *Sol. Stat. Electron.*, **48**, 1475.

4. Barba, D., Martin, F., Demarche, J., Terwagne, G., and Ross, G.G. (2012). *Nanotechnology*, **23**, 145701.

5. Baron, T., Pelissier, B., Perniola, L., Mazen, F., Hartmann, J.M., and Rolland, G. (2003). *Appl. Phys. Lett.*, **83**, 1444.

6. Blaser, J.M., Caragianis-Broadbridge, C., Walden, B.L., and Paine, D.C. (1996). *Proc. Mater. Res. Soc. Symp.*, **398**, 619.

7. Borany, J.V., Grotzschel, R., Heinig, K.H., Markwitz, A., Matz, W., Schmidt, B., and Skorupa, W. (1997). *Appl. Phys. Lett.*, **71**, 3215.

8. Brus, L.E. (1986). *IEEE J. Quantum Electron.*, **22**, 1909.

9. Busseret, C., Souifi, A., Baron, T., Guillot, G., Martin, F., Semeria, M.N., and Gautier, J. (2000). *Superlattices Microstruct.*, **28**, 493.

10. Caraganis-Broadbridge, C., Blaser, J.M., and Paine, D.C. (1997). *J. Appl. Phys.*, **82**, 1626.

11. Celler, G.K., Trimble, L.E., Sheng, T.T., Kosinski, S.G., and West, K.W. (1988). *Appl. Phys. Lett.*, **53**, 1178.

12. Cerdeira, F., Buchenauer, C.J., Pollack, F.H., and Cardona, M. (1972). *Phys. Rev. B*, **5**, 580.

13. Chang, S.-T., and Liao, S.-H. (2009). *J. Vac. Sci. Technol. B*, **27**, 535.

14. Chang, T.-C., Jian, F.-W., Chen, S.-C., and Tsai, Y.-T. (2011). *Mater. Today*, **14**, 608.

15. Chen, Z., Lei, Y., Chew, H.G., Teo, L.W., Choi, W.K., and Chim, W.K. (2004). *J. Cryst. Growth*, **268**, 560.

16. Chew, H.G., Choi, W.K., Foo, Y.L., Zheng, F., Chim, W.K., Voon, Z.J., Seow, K.C., Fitzgerald, E.A., and Lai, D.M.Y. (2006). *Nanotechnology*, **17**, 1964.

17. Chew, H.G., Zheng, F., Choi, W.K., Chim, W.K., Foo, Y.L., and Fitzgerald, E.A. (2007). *Nanotechnology*, **18**, 065302.

18. Chew, H.G., Choi, W.K., Liew, T.H., and Zheng, F. (2009). A novel process to fabricate nanoparticles and nanowires, US Patent Application US2009/0291311.

19. Chew, H.G., Zheng, F., Choi, W.K., Chim, W.K., Fitzgerald, E.A., and Foo, Y.L. (2009). *J. Nanosci. Nanotech.*, **9**, 1577.

20. Choi, W.K., Ng, V., Ng, S.P., Thio, H.H., Shen, Z.X., and Li, W.S. (1999). *J. Appl. Phys.*, **86**, 1938.

21. Choi, W.K., Kanakaraju, S., Shen, Z.X., and Li, W.S. (1999). *Appl. Surf. Sci.*, **144/145**, 697.

22. Choi, W.K., Thio, H.H., Ng, S.P., Ng, V., and Cheong, B.A. (2000). *Philos. Mag.*, **80**, 729.

23. Choi, W.K., Ho, Y.W., Ng, S.P., and Ng, V. (2001). *J. Appl. Phys.*, **89**, 2168.

24. Choi, W.K., Ng, V., Swee, V.S.L., Ong, S.C., Yu, M.B., Rusli and Yoon, S.F. (2001). *Scripta Mater.*, **44**, 2001.

25. Choi, W.K., Teh, L.K., Bera, L.K., Chim, W.K., Wee, A.T.S., and Jie, Y.X. (2002). *J. Appl. Phys.*, **91**, 444.

26. Choi, W.K., Chim, W.K., Heng, C.L., Teo, L.W., Ho, Vincent, Ng, V., Antoniadis, D.A., and Fitzgerald, E.A. (2002). *Appl. Phys. Lett.*, **80**, 2014.

27. Choi, W.K., Ho, V., Ng, V., Ho, Y.W., Ng, S.P., and Chim, W.K. (2005). *Appl. Phys. Lett.*, **86**, 143114.

28. Choi, W.K., Foo, Y.L., Ho, V., and Nath, R. (2005). *Chem. Phys. Lett.*, **416**, 381.

29. Choi, W.K., Chew, H.G., Ho, V., Ng, V., Chim, W.K., Ho, Y.W., and Ng, S.P. (2006). *J. Cryst. Growth*, **288**, 79.

30. Choi, W.K., Chew, H.G., Zheng, F., Chim, W.K., Foo, Y.L., and Fitzgerald, E.A. (2006). *Appl. Phys. Lett.*, **89**, 113126.

31. Cosentino, S., Ozen, E.S., Raciti, R., Mio, A.M., Nicotra, G., Simone, F., Crupi, I., Turan, R., Terrasi, A., Aydinli, A., and Mirabella, S. (2014). *J. Appl. Phys.*, **115**, 043103.

32. Cullis, A.G., Canham, L.T., and Calcott, P.D.J. (1997). *J. Appl. Phys.*, **82**, 909.

33. Das, K., NandaGoswami, M., Mahapatra, R., Kar, G.S., Dhar, A., Acharya, H.N., Maikap, S., Lee, J.–H., and Ray, S.K. (2004). *Appl. Phys. Lett.*, **84**, 1386.

34. Das, S., Das, K., Singha, R.K., Dhar, A., and Ray, S.K. (2007). *Appl. Phys. Lett.*, **91**, 233118.

35. Das, S., Singha, R.K., Manna, S., Gangopadhyay, S., Dhar, A., and Ray, S.K. (2011). *J. Nanopart. Res.*, **13**, 587.

36. Das, S., Singha, R.K., Dhar, A., Ray, S.K., Anopchenko, A., Daldosso, N., and Pavesi, L. (2011). *J. Appl. Phys.*, **110**, 024310.

37. Das, S., Aluguri, R., Manna, S., Singha, R.K., Dhar, A., Pavesi, L., and Ray, S.K. (2012). *Nanoscale Res. Lett.*, **7**, 143.

38. Dole, S.L., Hunter, O., and Wooge, C.J. (1977). *J. Am. Ceram. Soc.*, **45**, 452.

39. Dutta, A.K. (1995). *Appl. Phys. Lett.*, **68**, 1189.

40. Dutta, A., Hayafune, Y., and Oda, S. (2000). *Jpn. J. Appl. Phys. Part 1*, **39**, 855.

41. Elshocht, S.V., Caymax, M., Conard, T., Gendt, S.D., Hoflijk, I., Houssa, M., Leys, F., Bonzom, R., Jaeger, B.D., Steenbergen, J.V., Vandervorst, W., Heyns, M., and Meuris, M. (2006). *Thin Solid Films*, **508**, 1.

42. Fauchet, P., and Campbell, I.H. (1988). *Crit. Rev. Solid State Mater. Sci.*, **14**(Suppl 1), S79.

43. Freescale Semiconductor, www.freescale.com/files/32bit/doc/.../ BRKINETISELS.pdf

44. Fujii, M., Hayashi, S., and Yamamoto, K. (1991). *Jpn. J. Appl. Phys. Part 1*, **30**, 687.

45. Fukuda, H., Kobayashi, T., Endoh, T., and Ueda, Y. (1998). *Appl. Surf. Sci.*, **130/132**, 776.

46. Futatsugi, T., Nakajima, A., and Nakao, H. (1998). *Fujitsu Sci. Tech. J.*, **34**, 142.

47. Gay, G., Molas, G., Bocquet, M., Jalaguier, E., Gely, M., Masarotto, L., Colonna, J.P., Grampeix, H., Martin, F., Brianceau, P., Vidal, V., Kies, R., Baron, T., Ghibaudo, G., and De Salvo, B. (2012). *IEEE Trans. Electron Devices*, **59**, 933.

48. Gebel, T., Rebohle, L., Skorupa, W., Nazarov, A.N., Osiyou, I.N., and Lysenko, V.S. (2002). *Appl. Phys. Lett.*, **81**, 2575.

49. Grabert, H., and Devoret, M.H. (1992). *Single Charge Tunnelling: Coulomb Brockage Phenomenon in Nanostructures*, Plenum, New York.

50. Haesslein, H., Sielemann, R., and Zistl, C. (1998). *Phys. Rev. Lett.*, **80**, 2626.

51. Hall, R.N. (1952). *Phy. Rev.*, **87**, 387.

52. Hanafi, H.I., Tiwari, S., and Khan, I. (1996). *IEEE Trans. Electron Devices*, **43**, 1553.

53. Haselhoff, M., Reimann, K., and Weber, H.J. (1999). *Eur. Phys. J. B*, **12**, 147.

54. Heinig, K.H., Schmidt, B., Markwitz, A., Grötzschel, R., Strobel, M., and Oswald, S. (1999). *Nucl. Instrum. Methods Phys. Res. B*, **148**, 969.

55. Heng, C.L., Teo, L.W., Ho, V., Tay, M.S., Lei, Y., Choi, W.K., and Chim, W.K. (2003). *Microelectron. Eng.*, **66**, 218.

56. Ho, Y.W., Ng, V., Choi, W.K., Ng, S.P., Osipowicz, T., Seng, H.L., Tjui, W.W., and Li, K. (2001). *Scripta Mater.*, **44**, 1291.

57. Ho, V., Teo, L.W., Choi, W.K., Chim, W.K., Tay, M.S., Antoniadis, D.A., Fitzgerald, E.A., Du, A.Y., Tung, T.H., Liu, R., and Wee, A.T.S. (2003). *Appl. Phys. Lett.*, **83**, 3558.

58. Ho, V., Tay, M.S., Moey, C.H., Teo, L.W., Choi, W.K., Chim, W.K., Heng, C.L., and Lei, Y. (2003). *Microelectron. Eng.*, **66**, 33.

59. Hughey, M.P., and Cook, R.F. (2004). *Thin Solid Films*, **460**, 7.

60. Inokuma, T., Wakayama, Y., Muramoto, T., Aoki, R., Kurata, Y., and Hasegawa, S. (1998). *J. Appl. Phys.*, **83**, 2228.

61. International Centre for Diffraction Data, Card 01-081-0028.

62. International Centre for Diffraction Data, Card 01-004-0545.

63. International Technology Roadmap for Semiconductor Industry (ITRS) (2009). Semiconductor Industry Association, San Jose, CA.

64. Isikawa, Y., Wada, K., Cannon, D.D., Liu, J.F., Luan, H.C., and Kimmerling, L.C. (2003). *Appl. Phys. Lett.*, **82**, 2044.

65. Iyer, S.S., and Xie, Y.-H. (1993). *Science*, **260**, 40.

66. Jessensky, O., Müller, F., and Gösele, U. (1998). *Appl. Phys. Lett.*, **72**, 1173.

67. Jie, Y., Wee, A.T.S., Huan, C.H.A., Shen, Z.X., and Choi, W.K. (2011). *J. Appl. Phys.*, **109**, 033107.

68. Kahng, D., and Sze, S.M. (1967). *Bell Syst. Tech. J.*, **46**, 1288.

69. Kan, E.W.H., Choi, W.K., Leoy, C.C., Chim, W.K., Antoniadis, D.A., and Fitzgerald, E.A. (2003). *Appl. Phys. Lett.*, **63**, 2058.

70. Kan, E.W.H., Choi, W.K., Chim, W.K., Fitzgerald, E.A., and Antoniadis, D.A. (2004). *J. Appl. Phys.*, **95**, 3148.

71. Kan, E.W.H., Chim, W.K., Lee, C.H., Choi, W.K., and Ng, T.H. (2004). *Appl. Phys. Lett.*, **85**, 2349.

72. Kim, I., Han, S., Kim, H., Lee, J., Choi, B., Hwang, S., Ahn, D., and Shin, H. (1998). *Int. Electron Dev. Meet.*, 111.

73. Kim, I., Han, S., Han, K., Lee, J., and Shin, H. (1999). *IEEE Electron Dev. Lett.*, **20**, 630.

74. Kim, Y., Park, K.H., Chung, T.H., Bark, H.J., Yi, J.-Y., Choi, W.C., Kim, E.K., Lee, J.W., and Lee, J.Y. (2001). *Appl. Phys. Lett.*, **79**, 433.

75. King, Y.C., King, T.J., and Hu, C. (1998). *Int. Elecron Dev. Meet.*, 115.

76. King, Y.C. (1999). PhD Thesis, University of Berkeley, Berkeley, CA.

77. Klimenkov, K., Matz, W., and von Borany, J. (2000). *Nucl. Instrum. Methods Phys. Res. B*, **168**, 367.

78. Knotek, M.L., and Feibelman, P.J. (1978). *Phys. Rev. Lett.*, **40**, 964.

79. Koh, B.H., Chim, W.K., Ng, T.H., Zheng, J.X., and Choi, W.K. (2004). *J. Appl. Phys.*, **95**, 5094.

80. Koh, B.H., Kan, E.W.H., Chim, W.K., Choi, W.K., Antoniadis, D.A., and Fitzgerald, E.A. (2005). *J. Appl. Phys.*, **97**, 124305.

81. Kouvatsos, D.N., Ioannou-Sougleridis, V., and Nassiopoulou, A.G. (2002). *Appl. Phys. Lett.*, **82**, 397.

82. Lee, R.W. (1963). *J. Chem. Phys.*, **38**, 488.

83. Lee, S.J., Luan, H.F., Bai, W.P., Lee, C.H., Jeon, T.S., SEnzaki, Y., Roberts, D., and Kwong, D.L. (2000). *IEDM Tech. Digest.*, 31.

84. Lee, N., Choi, S., and Kang, S. (2006). *Appl. Phys. Lett.*, **88**, 073101.

85. Li, A.P., Müller, F., Birner, A., Nielsch, K., and Gösele, U. (1998). *J. Appl. Phys.*, **84**, 6023.

86. Likharev, K.K. (1998). *Appl. Phys. Lett.*, **73**, 2137.

87. Likharev, K.K. (1999). *Nanotechnology*, **10**, 159.

88. Liu, L., Teo, K.L., Shen, Z.X., Sun, J.S., Ong, E.H., Kolobov, A.V., and Maeda, Y. (2004). *Phys. Rev. B*, **69**, 125333.

89. Liu, J., Cannon, D.D., Wada, K., Ishikawa, Y., Danielson, D.T., Jongthammanurak, S., Michel, J., and Kimmerling, L.C. (2004). *Phys. Rev. B*, **70**, 55309.

90. Liu, H., Ferrer, D.A., Ferdousi, F., and Banerjee, S.K. (2009). *Appl. Phys. Lett.*, **95**, 203112.

91. Maeda, Y. (1991). *Phys. Rev. B*, **51**, 1658.

92. Maeda, Y., Tsukamoto, N., Yazawa, Y., Kanemitsu, Y., and Masumoto, Y. (1991). *Appl. Phys. Lett.*, **59**, 3168.

93. Maikap, S., Lee, H.Y., Wang, T.Y., Tzeng, P.J., Wang, C.C., Lee, L.S., Liu, K.C., Yang, R.J., and Tsai, M.J. (2007). *Semicond. Sci. Technol.*, **22**, 884.

94. Masuda, H., Yasui, K., Sakamoto, Y., and Nakao, M. (2001). *Jpn. J. Appl. Phys.*, **40**, 1267.

95. McWhorter, P.J., Miller, S.L., and Dellin, T.A. (1990). *J. Appl. Phys.*, **68**, 1902.

96. Mestanza, S.N.M., Doi, I., Swart, J.W., and Frateschi, N.C. (2007). *J. Mater. Sci.*, **42**, 7757.

97. Micron (2009). http://investors.micron.com/releasedetail.cfm?REleaseID=467770.

98. Min, K.S., Shcheglov, K.V., Yang, Y.M., Atwater, H.A., Brongersma, M.L., and Polman, A. (1996). *Appl. Phys. Lett.*, **68**, 2511.

99. Minke, M.V., and Jackson, K.A. (2005). *J. Non-Cryst. Sol.*, **351**, 2310.

100. Nayak, D.K., Kamjoo, K., Woo, J.C.S., Park, J.S., and Wang, K.L. (1990). *Appl. Phys. Lett.*, **56**, 66.

101. Nemanich, R.J., Solin, S.A., and Martin, R.M. (1981). *Phys. Rev. B*, **23**, 6348.

102. Ng, V., Ng, S.P., Thio, H.H., Choi, W.K., Wee, A.T.S., and Jie, Y.X. (2000). *Mater. Sci. Eng. A*, **286**, 161.

103. Ng, W.L., Lourenco, M.A., Gwilliam, R.M., Dedain, S., Shao, G., and Homewood, K.P. (2001). *Nature*, **410**, 192.

104. Ng, T.H., Chim, W.K., Choi, W.K., Ho, V., Teo, L.W., Du, A.Y., and Tung, C.H. (2004). *Appl. Phys. Lett.*, **84**, 4385.

105. Ng, T.H., Ho, V., Teo, L.W., Tay, M.S., Koh, B.H., Chim, W.K., Choi, W.K., Du, A.Y., and Tung, C.H. (2004). *Thin Solid Films*, **462/463**, 46.

106. Niquet, Y.M., Allan, G., Deletue, C., and Lannoo, M. (2000). *Appl. Phys. Lett.*, **77**, 1182.

107. Novikov, P., Heinig, K.H., Larsen, A., and Dvurechenskii, A. (2002). *Nucl. Instrum. Methods Phys. Res. B*, **191**, 462.

108. Okamoto, S., and Kanemitsu, Y. (1996). *Phys. Rev. B*, **54**, 16421.

109. Oku, T., Nakayama, T., Kuno, M., Nozue, Y., Wallenberg, L.R., Niihra, K., and Saganuma, K. (2000). *Mater. Sci. Eng. B*, **74**, 242.

110. Ostraat, M.L., De Blauwe, J.W., Green, M.L., Bell, L.D., Brongersma, M.L., Casperson, J., Flagan, R.C., and Atwater, H.A. (2001). *Appl. Phys. Lett.*, **79**, 433.

111. Paine, D.C., Caragianis, C., and Schwartwazman, A.F. (1991). *J. Appl. Phys.*, **70**, 5076.

112. Paine, D.C., Caragianis, C., Kim, T.Y., and Shigesato, Y. (1993). *Appl. Phys. Lett.*, **62**, 2842.

113. Peibst, R., de Sousa, J.S., and Hofmann, K.R. (2010). *Phys. Rev. B*, **82**, 195415.

114. Peibst, R., Erenburg, M., Bugiel, E., and Hofmann, K.R. (2010). *J. Appl. Phys.*, **108**, 054316.

115. Richter, H., Wang, Z.P., and Ley, L. (1981). *Sol. Stat. Commun.*, **39**, 625.

116. Rodríguez, A., Ortiz, M.I., Sangrador, J., Rodríguez, T., Avella, M., Prieto, A.C., Torres, Á., Jiménez, J., Kling, A., and Ballesteros, C. (2007). *Nanotechnology*, **18**, 065702.

117. Ruddy, D.A., Johnson, J.C., Smith, R.E., and Neale, N.R. (2010). *ACS Nano*, **4**, 7459.

118. Saeed, S., Buters, F., Dohnalova, K., Wosinski, L., and Gregorkiewicz, T. (2014). *Nanotechnology*, **25**, 405705.

119. Sharp, I.D., Yi, D.O., Xu, Q., Liao, C.Y., Beeman, J.W., Lilienal-Weber, Z., Yu, K.M., Zakarov, D.N., Ager, J.W. III, Chrzan, D.C., and Haller, E.E. (2005). *Appl. Phys. Lett.*, **86**, 063107.

120. She, M., and King, T.J. (2003). *IEEE Trans. Electron Devices*, **50**, 1934.

121. Shen, J.K., Wu, X.L., Tan, C., Yuan, R.K., and Bao, X.M. (2002). *Phy. Lett. A*, **300**, 307.

122. Shi, Y., Saito, K., Ishikuro, H., and Hiramoto, T. (1998). *J. Appl. Phys.*, **84**, 2538.

123. Shockley, W., and Read, W.T. (1952). *Phys. Rev.*, **87**, 835.

124. Takagahara, T., and Takeda, K. (1992). *Phys. Rev. B*, **46**, 15578.

125. Takeoka, S., Fujii, M., Hayashi, S., and Yamamoto, K. (1998). *Phys. Rev. B*, **58**, 7921.

126. Takeoka, S., Fujii, M., and Hayashi, S. (2000). *Phys. Rev. B*, **62**, 16820.

127. Takeoka, S., Toshikiyo, K., Fujii, M., Hayashi, S., and Yamamoto, K. (2000). *Phys. Rev. B*, **61**, 15988.

128. Tarachi, G., Saini, S., Fan, W.W., Kimmerling, L.C., and Fitzgerald, E.A. (2003). *J. Appl. Phys.*, **93**, 9988.

129. Teo, L.W., Choi, W.K., Chim, W.K., Ho, V., Moey, C.M., Tay, M.S., Heng, C., Lei, Y., Antoniadis, D.A., and Fitzgerald, E.A. (2002). *Appl. Phys. Lett.*, **81**, 3639.

130. Terasawa, N., Akimoto, K., Mizuno, Y., Ichimiya, A., Sumitani, K., Takahashi, T., Zhang, X.W., Sugiyama, H., Kawata, H., Nabatame, T., and Torumi, A. (2005). *Appl. Surf. Sci.*, **224**, 16.

131. Thielsch, R., Gatto, A., and Kaiser, N. (2002). *Appl. Opt.*, **41**, 3211.

132. Tiwari, S., Rana, F., Hanafi, H., Hartstein, A., Crabbé, E.F., and Chan, K. (1996a). *Appl. Phys. Lett.*, **68**, 1377.

133. Tiwari, S., Rana, F., Chan, K., Shi, L., and Hanafi, H.I. (1996). *Appl. Phys. Lett.*, **69**, 1232.

134. Wahl, J.A., Silva, H., Gokirmak, A., Kumar, A., Welser, J.J., and Tiwari, S. (1999). *Tech. Dig. Int. Electron Devices Meet.*, 375.

135. Wang, Y., Yang, Y., Guo, Y., Yue, Y., and Gan, R. (1996). *Mater. Lett.*, **29**, 159.

136. Wang, Y.Q., Chen, J.H., Yoo, W.J., Yeo, Y.C., Kim, S.J., Gupta, R., Tan, Z.Y.L., Kwong, D.L., Du, A.Y., and Balasubramanian, N. (2004). *Appl. Phys. Lett.*, **84**, 5407.

137. Weinberg, Z.A. (1977). *Sol. Stat. Electron.*, **20**, 11.

138. Wellner, A., Paillard, V., Bonafos, C., Coffin, C., Claverie, H., Schmidt, B., and Heinig, K.H. (2003). *J. Appl. Phys.*, **94**, 5639.

139. Wu, X.L., Gao, T., Bao, X.M., Yan, F., Jiang, S.S., and Feng, D. (1997). *J. Appl. Phys.*, **82**, 2704.

140. Yang, M., Chen, T.P., Wong, J.I., Liu, Y., Ding, L., Liu, K.Y., Zhang, S., Zhang, W.L., Gui, D., and Ng, C.Y. (2009). *J. Phys. D: Appl. Phys.*, **42**, 035109.

141. Zacharias, M., and Fauchett, P.M. (1997). *Appl. Phys. Lett.*, **71**, 380.

142. Zhang, J.-Y., Ye, Y.-H., and Tan, X.-L. (1999). *Appl. Phys. Lett.*, **74**, 2459.

143. Zhang, J.-Y., Ye, Y.-H., Tan, X.-L., and Bao, X.-M. (2000). *Appl. Phys. A*, **71**, 299.

144. Zheng, F., Choi, W.K., Lin, F., Tripathy, S., and Zhang, J.X. (2008). *J. Phys. Chem. C*, **112**, 9223.

145. Zheng, F., Choi, W.K., and Liew, T.H. (2008). *J. Appl. Phys.*, **104**, 084132.

146. Zhu, W.J., Ma, T.P., Tamagawa, T., Kim, J., and Di, Y. (2002). *IEEE Electron Device Lett.*, **23**, 97.

147. Chindalore, G., Yater, J., Gasquet, H., Suhail, M., Kang, S.-T., Hong, C.M., Ellis, N., Rinkenberger, G., Shen, J., Herrick, M., Malloch, W., Syzdek, R., Baker, K., and Chang, K.-M. (2008). *IEEE Symposium on VLSI Technology Digest*, 136.

Chapter 6

Nanographene Flash Memory

Jianling Meng,[a],* Rong Yang,[a],* and Guangyu Zhang[a,b,c],*

[a]*Beijing National Laboratory for Condensed Matter Physics and Institute of Physics, Chinese Academy of Sciences, Beijing 100190, China*
[b]*Collaborative Innovation Center of Quantum Matter, Beijing 100190, China*
[c]*Beijing Key Laboratory for Nanomaterials and Nanodevices, Beijing 100190, China*
gyzhang@iphy.ac.cn; ryang@iphy.ac.cn

Nanomaterials have novel physical and chemical properties as a result of the size-confinement effect. Metal- and semiconductor-based nanocrystals have been proved to be promising floating-gate charge-trapping media for nonvolatile flash memory. Miniaturization of memory devices requires one to reduce the size of such materials incessantly. However, due to the quantized energy levels associated with a size below the quantum limit, the performance of semiconductor nanocrystals is degraded. The instability and the fabrication process incompatibility also limit the development of metal nanocrystals for floating-gate memory. Newly emerging

---

*All authors contributed equally to this work.

Nanocrystals in Nonvolatile Memory
Edited by Writam Banerjee
Copyright © 2018 Pan Stanford Publishing Pte. Ltd.
ISBN 978-981-4774-73-4 (Hardcover), 978-1-351-20327-2 (eBook)
www.panstanford.com

graphene has attracted much research interest for flash memory materials. Compared with conventional materials, atomic-layer-thick graphene is chemically inert and thermally stable; meanwhile it has high electrical conductivity and high carrier mobility and is compatible with the conventional semiconductor fabrication processing. In this chapter, we give an overview of the synthesis, properties, and memory device applications of graphene and related graphene nanosheets. The promising findings as well as the future potential directions of this research field have been highlighted.

6.1   Introduction

6.1.1   Graphene Fundamentals

6.1.1.1   Structure and electronic properties of graphene

Ideally, graphene is an infinite 2D hexagonal structure consisting of sp^2-hybridized carbon atoms. The crystal structure of graphene can be thought of as two equivalent carbon triangular sublattices, labeled as A and B [1], as shown in Fig. 6.1. Figure 6.1b shows the band structure of graphene obtained by tight binding approximation. The valence and conduction bands of graphene intersect at the K and K' points, which makes this material a zero-bandgap semimetal with unique electronic properties. Near these crossing points, the electron energy is linearly dependent on the wave vector. This linear dispersion results in massless excitons, which are described by the Dirac equation [2]. Dirac fermions (electrons or holes) exhibit unusual properties compared with ordinary electrons. Furthermore, at the Dirac point, the wave function of the electron can be expressed as a linear combination of the two sets of A and B sublattice wave functions, which is similar to the spin wave function of quantum mechanics. Hence, the sublattice degeneracy and honeycomb symmetry lead to eigenstates that hold an additional quantum (Berry's) phase [3], associated with the so-called pseudospin quantum degree of freedom, making graphene contain rich and novel physical phenomena and properties. An anomalous integer quantum Hall effect [3], weak antilocalization [4] and minimum conductivity, Klein tunneling phenomenon [5], jittery motion of the wave function under confining potentials [6], and large mean free paths have been observed in graphene. For a comprehensive understanding of the properties of graphene, see Ref. [1].

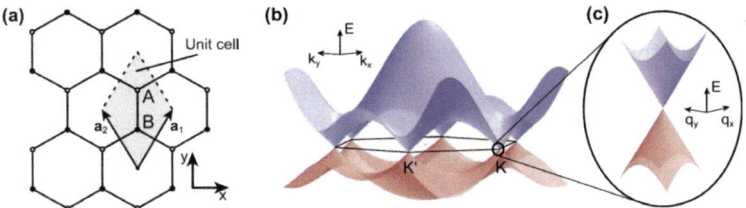

Figure 6.1 (a) Graphene lattice with two carbon atoms per unit cell, denoted by A and B. (b) Electronic dispersion of graphene. The π band (red) and the π^* band (blue) touch each other at singularity points (K and K' points). (c) The linear dispersion and the band structure at the Dirac point. Reprinted (figure) with permission from Ref. [1]. Copyright (2009) by the American Physical Society.

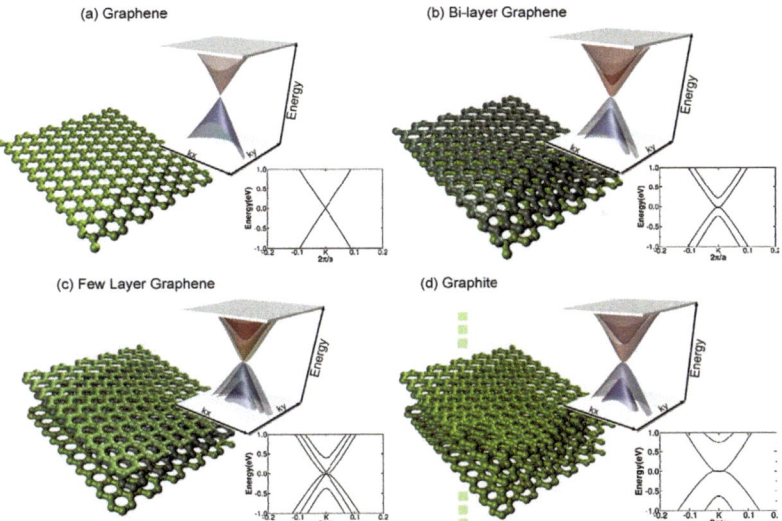

Figure 6.2 Low-energy DFT 3D band structure for (a) graphene, (b) bilayer graphene, (c) trilayer graphene, and (d) graphite. (a) The characteristic Dirac point of graphene. The Dirac point is lost in bilayer graphene (b) but appears again in trilayer graphene (c). (d) 3D graphite structure, which displays a semimetallic band structure with parabolic-like bands. The Fermi level has been set at zero in all cases. Reprinted from Ref. [7], Copyright (2010), with permission from Elsevier.

Strictly speaking, graphene is simply a single layer of graphite, making it only one atom thick. In fact, there is also bilayer, trilayer, and few-layer graphene, different from monolayer graphene or graphite. The electronic properties of graphene change with the number of layers and the stacking order (Fig. 6.2) [7]. Bilayer graphene with AB stacking shows parabolic bands (thus no Dirac electrons), which touch at the Fermi level. The bandgap can be opened in bilayer graphene with the symmetry being broken by an external electric field, which is of interest in technological applications [8]. Trilayer graphene shows an interesting band structure that looks like a combination of monolayer and bilayer graphene [9]. In general, for few-layer graphene with N layers (AB stacking), there will be a linear band (Dirac fermions) if N is odd [10]. As the number of layers increases, the band structure becomes more complicated: several charge carriers appear, and the conduction and valence bands start to overlap notably. As the number of layers increases, the limit of thickness at which graphene can still be considered as different from graphite is determined by the changed electronic structures [11].

As a 2D crystal, graphene is stable at room temperature and has many unique properties. Graphene has a highly symmetrical structure that is chemically inert. When a mechanical force is applied to graphene, the plane is very prone to bending and deformation due to its good flexibility. Regarding the mechanical strength of an individual graphene sheet (GS), graphene has a breaking strength 200 times larger than that of steel, with estimated Young's modulus values of 2.4 ± 0.4 TPa for monolayer graphene and 2.0 ± 0.5 TPa for bilayer graphene, respectively [12, 13]. A high transmittance of 97.7% per graphene layer is observed in visible light that is wavelength independent [14]. Carrier transport measurements have demonstrated that graphene is a half metal at room temperature and shows a bipolar effect. The charge carriers in graphene have zero effective mass, the free carrier concentration is as high as 10^{13} cm^2, and electron mobility has reached up to 15,000 cm^{2}V^{-1}S^{-1} with very little temperature dependency [15]. The low temperature dependency of the electron mobility of graphene indicates that electron movement is mainly affected by impurities or defect scattering rather than phonon scattering [16]. As a result, the carrier mobility of graphene can be improved to a value as high as 1,000,000 cm^{2}V^{-1}S^{-1} by increasing the crystal quality and decreasing the doping level [17]. Graphene has a theoretical resistivity of 10^{-6} Ωcm, which makes it the prominent material with the lowest resistivity at room temperature. Graphene also has extremely high thermal conductivity, ranging from $(4.84 \pm 0.44) \times 10^3$ to $(5.30 \pm 0.48) \times 10^3$ W/mK, higher than experimental values for carbon nanotubes (CNTs) and diamond [18].

6.1.1.2 Graphene nanostructures and graphene nanosheets

The extremely high carrier mobility, mechanical flexibility, optical transparency, and chemical stability of 2D Dirac material graphene provides a great opportunity for the development of high-performance electronic devices. However, one challenge in graphene electronics arises from its metallic nature. An effective way to open and control its bandgap is to tailor the 2D graphene crystals into nanostructures. Excitons in graphene have an infinite Bohr diameter. Thus, graphene fragments of any size will show quantum confinement effects (QCEs). When infinite graphene crystals become finite, edges and boundaries appear, and if the size is on the order of nanometers, graphene nanostructures exhibit different properties from those observed in bulk [19]. Typical graphene nanostructures are graphene nanoribbons (GNRs) (Fig. 6.3a) and nanosheets (also called nanographene [NG] or graphene quantum dots [GQDs]) (Fig. 6.3b). In analogy with the 2D counterpart, one can also have GNRs, bilayer nanoribbons, few-layer nanoribbons, and graphitic nanoribbons or nanocrystals. In general, a GNR could be defined as a 1D sp^2-hybridized carbon crystal with boundaries that expose non-three-coordinated carbon atoms and possess a large aspect ratio. Edge terminations could be armchair, zigzag, or a combination of both [20]. GNRs are commonly used as device/circuit architectures with width-edge-dependent bandgaps [21]. Theoretically, perfectly armchair edged GNRs show room temperature on-off property with high carrier mobility. NG has zero dimensionality and no periodicity is present [22]. NG, especially when it is below 20 nm in diameter, has a non-zero bandgap and excitation luminescence. The bandgap can be tuned by modifying the size and surface chemistry of the NG. Density functional theory (DFT) calculations have shown that the bandgap of NG increases to approximately 2 eV in a cluster consisting of 20 aromatic rings and 7 eV for benzene. Very little work has been done on the measurement of electronic properties of NG. The early work on NG was focused on the investigation of its theoretical spin properties due to the difficulty in sample preparation [22]. More recently, NG has been obtained in massive amounts, chemically modified, and used for the first time in energy conversion, memory, biology, and sensors (see Refs. [23, 24] for a comprehensive introduction). Here in this book, we will focus on the memory application of NG.

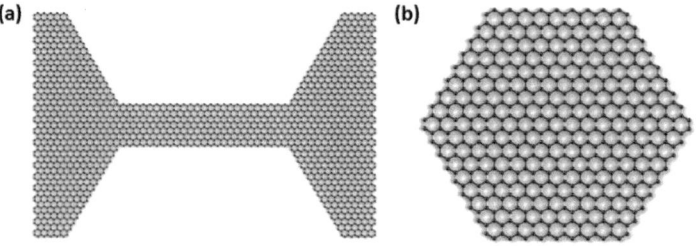

Figure 6.3 Sketch map of graphene nanoribbons (a) and nanographene (b).

6.2 Preparation/Synthesis of Graphene and Nanographene

6.2.1 Graphene Thin-Film Preparation

Since graphene was first isolated from graphite by a mechanical exfoliation method, many methods have been explored for graphene preparation, such as reduced graphene oxide (RGO), liquid exfoliation, thermal decomposition of SiC, epitaxial growth on transition metals, and chemical vapor deposition (CVD). The advantages and limitations of these methods have been discussed in previous review papers in detail [25, 26]. However, it remains a challenge to produce large-scale high-quality graphene. Here we focus on the promising RGO and CVD methods, which are inexpensive and feasible for large-area graphene production, thus providing technology compatibility and scale-up possibility for their application in memory devices.

6.2.1.1 Reduced graphene oxide

The key to the rapid development and application of a new material lies in the ability to produce high-quality materials on an industrial scale. The chemical reduction of graphene oxide (GO) is a promising route toward the large-scale production of graphene for commercial applications [27]. Figure 6.4 shows the synthesis process of graphene from GO.

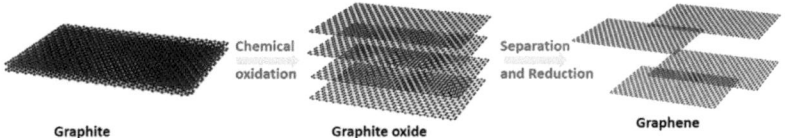

Graphite   **Graphite oxide**   **Graphene**

Figure 6.4   Wet-chemical synthesis of graphene from graphite.

GO is one type of graphene-based materials prepared by the chemical separation of graphite layers by Brodie's [28], Staudenmaier's [29], Hummers's [30], and improved Hummers's methods [31]. Firstly, natural graphite is treated with strong oxidative solutions for resulting graphite oxide. The selection of strong oxidative solutions has determined the efficiency of the oxidation process. Currently, Hummers's method ($KMnO_4$, $NaNO_3$, H_2SO_4) is the most common method used as the reaction is fast and no explosive gases are formed. Graphite oxide preserves the layered structure of the starting graphite, while the layers are heavily oxidized. Then, it is easy to separate graphite oxide into water or other polar solvents by mild sonication or stirring. The obtained solution is GO containing a majority of monolayer or few-layer flakes. Figure 6.5 shows the typical atomic force microscopy (AFM) topography images of GO spin-coated onto a silicon oxide substrate [31]. The root-mean-square roughness/mean roughness for Fig. 6.5a is 1.133/0.635 nm. The section analysis indicates (Fig. 6.5c) the thickness of the GO sheet is around 1 nm. Such thickness is larger than the theoretical value of 0.34 nm for perfectly flat graphene.

GO has an irregular structure with different oxygen-containing functionalities: hydroxyl and epoxy groups bonded to the basal plane and carboxylic or carbonyl groups attached to the edges of graphene (as shown in Fig. 6.6) [32].

Compared with highly ordered graphene with negligible defects and high conductivity, GO is heavily oxidized and electrically nonconductive. High-resolution transmission electron microscopy (TEM) results (Fig. 6.7b) have shown that GO has a highly inhomogeneous microstructure with holes and graphitic regions filled in the disorder oxidized regions that form a continuous network across the basal plane [33]. The property of GO is controlled by the degree of the oxidation process. Theoretical study indicates that GO sheets become conducting if their coverage with functional groups does not exceed 25%; otherwise they are insulators [34].

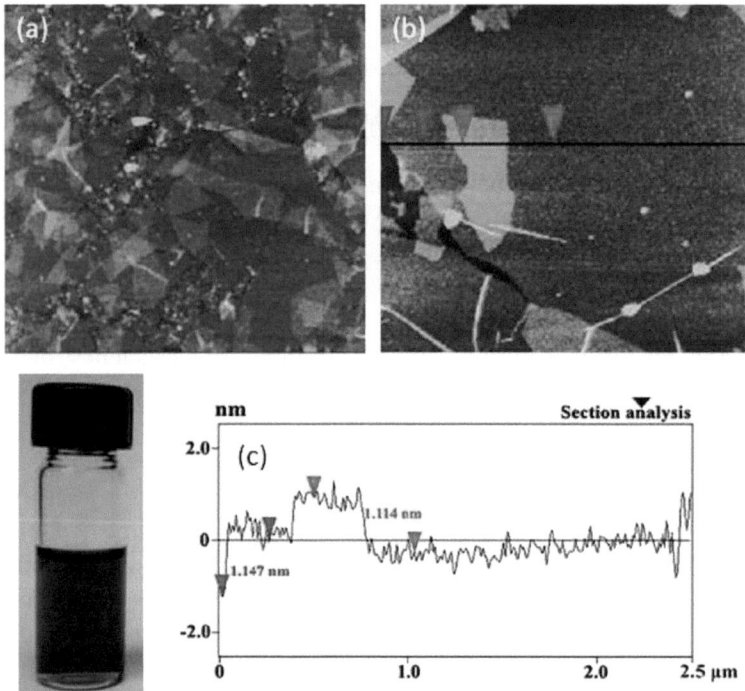

Figure 6.5 AFM images of graphene oxide and the corresponding section analysis. Adapted with permission from Ref. [31]. Copyright (2010) American Chemical Society.

GO has rather stronger oxidizing properties and can be usually chemical reduced by treatment with hydrazine [35], hydroxylamine [36], sodium borohydride [37], and other reducing agents [27]. Besides, GO can be reduced by high-temperature annealing in a reducing atmosphere (Ar/H_2) [38], by photothermal heating at room temperature [39], by a solvothermal approach [40], or by photocatalytic [41]. Recent studies suggest that high-quality GO can be obtained by high-temperature annealing or plasma treatment in a carbonaceous atmosphere [42–44]. These reducing methods remove the majority of the oxygen-containing functional groups in GO and the resulting product is called RGO, which has partially restored crystallinity and electrical conductivity. High-resolution TEM results in Fig. 6.7c have shown that RGO contains more holes and increased size of graphitic regions. The holes are expected as CO and CO_2 form during annealing [33]. The significant increase in graphitic regions results in a substantial restoration of sp^2 carbon networks. Disordered regions still exist due to the stubborn nitrogen and oxygen functionalities. Complete reduction of GO into graphene is very difficult, and 90% reduction of GO is relatively easy.

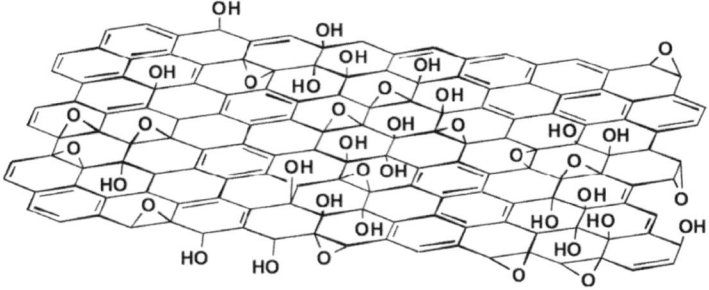

Figure 6.6 Proposed schematic (Lerf–Klinowski model) of the graphene oxide structure. Reprinted from Ref. [32], Copyright (1998), with permission from Elsevier.

Figure 6.8 shows the electrical properties of a typical electronic device based on RGO that was obtained by reduction of GO using CH_4 plasma, which is a high-efficiency method that can restore high-quality graphene from GO through defect repair [42]. The conductivity of the RGO is found to be over 4 orders of magnitude higher than that of as-made GO. The sheet resistance RS is 9–20 kΩ, which is very close to that of pristine graphene. This achieved best graphene device also gives a very high conductivity, with σ around 1590 S/cm. The RGO behaves as p-type doping in air and recovers slightly ambipolar in vacuum after desorption of atmospheric absorbates. Temperature-dependent electrical measurements were carried out in vacuum from room temperature down to 4 K. RGO devices also show p-type behavior in low-temperature measurements, which can be explained by the presence of persistent positively charged impurities and substrate charge transfer. The

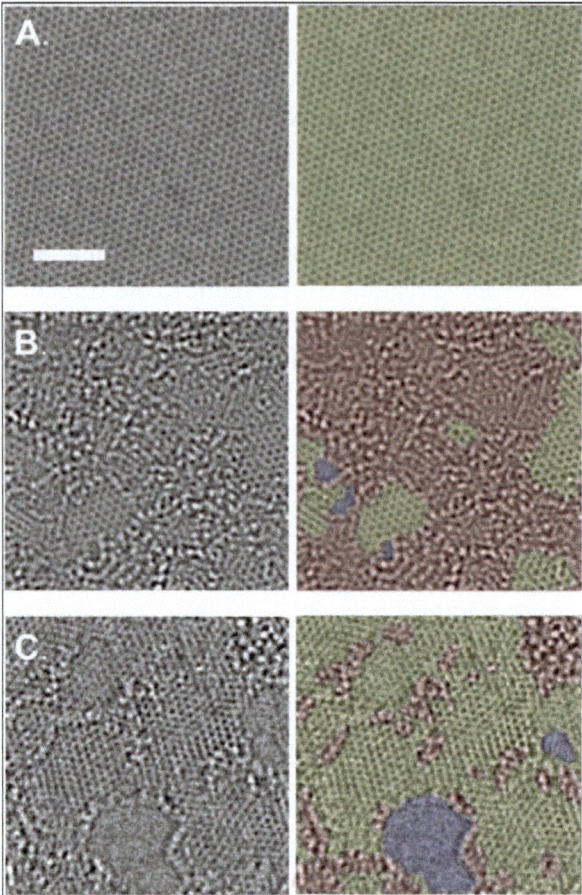

Figure 6.7 Aberration-corrected TEM images of the following materials (scale bar, denoting 2 nm, valid for all images): (A) a single suspended sheet of graphene; on the right, the graphitic area indicated in yellow; (B) a single suspended sheet of graphene oxide; on the right, holes indicated in blue, graphitic areas in yellow, and high-contrast, disordered regions, indicating oxygen functionalities, in red; (C) a suspended monolayer of RGO; on the right, holes indicated in blue, graphitic regions in yellow, and areas with remaining functionalities in red. Adapted with permission from Ref. [33]. Copyright © 2010 WILEY-VCH Verlag GmbH & Co. KGaA, Weinheim.

resistance of RGO increases with decreasing temperature, which is different from the behavior of graphene, which shows temperature independence. However, the superior conduction performance of RGO is also supported by its high conductivity, even at low temperature, with only a slight rise in resistance upon cooling down. Compared with pristine graphene, RGO has larger sheet resistance due to its corrugated surface and residual defects and its electrical conductance can be described by the 2D variable range hopping (VRH) transport mechanism [44, 45].

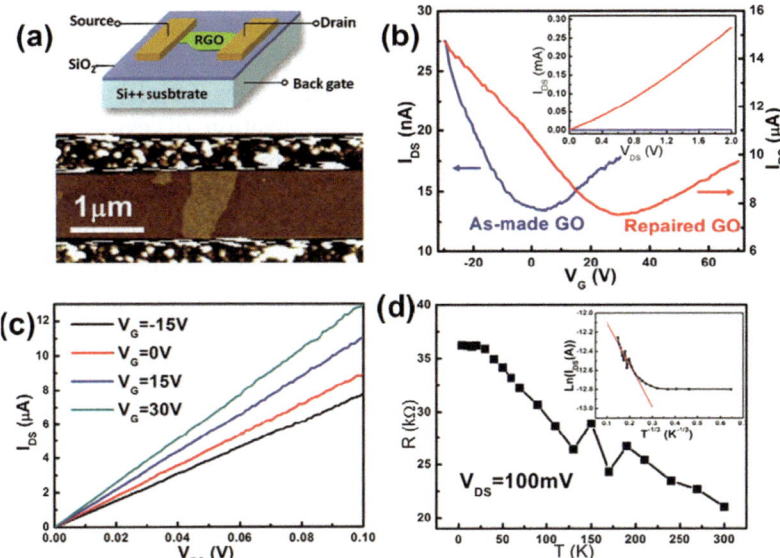

Figure 6.8   Electrical properties of an RGO sheet. (a) Schematic diagram (top) and a typical AFM image (down) of an RGO sheet device. (b) $I–V_G$ curves for as-made graphene oxide (blue) and repaired graphene oxide (red) (V_{bias} = 0.1 V), whose $I–V$ curves are showed in the respective inset diagrams. (c) $I–V$ curves for different gate voltages that show the tunability of graphene-based material and the linear relation of $I–V$. (d) Resistance at different temperatures of a repaired graphene oxide device. The inset is a plot of current against $T^{-1/3}$ partly linearly fit.

Although the quality of RGO is far less than that of pristine graphene, the facile, cheap, and compatible preparation techniques make RGO easy to produce on a massive scale, satisfying the requirement of large-scale commercial applications.

6.2.1.2   Chemical vapor depositions

Similar to CNTs, CVD is the most promising technique to synthesize large-area uniform graphene films at a relatively low cost. The first CVD growth of graphene was reported in 2008 by using Ni as the catalyst substrate [46]; then an explosion of research works followed, using various kinds of transition metal substrates, for example, Cu, Ir, Co, Pt, Fe, Au, Rh, Ru, Cu-Ni, Au-Ni, and Ni-Mo alloys, which show different carbon solubility and catalytic effects (see Refs. [47–49] for a comprehensive introduction). Single-layer and multilayer graphene continuous films are formed on these polycrystalline metals and metal alloys through catalytic pyrolysis of hydrocarbons (such as methane, ethylene, and acetylene). The growth process of graphene involves three stages [50]: (i) annealing the metal substrates to induce its recrystallization, (ii) exposing to a diluted hydrocarbon source, and (iii) cooling the metal substrates. Any steps can affect the crystal quality and layer thickness of graphene. The growth mechanisms include two broad factors (Fig. 6.9): surface segregation of carbon from metal substrates and surface decomposition and deposition of hydrocarbon. During the reactions, transition metal substrates play a dominant role that not only works to catalyze the dehydrogenation of hydrocarbon but also determines the graphene growth mechanism, which ultimately affects the quality of as-grown graphene. Among so many optional metal substrates, low-cost Ni and Cu are two dominant substrates that can achieve reproducible growth of high-quality, single-/few-layer graphene at ambient pressure and a low temperature with a simple post-transfer process. In the following sections, we will focus on graphene growth on Ni and Cu.

6.2.1.2.1 *CVD synthesis of graphene on Ni*

Usually, thin (200–500 nm) polycrystalline Ni films deposited by e-beam evaporation or sputtering on oxidized silicon substrates are used as metal catalysts. In a typical growth process [51, 52], polycrystalline Ni films are first annealed in a protective gas (Ar/H_2) at ~1000°C to increase grain size and then exposed to a H_2/CH_4 gas mixture. At this time, the hydrocarbon decomposes and carbon atoms dissolve into the Ni film to form a solid solution because Ni has relatively high carbon solubility at elevated temperatures. After a short time, Ni films are cooled down in argon gas. During the cooling, dissolved carbon atoms start to diffuse out and then segregate onto the Ni surface to form graphene films as the Ni (111) surface has an excellent lattice match with hexagonal graphene, which can serve as an ideal template for graphene growth. Finally, continuous graphene films with monolayer and few-layer regions mixed can form on Ni substrates (Fig. 6.10) [51]. Most of the multilayer nucleation occurs at Ni grain boundaries. The growth of graphene on Ni follows the rules of carbon segregation and precipitation process. Hence, controlling the carbon dissolution and segregation process on Ni is the key step to achieving high-quality graphene with a uniform thickness [52]. The growth time and hydrocarbon concentration may affect the graphene thickness due to different amounts of carbon dissolved in Ni films. High temperature, high hydrogen concentration, and short growth time are necessary parameters for producing few-layer graphene with minimal defects [46, 51]. The cooling-rate-dependent segregation behavior strongly affects the thickness and quality of graphene films. Medium cooling rates are demonstrated to induce optimal carbon segregation and produce few-layer graphene with fewer defects. Besides, the crystal structure of Ni films also plays an important role in the morphology formation of the graphene [53]. A single-crystal Ni (111) substrate with a smooth surface might be the best choice for possible uniform monolayer graphene growth [54].

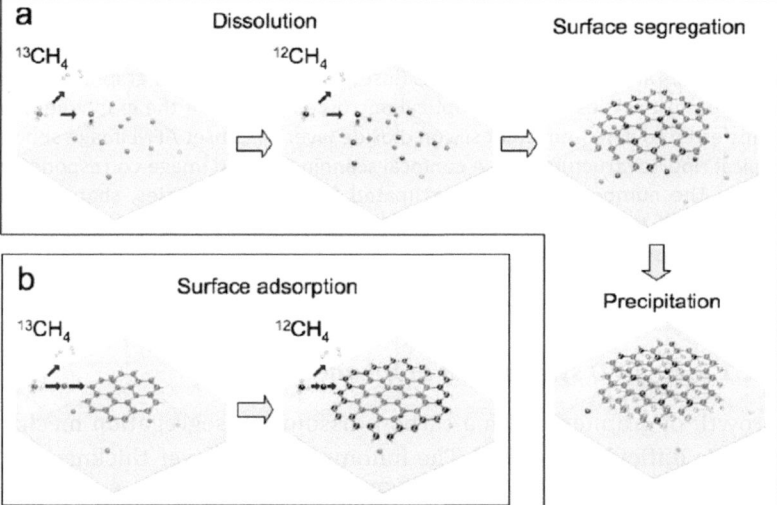

Figure 6.9 Schematic diagrams of the possible growth mechanisms of CVD graphene. (a) Surface segregation and precipitation after dissolution of carbon in the metal. (b) Surface adsorption on metal surface. Adapted with permission from Ref. [50]. Copyright (2009) American Chemical Society.

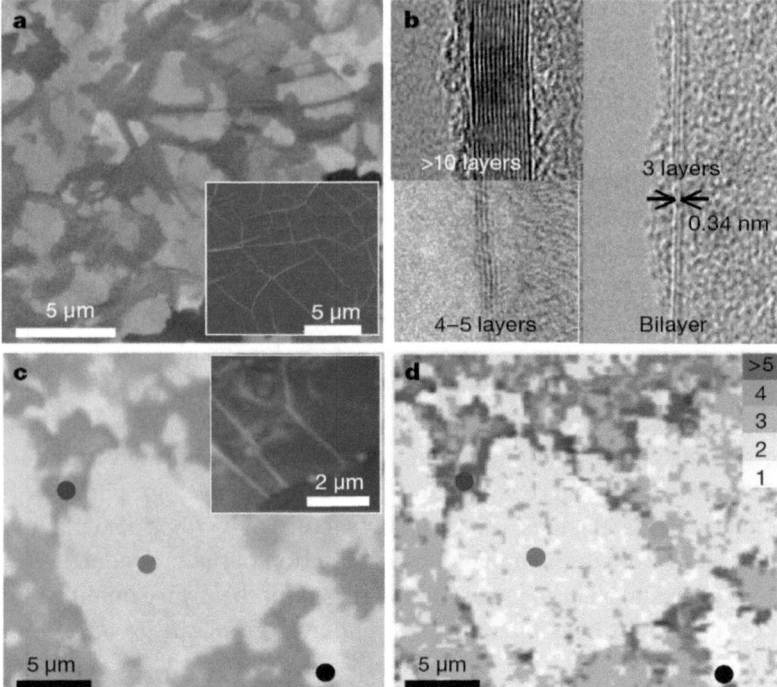

Figure 6.10 (a) SEM images of as-grown graphene films on thin (300 nm) nickel layers and thick (1 mm) Ni foils (inset). (b) TEM images of graphene films of different thicknesses. (c) An optical microscope image of the graphene film transferred to a 300-nm-thick silicon dioxide layer. The inset AFM image shows typical rippled structures. (d) A confocal scanning Raman image corresponding to (c). The number of layers is estimated from the intensities, shapes, and positions of the G band and 2D band peaks. Adapted by permission from Springer Customer Service Centre GmbH: Springer Nature, *Nature* (Ref. [51]), Copyright (2009).

6.2.1.2.2 *CVD synthesis of graphene on Cu*

Growth of graphene with a carbon-dissolution-segregation mechanism is difficult to control. The inhomogeneous layer thickness reduces the property of graphene [50]. Direct deposition of graphene with CVD without dissolution-precipitation can be accomplished by using metallic substrates that have negligible carbon solubility at an elevated temperature. Cu is just the substrate. A Cu foil is first annealed in a hydrogen atmosphere at 1000°C, and then a mixture of

H_2/CH_4 is introduced into the system to initiate graphene growth. After a continuous graphene layer is formed on the Cu foil, the system is cooled down to room temperature. Compared with graphene growth on Ni, graphene growth on Cu follows the surface decomposition and deposition process. Due to the ultralow carbon solubility in Cu, there is only a small amount of carbon dissolved in Cu, even at high hydrocarbon concentrations or when the growth time is long [55]. Most of the carbon source for graphene formation is the hydrocarbon that is catalytically decomposed on the Cu surface. After the first layer of graphene is deposited, the Cu surface is fully covered and there is no catalyst exposed to hydrocarbon to promote decomposition and growth. Thus, graphene growth on Cu is a surface reaction process. About 95% is monolayer, with small regions of bi- and multilayer graphene. Graphene growth is found to be self-limiting and robust at producing similar films no matter how short or long the methane exposure [56]. This mechanism is experimentally proved by using isotopic labeling of hydrocarbon precursors combined with Raman spectroscopic mapping. Both the surface microstructure of Cu and the hydrocarbon concentration can impact graphene uniformity. The hydrocarbon concentration can affect the number of graphene nucleation sites. When electropolished Cu foils are used, 95% monolayer coverage can be obtained under low methane concentrations, while increasing the methane concentration affords thicker graphitic regions [57]. Furthermore, chemical mechanical polishing of Cu foils results in improved graphene films compared with unpolished samples. This can be attributed to the reduction in graphene nucleation sites on the smoother metal surfaces. At lower hydrocarbon concentrations, there are few nucleation sites, which leads to the production of more uniform graphene. Growth of graphene with average domain sizes of $\sim 140\ \mu m^2$ has been reported for low-pressure CVD using carefully controlled hydrocarbon exposures [58]. Methane exposure is kept low at the beginning of the process to reduce the number of graphene nucleation sites and then increased during the reaction to promote continuous graphene coverage.

6.2.1.2.3 *Transfer of CVD graphene*

For most device applications, the obtained CVD graphene on transition-metal substrates needs to be transferred on the dielectric substrates. The isolation of graphene relies on chemical etching of metallic substrates, which is usually done by a wet-etch process [59, 60]. A schematic diagram of the typical transfer process is shown in Fig. 6.11. Before etching, graphene was first coated by a thin, robust, and flexible layer to give mechanical support to the graphene layer and allow for handling after removing the metallic film. Usually, this layer is polymethyl methacrylate (PMMA). PMMA is widely used as a positive e-beam resist in lithography, and it can be easily dissolved in acetone. After baking at 120°C to evaporate the solvent in PMMA and harden the PMMA, the metal catalyst was then removed by Ni or Cu etchant, leaving only the PMMA/graphene film. The film was cleaned by deionized (DI) water and then transferred onto a targeting substrate. After evaporating the water vapor, PMMA was removed by acetone, leaving a graphene film on top of the targeting substrate. Direct transfer of graphene is also possible by replacing the support layer with other target substrates, for example, polydimethylsiloxane (PDMS).

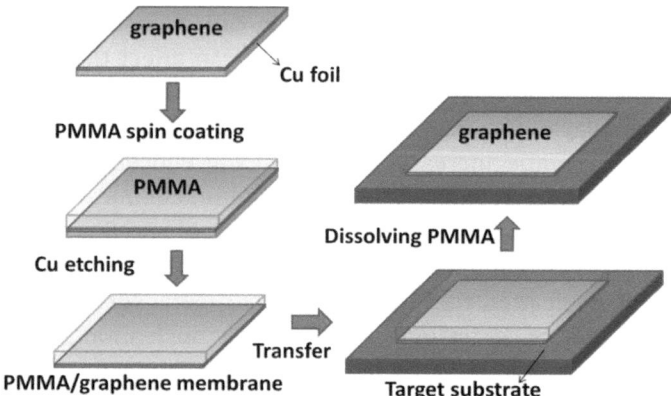

Figure 6.11 Schematic illustration of a wet-transfer process of Cu-CVD graphene onto a target substrate by a PMMA mask.

The scalable transfer of graphene can also be achieved by a "roll-to-roll" process compatible with industry techniques [55], where graphene grown on flexible metal films is rolled onto flexible substrates, followed by removal of the metal layer to yield bare graphene on the substrate surface (Fig. 6.12). The transfer of 30-inch graphene films grown on a Cu foil has been demonstrated. Firstly, the graphene film is attached to a thermal release tape through the application of soft pressure (0.2 MPa) between two rollers. Then, the copper foil is etched in a plastic bath filled with a 0.1 M ammonium persulphate aqueous solution. After etching, the transferred graphene film on the tape is rinsed with DI water to remove residual etchant. Lastly, the graphene film on the thermal release tape is inserted between the rollers together with a target substrate and exposed to mild heat (90–120°C) to release the adhesive tape while leaving the graphene to the target surface. The transfer rate is 150–200 mm/min.

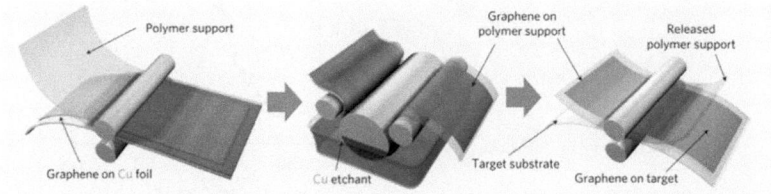

Figure 6.12 Schematic of the roll-based production of graphene films grown on a copper foil. The process includes adhesion of polymer supports, copper etching (rinsing), and dry transfer-printing on a target substrate. A wet-chemical doping can be carried out using a setup similar to that used for etching. Adapted by permission from Springer Customer Service Centre GmbH: Springer Nature, *Nature Nanotechnology* (Ref. [55]), Copyright (2010).

Large-scale CVD graphene films are typically polycrystalline, with domain boundaries, which tends to degrade the quality of graphene. Hence, the electronic quality of graphene is dependent on the grain size of single domains. Figure 6.13 shows the transport properties of CVD graphene with different average domain sizes of 6 μm and 20 μm linear length and compared with exfoliated graphene [58]. The device layout of a back-gated field-effect transistor (FET) with Ni contacts is showed in Fig. 6.13a. The effective mobility extracted from the FET device model for films with small domains, 6 μm, is in the range of 800–7000 $cm^2V^{-1}s^{-1}$. In contrast, devices with larger graphene domains, 20 μm, have a much higher range of mobility, 800–16,000 $cm^2V^{-1}s^{-1}$. The mobility is 2500 to over 40,000 $cm^2V^{-1}s^{-1}$ for exfoliated graphene. It can be seen that CVD graphene films with large domains show higher mobility. However, the quality of CVD graphene is still not as good as that of exfoliated graphene.

The lower mobility for CVD graphene films may be associated not only with the presence of domain boundaries but also with the presence of wrinkles and other defects, some of which are induced by the transfer process [60]. Besides, it has also been found that the substrate has a large effect on the transport properties; specifically, the mobility of CVD graphene transferred onto a flat hexagonal boron nitride (h-BN) substrate is 10^4 to 3×10^4 cm^{2}V^{-1}s^{-1} at room temperature, which is comparable to that of exfoliated graphene [61, 62].

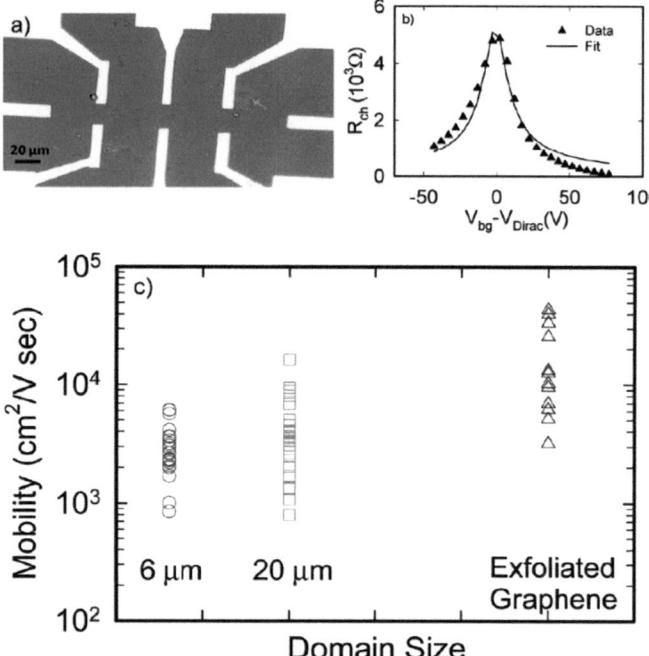

Figure 6.13 (a) Optical micrograph of a FET device (top view). (b) A typical plot of the normalized channel resistance (R_{ch}) as a function of applied back gate voltage (V_{bg}). (c) Carrier mobility as a function of graphene domain size in comparison with exfoliated graphene. Adapted with permission from Ref. [58]. Copyright (2010) American Chemical Society.

6.2.2 Synthetic Strategies for Nanographene

In general, the synthesis of NG can be completed by top-down and bottom-up approaches. The top-down methods are the cleavage or cutting of sp^2 carbon containing materials, such as graphene, GO, fullerene (C$_{60}$), and glucose, by plasma etching, chemical oxidation, electrochemical oxidation, or hydrothermal processes. The bottom-up methods usually control the structure of NG precisely from micromolecules in organic solvents or carbon active groups in CVD ways. The synthetic methods determine the physicochemical properties of the NG, such as size, density, edge termination, stability, and crystal quality.

6.2.2.1 Top-down methods

6.2.2.1.1 *Plasma etching*

It is a straightforward way to combine ultrahigh-resolution-electron-beam-lithography-defined masks and O$_2$-plasma etching for graphene nanostructure fabrication. Although this method offers high precision and reproducibility, it requires very specialized equipment and the yield is low. Alternatively, self-assembled block copolymer (BCP) nanospheres are used as a simple but effective mask to fabricate periodic aligned NG by oxygen-plasma etching (as shown in Fig. 6.14) [63]. Firstly, monolayer graphene grown by CVD was transferred to a SiO$_2$/Si substrate. Then a self-aligned monolayer of BCP nanospheres was spin-coated onto the graphene surface. After thermal annealing and washing processes, well-ordered silica nanospheres in a polystyrene (PS) matrix were formed as a masking layer. Followed by O$_2$-plasma etching, the graphene under the interval of nanospheres was etched away. Finally, nanospheres were removed and ordered NG was obtained. Although adjacent nanospheres are very closely packed, NG is overetched depending on the condition of oxygen plasma treatment. The size of the NG (GQDs) can be tuned by the diameter of the initial BCP nanospheres and the plasma etching time with narrow size distribution. The minimum diameter of the GQDs is 10 nm.

6.2.2.1.2 *Chemical method*

The chemical method mainly uses oxidation reagents to break up carbon–carbon bonds in the starting material, and NG is exfoliated when the oxygen-containing functional groups line up. The starting

GO material actually has been achieved by oxidation of graphite using modified Hummers's method. Then GO is oxidized/cut into small pieces of GQDs by $H_2SO_4/KMnO_4$ (Fig. 6.15). The oxidation process is similar to that in Hummers's method used to acquire GO. GO nanocrystals have some functional groups, including hydroxyl, epoxy, and carboxylic acid groups. Figure 6.16 shows the achieved controllable sizes 3–5 nm and 2–4 nm by different oxidation levels and duration [64]. Also, the GO nanocrystals consist of five to seven layers of graphene. Apart from the strong acid reagents, a nontoxic oxone oxidant is also suitable for the oxidation. For example, acid-free potassium mono persulfate was used to oxidize GO, which prevents subsequent neutralization [65]. Through sonication or UV irradiation, the process of tailoring GO has been accelerated compared with Hummers's method. The lateral size of GO nanocrystals is 2–6 nm, and the height is one to three layers. Apart from GO, the original material can also be carbon fibers [66]. Due to the sp^2 graphene domain in a carbon fiber, it is susceptible to chemical cleavage for NG 1–4 nm in size and one to three layers thick.

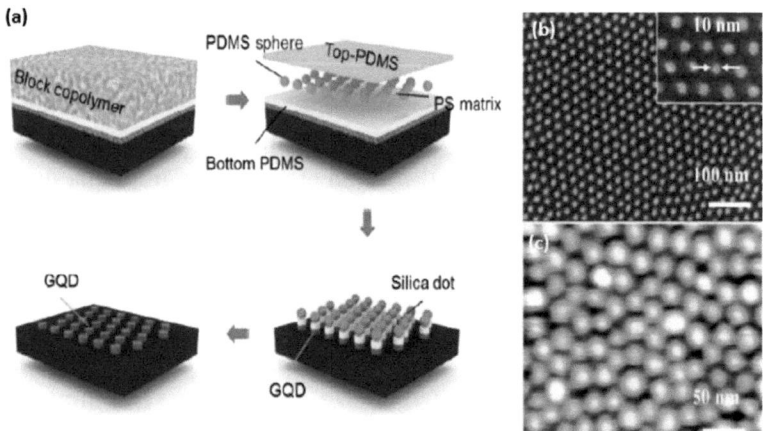

Figure 6.14 (a) Schematic illustration of the fabrication of GQDs, including the spin coating of BCP, formation of silica dots, and etching process by O_2 plasma. (b) SEM image of self-assembled 10 nm silica nanodots on graphene. (c) AFM image of 10 nm GQDs on a mica substrate. Adapted with permission from Ref. [63]. Copyright (2012) American Chemical Society.

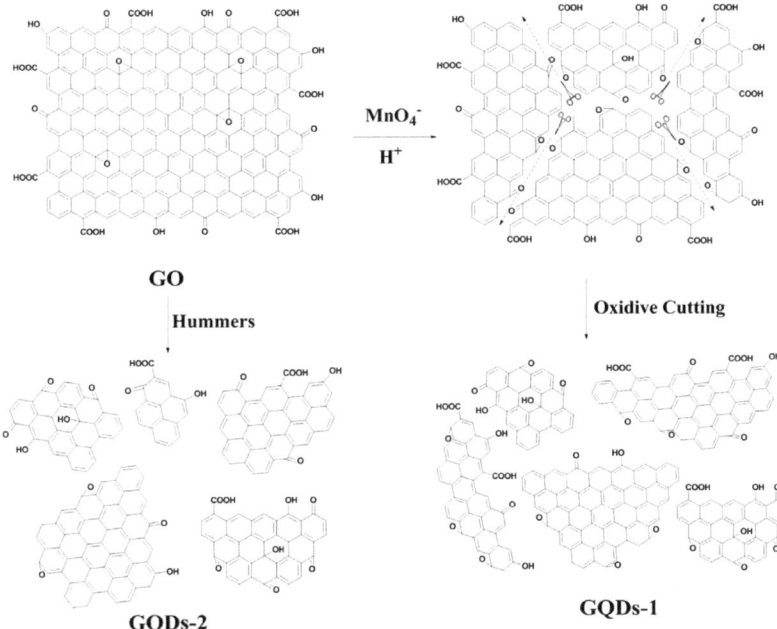

Figure 6.15 Mechanism of fabricated nanographene (GQDs) for cutting graphene oxide by modified Hummers's method [64].

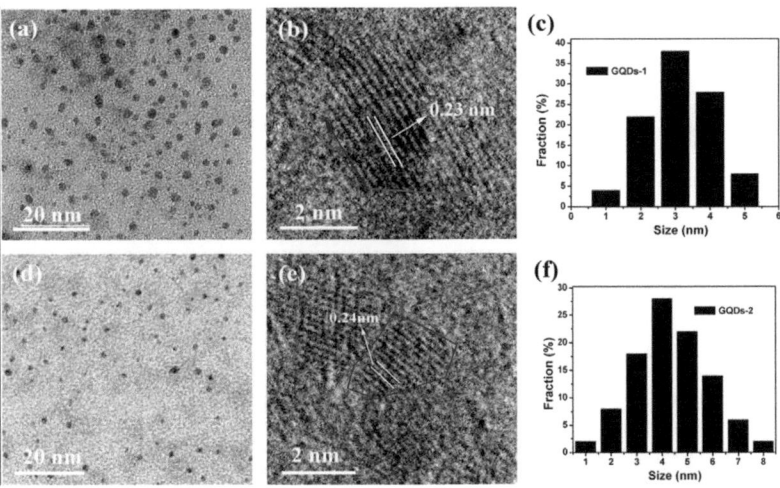

Figure 6.16 (a) TEM images of GNC-1 prepared by Hummers's method. (b) The HRTEM images of GNC-1 with measured lattice spacing and edge structures. (c) Diameter distribution of GNC-1. (d) TEM images of GNC-2 by modified Hummers's method. (e) The HRTEM images of GNC-2 with measured lattice spacing and edge structures. (f) Diameter distribution of GNC-2 [64].

6.2.2.1.3 *Electrochemical method*

Monodispersed NG with a lateral dimension of 3–5 nm and a height of 1–2 nm can also be achieved by an electrochemical approach, as shown in Fig. 6.17 [67]. A filtration-formed graphene film is used as the working electrode and a phosphate buffer solution (PBS) as the electrolyte. The obtained NG is soluble in water with abundant oxygen functional groups. An alternative electrochemical method is to use multiwalled carbon nanotubes as the working electrode [68]. The electrolyte is nonaqueous propylene carbonate/LiClO$_4$. The method consists of two steps, electrochemical oxidation and electrochemical reduction. The electrolyte oxidized from the C–C bond and Li ion is intercalated to reduce GQDs for exfoliating. Additionally, the controllable size of GQDs can be tuned by the parameters of electrochemical experiments. One parameter is oxidation duration, and GQDs of a smaller size are achieved when the oxidation duration is longer. The smallest GQDs are about 3 nm and 1–3 nm in height. Another parameter is the temperature of the electrolyte. A high temperature (~90°C) can accelerate the redox process thoroughly; thus smaller GQDs can be obtained at a lower temperature (30°C).

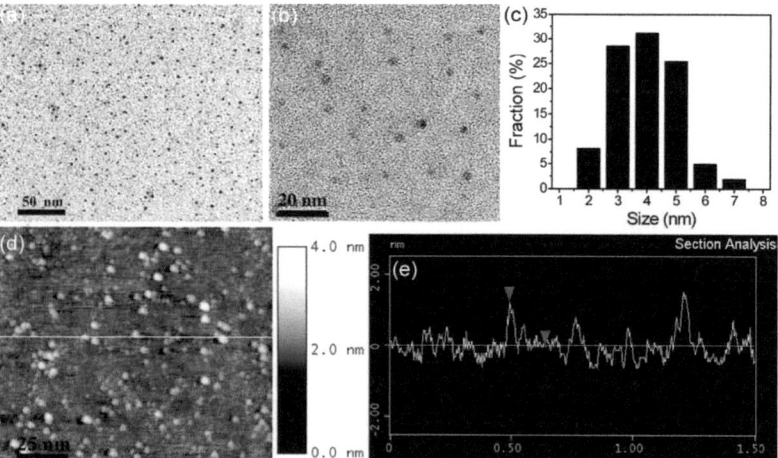

Figure 6.17 (a, b) TEM images of as-prepared GQDs with different magnifications achieved by electrochemical cutting. (c) Size distribution of GNDs. (d) AFM image of the GNDs on a Si substrate. (e) The height profile along the line in (d). Adapted with permission from Ref. [67]. Copyright © 2010 WILEY-VCH Verlag GmbH & Co. KGaA, Weinheim.

6.2.2.1.4 *Hydrothermal method*

Hydrothermal cutting [69], as shown in Fig. 6.18, is another method for NG fabrication and uses dehydration of materials containing C–H–O under a high temperature.

Figure 6.18 Schematic diagram of the hydrothermal deoxidization of graphene sheets: a mixed epoxy chain composed of epoxy and carbonyl pair groups (left) is converted into a complete cut (right) under hydrothermal treatment.

Micrometer-sized GSs are acquired by thermal reduction of GO sheets [69]. Then GSs are oxidized in H_2SO_4 and HNO_3. During the oxidation, C=O/COOH, OH, and C–O–C are introduced in the edge and the basal plane. The epoxy groups tend to appear on the edge. Once an epoxy appears, one pair of epoxy group forms. Finally, the lined epoxy groups are transformed to stable carbonyl pairs and thus the C–C bonds are broken. The GSs then are hydrothermally treated at 200°C in a poly (tetrafluoroethylene) (Teflon)-lined autoclave (50 ml). The NG is 5–13 nm in size and one–three layers in height. Of course, GO sheets can be used as the starting material directly [70]. A mixture of GO sheets and polyethylene glycol (PEG) is heated at 200°C for 24 h and the resulting GO nanocrystals 5–20 nm in size are passivated by PEG. The oxidized GO sheets are reduced at 250°C in an argon atmosphere. Micrometer-sized NG with a controllable size is achieved by filtering and dialysis, as shown in Fig. 6.19. Smaller size (~6 nm) and height (~5 nm) are achieved by a repeated oxidation/reduction process [71].

Apart from GO, glucose can also be used as the carbon source [72]. Actually, most of the carbohydrates that contain C, H, and O in the ratio of ~1:2:1 can be used as a carbon source for NG synthesis. The glucose solution and DI water mixture are heated using microwave methods. Figure 6.20 shows the schematic of the forming process of NG. The glucose is dehydrated into small pieces and then small dots are nucleated. On increasing the heating time, the size of the NG increases. NG as small as 1.65 nm (~5 layers) can be prepared by this method.

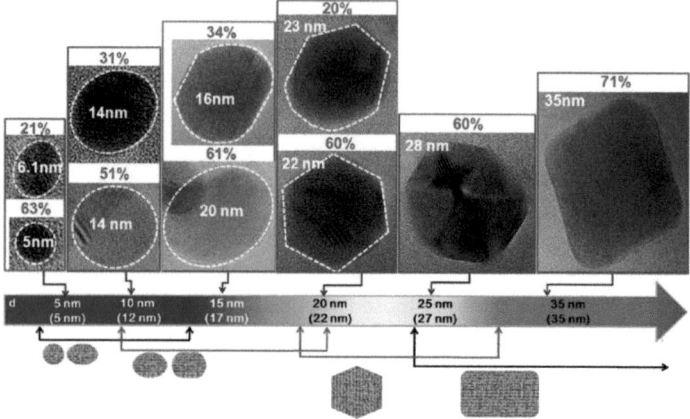

Figure 6.19 HRTEM images of GQDs for their major shapes and corresponding populations (*p*) with increasing average size of GQDs. Average sizes (*d*$_a$) of GQDs estimated from the HRTEM images at each *d* are indicated in the parentheses at the bottom. The connected arrows indicate the range of the average size in which GQDs with particular major shapes are found. Adapted with permission from Ref. [71]. Copyright (2012) American Chemical Society.

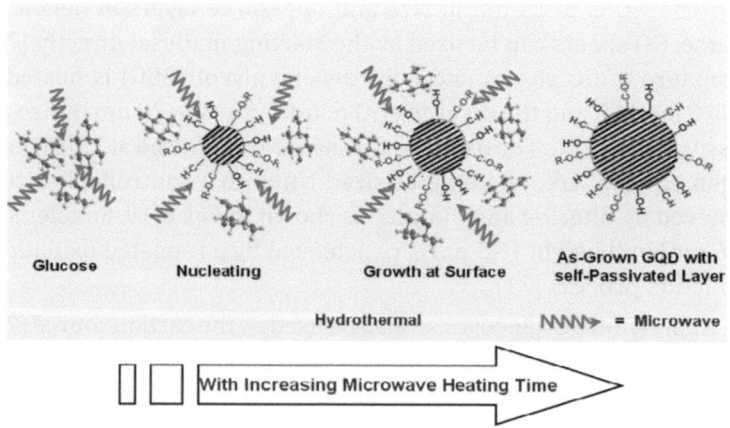

Figure 6.20 Schematic diagram of GQDs fabricated from glucose by the hydrothermal route. Adapted with permission from Ref. [72]. Copyright (2012) American Chemical Society.

6.2.2.1.5 *Other methods*

NG synthesis through cage opening of C_{60} on ruthenium (Ru) has been realized [73]. The bottom hexagon ring of C_{60} and the closest Ru atoms are bonded initially. Then, the carbon atoms diffuse and combine into quantum dots by heating. Note that it is unique for C_{60} to produce GQDs. The reasons for the formation of GQDs are the limited mobility of its cluster and surface-mediated growth. The size of GQDs is 5 nm, they are hexagonal in shape, and they are monolayer.

Another way to synthesize GQDs is combustion oxidation, and the produced carbon soots contain graphite nanocrystals. The method is similar to that used by James Tour's group, where graphene was synthesized from food, insects, and waste or solid carbon sources at high temperatures [74, 75]. The average size is 6 nm.

6.2.2.2 Bottom-up methods

6.2.2.2.1 *Direct organic synthesis*

Usually the bottom-up methods use organic reagents. Four kinds of NG ($C_{42}H_{18}$, $C_{96}H_{30}$, $C_{132}H_{34}$, and $C_{222}H_{42}$) have been synthesized using organic methods. The synthesis processes of the four kinds of NG are similar, and we will introduce the method for synthesizing $C_{42}H_{18}$. First, diphenylacetylene (1 g) and tetraphenylcyclopentadienone (2.6 g) are dissolved in o-xylene (40 ml) in an argon atmosphere to obtain hexaphenylbenzene by purifying the mixture after heating and cooling. Then, hexaphenylbenzene (0.7 g) is dissolved in dry dichloromethane (200 ml). The reaction is kept for 12 h before dry iron (III) chloride (4.6 g), dissolved in nitromethane (30 ml), is added. Then, the reaction is quenched by adding methanol (350 ml). By filtration, washing, and drying, a dark-brown solid material is collected. Then GQDs are obtained after purification by chromatography in silica gel [76].

Hexa-peri-hexabenzocoronene (HBC) can also be used as an organic carbon source. The powder of HBC is pyrolyzed to graphite at 1200°C. Then, the graphite is oxidized and exfoliated using a modified Hummers's method. Subsequently, aqueous solutions of the resultant GO nanocrystals are heated to reflux for 48 h with oligomeric poly (ethylene glycol) diamine (PEG1500N) and then reduced with hydrazine. The obtained GQDs have a uniform morphology of ∼60 nm size and 2–3 nm thickness (Fig. 6.21) [77].

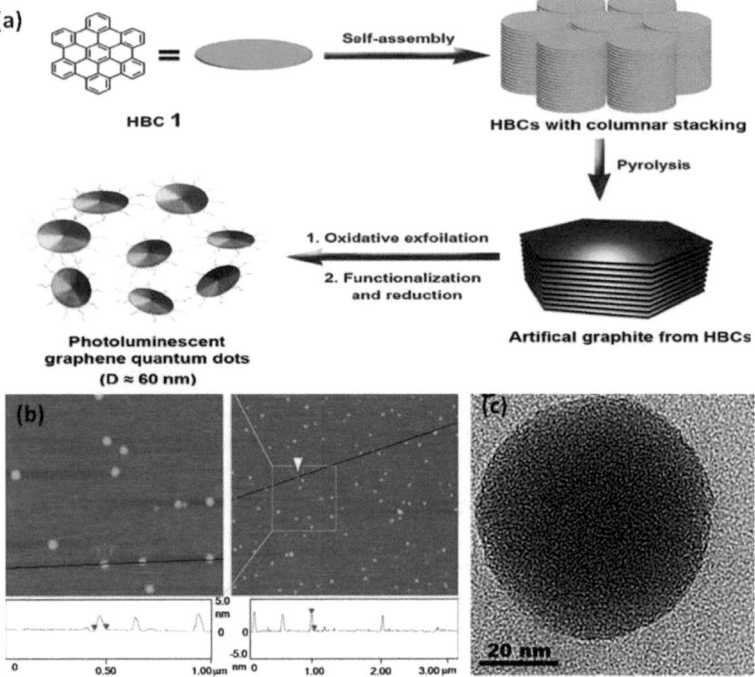

Figure 6.21 (a) Processing diagram for the preparation of GQDs by using HBC as a carbon source. (b) AFM topography images of GQDs on mica substrates, with the width and height profiles along the line in the images. (c) TEM image of a GQD. Adapted with permission from Ref. [77]. Copyright (2011) American Chemical Society.

6.2.2.2.2 *Chemical vapor deposition*

CVD growth of graphene on metallic substrates requires a post-transfer process and usually needs a high temperature, which is incompatible with the traditional semiconductor technology. However, NG can be grown directly on the insulating substrates without the need of a metallic catalytic substrate by a plasma-enhanced CVD (PECVD) process (Fig. 6.22) [78]. The growth is done at a low temperature (550°C) using methane as a precursor. The plasma produces C_xH_x radical species that play an important role in the formation of NG. PECVD growth of NG follows a van der Waals growth mode, with only weak van der Waals interaction between the grown layered materials and the substrate. Compared with 3D materials grown via CVD, quasi–van der Waals growth of layered 2D material allows precise control of the nucleation and formation of a monolayer. The substrate roughness affects the growth of NG. Oxide substrates with large surface roughness are preferred for the growth of NG crystals. The density of NG is decided by the density of the nuclei. During a thermal kinetic growth process, the density of the nuclei has a saturated maximum value when the density of the metastable nuclei reaches equilibrium within a certain time. The saturated density is independent of the growth rate while dependent on the growth temperature.

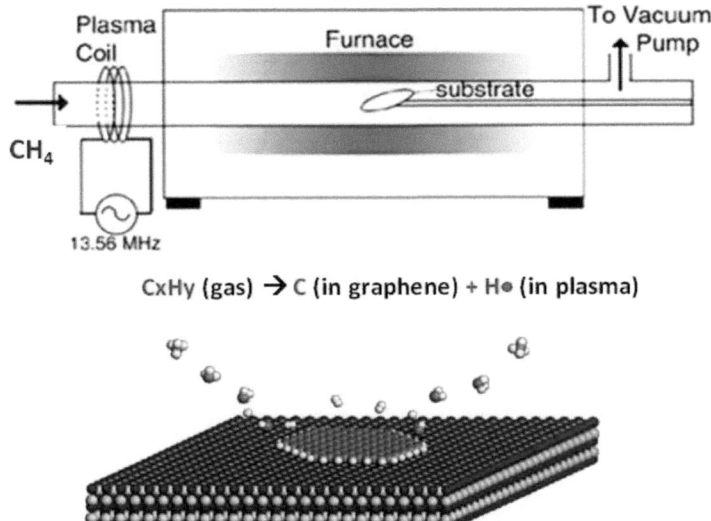

Figure 6.22   Setup and mechanism of nanographene growth by PECVD.

The density and size can be tuned by controlling the growth time and growth temperature. Under an optimal growth environment, NG can be grown at a high density and with good uniformity over a relatively wide area (~4-inch wafer). The average size of the as-grown NG samples (Fig. 6.23) is 6–100 nm, with a saturated density of more than 2×10^{11} cm^{-2} to 1×10^{10} cm^{-2}, and the height of the NG samples is 0.4–0.7 nm, which corresponds to a monolayer thickness [78]. Figure 6.23b shows Raman spectra of these samples. The high D peaks and the appearance of D′ peaks (1620 cm^{-1}) reveal strong defect scattering at abundant edges of the NG samples. Small-sized

NG samples with higher densities show a larger I_D/I_G ratio and a lower effective area ratio than large-sized NG samples with lower densities. Hence, the properties of NG can be modulated by these edges.

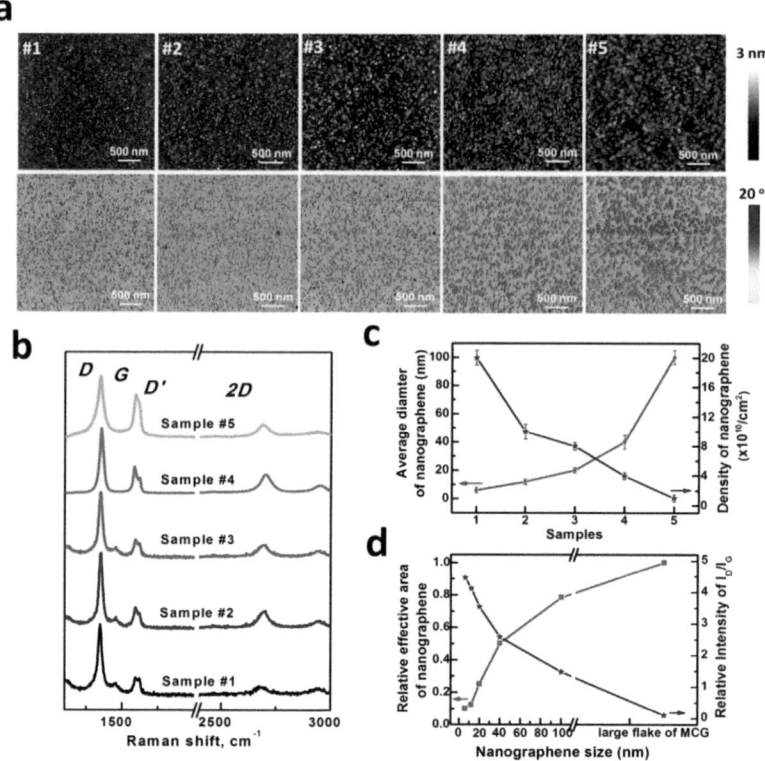

Figure 6.23 (a, b) AFM height/phase images and Raman spectra of nanographene samples with controlled sizes and densities by controlled growth conditions. (c) Corresponding size and density of samples. (d) Size dependence of effective area ratio and Raman I_D/I_G ratio, which reveals the edge proportion in nanographene. Scale bar: 500 nm.

Through the above introduction, we can see that the preparation technology for graphene and its nanostructures is relatively mature. Large-scale production of these materials can be achieved through simple, effective, and inexpensive ways, which is one of the most favorable factors for the promoted application of graphene. However, there remains a large scope in the preparation of NG with various properties, controllable assembly, and extended applications.

6.3   Graphene-Based Flash Memory

The rapid development of flash memory is accompanied by the prominent applications of modern portable electronic products. As the market for portable electronic equipment, such as smart phones, iPads, tablet PCs, and flexible memory, has increased, huge attention has been focused on increasing the storage intensity and decreasing the node size. According to the survey of the International Technology Roadmap for Semiconductors (ITRS), the requirement for nonvolatile memory is approaching 128 Gb per chip, or approximately 10 nm^2 per device cell [79]. Generally, flash memory can be classified into two types according to the structure and material of the charge storage layer: floating-gate flash memory, which is the current industry standard, and charge-trap flash memory, which is the focus of emerging technologies. Floating-gate memory contains two gates: a control gate, which is used to modify electrons, and a floating gate, which is used to store the electrons. After increasing the density of storage elements, a floating gate confronts several challenges that would normally jeopardize device performance. One of the challenges is the increased parasitic capacitance between the neighboring cells, which induces cell-to-cell interference [80]. The shift of one cell's threshold voltage (V_{th}) has been reportedly proportional to the V_{th} of the adjacent cell. Also, the parasitic capacitor is adverse to the programming speed of the cell. Additionally, the gate coupling ratio is lowered since the capacitive coupling between the drain electrode and the floating gate is influenced by the cross talk. Charge-trap flash memory has an insulating storage layer and a single gate to control the channel directly. However, in terms of scaling up production, charge-trap flash memory can be hindered by variability in V_{th} from the implementation limit of the trap density ($\sim 10^{19}$ cm^{-3}) and uniformity and the short retention times due to shallow trap energy levels, which can promote trap-assisted Poole–Frenkel conduction during the retention state.

Height reduction is an effective solution to reduce the parasitic capacitive coupling. However, the requirements of charge retention limit the thickness of the tunneling oxide and the blocking oxide and the limit for the minimum tunneling oxide is about 5 nm. So, a lot of effort has been made to search for a good floating-gate material. Earlier, polysilicon has been used as a floating-gate material and when the polysilicon is scaled down to 7 nm, the ballistic current that tunnels from the substrate to the control gate directly increases [81]. A thin metal film can decrease the ballistic ratio due to an increase in the elastic scattering of electrons in metals. However, a metal film encounters a thermal stability problem. Graphene is currently being investigated as the floating-gate material due to its unique properties: (i) an atomically thin nature, which can reduce the vertical thickness; (ii) metallic characteristics, which can decrease ballistic current; and (ii) high thermal stability, which can overcome the drawbacks of metal [82].

The first graphene floating-gate memory was reported in 2011 [83]. Graphene as a floating gate is transferred from graphene grown on Cu foils by CVD. Both monolayer graphene and multilayer graphene are fabricated into the device. The structure uses a heavily p-doped Si substrate as the channel, 5 nm SiO_2 as the tunneling layer, transferred graphene as the floating gate, 35 nm Al_2O_3 as the blocking layer, and Ti/Al/Au (10 nm/500 nm/50 nm) as the top electrode. The specific fabrication procedure of the flash memory is shown in Fig. 6.24.

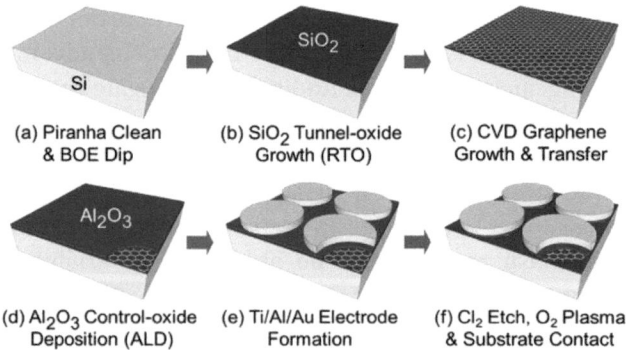

(a) Piranha Clean & BOE Dip   (b) SiO_2 Tunnel-oxide Growth (RTO)   (c) CVD Graphene Growth & Transfer

(d) Al_2O_3 Control-oxide Deposition (ALD)   (e) Ti/Al/Au Electrode Formation   (f) Cl_2 Etch, O_2 Plasma & Substrate Contact

Figure 6.24 The fabrication procedure of the graphene floating-gate memory. Adapted with permission from Ref. [83]. Copyright (2011) American Chemical Society.

The charges are stored in the graphene layers. The control structure without graphene shows almost no hysteresis, while the single-layer graphene shows ~2 V with a sweep voltage of ±7 V. This demonstrates that monolayer graphene can store charges. A multilayer graphene device indicates a ~6 V memory window with a sweep voltage of ±7 V (Fig. 6.25). The wider memory window of the multilayer graphene compared to that of single-layer graphene confirms that the charge is stored in the graphene layer. This can be explained by the fact that the thickness of single-layer graphene is about 0.35 nm and the interlayer screening length of the multilayer graphene is ~0.8 nm, which can store more charge. It is noted that the curve of single-layer-graphene flash memory is initially a negative shift, which can be explained by the defects or contaminations that contribute the p doping. The low program/erase (P/E) voltage can reduce the energy consumption because its density of states (DOS) is sensitive to gate voltage. Additionally, the high work function (~4.6 eV) of multilayer graphene is stable, which is not sensitive to the graphene layers. Also, the low electricity along the c axis can decrease the ballistic current.

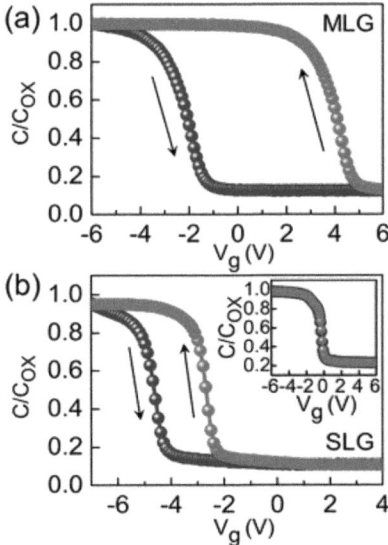

Figure 6.25 The memory window of multilayer- and single-layer-graphene-based flash memory. Adapted with permission from Ref. [83]. Copyright (2011) American Chemical Society.

The graphene floating-gate memory also shows excellent retention stability. After 10 years, only 8% of electrons are lost. This is also due to the high work function of graphene (~4.6 eV), which makes the electrons harder to tunnel from the charge-trapping layer to the substrate. Direct tunneling is estimated to be the main reason of charge losses.

Additionally, the graphene thinness has more advantages than traditional polysilicon flash memory in that there is a decrease in cell-to-cell cross talk. The simulations of the interference of the neighboring cells are done on polysilicon and graphene-like flash memory with the assumption that the adjacent cells are programmed while the middle cell is unprogrammed. The simulation results as shown in Fig. 6.26 indicate that polysilicon flash memory has significant cross talk below 25 nm while graphene flash memory can survive less than 10 nm [83]. The reduction in the cross talk in graphene flash memory can improve the density twice compared to the current polysilicon flash memory.

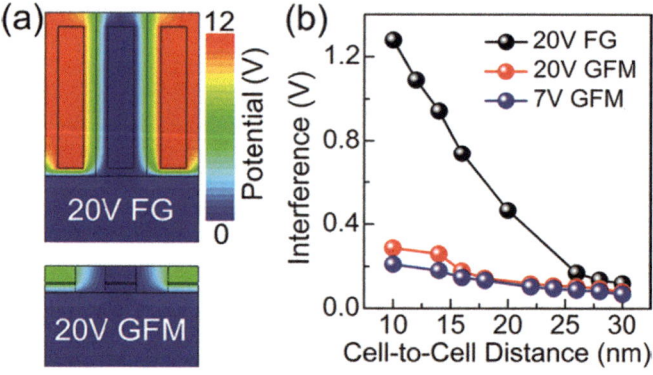

Figure 6.26 Cell-to-cell cross talk in polysilicon flash memory (FG) and graphene flash memory (GFM). Adapted with permission from Ref. [83]. Copyright (2011) American Chemical Society.

Excepting for CVD transferred graphene, alternative thermal or chemical reduction of GO is also used as a charge storage layer [84, 85]. The capacitor structure of the RGO flash memory is similar to the first-reported graphene flash memory. The thickness of the RGO is 2–3 nm and the corresponding layer is 6–7 nm.

Compared with the control sample without a graphene layer, the RGO flash memory shows an obvious memory window at ±8 V (Fig. 6.27). A large operation voltage compared with the above-mentioned flash memory may be due to the different fabrication process and the different graphene synthesis method. Another merit of the memory-P/E transition is also characterized. The transient time is about 1 s. Note that the same group further improves the capacitor structure to a transistor structure and the P/E response time improves to 1 μs. The retention characteristics are further investigated. At 150°C and room temperature, the extrapolation of the P/E state retention indicates 74% charge retention and 30% charge retention, respectively, after 10 years. Endurance characteristics are also performed, and the 1000 P/E cycles show robust characteristics.

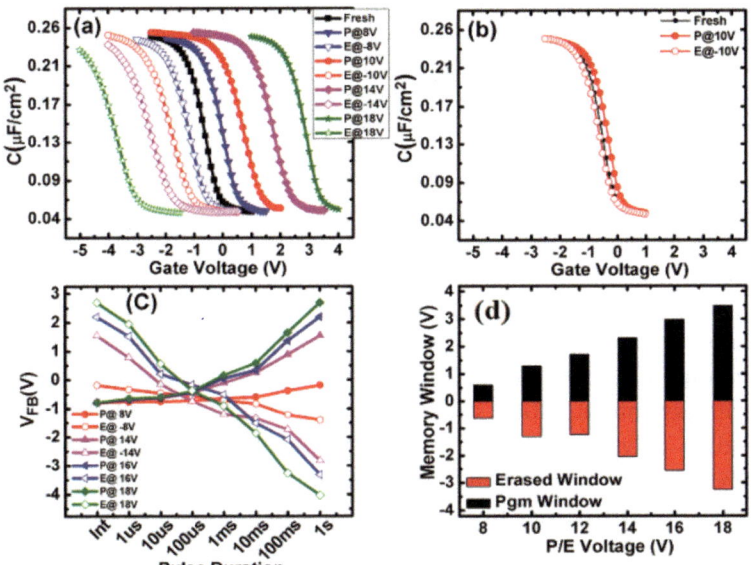

Figure 6.27 Characteristics of the RGO floating-gate memory. (a) *C–V* plots of the flash memory capacitors after successive programming and erasing. (b) *C–V* plots for a control sample (without monolayer graphene sheets) programmed and erased at ±10 V for 1 s. (c) Program/erase transients of the monolayer graphene flash devices. (d) Program/erase memory window at each P/E voltage used. Numbers above bars indicate the number of stored electrons per cm^2 in the floating gate of the programmed device. © [2012] IEEE. Reprinted, with permission, from, Ref. [84].

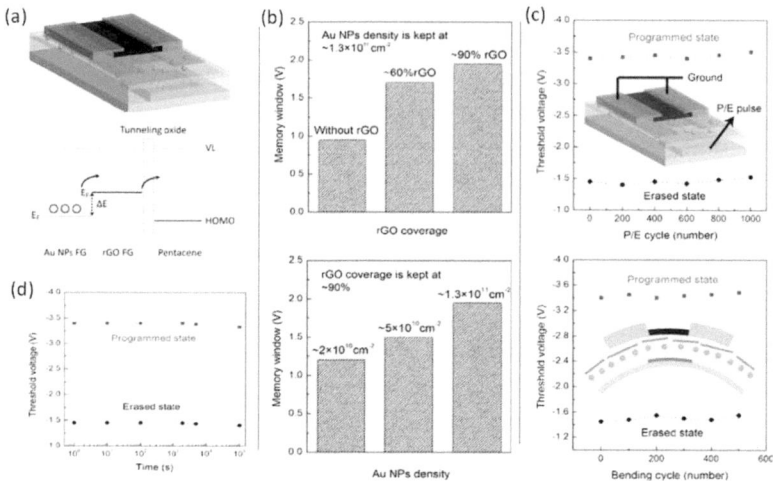

Figure 6.28 (a) Schematic diagram depicting RGO-Au NP hybrid double-floating-gate (DFG) memory device. (b) Memory window with respect to different densities of RGO, while the density of the Au NPs is kept at ~1.3 × 10 ^{11} cm^2 with respect to different densities of Au NPs, and the RGO coverage is kept at ~90%. (c) Endurance characteristics of a DFG memory device and mechanical-stability test of flexible DFG memory device. (d) Data retention capability of a DFG memory device. Reprinted from Ref. [89], with permission of AIP Publishing.

On the basis of graphene floating-gate flash memory, a graphene-based double-floating-gate (DFG) structure for low-voltage flash memory has been developed [86]. Figure 6.28 shows layer-by-layer-assembled RGO sheets/Au nanocrystal hybrid DFGs on flexible poly(ethylene terephthalate) (PET) substrates. More than 90% coverage of monolayer RGO sheets and Au nanocrystal arrays with a saturated density (1.3 × 10^{11} cm^2) act as the upper and lower floating gates, respectively. The self-assembled large-area RGO monolayer almost fully covers the Au nanocrystal array and overcomes the problem of a pair of charge-trapping layers along the vertical direction of the nanocrystal DFGs. Compared with single-Au-nanocrystal nano-floating-gate memory devices, the memory characteristics, including the memory window and the retention capability, are significantly improved in RGO/Au nanocrystal double-trapping-layer memory devices. The high DOS of RGO induces an enhancement of the memory window. Meanwhile, RGO has a smaller work function than Au nanocrystals and a Fermi-level offset is created between the hybrid DFGs. Therefore, an energy barrier between the Au nanocrystal charge-trapping layer and the semiconductor channel of the memory device is achieved eventually, making it harder for trapped charge carriers to leak out from the lower floating gate, consequently enhancing the retention time. In addition, the P/E speed, the endurance properties, and the bending cycles of the hybrid DFG memory device do not show significant degradation in comparison with single-floating-gate Au nanocrystal memory devices.

6.4   Graphene Nanostructures Flash Memory

Graphene flash memory has the potential to exceed the performance of current flash memory technology by utilizing its intrinsic properties. However, the performance of graphene flash memory will be severely degraded when the size decreases. Since these flash memory structures use transferred CVD graphene as the floating gate, these floating-gate type of memory structures face density and retention problems. In practice, such continuous and conducting graphene as a floating gate is less efficient as a single pinhole in a device would cause device failure. Although nearly insulated GO flakes can also be used as storage nodes for charge-trapping flash memory and a large memory window with low power consumption can be achieved, the poor thermal stability makes GO less favorable for high-performance devices. To solve the problem, a charge trap memory structure based on discrete storage sites has been developed [87]. The nanocrystal-based charge-trapping memory (CTM) device (which is also called nano-floating-gate memory) has advantages over other two typical types of flash memory devices. The charge-trapping sites and trapped energy levels are controllable and can be tuned by varying the size, density, and work functions of nanocrystals. In addition, the isolated storage structure can prevent electrons' lateral movements even if some defects exist, thus enhancing the retention character. A lot of effort has been made on nanocrystal flash memory by employing semiconductor nanocrystals or metal nanoparticles. These devices show massive memory capacity, better endurance, smaller devices size, and lower operation voltage as compared with conventional flash memory devices [88–95]. Semiconductor nanocrystals are more favorable in terms of the fabrication compatibility with complementary-metal-oxide-semiconductor (CMOS) technologies; however, their relatively large sizes limit their use for miniaturized device applications as very small semiconductor nanocrystals are not good for charge storage due to the quantum confinement effect (QCE). In contrast, metal nanoparticles, usually expensive and high-work-function metals such as Au and Pt, can be very small while keeping high charge storage capacity; however, the fabrication process is not compatible with CMOS technologies. Besides, it is also challenging to control the sizes, shapes, and configurations of these metal nanoparticles, thus fluctuations of the device performance among individual devices become severe, especially when reducing the devices' sizes to nanoscale. Memory cell size is likely to reduce to 1000 nm^2 by 2020 [89, 90, 96, 97] and only ~10 nanocrystals could be included in an individual memory cell on such a small scale. Hence, it becomes increasingly challenging to synthesize suitable materials with large charge-trapping capacity while keeping their sizes and shapes uniform and assembling them into a well-ordered nanocrystal matrix.

The arising graphene nanocrystal has characteristics similar to those of metal nanocrystal. Besides, NG also has unique characteristics more favorable for flash memory. Firstly, graphene nanocrystal has thermal stability that can survive even at 1500°C and thus is compatible with current CMOS technology [82]. Additionally, the atomic thinness has the potential for transparent flexible and folded memory electronics [96–98]. The detailed characteristics of NG have been addressed above. Although graphene nanocrystal flash memory is promising, the research is at its early stage. Next, an overview of reported nanocrystal flash memory is presented and the preliminary conclusions and challenges will be discussed in detail. The basic characteristics of a nonvolatile memory device, including the classification of memory window, P/E transition, endurance, and retention, will also be addressed.

6.4.1 Memory Window

A memory window that is characterized by shifts of the threshold voltage indicates the capacitance of the electron/hole storage. The wider the memory window, the more is the number of electrons/holes trapped. Additionally, the direction of the hysteresis curve has different meanings. An anticlockwise curve indicates that the trapped electrons/holes are tunneled from the substrate. A clockwise curve is attributed to electrons/holes tunneling from the gate electrode, which decreases the stored charge and should be avoided since it can decrease the memory window. Two memory structures are fabricated to characterize the memory window. One of the structures is the capacitor, which is usually composed of a tunneling layer, a charge-trapping layer, and a blocking layer sandwiched between the top gate and the bottom channel. The other structure is the transistor, which has typically source and drain electrodes and a bottom channel. Two memory structures based on graphene nanocrystals will be introduced below.

The fabrication process of the capacitor structure with NG is as follows [78, 97]:

1. A 10 nm SiO_2 tunneling layer is deposited by ion-beam sputtering on a p-type Si wafer.

2. Synthesized NG is coated on the 10 nm SiO_2/p-type Si. It is noted that the NG is obtained through hydrothermal method or by PECVD growth, which have been discussed in Section 6.2.2.

3. A 20 nm SiO_2 blocking layer is deposited on the NG using the same ion-beam sputtering method.

4. Aluminum is deposited as the top electrode. A typical Raman G-peak of NG before and after rapid thermal annealing (RTA) at 1000°C confirms that the NG encapsulated in the oxide layer is stable even at high temperatures.

The schematic diagram of the GQD memory capacitors has been shown in Fig. 6.29a [97].

Except for silicon dioxide, a high-dielectric-constant (high-k) material (Al_2O_3, HfO_2) is often chosen as the tunneling and blocking material (as shown in Fig. 6.29b) [78]. HfO_2 and Al_2O_3 are always deposited by atomic layer deposition (ALD). Additionally, Al_2O_3 can be achieved by oxidation of a thin Al layer. The high dielectric constant is advantageous for decreasing the operation voltage.

Also, the graphene nanocrystal as a charge-trapping element can be enhanced by the addition of a thin-layer ZnO dielectric [99].

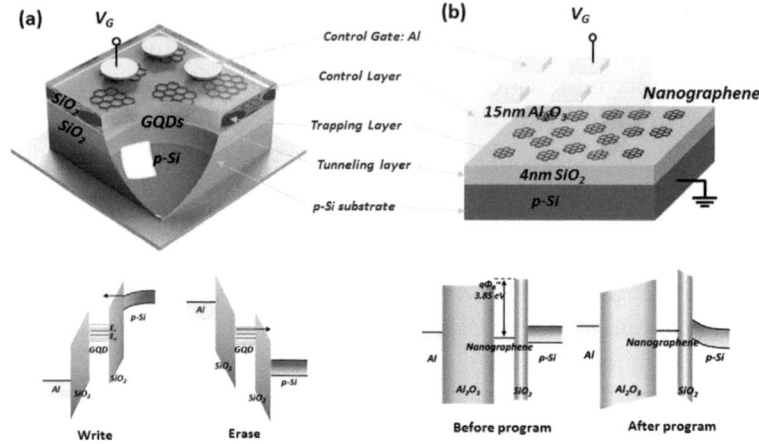

Figure 6.29 The schematic diagrams of the capacitor structure with nanographene as charge-trapping memory.

The transistor structure of NG has a similar fabrication process. A typical high-performance floating-gate transistor is fabricated using solution-based NG as the charge-trapping elements and high-k Ta$_2$O$_5$ as a controlled gate dielectric [100]. Figure 6.30a shows the schematic of the floating-gate transistors. A 30-nm-thick Ta$_2$O$_5$ dielectric layer with a wide bandgap of 4.85 eV is deposited by thermal evaporation onto the indium tin oxide (ITO)-coated substrates to serve as the gate dielectric and then is subjected to annealing treatment at 550°C for 20 min. The synthesized NG solution is spin-coated to form the carrier trapping layer. Next, a 10 nm poly(4-vinylphenol) (PVP) dielectric with a wide bandgap of 4.77 eV is spin-coated onto the NG so that the NG is loaded in the PVP/Ta$_2$O$_5$ bilayer. Next, highly conjugated conducting poly(3-hexylthiophene) (P3HT, Sigma-Aldrich) with 98% head-to-tail linkages is deposited by spin coating in a nitrogen glove box at a deposition rate of 1000 rpm. The coated P3HT film is annealed at 70°C for 30 min. and at 120°C for 20 min. Finally, to form the source and drain electrodes, a 100-nm-thick gold layer with a channel length (L) of 100 mm and a width (W) of 1 mm is deposited using a shadow mask and a sputtering system at room temperature.

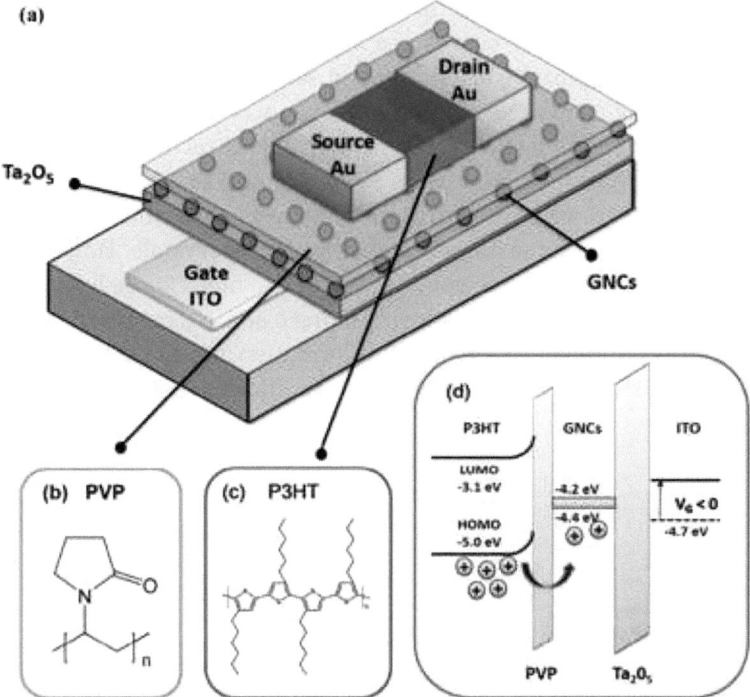

Figure 6.30 (a) Schematic of the high-performance organic nano-floating-gate memory (NFGM) device fabricated by integrating nanographene as charge-trapping elements and high-k Ta$_2$O$_5$ as a controlled gate dielectric. (b) Molecular structure of the polymer dielectric PVP. (c) Molecular structure of the polymer semiconductor P3HT. (d) Schematic illustration of the operating mechanism of the organic NFGM device shown in (a). Adapted from Ref. [100] with permission of The Royal Society of Chemistry.

The characteristics of the memory window are measured by a semiconductor parameter analyzer. For the capacitor structure, the memory window is characterized by the hysteresis of the capacitance–voltage curve. Usually, the analyzer is worked at a high frequency (0.1–1 MHz). On the other hand, for the transistor structure, the memory window is characterized by a shift of the transfer curve.

Figure 6.31a shows the high-frequency (100 kHz) capacitance–voltage (C–V) characteristics of a typical graphene nanocrystal device (Al$_2$O$_3$/8 nm NG/SiO$_2$, metal/aluminum oxide/nanographene/

silicon oxide/silicon structure [MANGOS]) under lower sweep voltages [78]. The *C–V* curves at a high sweep voltage are shown in Fig. 6.31b. Compared with the control sample, showing no memory window, *C–V* curves of a graphene nanocrystal device reveal a large flat-band voltage (V_{fb}) shift under dual-direction gate voltage sweeping (V_{fb} = 4.52 V at ±8 V, V_{fb} = 8 V at ±13 V). High-density charge storage and less fluctuation between individual cells can be achieved by increasing the density of nanocrystals. However, a high-density nanocrystal would also lead to a high charge leakage or direct tunneling through the surrounding oxide. NG memory devices might be improved to overcome this limitation as a single NG 6–8 nm in size can store over 15 electrons with sufficient spacing gaps.

(a)   (b)

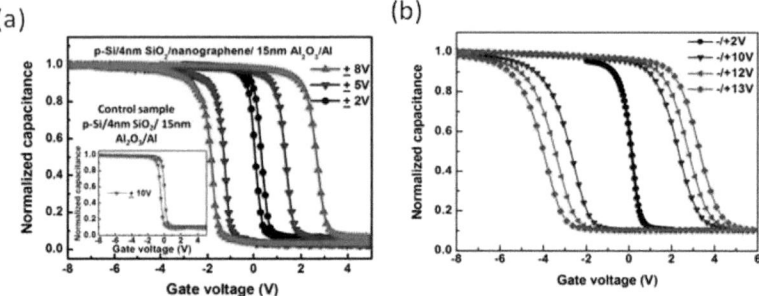

Figure 6.31 High-frequency *C–V* characteristics of MANGOS under different gate voltage sweepings.

Usually, the characteristics of nanocrystal memory are directly related to the nanocrystal size, density, and configuration. Thus it is important to investigate these structural effects. The first factor is the size of NG. NG samples of 6 nm, 12 nm, and 27 nm are chosen as a charge trap layer, and the corresponding *C–V* plots indicate the obvious hysteresis compared with a few millivolt memory windows of metal-oxide-semiconductor (MOS) structures without NG [97]. It is demonstrated that the memory window increases with the increase of NG average size. Additionally, at a large sweep voltage the memory window deflects the linear relation between memory window and sweep voltage. The memory window decreases, saturates, and further increases after reaching ∼3 V, ∼4 V, and ∼8 V for 6 nm, 12 nm, and 27 nm GQDs, respectively. The increasing and saturation of memory increasing are reasonable since they are dependent on whether the trap sites are saturated or not. However, the decreasing of the memory window is odd and still unexplained (Fig. 6.32).

Actually, the amount of charge stored in GQDs can be estimated by the formula

$$\Delta n = (\Delta V_{th} \times C)/q, \tag{6.1}$$

where C is the memory capacitance density, ΔV_{th} is the memory window, and q is the elementary charge. Here, the density of the stored charges is 2.8×10^{11}, 2.2×10^{11}, and 2.6×10^{11} cm^{-2} for $d = 6$ nm, 12 nm, and 27 nm, respectively, in the linear region of the C–V curve. It is interesting to find that the small-sized NG has the largest charge storage capacity, which can act as a guide for improving the memory window.

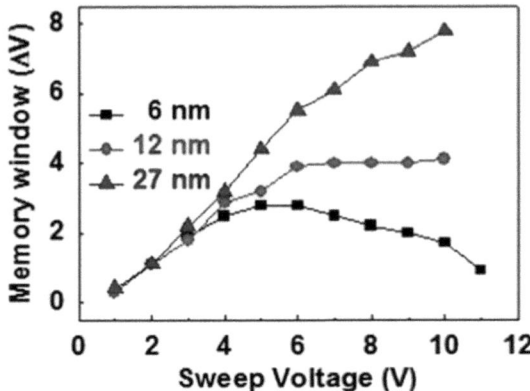

Figure 6.32 Memory window versus sweep voltage for GQD-based CTMs of different average sizes. Reprinted from Ref. [97]. © IOP Publishing. Reproduced with permission. All rights reserved.

Another factor is the density of graphite nanocrystals. There is limited reporting about the direct relation between the density of NG and the memory windows. However, we can get a glimpse of the relation from the reported work about the controllable density of graphite nanocrystal flash memory. The threshold voltage of a memory window increases with the increase of graphite nanocrystal density until it is saturated (see Fig. 6.33). The results demonstrate that high-density graphite nanocrystals have more trap sites.

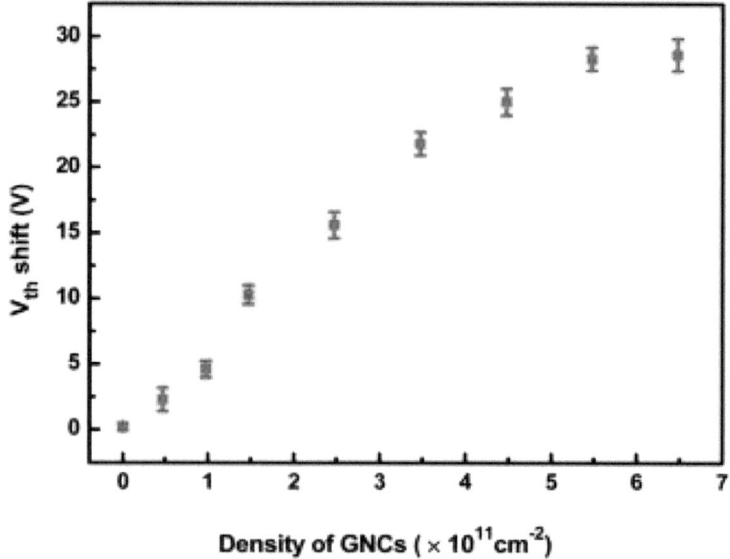

Figure 6.33 Threshold voltage versus density of graphite nanocrystals (GNCs). Adapted from Ref. [100] with permission of The Royal Society of Chemistry.

The combined effects of size and density on the charge-trapping capacity of NG have been demonstrated by scanning Kelvin probe microscopy (SKPM) measurements, as shown in Fig. 6.34 [78]. Without the blocking high-k oxide layer, the variation in the surface potential is determined by the carrier concentration variation in NG. The change of carrier concentration is ascribed by the Fermi-level shift due to the electric-field-effect-induced charge doping, which is well quantified by the DOS of graphene. The measured contact potential difference (CPD) values were used to estimate the charge-trapping capacity of NG. Figure 6.34 presents the CPD values and corresponding calculated electron densities extracted from the SKPM measurements of NG of different sizes and densities (as shown in Fig. 6.23) after low electrical field charging (E, –5 MV/cm; charged area, 1 × 1 um^2), together with exfoliated large graphene flakes and silicon nanocrystals for comparison. The total trapped electron densities in smaller-sized NG samples are obviously higher than those in bigger-sized NG samples. The size-dependent total charge trap density agrees well with the measured edge densities by Raman scattering. When the size decreases, the saturated density increases, and the available edge states in NG would increase and thus supply more trapping sites for charges. These results reveal that smaller NG with a high density is more favorable for a higher trapping density than graphene flakes. We deduce that the edge structure plays an important role in contributing to the charge-trapping performance of NG.

Note that the size-dependent charge-trapping capacity of NG, as depicted in Fig. 6.34 with dotted marks, has an opposite trend as compared to that of silicon nanocrystals, which has been well explained by the size-related-quantum-effect-resulting charge loss in semiconductor nanocrystal flash memory. As the size of the semiconductor nanocrystal decreases down to its Bohr radius (e.g., Si has a Bohr radius of ~11 nm), the total quantum confinement energy increases and available DOS decreases, leading to less charge being injected and stored. Unlike semiconductor nanocrystals, NG is a zero-bandgap semiconductor with a Bohr radius of around 0.74 nm (which is comparable to that of noble metals, for example, Au, which has a Bohr radius of ~0.5 nm) and shows an ambipolar conducting behavior with weak gate modulation [102]. Besides, NG preferentially has zigzag-dominated edges; thus there is no bandgap or discrete energy induced by the quantum size effect when the size of NG shrinks to a few nanometers, and the increased proportion of the metallic edge states would increase the charge storing. Hence NG is a favorable charge-trapping material due to its high DOS, high work function, and small Bohr radius.

Through the above introduction, we can see that high-density, small-sized NG with abundant defects has massive charge-trapping sites for a wider memory window in flash memory. However, whether we use a top-down approach or a bottom-up approach, there is no available fabrication technology to obtain NG with a density higher than 10^{12} cm^{-2}, the same problem as in other nanocrystal-based memory devices. There is a potential solution to solve this problem, a two-step fabrication approach: (i) growth of continuous polycrystalline films on silicon dioxide substrates in a remote-plasma-enhanced chemical vapor deposition (rPECVD) system [101, 102] and (ii) Ar- or O$_2$-plasma etching of the as-grown films for high-density and isolated NG. Usually isolated NG directly grown by rPECVD has a typical density saturated at ~10^{11} cm^{-2}.

Through the above two-step process, isolated NG with a density of $\sim 10^{12}$ cm^{-2} can be achieved [103], which is an improvement of ~ 10 times (Fig. 6.35). The obtained high-density NG is then used as the charge-trapping material in a flash memory device with a typical geometry of Si (p++)/SiO$_2$ (4 nm)/NG/Al$_2$O$_3$ (15 nm)/Ti-Au [103]. The wide memory windows of the sample clearly demonstrate the capacity of high-density NG. Figure 6.35 depicts *C–V* curves of the best device with a large memory window of 9 V at a sweep voltage of ±8V, the largest ever reported in a graphene CTM so far. Plasma etching of as-grown continuous NG films for high-density NG is efficient in two aspects. First, by precisely control of the etching time, grain boundaries can be etched off and the continuous film is changed to isolated graphene nanosheets. Second, plasma etching is also capable of creating additional defects within these graphene nanosheet basal planes, thus providing more charge-trapping sites. This two-step method for high-density NG makes graphene an outstanding alternative for downscaling technology beyond the current flash memory.

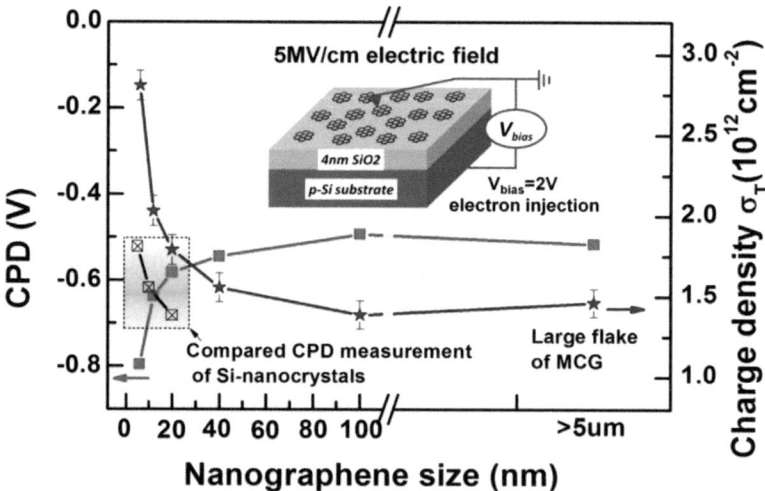

Figure 6.34 Size-/density-dependent charge-trapping capacity in nanographene. The CPD values and corresponding electron densities extracted from the SKPM measurements of samples with different sizes, together with exfoliated large graphene flakes and silicon nanocrystals for comparison. The inset is a sketch map of a local back-gated nanographene device for surface potential measurement.

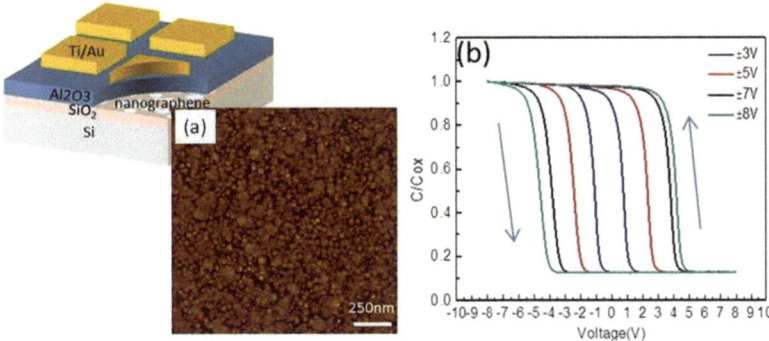

Figure 6.35 (a) AFM image of etched high-density nanographene film as well as a schematic diagram of the structure of Si (p++)/SiO$_2$ (4 nm)/GNC/Al$_2$O$_3$ (15 nm)/Ti-Au metal gate flash memory device. (b) Capacitance–voltage curve of nanographene flash memory. The capacitance has been normalized with respect to the capacitance of the control oxide.

Besides, the memory window can also be enhanced by changing the structures of the storage layer. NG sandwiched between ZnO double layers as charge-trapping elements in 15 nm Al$_2$O$_3$/2 nm ZnO/NG/2 nm ZnO/3.6 nm Al$_2$O$_3$ has been investigated (graphene nanoplatelets in ZnO [GNIZ]) [99]. A control sample using NG with 5-nm-thick ZnO (or only 4 nm ZnO) is also fabricated. The memory windows versus sweeping voltage for these three compared memory structures are plotted (see Fig. 6.36). The obvious memory window contrast with only graphene nanoplatelets and only the ZnO structure indicates the remarkable electron storage in a NG-sandwiched structure. It indicates that ZnO layers can effectively reduce the leakage of electrons from NG.

Table 6.1 summarizes the memory characteristics of NG synthesized by various methods. It can be deduced that the crystal quality, size, and density of NG play an important role in the charge storage capacity. Hence, to further improve the storage performance of the memory device, the most important and critical thing is to get a proper method to produce NG of a much higher density and much smaller size.

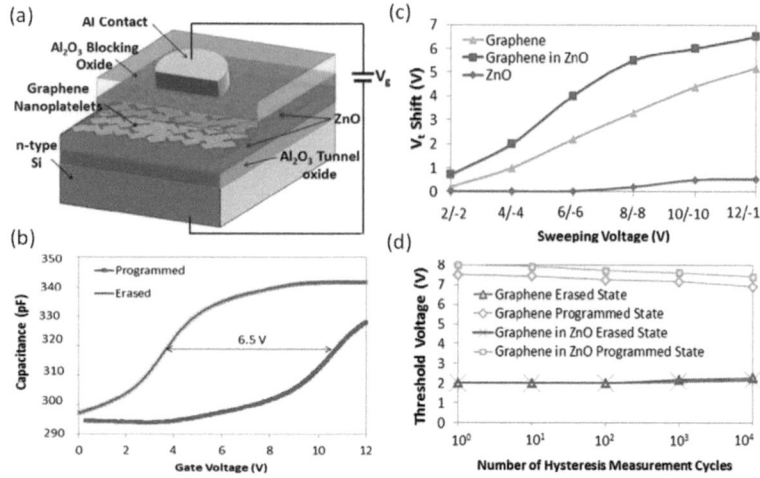

Figure 6.36 (a) Cross-sectional illustration of the fabricated MOS memory with GNIZ. (b) *C–V* measurement of the memory with GNIZ. (c) Threshold voltage versus sweeping voltage for three compared memory structures. (d) Endurance measurement showing threshold voltage shift versus number of hysteresis measurement cycles. Adapted from Ref. [99], with the permission of AIP Publishing.

Table 6.1 Memory characteristics of nanographene synthesized by various methods

Synthesis methods	Size and density of NG	Tunnel dielectric	Sweep voltage (V)	Memory window (V)	Reference
Chemical cutting RGO	6–30 nm, $<6 \times 10^{11}$ cm^{-2}	SiO_2	10 to –10	8	[97]
Chemical cutting RGO	100–500 nm, $<10^9$ cm^{-2}	Al_2O_3	6 to –6, –12 to 2	4, 6.5	[99]
Combustion oxidation method	3–9 nm, 1.5×10^{11} cm^{-2}	Ta_2O_5	4 to –4	3.3	[100]
CVD growth	6–20 nm, $\sim 2 \times 10^{11}$ cm^{-2}	SiO_2	8 to –8, –13 to 13	4.5, 8	[78]
CVD growth and plasma etching	6–8 nm, $\sim 10^{12}$ cm^{-2}	SiO_2	8 to –8	9	[103]

6.4.2   P/E Transient Time

Improving the P/E speed can effectively reduce the energy consumption of a CTM. Especially, the transistor structure has a faster P/E speed than a capacitor. This is mainly due to the faster detection of the shift of threshold after a P/E pulse in the transistor structure than that in the capacitor structure. As we know, the fastest P/E transient time is ~10 ns in a NG-based flash memory transistor structure (Fig. 6.37) [100].

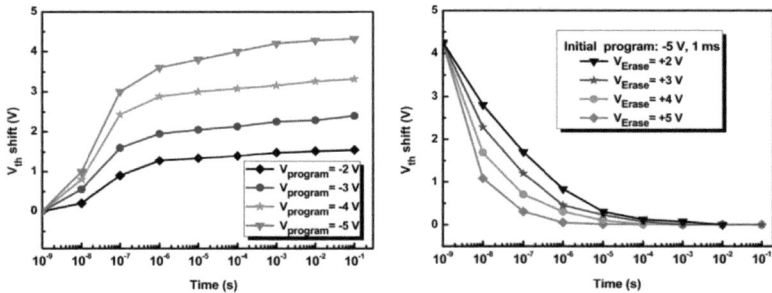

Figure 6.37   Programming and erasing speeds of the nano-floating-gate memory (NFGM) devices with 1.5×10^{11} cm^{-2} nanographene under different voltage pulses. Adapted from Ref. [100] with permission of The Royal Society of Chemistry.

Since it is less meaningful to compare the P/E speed of flash memory devices with different structures, the parameters affecting the P/E transition time are discussed. To test the P/E speed, the flash device should be totally erased or programmed. For NG nano-floating-gate memory devices with different nanocrystal sizes (6 nm/12 nm/27 nm), a pulse is applied to initialize the P/E state. Different pulses with various pulse widths are applied to test the shift of flat-band voltage (Fig. 6.38) [97]. Overall, a program operation is faster than an erase operation. Also, the program is the fastest while the erase is the slowest for a 12 nm NG device and it is vice versa for a 27 nm NG device. Two unique characteristics of NG affect the speed: the QCE and the edge effect. It is QCE that makes a 12 nm NG have a faster program speed and a slower erase speed. A 12 nm NG has a deeper conductance band than 6 nm due to the QCE. Thus electrons are easily trapped and find it hard to escape. However, the situation is complex for a 27 nm NG since the edge effect is prominent at

~17 nm. The mostly armchair edge effects make electron tunneling to the NG hard and make it easy to erase them.

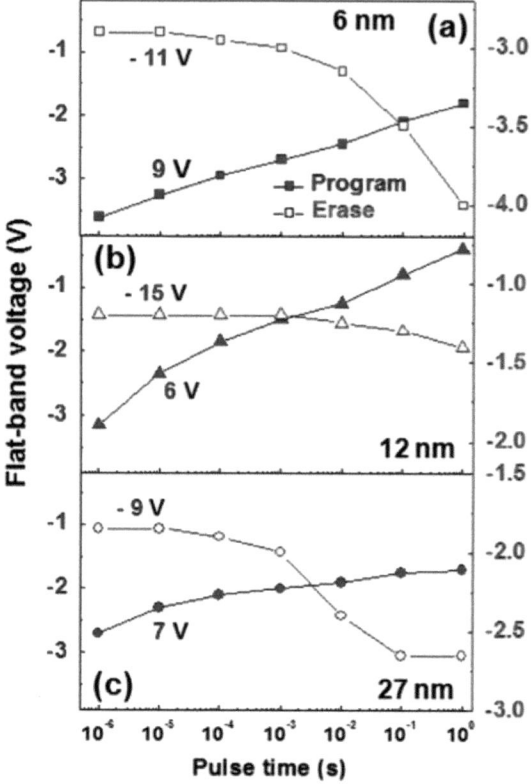

Figure 6.38 P/E transient time for nanographene of 6 nm, 12 nm, and 27 nm. Reprinted from Ref. [97]. © IOP Publishing. Reproduced with permission. All rights reserved.

6.4.3 Retention Characteristics

It is well known that retention characteristic is a key parameter for flash memory. The memory device should retain stored information after a certain time with as small a charge loss as possible. The retention characteristics of NG-based flash memory largely depend on the device structure. Loss of storage is typically a result of charge tunneling through the tunneling oxide, and the tunneling rate depends on the thickness and the barrier height of the oxide.

Figure 6.39 shows the cycled retention characteristic of a typical graphene nanocrystal device (Al_2O_3/GNC/SiO_2) at both room temperature and a high temperature [78, 103]. Generally, a retention time of more than 10 years is required before the device loses 50% of the stored charge. The large barrier height (~3.85 eV) of NG/SiO_2 and the low electric field across the tunneling layer under retention state make direct tunneling the most probable charge loss mechanism. A charge loss of 45% in Al_2O_3/NG/SiO_2 after 10 years' operation is predicted from experimental data. However, due to some overlap of the NG dots as well as the quality of the blocking Al_2O_3 layer, the high-temperature retention after 1000 cycled operations shows a charge loss of 55%. Further optimization is needed to improve the retention characteristics by improving the quality of the high-κ layers. Due to the high dielectric constant of the tunneling oxide, the HfO_2/NG/HfO_2 structure shows obviously improved retention, with a charge loss of 28% after 10 years, as shown in Fig. 6.39.

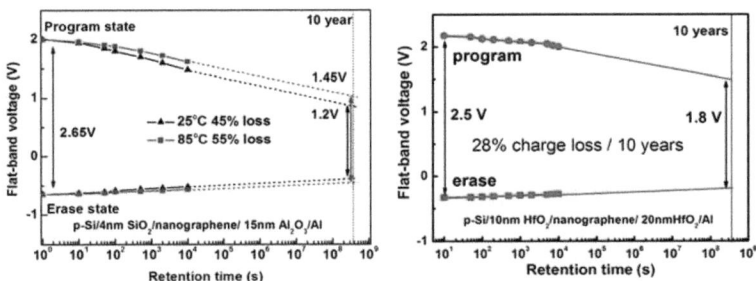

Figure 6.39 Cycled retention characteristics of a MANGOS device.

We already know that a graphene nanocrystal in ZnO (GNCIZ) enlarging the memory window is attributed to the reduction of tunneling current leakage. Hence, there is no doubt that GNCIZ-based flash memory can have a longer retention time than only a NG-based device. Here, the GNCIZ-based flash memory has only a 3.6 nm tunneling oxide layer while a NG has a 5 nm tunneling oxide layer. Though it has a thinner tunneling layer, GNCIZ-based flash memory has a 25% charge loss after 10 years while NG-based flash memory has a 29% charge loss [99]. This again demonstrates that GNCIZ has potential for future scaling technology without degrading the charge storage retention.

For an RGO-based flexible flash memory structure, the data retention is relatively worse. The memory window is large after a time elapse of one year. The work did not show data for a longer time span. However, further effort is needed to enhance the data storage time.

An upward shift of the retention curve is also observed. Flat-band voltage versus retention time for graphene of different sizes is shown in Fig. 6.40. Different from the common downward shift of the flat-band voltage, the upward shift of the curve is due to charge leakage through the blocking layer from the control gate [97]. The phenomenon should be avoided during memory device fabrication.

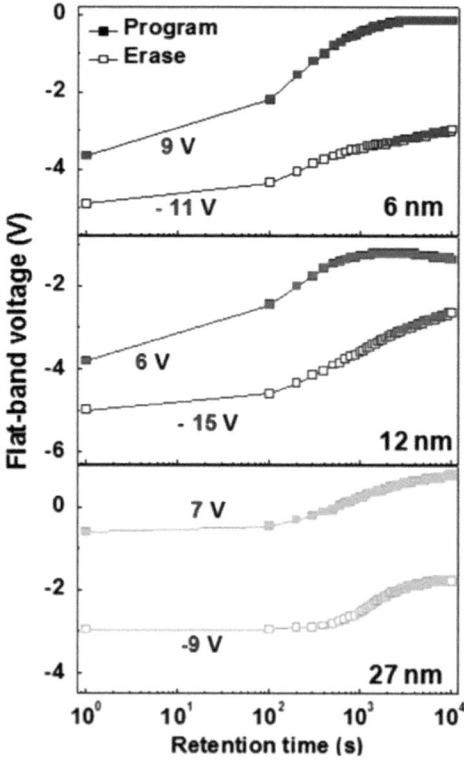

Figure 6.40 Retention characteristics of nanographene of different sizes. Reprinted from Ref. [97]. © IOP Publishing. Reproduced with permission. All rights reserved.

6.4.4 Endurance Cycles

Endurance cycling tests characterize device reliability. Overall, NG-based flash memory devices show robust endurance. Here, the shift of threshold voltage versus cycles is typically presented (see Fig. 6.41). The memory window has insignificant change after 10^4 P/E cycles [99].

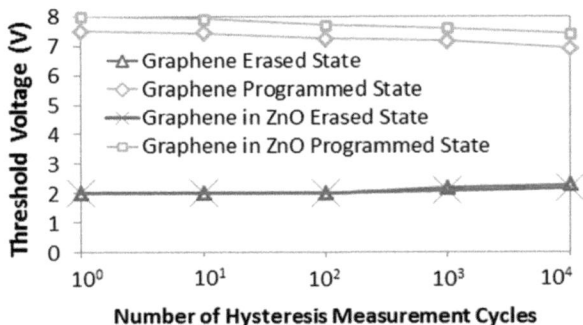

Figure 6.41 Endurance cycles of nanographene-based flash memory. Adapted from Ref. [99], with the permission of AIP Publishing.

Overall, a graphene nanocrystal not only possesses the advantages of metal nanocrystals but also shows its superiority in terms of low cost, high charge storage capacity, high temperature stability, and wafer-scale homogeneity, which can satisfy the nano-miniaturized requirement compared with the existing semiconductor-nanocrystal- or metal-nanoparticle-based CTM device. And still the NG device needs to be optimized to improve the retention and endurance performance.

6.5 Graphene Memory Hybrids

Besides the above-mentioned topics in this field, graphene nanosheet flash memory combined with the following aspects needs to be further investigated for broadening their future applications.

6.5.1 Flexible Transparent Flash Memory

The future trend in flash memory is flexibility and transparency, which are highly required in wearable or portable electronics. Organic materials have been widely explored in the last few decades [104]. The easy mass fabrication of polymers (via spin coating, dip-drop, or printing) can reduce the costs effectively. In the active layer of flash memory, the polymer P3HT is usually used as a p-type semiconductor [105–107].

It has been noted that P3HT is a unipolar material. Lately, the organic material poly-diketopyrrolopyrrole thiophene-benzothiadiazole-thiophene (PDPP-TBT) has been demonstrated as an ambipolar semiconductor material that can provide a wide range of electron and hole charge, thus enhancing the memory window of the flash memory [108]. Hence, it is advantageous for future multibit charge trapping. Besides polymers, graphene is suitable as an active layer material due to its ambipolar electrical property. Related works have been reported using transferred Cu-CVD monolayer graphene as the channel [109]. The specific schematic diagram of the floating-gate memory is shown in Fig. 6.42. The flexible and transparent substrate is polyethylene naphthalate (PEN). The charge storage module is $Al_2O_3/AlO_x/Al_2O_3$, and the top gate is ITO. For a graphene channel, the complicated source and drain electrode fabrication can be eliminated. The transfer curve demonstrates the feasibility of a graphene channel (Fig. 6.42b). Also, the as-made floating-gate memory has >80% optical transmittance in the visible range (Fig. 6.42c). Figure 6.44d shows a prototype of large-area, high-density graphene-based flash memory arrays.

Another fabrication process is opposite to that mentioned above [110]. ITO is firstly deposited on the PEN substrate. Then Al_2O_3, Au nanoparticle, and cross-linked poly(4-vinylphenol) (cPVP) are integrated. Finally, prepatterned graphene is transferred from a Cu coil to the device. However, the trapped graphene is p-doped, which is due to the residual peroxy-disulfate during the copper etching. This can degrade the memory window. So, moderate n-doped graphene is prepared by immersing the device into a poly(ether imine) (PEI) solution. Figure 6.43b shows that the memory window increases after the device is dipped in the PEI solution. The result confirms mechanical stability even after 500 bending cycles.

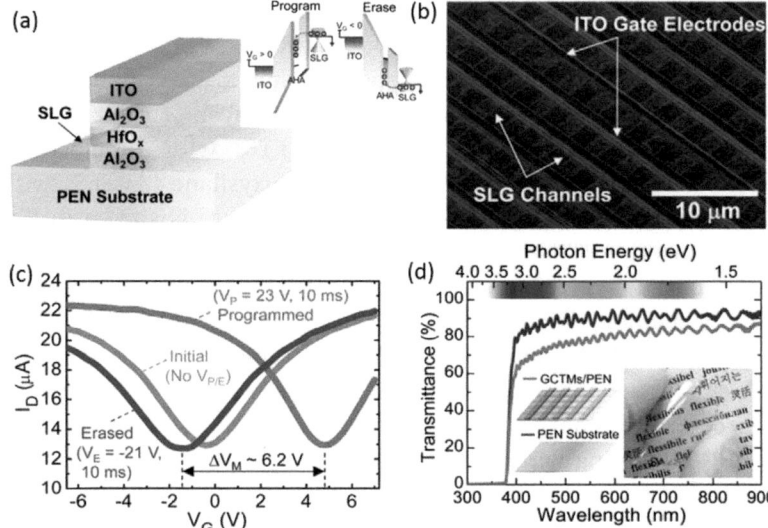

Figure 6.42 Graphene-based flexible flash memory. (a) Schematic diagram of the structure of graphene-based flexible flash memory. (b) Optical images of graphene-based flash memory arrays. (c) Transfer curve of the graphene-based flexible flash memory. (d) Optical transmittance of graphene-based flash memory. Adapted with permission from Ref. [109]. Copyright (2012) American Chemical Society.

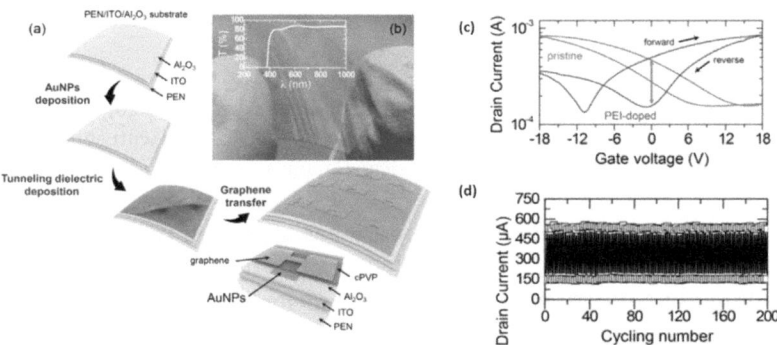

Figure 6.43 Graphene nano-floating-gate transistor memory on plastic. (a) Fabrication process of the device. (b) Optical transmittance of the device. (c) Transfer curve of the pristine graphene and PEI-doped graphene. (d) Cycle tests of the device. Adapted from Ref. [110] with permission of The Royal Society of Chemistry.

Preliminary research on graphene charge elements has been done using RGO as the charge-trapping layer [86, 111, 112]. RGO has charge capacitance for both electrons and holes. RGO can be doped utilizing alkylsilane self-assembled monolayer, (tridecafluoro-1,1,2,2,-tetrahydrooctyl) trichlorosilane (FTS) with an electron-withdrawing group, or aminopropyl triethoxysilane (APTES) with an electron-donating group. The device structure is shown in Fig. 6.44a (insert APTES-RGO into polyethylene terephthalate/25 nm Ag/15 nm Al_2O_3 and 5 nm Al_2O_3/PDPP-TBT) [111]. The device shows a 1.6 V memory window, $>10^5$ retention time, and >500 bending cycles at the electron enhancement mode, which demonstrates that it is has great potential for future flexible electronics. To further improve the memory window controllability, tuning work function of RGO by doping with gold chloride ($AuCl_3$) is demonstrated (Fig. 6.44b) [112]. To improve the charge capacitance, an RGO and gold nanoparticle DFG can be used [86].

(a)   **(b)**

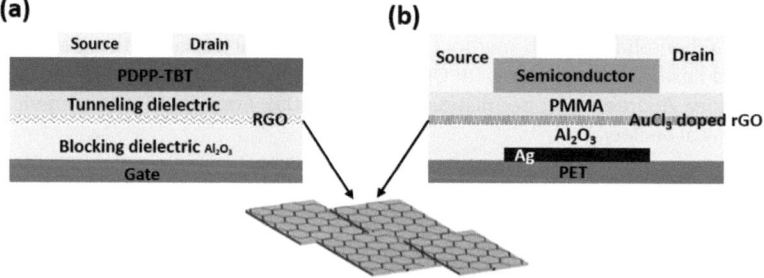

Figure 6.44 (a) Schematic diagram of RGO-based flash memory. (b) Schematic diagram of the $AuCl_3$-doped RGO-based flash memory. Adapted with permission from Ref. [111]. Copyright (2015) American Chemical Society.

RGO-based flexible flash memory has also been reported [113] by using a mixture of poly(3,4-ethylenedioxythiophene:polystyrene sulfonate) (PEDOT:PSS) and dimethyl formaldehyde (DMF) as the conductive gate that is spin-coated on the flexible polyethersulfone (PES) substrates. A 400 nm layer of cross-linked PVP is spin-coated on the gate, which is used as a blocking layer. The charge-trapping layer is chemical RGO from graphite. Next, a 10 nm layer of PVP is spin-coated as a tunneling layer. Finally, 60 nm of an organic pentacene material is deposited as the channel and gold layers are deposited as source and drain. The schematic illustration of an RGO-based organic transistor memory device is shown in Fig. 6.45. Here, we note that all the material is organic except for the gold electrode. Of course, SiO_2 or another high-k material can also be chosen as a tunneling and blocking material. However, the channel material is always organic, such as P3HT and indium zinc oxide (IZO).

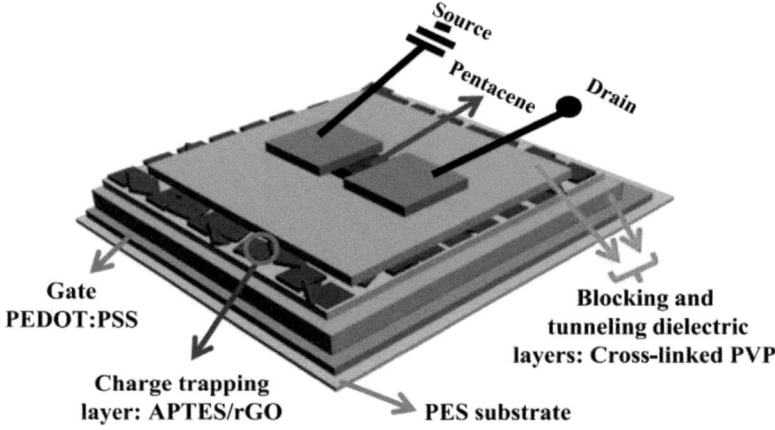

Figure 6.45 Schematic diagram of the RGO-based flexible organic flash memory. Adapted from Ref. [113], with the permission of AIP Publishing.

The all-organic flash memory seems to have a huge commercial market in flexible transparent electronics. These pioneering works demonstrate the practicability of graphene incorporation into organic transistors. Further work is still needed to optimize the process and improve the performance. The flexible mechanical properties of an organic material attached to a plastic substrate can be used for lightweight and wearable products.

6.5.2 3D Stacking

Besides 2D graphene, other arising 2D transition metal dichalcogenide (TMD) materials are also stacked into the charge-trapping structure. MoS_2, a typical TMD material, as a charge-trapping layer has been reported. Plasma-treated multilayer MoS_2 was used as a field transistor for multibit data storage (shown in

Fig. 6.46a) [114]. O_2 plasma was used to treat the top layer of MoS_2 and induced ripples up and down. The treated MoS_2 served as a charge-trapping site, and the rippled space served as a tunneling layer. The memory window and retention characteristics were investigated. Metallic MoS_2 nanoflakes fabricated by lithium intercalation was also used for charge storing in organic memory (Fig. 6.46b) [115]. The device shows sufficient memory used for multilevel memory cells, robust endurance, and quasi-permanent charge storing. Also, MoS_2 consisting metal nanoparticles are demonstrated as a charge-trapping layer [116].

(a)   **(b)**

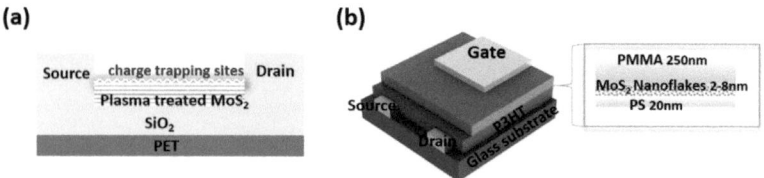

Figure 6.46 (a) Schematic diagram of O_2-treated MoS_2-based flash memory. (b) Schematic diagram of the MoS_2-nanoflake-based flexible flash memory. (c) Fabrication process of MoS_2-metal-particle-based flash memory.

Actually, similar to graphene, a TMD is suitable for a channel due to its atomic thickness and pristine surface. One example has been shown by using mechanically exfoliated few-layer MoS_2 transferred on Si/SiO_2 as channel and $Al_2O_3/HfO_2/Al_2O_3$ as charge storage element [117]. The corresponding memory structure is shown in Fig. 6.47.

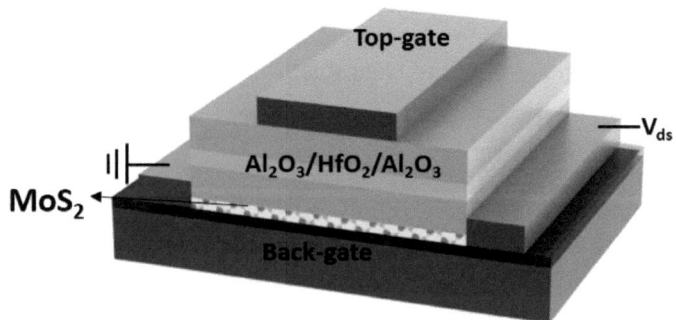

Figure 6.47 Schematic diagram of the structure of MoS_2-channel-based floating-gate memory.

The simulation has also demonstrated that monolayer-graphene-and TMD-based floating-gate memory are options for future scaling memory [118]. Firstly, the atomic height limit can reduce the gate channel. Second, cell-to-cell interference can be reduced. Most importantly, the stacking heterostructure can extend the retention of trapping charges. Graphene and TMD material have good affinity, and an offset can form a barrier that reduces the leakage current through F-N tunneling. The Fermi level of graphene can be tuned by doping to increase the barrier height. The pioneer work uses MoS_2 as the channel, insulating boron nitride (BN) as the tunneling and blocking layer, and graphene as the charge-trapping layer [119]. The corresponding memory window and retention characteristics confirm the practicability of graphene/MoS_2 shown in Fig. 6.48. It is also noted that 2D materials have the competitive superiority to replace the traditional materials of the CTM structure due to their unique properties. For example, an insulating tunneling or blocking layer material can be replaced by insulating thin BN. Graphene or MoS_2 can be used as a channel or a charge storage material. The conductive electrode can be conductive graphene [61]. The 2D-material-based floating-gate memory is compatible with the current CMOS technology.

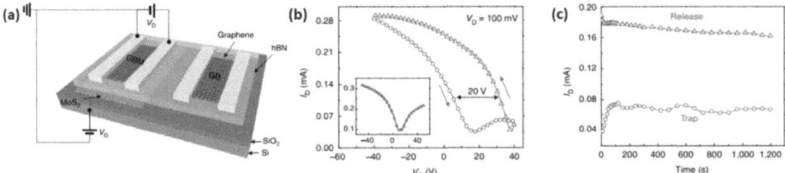

Figure 6.48 (a) Schematic diagram of MoS_2/h-BN/graphene-based flash memory. (b) Memory window of the flash memory. (c) Retention time of the flash memory. Adapted by permission from Springer Customer Service Centre GmbH: Springer Nature, *Nature Communications* (Ref. [119]), Copyright (2013).

6.6 Conclusion and Prospects

Nano-floating-gate memory devices based on FETs with semiconductors or metal nanocrystals as charge storage elements are promising candidates for the next-generation miniaturized flash memory device. This type of device structure can provide single-transistor memory, nondestructive read-out, high density, low power, better endurance, and compatibility with current CMOS technology. In particular, the deposition methods of nanocrystals can usually facilitate fast, low-cost, roll-to-roll printing processes, which is a significant potential advantage for flexible electronic devices. Although these studies have shown considerable advancements in the development of nonvolatile memory devices, the charge capture efficiency still needs a lot of improvement. In this regard, as an alternative approach, flash memory devices based on graphene-based materials have been proposed and exhibit a wide memory window at a low operation voltage. Graphene is a unique type of 2D semiconductor with a zero bandgap and a zero effective mass of the charge carriers. Thus, in nanometer-sized NG, many interesting phenomena have been observed that are different from those in any other semiconductor nanocrystals. NG has been demonstrated to possess many unique properties, such as low dimensionality, strong quantum confinement, edge effects, high thermal stability, chemical inertness, and a high DOS. More importantly, most of the fabrication methods for NG are controllable, scalable, cheap, and compatible with the traditional fabrication technology. In terms of these unique advantages, NG are very favorable as the charge storage nodes in memory devices. High-performance-graphene-nanocrystal-based CTM devices were developed with large memory windows at low operation voltages, good retention, high temperature tolerance, good endurance, high speed, and tunable memory performance.

Though NG-based CTMs show good performance, the research is still at the beginning. For practical applications, the corresponding technology should catch up. In the prototype of the NG-based CTM, some challenges remain:

- **Reproducibility:** The threshold voltage of a graphene-nanocrystal-based CTM is dependent on the size, density, and layers of NG. Thus, a controllable fabrication method of NG needs further investigation to guarantee the stability. By the year 2020, only 10 nanocrystals will be probably allowed in a single memory cell. The current bottom-up and top-down methods are far from the standard. This requires that the fabrication process of nanocrystals be uniform, controllable, and well-ordered.

- **Reliability:** Discrete nanocrystals can effectively decrease the lateral charge leakage that happens in traditional floating-gate memory. However, when the density of nanocrystals is increased, the CTM faces an unstable shift of the threshold voltage. The shift is from the leakage path of the nanocrystals into the surrounding oxide or the unstable trap level.

- **Mechanism:** Charge stored in graphene itself or the edges or the functional sizes is confirmed, but exactly what percentage of charge is stored is still unclear.

If these problems could be solved, such memory devices are promising for nanoscale-integrated high-performance and low-cost nonvolatile CTM applications, which even have the potential to exceed the performance of current flash memory technology.

References

1. Castro Neto, A.H., Novoselov, K.S., Geim, A.K., et al. (2009). *Rev. Mod. Phys.*, **81**, 109–162 (2009).

2. Novoselov, K.S., Geim, A.K., Morozov, S.V., et al. (2005). *Nature*, **438**, 197–200.

3. Zhang, Y.B., Tan, Y.W., Horst, L.S., and Philip, K. (2005). *Nature*, **438**, 201–204.

4. Morozov, S.V., et al. (2006). *Phys. Rev. Lett.*, **97**, 016801.

5. Katsnelson, M.I., et al. (2006). *Nat. Phys.*, **2**, 620–625.

6. Young, A.F., and Kim, P. (2009). *Nat. Phys.*, **5**, 222–226.

7. Terronesa, M., et al. (2010). *Nano Today*, **5**, 351–372.

8. Zhang, Y., Tang, T., Wang, F., et al. (2009). *Nature*, **459**, 820–824.

9. Partoens, B., and Peeters, F.M. (2007). *Phys. Rev. B*, **75**, 193402.

10. Partoens, B., and Peeters, F.M. (2006). *Phys. Rev. B*, **74**, 075404.

11. Luk'yanchuk, I.A., and Kopelevich, Y. (2006). *Phys. Rev. Lett.*, **97**, 256801.

12. Lee, C., Wei, X., Kysar, J.W., and Hone, J. (2008). *Science*, **321**, 385–388.

13. Bunch, J.S., et al. (2008). *Nano Lett.*, **8**, 2458–2462.

14. Nair, R.R., Blake, P., Grigorenko, A.N., et al. (2008). *Science*, **320**, 1308–1311.

15. Nikolaos, T., Csaba, J., Mihaita, P., Harry, T.J., and Bart, J.W. (2007). *Nature*, **448**, 571–574.

16. Chen, J., Jang, C., Xiao, S., Ishigamin, M., and Fuhrer, M.S. (2008). *Nat. Nanotechnol.*, **3**, 206–209.

17. Dean, C.R., Young, A.F., Hone, J., et al. (2010). *Nat. Nanotechnol.*, **5**, 722–726.

18. Balandin, A.A., et al. (2008). *Nano Lett.*, **8**, 902–907.

19. Molitor, F., et al. (2011). *J. Phys.: Condens. Matter*, **23**, 243201.

20. Nakada, K., et al. (1996). *Phys. Rev. B*, **54**, 17954–17961.

21. Han, M.Y., Özyilmaz, B., Zhang, Y., and Kim, P. (2007). *Phys. Rev. Lett.*, **98**, 206805.

22. Sepioni, M., Nair, R.R., and Rablen, S., et al. (2010). *Phys. Rev. Lett.*, **105**, 207205–207208.

23. Shen, J., Zhu, Y., Yang, X., and Li, C. (2012). *Chem. Commun.*, **48**, 3686–3699.

24. Bacon, M., Bradley, S.J., and Nann, T. (2014). *Part. Part. Syst. Char.*, **31**, 415–428.

25. Fiori, G., Bonaccorso, F., Iannaccone, G., et al. (2014). *Nat. Nanotechnol.*, **9**, 768–779.

26. Novoselov, K.S., et al. (2012). *Nature*, **490**, 192–200.

27. Chun, K., and Martin, P. (2014). *Chem. Soc. Rev.*, **43**, 291–312.

28. Brodie, B.C. (1859). *Philos. Trans. R. Soc. London*, **149**, 249–259.

29. Staudenmaier, L. (1898). *Ber. Dtsch. Chem. Ges.*, **31**, 1481–1487.

30. Hummers Jr., W.S., and Offeman, R.E. (1958). *J. Am. Chem. Soc.*, **80**, 1339–1339.

31. Marcano, D.C., Kosynkin, D.V., and Tour, J.M. (2010). *ACS Nano*, **4**, 4806–4814.

32. He, H., Klinowski, J., Forster, M., and Lerf, A. (1998). *Chem. Phys. Lett.*, **287**, 53–56.

33. Erickson, K., Erni, R., Lee, Z., Alem, N., Gannett, W., and Zettl, A. (2010). *Adv. Mater.*, **22**, 4467–4472.

34. Boukhvalov, D.W., and Katsnelson, M.I. (2008). *J. Am. Chem. Soc.*, **130**, 10697–10701.

35. Li, D., Müller, M.B., Gilje, S., Kaner, R.B., and Wallace, G.G. (2008). *Nat. Nanotechnol.*, **3**, 101–105.

36. Zhou, X., Zhang, J., Wu, H., Yang, H., Zhang, J., and Guo, S. (2011). *J. Phys. Chem. C*, **115**, 11957–11961.

37. Gao, W., Alemany, L.B., Ci, L., and Ajayan, P.M. (2009). *Nat. Chem.*, **1**, 403–408.

38. Yang, D., et al. (2009). *Carbon*, **47**, 145–152.

39. Cote, L.J., Cruz-Silva, R., and Huang, J. (2009). *J. Am. Chem. Soc.*, **131**, 11027–11032.

40. Dubin, S., et al. (2010). *ACS Nano*, **4**, 3845–3852.

41. Williams, G., Seger, B., and Kamat, P.V. (2008). *ACS Nano*, **2**, 1487–1491.

42. Cheng, M., Yang, R., Zhang, G.Y., et al. (2012). *Carbon*, **50**, 2581–2587.

43. Baraket, M., Walton, S.G., Wei, Z., et al. (2010). *Carbon*, **48**, 3382–3390.

44. Lopez, V., Sundaram, R.S., Gomez-Navarro, C., et al. (2009). *Adv. Mater.*, **21**, 4683–4686.

45. Kaiser, A.B., Gomez-Navarro, C., Sundaram, R.S., Burghard, M., and Kern, K. (2009). *Nano Lett.*, **9**, 1787–1792.

46. Reina, A., Jia, X., Kong, J., et al. (2009). *Nano Lett.*, **9**, 30–35.

47. Muñoz, R., and Gómez-Aleixandre, C. (2013). *Chem. Vap. Deposition*, **19**, 297–322.

48. Zhang, Y., Zhang, L., and Zhou, C. (2013). *Acc. Chem. Res.*, **46**, 2329–2339.

49. Edwards, R.S., and Coleman, K.S. (2013). *Acc. Chem. Res.*, **46**, 23–30.

50. Li, X., Cai, W., Colombo, L., and Ruoff, R.S. (2009). *Nano Lett.*, **9**, 4268–4272.

51. Kim, K.S., Zhao, Y., Kim, P., and Hong, B.H. (2009). *Nature*, **457**, 706–710.

52. Yu, Q.K., Lian, J., Siriponglert, S., et al. (2008). *Appl. Phys. Lett.*, **93**, 113103.

53. Chae, S.J., Guenes, F., and Kim, K.K. (2009). *Adv. Mater.*, **21**, 2328–2333.

54. Zhang, Y., Gomez, L., and Zhou, C. (2010). *J. Phys. Chem. Lett.*, **1**, 3101–3107.

55. Bae, S., Kim, H., Iijima, S., et al. (2010). *Nat. Nanotechnol.*, **5**, 574–578.

56. Li, X., Cai, W., and Ruoff, R.S. (2009). *Science*, **324**, 1312–1314.

57. Luo, Z., Lu, Y., Singer, D.W., et al. (2011). *Chem. Mater.*, **23**, 1441–1447.

58. Li, X., Magnuson, C.W., and Ruoff, R.S. (2010). *Nano Lett.*, **10**, 4328–4334.

59. Liang, X., Sperling, B.A., Calizo, I., et al. (2011). *ACS Nano*, **5**, 9144–9153.

60. Liu, N., Pan, Z., Fu, L., Zhang, C., Dai, B., and Liu, Z. (2011). *Nano Res.*, **4**, 996–1004.

61. Dean, C.R., Young, A.F., Hone, J., et al. (2010). *Nat. Nanotechnol.*, **5**, 722–726.

62. Yankowitz, M., Xue, J., Cormode, D., et al. (2012). *Nat. Phys.*, **8**, 382–386.

63. Lee, J., Kim, K., Park, W.I., et al. (2012). *Nano Lett.*, **12**, 6078–6083.

64. Fan, T.J., Zeng, W.J., Tang, W., et al. (2015). *Nanoscale Res. Lett.*, **10**, 1–8.

65. Shin, Y., Park, J., Hyun, D., et al. (2015). *New J. Chem.*, **39**, 2425–2428.

66. Peng, J., Gao, W., Gupta, B.K., et al. (2012). *Nano Lett.*, **12**, 844–849.

67. Li, Y., Hu, Y., Zhao, Y., et al. (2011). *Adv. Mater.*, **23**, 776–780.

68. Zhou, J., Booker, C., Li, R., et al. (2007). *J. Am. Chem. Soc.*, **129**, 744–745.

69. Pan, D., Zhang, J., Li, Z., et al. (2010). *Adv. Mater.*, **22**, 734–738.

70. Shen, J., Zhu, Y., Yang, X., et al. (). *New J. Chem.*, **36**, 97–101.

71. Kim, S., Hwang, S.W., Kim, M., et al. (2012). *ACS Nano*, **6**, 8203–8208.

72. Tang, L., Ji, R., Cao, X., et al. (2012). *ACS Nano*, **6**, 5102–5110.

73. Lu, J., Yeo, P.S.E., Gan, C.K., et al. (2011). *Nat. Nanotechnol.*, **6**, 247–252.

74. Ruan, G., Sun, Z., Peng, Z., et al. (2011). *ACS Nano*, **5**, 7601–7607.

75. Sun, Z., Yan, Z., Yao, J., et al. (2010). *Nature*, **468**, 549–552.

76. Zhu, S., Wang, L., Li, B., et al. (2014). *Carbon*, **77**, 462–472.

77. Liu, R., Wu, D., Feng, X., et al. (2011). *J. Am. Chem. Soc.*, **133**, 15221–15223.

78. Yang, R., Zhu, C., Meng, J., et al. (2013). *Sci. Rep.*, **3**, 2126.

79. The International Technology Roadmap for Semiconductors (ITRS) (2011). Semiconductor Industry Association, International Sematech, Austin, TX.

80. Lee, J.-D., Hur, S.-H., and Choi, J.-D. (2002). *IEEE Electron Device Lett.*, **23**, 264.

81. Raghunathan, S., Krishnamohan, T., Parat, K., and Saraswat, K. (2009). *IEEE International Electron Devices Meeting (IEDM)*, Baltimore, MD, 1–4

82. Delgado, J.C.D., Kim, Y.A., Hayashi, T., et al. (2009). *Chem. Phys. Lett.*, **469**, 177.

83. Augustin, J.H., Emil, B.S., Hyung, S.Y., et al. (2011). *ACS Nano*, **5**, 7812.

84. Mishra, A., Kalita, H., Waikar, M., et al. (2012). *IEEE Proceedings*.

85. Mishra, A., Janardanan, A., Khare, M., et al. (2013). *IEEE Electron Device Lett.*, **34**, 1136.

86. Han, S.-T., Zhou, Y., Wang, C., et al. (2013). *Adv. Mater.*, **25**, 872–877.

87. Hanafi, H.I., Tiwari, S., and Khan, I. (1996). *IEEE Trans. Electron Devices*, **43**, 1553.

88. King, Y.C., King, T.J., and Hu, C.M. (2000). *IEEE Electron Device Lett.*, **21**, 543.

89. Wang, C.-C., Tseng, J.-Y., Wu, T.-B., et al. (2006). *J. Appl. Phys.*, **99**, 026102.

90. Kouvatsos, D.N., Ioannou-Sougleridis, V., and Nassiopoulou, A.G. (2003). *Appl. Phys. Lett.*, **82**, 397.

91. Shi, Y., Saito, K., Ishikuro, H., et al. (1998). *J. Appl. Phys.*, **84**, 2358.

92. Guan, W.H., Long, S.B., Liu, M., et al. (2007). *J. Phys. D: Appl. Phys.*, **40**, 2754.

93. Panda, D., Dhar, A., and Ray, S.K. (2009). *Semicond. Sci. Technol.*, **24**, 115020.

94. Ramalingam, B., Zheng, H., and Gangopadhyay, S. (2014). *Appl. Phys. Lett.*, **104**, 143103.

95. Choi, H.J., Choi, B.-S., Kim, T.-W., et al. (2008). *Nanotechnology*, **19**, 305704.

96. Kim, K.S., Zhao, Y., Jang, H., et al. (2009). *Nature*, **457**, 706–710.

97. Joo, S.S., Kim, J.K., Kang, S.S., et al. (2014). *Nanotechnology*, **25**, 255203.

98. Nair, R.R., Blake, P., Grigorenko, A.N., et al. (2008). *Science*, **320**, 1308–1308.

99. Nazek, E.A., Furkan, C., Sabri, A., et al. (2014). *Appl. Phys. Lett.*, **105**, 033102.

100. Dai, M.-K., Lin, T.-Y., Yang, M.-H., et al. (2014). *J. Mater. Chem. C*, **27**, 5342.

101. Zhang, L., Shi, Z., Wang, Y., et al. (2011). *Nano Res.*, **4**, 315.

102. Yang, W., He, C., Zhang, L., et al. (2002). *Small*, **8**, 1429–1435.

103. Meng, J., Yang, R., Zhang, G.-Y., et al. (2015). *Nanotechnology*, **26**, 455704.

104. Howard, E.K., Hong, X.M., Ananth, D., et al. (2002). *J. Appl. Phys.*, **91**, 1572 (2002).

105. Paul, H., Gerwin, H.G., Robert, M., et al. (2011). *Chem. Mater.*, **23**, 341.

106. Ling, Q.-D., Liaw, D.-J., Zhu, C.X., et al. (2008). *Prog. Polym. Sci.*, **33**, 917.

107. Tadanori, K., Tomoya, H., Mitsuru, U., et al. (2013). *Polym. Chem.*, **4**, 16.

108. Hou, Y., Han, S.-T., Sonar, P., et al. (2013). *Sci. Rep.*, **3**, 2319.

109. Sung, M.K., Song, E.B., Sejoon, L., et al. (2012). *ASC Nano*, **6**, pp.7879.

110. Jang, S., Hwang, E., Cho, J.H., et al. (2014). *Nanoscale*, **6**, 15286–15292.

111. Han, S.-T., Zhou, Y., Sonar, P., et al. (2015). *ACS Appl. Mater. Interfaces*, **7**, 1699–1708.

112. Han, S.-T., Zhou, Y., Yang, Q.D., et al. (2014). *ACS Nano*, **8**, 1923–1931.

113. Rani, A., Song, J.-M., Jung Lee, M., et al. (2012). *Appl. Phys. Lett.*, **101**, 233308.

114. Chen, M., Nam, H., Wi, S., et al. (2014). *ACS Nano*, **8**, 4023–4032.

115. Kang, M., Kim, Y.-A., Yun, J.-M., et al. (2014). *Nanoscale*, **6**, 12315–12323.

116. Han, S.-T., Zhou, Y., Chen, B., et al. (2015). *Nanoscale*, **7**(41), 17496–17503.

117. Zhang, E., Wang, W., Zhang, C., et al. (2015). *ACS Nano*, **9**, 612–619.

118. Wei, C., Jiahao, K., Bertolazzi, S., et al. (2014). *IEEE Trans. Electron Devices*, **61**, 3456–3464.

119. Sup, C.M., Lee, G.-H., Yu, Y.-J., et al. (2013). *Nat. Commun.*, **4**, 1624.

Chapter 7

Data Recovery of Flash Memory

Bernard Kasamani Shibwabo* and Ismail Ateya Lukandu*

Faculty of Information Technology, Strathmore University,
Ole Sangale Road, Madaraka Estate, PO 59857-00200 Nairobi, Kenya
bshibwabo@strathmore.edu; bernardshib@gmail.com

With the usage of flash memory increasing, coupled with portable flash drives, and secure digital (SD) cards and other flash media becoming ubiquitous, all sorts of important information is often stored on them, including important and sensitive documents. Loss of data can often cause a heartache, and it can also cause practical problems, especially if these devices store important or sensitive information. Even for the case of NAND flash memory, which is referred to as a nonvolatile memory, the nonvolatility of the data stored on it is guaranteed only for a specified retention time. There is need to formulate techniques of data recovery in flash. Data recovery is the process of reclaiming data from damaged, failed, corrupted, or inaccessible secondary storage media when it cannot be accessed normally. The reasons for data recovery could

---

*All authors contributed equally to this work.

Nanocrystals in Nonvolatile Memory
Edited by Writam Banerjee
Copyright © 2018 Pan Stanford Publishing Pte. Ltd.
ISBN 978-981-4774-73-4 (Hardcover), 978-1-351-20327-2 (eBook)
www.panstanford.com

be physical damage to the storage device or logical damage to the file system that stops it from being mounted by the host operating system. This chapter presents a comprehensive analysis of flash memory and proceeds to describe a set of high-level and low-level approaches for data recovery of flash memory. Artifacts, triggered by flash-specific operations, including block erasing and wear leveling, are discussed and directions are given for enhanced data recovery from and analysis of data stored on flash memory. This chapter mainly consists of four parts. The first part discusses how computers store information. The second part discusses in-depth concepts relating to flash memory. The third part focuses on data recovery concept, and the fourth part presents various techniques of data recovery in flash memory. Therefore, we discover that, despite data loss being a problem, the good news is that most of the time data can be recovered.

Memory . . . is the diary that we all carry about with us.

—Oscar Wilde

Time moves in one direction, memory in another.

—William Gibson

Without memory, there is no culture. Without memory,
there would be no civilization, no society, no future.

—Elie Wiesel

7.1 Introduction

This chapter presents as a discussion the vital topic of how data recovery can be achieved in flash memory. At any given time, the human mind's content is described by the nature or kind of things the mind interacts with on a daily basis. The interaction in the physical world is commonly achieved through the senses. What we perceive through our senses is categorized in the mind as data, information, knowledge, understanding, or wisdom. Of all these categories, wisdom is the only one that is not purely a result of what was recorded in the past. In the present world of technology, computer systems have the capability to process and represent data, information, and knowledge. Figure 7.1 presents the noise, data, information, knowledge, and wisdom (NDIKW) pyramid.

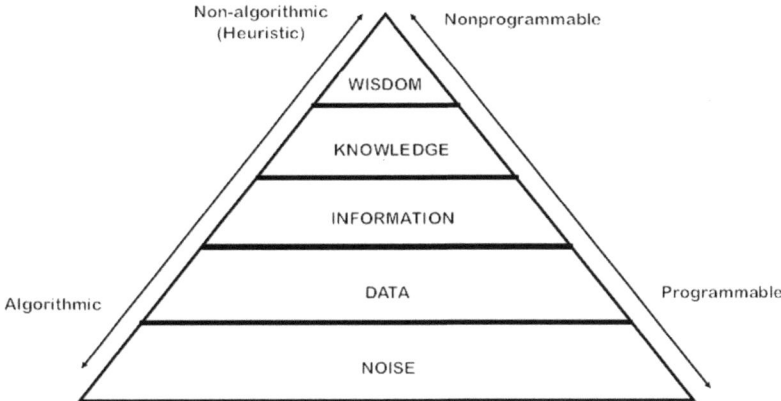

Figure 7.1   The NDIKW pyramid.

Data and information have become very important resources toward achieving any objective. Data is defined as a set of facts. In computing, data consists of characters, symbols, or quantities. This is what is usually transmitted, stored, operated on, or reasoned upon. Data is an irreplaceable strategic asset, and therefore any loss or destruction of data can have horrific financial and even social consequences for any party. Information, on the other hand, is defined as facts provided, obtained, or learned about a particular domain. The domain in this case can be an event, a subject, a news item, a crime, or a party, among other areas. Hence the domain is an area of interest. As an example, if we take a picture of a person, the photograph is information but what that persion looks like is purely data. Table 7.1 shows the distinguishing aspects about data, information, knowledge, and wisdom.

Information is needed in various daily activities, such as planning, monitoring, controlling, measuring, and decision making, where these needs make information critical to the survival and competitiveness of the business. Unfortunately, the same information can easily be corrupted, lost, or stolen due to disasters and events such as equipment failure, virus attacks, human error, theft, or natural disasters. Hence, it is now common for individuals and enterprises to find themselves storing data continuously in various storage locations around the world. Data may be stored in any data storage medium or could be in form of data packets in transit to a destination. With data storage cost per megabit reducing with improvement in technology, data is presently being stored at a higher rate than ever before. Additionally, enterprises are increasingly interconnected, exposing information to a growing number and a wider variety of threats and vulnerabilities.

Table 7.1   Data versus information versus knowledge versus wisdom

Term	Characteristics	Examples
Data	Uninterpreted raw signal	It is raining.
Information	Data + context meaning	• SOS. • The temperature dropped 24 degrees and then it started raining.
Knowledge	General awareness, understanding or possession of information, facts, ideas, truths, or principles, often gained through experience or study	• Emergency! Start rescue. • If the humidity is too high and the temperature drops substantially, the atmospheres is often unlikely to be able to hold the moisture. So it rains. • Knowing that a tomato is a fruit.
Wisdom	High level of predictability as to what is described or what will happen next	• It rains because it rains. • Knowing not to put a tomato in a fruit salad.

Flash memory is now widely known as a type of electronic (solid-state) nonvolatile memory that erases data in units referred to as blocks, hence the use of the term "flash." Therefore, a block stored on a flash memory chip must be electrically erased before any data can be written or reprogrammed onto the microchip. Flash memory retains data for a prolonged period of time whether a flash-equipped device is powered on or off. Flash memory was introduced by Toshiba in 1984, apparently as an invention credited to Dr. Fujio Masuoka. The term "flash" was coined because the process of erasing all the data from a semiconductor chip is reminiscent of the flash of a camera.

Flash memory was developed from EEPROM (electrically erasable programmable read-only memory). There exist two major types of flash memory, which are named after the NAND and NOR logic gates. The separate flash memory cells have internal characteristics that resemble those of the corresponding gates.

Flash memory has made a major transition from embedded devices to mobile phones and devices, laptops, desktops, and even big data centers. The memory offers enormous performance gains and power savings compared to the disk [1] while being denser and less power hungry than dynamic random access memory (DRAM). However, in order to fully exploit the advantages of flash memory, there is a need to overcome some of the issues—its durability is limited; it suffers from data integrity problems; and its read, program, and erase operations function at unequal granularities and have huge variety of latencies. Organizations, in accordance with the ISO/ IEC 17799:2005 standards, do use different ways to recover any lost important information—a process referred to as data recovery. They have the option to use some data recovery software in an attempt to retrieve lost data, use some data software to repair corrupted data, or use external data recovery services or even try restoring data from existing data backup.

Data recovery software in itself is not sufficient to restore lost or corrupt data because it does not fully guarantee total recovery of the lost data in addition to the fact that it may end up being a long and very tedious exercise trying to restore data using data recovery software, yet it could just be achieved by simply copying the lost or corrupted data back to the computer. Therefore, it is important and easier to always maintain a backup of your important data. Due to an ever-increasing necessity of data and the dependence of organizations on data, in addition to increased competitive and regulatory pressures in businesses, there is a need for high data availability. Achieving 100% data availability can never be guaranteed, since there always exist threats to data and therefore an organization will at some time or another lose some critical data. Due to the expected chances of data loss, there is a need to maintain backups and formulate means to recover the data in a timely fashion.

7.2 How Computers Store Information

7.2.1 Kinds of Computer Memory

Over time, computers have been known as devices that can store and process information, with their usefulness derived from their speed, ability to manipulate information, and ability to store large amounts of information. During an ordinary telephone conversation, the parties speak and hang up the phone; nothing is stored. However, this is not sufficient for an answering machine or a computer, both of which need to store information. An answering machine answers the phone and stores the information given by the caller. Since a computer consists of a processor separate from the main storage, it is necessary to store information so that the processor can work on it. The information that a computer stores could be what comes from the user or a set of instructions. Figure 7.2 presents the main hardware components of a computer based on the von Neumann architecture or model.

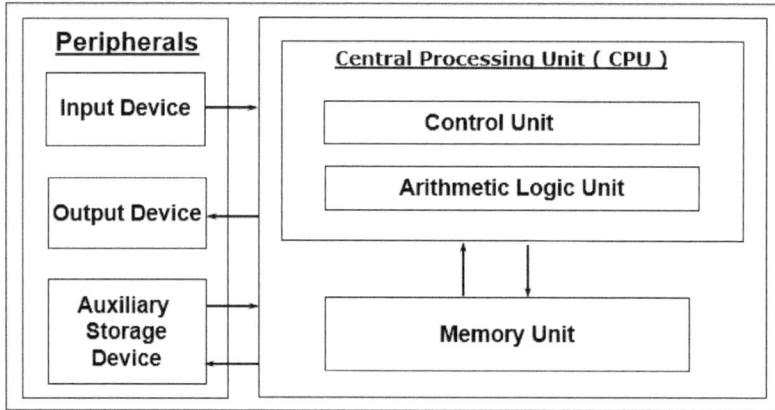

Figure 7.2 Main hardware components of a von Neumann architecture computer.

Computers use two kinds of storage to store information available. One is the short-term or temporary storage, with the information stored being actively used by the processor. A good example of temporary storage is random access memory (RAM).

RAM will mainly accept new information often from long-term storage devices connected to the computer. The second kind is long-term storage—the kind of information that is repeatedly used by computers, such as the instructions the computer prepares itself with every time you turn it on. These instructions are stored in read-only memory (ROM), a type of memory that does not accept new information.

Additionally, computers use a variety of devices to store information that is not actively being used for processing. These devices include hard drives, optical discs, storage, and removable media. These devices are also referred to as secondary storage devices. Secondary storage holds permanent or semipermanent data on some external magnetic or optical medium. The physical attributes of secondary storage devices determine the way data is organized on them. In reality, information on a disk must be copied to electronic memory (primary/temporary memory) before the computer can do anything with it, and data in memory must be copied back to the disk if it is to be preserved after the computer is shut off. This information is stored in the working memory of the computer in the state of electrical switches and in external memory as magnetic images.

7.2.2 Bits and Memory

Computers store information by using switches (on-off switches, as in the case of turning lights on/off, but tiny and implemented with electronic chips, that is, no parts are in motion). Regardless of whether information is stored on secondary or primary storage or not, information is stored by the fact that a given electrical switch can be in either of two states (or a given area of storage can be magnetized or not). Imagine storing information by using a long series of switches each of which can be on or off. When a switch is on, we can represent it by the numeral 1, and when it is off, we can represent that by the numeral 0. This leads to a digital format where something can have only one of a set number of possible states, and in this case, it is binary, meaning it can have only two such possible states.

Since in computing we have binary storage devices, there is a need to comprehend how to store information on them. In binary, we have two values to use, 0 and 1. Therefore, one switch can store two values. These values can be coded using 1 and 0 as follows for the example of a bulb: 0 = It is on; 1 = It is off (or vice versa).

In modern computers, switches es are arranged in groups of 8, with 256 possible combinations (since each switch can represent 2 values, the total number = 2^8 = 256). This particular sequence of 8 switches is what is called a byte (the information stored in 8 switches is called a byte). One individual switch would be a bit (the information stored in 1 switch is called a bit). The groups of 8 (1 byte) would appear like this: 00000000, 00000001, 00000010, 00000011, and so on.

Since the 8-group switches have 256 possible combinations, they can be used to represent integers ranging from 0 to 255. Or any given switch combination can be assigned a symbol to represent it (sometimes called a "code"). An example of a code used in many applications in computing is the ASCII code. The other essential code is Unicode. Unicode uses 2 bytes to represent 1 character. Since it uses 2 bytes, it has 65,536 possible combinations and so can be used to represent far more characters than the ASCII code can represent. In Table 7.2, let us see how the combinations are derived given the number of switches:

Table 7.2　Possible combinations given the number of switches

Number of switches	Possible combinations
1	2 (i.e., on/off)
2	4 (i.e., on,on/off,off/off,on/off,on = both on, both off, switch 1 off while switch 2 is on, switch 2 off while switch 1 is on).
3	8
4	16
5	32
6	64
7	128
8	256

In case you start a text editor like Notepad or any other text editor and type text in it, it stores each letter as a byte in the computer's memory, or RAM. Computer memory is a set (array) of millions of electronic switches. The letter is stored by using the ASCII code. When you later save the text to your hard disk, the program (in conjunction with the operating system) copies the bytes from memory and writes them on to the hard disk. A word, on the other hand, would typically be 16, 32, or 64 bits depending on the computer and its capability for handling information. The word size is the computer's preferred size for moving units of information around: technically it is the width of a computer's processor registers, which are the holding areas the processor uses to do arithmetic and logical calculations. When you hear people talking about computers having bit sizes (referring to them as 32-bit or 64-bit computers), this is what they mean. Most computers nowadays have a word size of 64 bits. Until recently (in the early 2000s) many computers had 32-bit words. The old 286 machines, during the 1980s, had a word size of 16. Old-style mainframes often had 36-bit words.

The hard disk is simply a data storage device that has a group of tiny spots on it that are either magnetized or not. A magnetized spot is used to represent the integer 1 and a demagnetized spot represents 0. Thus, the same information can comfortably be represented on the disk that was represented by the electronic switches that form the computer's memory. It is just being represented in a different manner. Magnetic charges on the hard disk persist even after the computer is turned off. In contrast, the electronic switches all reset to 0 when the computer is powered off. The operating system assigns each collection of bytes a name and keeps a record of the name and the matching bytes belonging to the name. This named collection of bytes is called a file. Files are often grouped into folders. A folder is simply a special type of file that has a list of files in it.

7.3   Flash Memory

7.3.1   Introduction to Flash Memory

Picture a scenario whereby your memory works only while you were awake. The moment you go to sleep, your memory gets cleared! That implies that every time you wake up, you have to relearn everything that you ever knew so as to be as you are now. That can be scary indeed, but computers are faced with this challenge. Ordinary computer chips lose/forget everything (their entire contents) when the computer that they are attached to is powered off. Computers address this by having magnetic storage called hard disks for storage. The contents on these disks can be retrieved irrespective of whether the machine is switched off or switched on. We also have smaller, more portable devices, such as digital cameras and MP3 players, which need smaller and more portable memory. These devices use special chips called flash memory to store information permanently. The kind of memory that loses/forgets everything when computer is powered off is called RAM. There is, however, another kind of memory, called ROM, that does not lose information when power goes off. ROM chips are usually prestored with information during their manufacture, and they retain that information irrespective of whether the power is switched on or switched off.

A major problem with ROM is that the information it stores is there permanently: ROM can never be rewritten. In reality, a computer uses a combination of different kinds of memory for different purposes. The data/information it needs to remember all the time, for example, the tasks to perform when the computer is just powered on (booting process information) is stored on ROM chips. During the time that you are using a computer (e.g., typing a text document) and it needs temporary memory for processing what you are typing, it uses RAM chips; it does not matter that this information is lost later. Information that needs to be remembered indefinitely is stored on the hard disk. However, it takes longer to read and write information from hard disks than from memory chips; thus, hard disks are not generally used as temporary memory. That explains why flash memory is used instead of hard disks in gadgets like digital cameras and small MP3 players. Flash memory shares some similarities with both RAM and ROM. Like ROM, it remembers information when the power is off; like RAM, it can be erased and rewritten more or less as many times as necessary.

7.3.2 The Features of Flash Memory

Flash memory is an electronic nonvolatile computer storage medium that can be electrically erased and reprogrammed. Flash memory is an advancement from erasable programmable read-only memory (EPROM) and EEPROM. The technology industry strictly uses the term "EEPROM" for byte-level erasable memory and applies the term "flash memory" to larger block-level erasable memory. Flash memory is extensively used for data storage as well as data transfer in end user devices, enterprise systems, and industrial applications. Flash disks are common devices that use flash memory. Figure 7.3 shows sample USB flash disks that utilize flash memory technology.

Figure 7.3 Sample USB flash disks in the market that use flash memory technology.

Before discussing more about flash memory, it is important to introduce transistors. Transistors have significantly revolutionized electronics since they there invention at Bell Labs over half a century ago (the year 1947) by John Bardeen, Walter Brattain, and William Shockley. The three inventors were jointly awarded the 1956 Nobel Prize in Physics for their achievement. The invention of the first transistor was named an IEEE Milestone in 2009. Many consider it to be one of the greatest inventions of the 20th century.

7.3.3 Transistors

A transistor is defined as a semiconductor device used to amplify and switch electronic signals and electrical power. Transistors consist of semiconductor material with at least three terminals for connection to an external circuit, for example, the motherboard. A voltage or current applied to one pair of the transistor's terminals changes the current through another pair of terminals. Transistors can be acquired individually or more often found embedded in integrated circuits. The transistor is therefore considered the fundamental building block of modern electronic devices and is ubiquitous in modern electronic systems. The human brain contains around 100 billion cells called neurons—which are tiny switches that let any human think and remember things. Analogously, computers contain billions of miniature "brain cells" as well, which we refer to as transistors. Transistors are made from silicon, a chemical element commonly found in sand and that normally does not conduct electricity. Silicon is actually a semiconductor, which means it is neither a conductor nor an insulator.

By treating silicon with impurities (a process known as doping), we can make it behave in a different way. Silicon can be doped with the chemical element arsenic, phosphorus, or antimony, which makes it gain some extra "free" electrons—which can carry an electric current. Since electrons have a negative charge, silicon that is treated this way is called n-type (negative type). Alternatively, we can dope silicon with impurities such as boron, gallium, and aluminum. Any silicon treated this way has fewer of those "free" electrons; therefore, electrons in nearby materials tend to flow into it. Silicon that is treated this way is called p-type (positive type). It is important to note that neither n-type nor p-type silicon actually has a charge in itself: both are electrically neutral. The most advanced transistors work by controlling the movements of individual electrons, so you can picture just how small they are. In a modern computer chip, the size of a fingernail, you will possibly find between 500 million and 2 billion separate electrons. Flash memory has a grid of columns and rows, with a cell that has two transistors at each intersection. Figure 7.4 shows a representation of a sample of transistors.

A standard transistor has three connections (control wires): source, drain, and gate. Imagine a transistor to be a pipe through which electricity can flow in the same way as water. One end of the pipe would have the water flowing in, called the source, for example, a dam. The other end of the pipe is called the drain (where water flows out). In between the source and the drain, blocking the pipe, there is a gate. For a transistor, any time that the gate is closed, the pipe is shut off and no electricity can flow; therefore, the transistor is off. In this case, the transistor stores a value of 0 for the state. Any time the gate is opened, electricity flows in, the transistor is on, and it therefore stores a value of 1. However, when the power is turned off, the transistor switches off too. When you switch the power back on, the transistor is still off, and since it is impossible to tell whether it was on or off before the power was removed, you can see why we say it forgets/loses any information it stores.

Figure 7.4 Sample transistors available on the technology market.

Flash memory consists of transistors. These transistors can be triggered to change their state (from a value of 1 to a value of 0 and vice versa) with electrical current but will retain that state in the event the electrical current is gone. Flash functions by using a totally different kind of transistor, one that stays switched on (or switched off) even when the power is turned off. The same way the EEPROM chips used to store computer motherboard basic input/ output system (BIOS) information, flash memory requires electricity in order to write or read data from its storage but retains the data when power is turned off. A flash transistor is different from a standard transistor because it has a second gate, above the first one. Figure 7.5 shows a representation of a basic transistor.

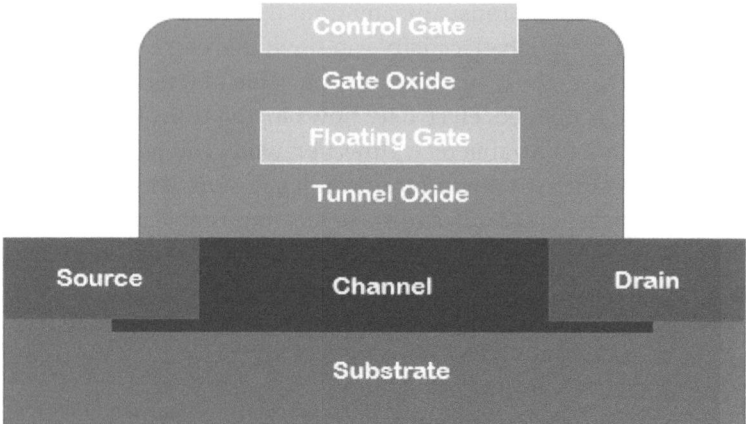

Figure 7.5 A basic transistor.

On opening the gate, electricity seeps up the first gate and is held between the first gate and the second one, changing the state to 1. The electricity stays between the two gates even when electricity is turned off. That explains the fact that the transistor stores its information whether the power is on or off. The stored information can be erased by making the "held electricity" drain back down again. The two gates found in flash memory are referred to as control gate and floating gate [2, 3]. These gates are separated by oxide layers through which current cannot normally pass. The floating gate stores the electrical charge and controls the flow of the electrical current.

The floating gate is insulated from all sides by an oxide layer. In the state represented in Fig. 7.5, the transistor is switched off—and effectively storing a 0 value. To understand how we switch it on, both the source and the drain regions are rich in electrons (because they are made of n-type silicon). However, electrons cannot flow from source to drain because of the electron deficient, p-type material that exists between them. However, if we apply a positive voltage to the transistor's two contacts, called the bitline and the wordline, electrons get pulled in a rush from source to drain. A few also manage to twist and turn with quick twisting movements through the oxide layer by a process called tunneling and get stuck on the floating gate.

The floating gate stores a value of 1 due to the presence of electrons on the gate. Those electrons stay there indefinitely, even when the positive voltages are removed and whether the circuit is supplied with power or not. The electrons can be flushed out by putting a negative voltage on the wordline. This has the effect of repelling the electrons back the way they came, clearing the floating gate and making the transistor store a value of 0 again. Due to the fact that the floating gate is electrically isolated by its insulating layer, electrons that are placed on it are held until they are removed by another application of electric field. Electrons are either added to or removed from the floating gate in order to change the storage transistor's threshold voltage so as to program (set) the cell to be a 0 or a 1. It is also possible to later remove the electrons from the floating gate with a process called field emission (also called Fowler–Nordheim tunneling), which removes electrons from the floating gate. Figure 7.6 showing the contacts of the transistor.

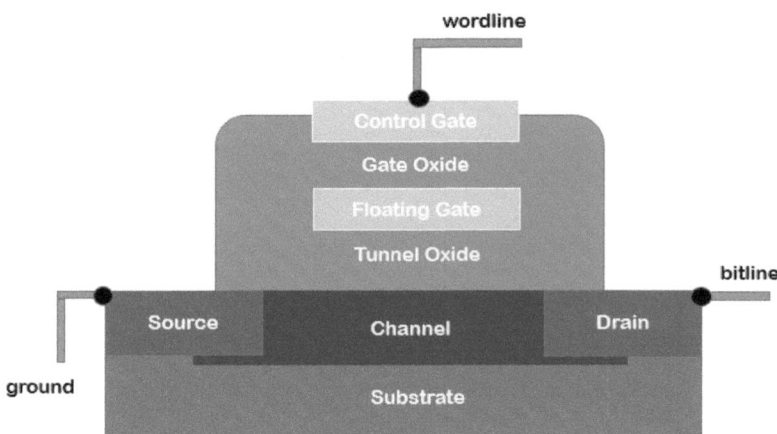

Figure 7.6 Transistor contacts (wordline and bitline).

Field emission is the process whereby electrons tunnel through a barrier in the presence of a high electric field. Either field emission or a phenomenon called channel hot-electron injection holds the electrons in the floating gate. Channel hot-electron injection originates from the heating of the channel carriers traveling from the source to the drain, due to the lateral electric field EII produced by the applied drain-source voltage $V_{DS} = V_D - V_S$.

When removing electrons through field emission, a strong negative charge on the control gate forces electrons off the floating gate and into the channel, where a strong positive charge exists. To hold (trap) electrons in the floating gate, the reverse of this process happens when using field emission. Electrons seep through the thin oxide layer to the floating gate due to the presence of a high electric field, with a strong negative charge on the cell's source and the drain and a strong positive charge on the control gate. This is illustrated in Fig. 7.7—a further representation of how field emission really works.

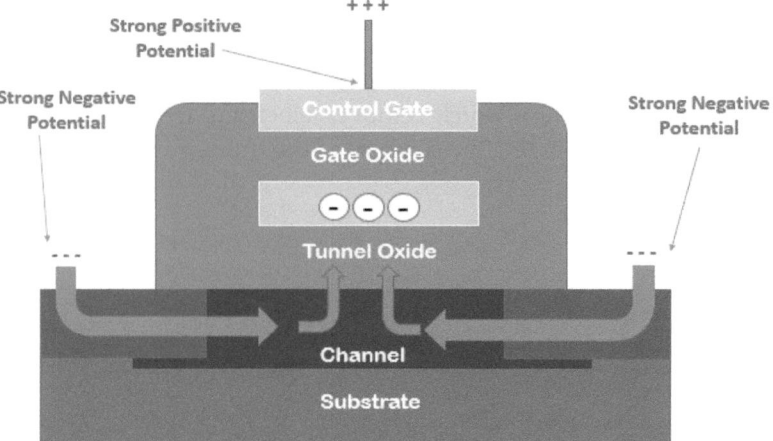

Figure 7.7 Demonstration of field emission (Fowler–Nordheim tunneling).

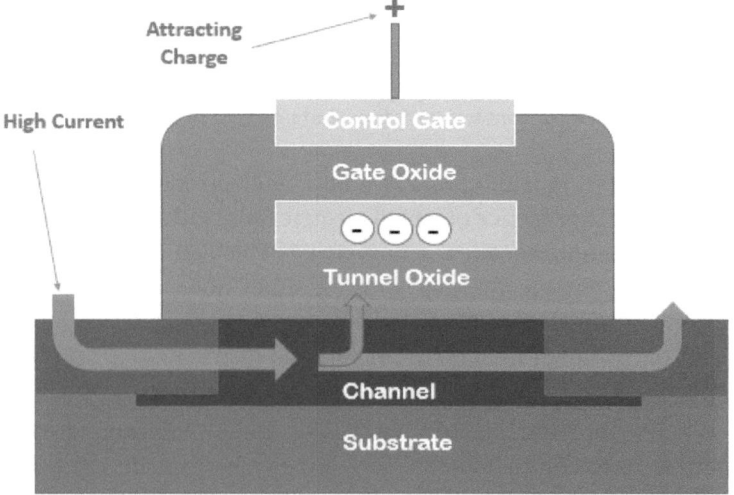

Figure 7.8 Hot-electron injection.

Alternatively, when using channel hot-electron injection (or hot-carrier injection), electrons gain sufficient energy from the high current in the channel and attract a charge on the control gate to break through the gate oxide and change the threshold voltage of the floating gate, as represented in Fig. 7.8.

7.3.4 NAND and NOR Flash Memory

NAND and NOR are two key architectures of flash memory [4]. NOR is easy to implement and requires minimal ongoing management due to the underlying cell structure. However, NOR flash is generally more expensive to produce than NAND flash and tends to be used primarily for reliable code storage (boot, application, OS, and execute-in-place [XIP] code in an embedded system). Almost all modern flash memory devices use NAND flash memory [5]. NOR flash uses no shared components and can connect individual memory cells in parallel, enabling random access to data. NOR flash is fast on data reads, but it is typically slower than NAND on erases and writes. The two main types of NOR flash memory are parallel and serial. The parallel interface was the original available interface on a NOR flash. Parallel NOR offers high performance [6] and security, among other additional features. It is primarily used for automotive, industrial, and networking purposes and telecom systems and equipment.

Serial NOR flash, on the other hand, has lower pin counts and smaller packaging and is generally less expensive than parallel NOR. The serial NOR usage areas include personal and ultrathin computers, servers, hard disks, printers, digital cameras, modems, and routers. A typical NAND flash memory device will contain one or more memory chips. NAND flash devices are used to store data and code. A NAND flash cell is more compact in size, with fewer bitlines, and consists of a set of floating-gate transistors to achieve greater storage density. NAND flash is better suited to serial rather than random data access. It is more suitable for data storage in consumer devices and enterprise server and storage systems due to its lower cost per bit to store data, greater density, and higher programming and erase speeds. NAND flash uses less power than NOR flash for write-intensive applications. There has been continuous effort to reduce the cost per gigagyte of NAND devices; therefore, device life cycles tend to be shorter.

NAND requires a controller and specific firmware for error code correction (ECC), bad block management, and wear leveling. For example, SD cards include controller circuitry to perform bad block management and wear leveling. During access of a logical block by high-level software, it is mapped to a physical block by the device driver or controller. Additionally, NOR flash usually programs data at the byte level, while NAND flash programs data in pages. Pages are larger than bytes but smaller than blocks [7]. For instance, a page might be 8 kilobytes (KB), while a block might be 128 KB to several megabytes in size. While NAND flash reading and programming are performed on a page basis, erasure can only be performed on a block basis. Devices such as a camera phone may use both NOR and NAND flash in addition to other memory technologies to facilitate code execution and data storage. Table 7.3 presents the major differences between NOR and NAND flash memory types.

Table 7.3   Major differences between NOR and NAND

Aspect	NOR	NAND
Performance	Very slow erase Slow write Fast read	Fast erase Fast write Fast read
Reliability	Standard: Bit-flipping issues reported Less than 10% the life span of NAND	Low: Requires at least 1 bit for error management (bit-flipping issue) Bad block management required
Life span	Less than 10% the life span of NAND	Over 10 times more than NOR
Execute-in-place (XIP) capabilities	Yes	None
Access method	Random	Sequential
Data programming	Byte level	Pages
Price	High	Low
Ideal usage	Code storage—limited capacity due to cost of high capacity May save limited data as well	Data storage only—due to complicated flash management Code usually not stored in raw NAND flash

Each gate in flash memory can store 1 bit or more of information depending on whether it is a single-level cell (SLC), a multilevel cell (MLC), a triple-level cell (TLC), or 3D [8, 9]. SLC can hold only 1 bit per cell. It is the most expensive type of flash medium, uses less power, has faster write speeds, and is used for industrial applications as well as storing critical data. MLC holds 2 bits per cell and stores two times more information. It is typically used in consumer products. TLC holds 3 bits per cell. It is the cheapest type of flash medium and has a higher density. It also has slower read and write speeds as well as much lower endurance. It is only usually found in low-end products, and it is not recommended for storing critical data. The internal structure of a flash memory chip consists of one or more planes or banks. Depending on the device, the planes are generally independent of each other, contain local buffering for read and program data, and perform a set of operations in parallel. Each plane consequently contains a set of blocks, each made up of 64 (SLC) or 128 (MLC) pages. Each page contains between 2112 and 8448 bytes. This includes a 2 to 8 KB primary data area as well as an out-of-band (OOB) data area used to store bad block information, error correction code (ECC) [10], and other metadata. NAND flash memory devices support three primary operations: erase, program, and read. Erase operates on entire blocks and sets all the bits in the block to 1. Program, on the other hand, writes entire pages at once and can only change 1s to 0s, meaning that an erase operation on a block only serves to arbitrarily modify the page's contents. Read operations read an entire page in parallel.

7.4 Data Recovery

7.4.1 Introduction to Data Recovery

Data recovery is the process whereby any corrupted or inaccessible data is retrieved from a damaged or corrupted digital medium when it cannot be accessed normally. The term is used when referring to circumstances when the data needs to be recovered from such devices as CDs, DVDs, floppy disks, hard disk drives, Xboxes, mobile phones, and memory cards. Some causes of data loss that lead to the need for data recovery are device mechanical failure (hardware failure/malfunction), damage to the device (e.g., due to natural disaster), human error, power surges, and software viruses. Most of the data losses are largely due to device mechanical failure and human error, in that order. In general, most types of data media failure (data loss) will fall under two categories: logical failures and physical failures. Physical failure simply means that the hardware itself has suffered physical damage and is normally caused by and related to malfunctions due to electrical, mechanical, and material deterioration. This is what led to the coining of the most commonly used term, "disk crash."

On the other hand, the meaning of logical failure is that the actual data (the bits and bytes) have been altered or is in some inconsistent state that hinders the normal data access. Logical failure can be caused by file system corruption, operating system malfunction, serious conflicts with a recently installed hardware or software, and virus /malware infections. In general, data that is unavailable due to these logical cases is easier to recover as long as it has not been overwritten by frequent usage [11]. Let us use a simple analogy of a plastic or glass water bottle. If you drop it and it breaks, it is referred to as a physical failure. However, if you drop it and it bends, it is a logical failure. A data medium can suffer both logical and physical damage simultaneously. For instance, physical damage may directly cause logical failure. This then causes the data in the medium to become inaccessible due to hardware failure.

7.4.2 The Need for Data Recovery

Information is an asset and just like any other important assets, is essential to an individual's or organization's operation and consequently needs to be appropriately stored and available when needed. This is particularly vital in the increasingly interconnected computing environment. Due to this increasing interconnectivity, information is now even more exposed to a growing number and a wider variety of threats and vulnerabilities than before. These threats often lead to loss or distortion of information, apart from other issues, like invasion of privacy and denial of service. Because data is an irreplaceable strategic asset, loss or destruction of data can result in dreadful social, technical, financial, environmental, social, and political consequences for an individual or organization. In today's world of tight competition, entities cannot afford to lose critical data and information because such loss can be too costly to any desired smooth progression.

Entities globally use various ways to recover important information that was lost. One way is to use some data recovery software to attempt to retrieve the lost data and later use some data repair software to repair corrupted data. Another possible way is to use external data recovery services. The third way is to restore data from a data backup. Data recovery software alone is often fronted as not being enough to restore lost or corrupt data because it does not fully guarantee that the lost data will be totally recovered. It may also end up being a long and very tedious process attempting to restore data using recovery software.

7.4.3   Data Extraction/Acquisition

It is critical to draw a distinction between data recovery and data extraction. Although data recovery might include data extraction, the two concepts are fundamentally different. The focus of the authors in this chapter is primarily on data recovery. Data extraction is drawing out something from any substance. In this case we extract data from a flash chip or device. This may not necessarily imply that the device has failed but may mean the board on which it is attached or even its connector has failed. There are several possible data extraction/ acquisition approaches for obtaining a full copy of flash memory data. The key approaches to be discussed are data extraction tools and physical extraction.

7.4.3.1   Data extraction tools

The easiest and a noninvasive way to read flash is simply using a simple hardware interface and software that copies all flash memory data from the target system to another system for further presentation or analysis. Unfortunately, there is a lack of a general method for this procedure because every system can have its own dedicated interface to data stored in flash memory chips. There is also a limitation on a standard operating system documentation on low-level flash memory access functions across multiple devises.

Nevertheless, memory copying tools specifically targeted toward certain devices or operating systems can sometimes be used for data extraction. These tools mainly originate from manufacturers or service centers that use them for debugging and diagnostics; in field software updates among other uses; and from hackers as well, who use them for checking and altering device functionality. One should therefore be careful when using these tools.

7.4.3.2 Physical extraction

Another approach to producing an image of flash memory is to physically remove a flash memory chip from a printed circuit board (PCB) and read this flash memory chip with a memory chip programmer or reader. This method can be used when software tools cannot be used. With this method, the chip has to be desoldered from a PCB and then prepared for further processing (cleaning and restoring connections) after removing. Special desoldering equipment is needed for desoldering in order to prevent damage to the chips and therefore loss of data. A flash memory chip can be read with a commercially available flash memory chip programmer or reader. A challenge is that a driver is required for each type of memory chip. In case a driver for a certain type of chip is unavailable, the manufacturer or sometimes the programmer has to program the required driver. This can be time consuming and is not always possible, for example, when a datasheet is not available. Another suitable solution can be to use a universal flash chip reader.

7.5 Data Recovery in Flash Media

7.5.1 Data Loss on Flash Media

Flash memory reliability is approaching its ultimate physical limitations. As a key attribute, one of the leading considerations of working with a flash medium is its reliability. A flash medium is the preferred storage solution for systems that require a very long life span, and huge mean time between failures (MTBF) rates, for example, military applications, consumer appliances, and mobile devices. However, flash memory can fail in a number of different ways. The most inevitable way is that the devices wear out with use.

Flash storage cells by default have the state "erased" (which matches the logical bit 1). Once a bit gets flipped to 0, you can only get it back to 1 by erasing the entire block. Following numerous repetitions, the erase and program process can cause cells to become unreliable because of charge trapping in the gate oxide. Remember that the physical mechanism of storing data in flash memory is based on the storage of electrical charge in a floating gate of a transistor. This stored charge can be trapped there for extended periods of time without using an external power supply. However, it will gradually leak away due to physical effects.

The data retention period specifications for current flash memory are between 10 and 100 years. Recall that flash memory can be written at the byte level but it has to be erased in blocks at a time before it can be rewritten [12]. Erasing a block causes all its bits to be totally filled with 1s. Moreover, erasing a block causes a block to gradually deteriorate. Typically, a block can be erased between 10,000 and 1,000,000 times before bits in the block begin to get inerasable. Such a block is referred to as a bad block. It is interesting to note that NAND flash memory usually already consists of bad blocks when leaving the factory. The initial bad blocks are marked as such using metadata in the spare area. Unfortunately, how this information is stored on flash memory is vendor/product specific. The guaranteed minimal number of good blocks on NAND flash memory is often specified in the accompanying documentation or datasheets. In a typical case, at least 98% of the blocks are guaranteed to be in working order.

To evenly spread the erasing of blocks over the full range of physical blocks, flash memory manufacturers have formulated what is referred to as "wear leveling" algorithms. The concept behind the algorithms is that spreading the wear, caused by erasing a block, as much as possible over the entire capacity of the flash memory will increase its overall lifetime. The wear leveling algorithm can often be very sensitive intellectual property for most manufacturers of flash memory devices; therefore, any detailed information sought about the algorithm used will often not be answered. However, for purposes of reconstruction of data in a flash memory, it is not necessary to know how the wear leveling process created the physical image copied in a flash chip. What is important is that one needs to know how to recreate the correct order of physical blocks in order to create a logical copy of the higher-level file system. In simpler terms, wear leveling can be considered a dynamic process that rearranges pages and/or blocks over and over again in order to extend flash lifetime. When comparing the reliability of flash media, four key factors must be considered: (i) bit flipping, (ii) bad block handling, (iii) life span (number of erase cycles allowed)/endurance, and (iv) retention.

7.5.1.1 Bit flipping

All existing flash architectures suffer from bit flipping, when a bit either gets reversed or is reported reversed. Cells that are not meant to be accessed during a specific read or write operation can change contents due to read and write activities in adjacent cells or pages. Bit flipping and associated problems are more common with NAND devices than with NOR devices; hence, an error code detection/error code correction (ECD/ECC) algorithm is very much recommended for NAND devices. MLC NAND is also far more susceptible to bit flips than is SLC NAND. The bit-flipping problem is not as critical in case NAND is used to store multimedia information. However, when flash memory is used as a local storage device that is storing the operating system, configuration files, or any sensitive information, an EDC/ECC mechanism must be implemented so as to ensure reliability. Bit flipping could be due to read disturb or write disturb [13].

A read disturb occurs when a cell that is not currently being read receives elevated voltage stress. Typically, stressed cells always exist in the block that is being read and are always on a page that is currently not being read. A read disturb is much less common than a write disturb. A write disturb occurs when a cell that is not currently being programmed receives elevated voltage stress. Stressed cells also always exist in the block that is being programmed and can be either on the page that is programmed (but the cell was not selected) or on any page within the same block. The cell is reset to its original state by being erased, which eliminates the data and, subsequently, the data errors that resulted from the read or write disturbs. An ECC mechanism in the data flow path detects bit flips and corrects them before providing the data to the host. As flash cell sizes decrease and more cells are placed onto wafers, there is an increased probability of errors and bit flips and consequently, NAND flash controllers require more powerful error detection/correction (EDC/ECC) algorithms.

7.5.1.2 Bad block handling

NAND devices are generally manufactured and shipped with bad blocks randomly scattered within them. There was an early attempt to manufacture NAND devices without any bad blocks, and it was found not to be economical due to the very high price tag caused by low-production yield rates. NAND devices require that first, the medium is scanned for bad blocks, which are mapped as unusable. In case this process is not performed in a reliable way, it is expected that the finished devices will have high failure rates.

7.5.1.3 Life span/endurance

As mentioned earlier, a flash block must be erased before it can be written to. However, the number of times that it can be erased is limited. Endurance refers to the maximum possible allowed number of erase/write cycles for a memory device [14]. NAND devices offer up to 10 times the life span of NOR devices. The typical maximum allowed number of erase cycles for a NAND device is higher (1,000,000 cycles) compared to a NOR device (100,000 cycles). This number does not imply that the erase block will not be operational after this threshold is reached, but it is simply a signal that the flash cells have aged and may therefore start showing signs of wear more rapidly. Various flash blocks can live much longer than their specified program/erase cycle limit. In addition to the 10-to-1 block-erase-cycle advantage of NAND memory devices, the typical NAND block size itself is usually about 8 times smaller than a NOR block size, which means that each NOR block will be erased relatively more times over a given period of time (this is particularly significant when working with small files) than each NAND block, which further widens the gap in favor of NAND flash.

7.5.1.4 Retention

Data retention refers to the memory device's ability to retain data. Data retention failure is when there is at least 1 bit of data that either cannot be read or is read incorrectly. Just like endurance, data retention is also a function of the number of erase/write cycles.

Flash memory management techniques, such as wear leveling, error correction, and bad block management, are needed in order to overcome and manage the flash wear-out limitation.

7.5.2   Bad Blocks

The two types of bad blocks in a NAND flash device are:

- Initial bad blocks: These are due to production yield constraints and the pressure to keep production costs low. NAND flash manufacturers specify that up to 2% of the SLC flash can contain bad blocks; the number for MLC flash is about 5%.

- Accumulated bad blocks: These are due to numerous write/ erase cycles, as a result of which the trapped electrons in the floating gate cause a permanent shift in the voltage levels of the cells. When the voltage level shifts enough, this will be observed as a read, write, or erase failure.

A proper bad block management mechanism is required to map out both the initial bad blocks and the bad blocks that were accumulated during device operation. Marking a bad block means adding this bad block to a table of bad blocks (which mostly is set at the end of the flash). Due to the fact that this table sits within blocks on the same flash, which might get bad as well, this table is usually redundant. Bad blocks happen, especially on NAND flash, including never used NAND flash right from the factory. Some available NAND chips can do bad block management on their own. Removable flash memory cards and USB flash drives have built-in controllers to perform wear leveling and error correction. Embedded flash memory that does not have a controller may use a flash file system. The basic concept behind flash file systems is that when the flash store is about to be updated, the file system will write a new copy of the changed data to a fresh block, remap the file pointers, and then erase the old block later, when it has time.

In case you are using a flash controller to deal with bad blocks, it needs to be able to read and write the type of format of the often-preinstalled bad block information that is stored in it. The controller can be hardware or software. The controller needs to be aware about whether the flash has any OOB areas or not and how they are organized. Therefore, this is not guaranteed. The OOB area is usually is a fraction of the block size. It is found in NAND flash. It is dedicated for metainformation (e.g., bad blocks, ECC data, and erase counters) and is therefore not supposed to be used for your actual data payload. NOR flash is by far much predictable than NAND flash. NAND flash blocks can get bad in many circumstances (e.g., humidity and temperature among others). Reading NAND flash is a stressful operation as well (it actually contributes to bad blocks!) Even though this happens less often than due to write operations (and much less on SLC than on MLC flash memory), it is a fact that we have to deal with.

Although the number of expected erase cycles is documented by manufacturers of NAND flashes, the number of read cycles is not. However, it is usually about 10 to 100 times the erase cycles. Consider a scenario whereby you want to boot from your NAND flash and the boot-loader (which does not currently yet deal with bad blocks) resides within blocks that are either bad or react unpredictably. That will be a big problem! That is the reason why it is common now that the first few blocks of NAND flash are guaranteed to be safe for the first N erase cycles. Additionally, the first physical block (block 0) is always guaranteed by manufacturers to be readable and free from errors. One should therefore make sure the boot-loader fits into those safe blocks in case one wants to avoid such a big problem.

7.5.3 File Systems

A file system organizes data that needs to be retained after a program stops executing by providing procedures to store, retrieve, as well as update data, in addition to managing the available space on the devices that contain the data. In the absence of a file system, programs cannot access data by file name or directory and would therefore need to be able to directly access data regions on a storage device. A file system is used on data storage devices, including flash memory, to maintain the physical locations of the computer files. It organizes data in an efficient way and is configured to the specific characteristics of the target device. There is a usually a very high degree of mapping (working together) between the operating system and the file system. For example, the two file systems used in Microsoft Windows operating systems are NTFS and FAT32. NTFS is considered as the most secure and robust file system for Windows 7, Vista, and XP. It offers security via supporting access control and ownership privileges and, therefore, users can set permission for groups or individual users to access certain files. It also supports compression, encryption, disk quotas, and even recoverability by facilitating undo or redo operations for operations that faced problems such as system failure or power loss. FAT32 is the file system used in some older versions of Windows. It is still relevant since it supports disk partitions as large as 2 TB as well as the fact that it wastes much less disk space on large partitions.

On a UNIX system, everything is a file; if something is not a file, then it is a process. A Linux system, just like UNIX, does not distinguish between a file and a directory, since a directory is just a file containing names of other files. Programs, services, texts, images, video, etc., are all files. Both input and output devices (generally all devices) are considered to be files according to the system. For purposes of managing all those files in an orderly fashion, they are often considered to be in an ordered tree-like structure on the hard disk. On Linux systems, all partitions are attached to the system via a mount point. The mount point defines the location of a particular dataset in the file system. Generally, all partitions are connected through the root partition. On the root partition, which is indicated with a slash (/), directories are created. Some Linux file system types are ext, ext2, ext3, ext4, hpfs, iso9660, JFS, minix, msdos, ncpfs nfs, ntfs, proc, Reiserfs, smb, sysv, umsdos, vfat, XFS, and xiafs. The ext2 is the high-performance disk file system used by Linux for fixed disks as well as removable media.

7.5.4   File Attributes

When dealing with files, it is important to understand the set of file attributes. In reality, one of the various characteristics stored for each file is a set of file attributes that provide the operating system and application software further information about the file and how it can be used. These file attributes are:

- Permissions (especially read-only)
- Hidden
- System

- Volume Label
- Directory
- Archive

For a file that is marked as read-only, most software that gets it will refuse to delete or modify it. For example, on Windows Systems, DOS will indicate "Access denied" if you try to delete a read-only file. However, Windows Explorer will delete it without any problem. Some software will choose the middle ground by letting you modify or delete the file but only after asking for your confirmation. The same applies to Linux for the permissions set for the owner, group, and others.

In case a file is marked as hidden, then under normal circumstances it is hidden from view. In both DOS and Linux, the file will not be displayed unless a special flag is used, for example, ls –a in Linux. The system flag is used to tag special files that are used by the system and should not be altered or removed from the flash/disk. Think of this as a "more serious" read-only flag.

The volume label is often assigned to every disk volume as an identifying label. This can be done either when it is formatted or later, through various tools, such as the DOS command LABEL. The volume label is stored in the root directory as a file entry with the label attribute set.

Directory is the bit that distinguishes between entries that describe files and those that describe subdirectories within the current directory. Theoretically, you can convert a file to a directory by changing this bit. Of course in practice, trying to do this would result in a mess because the entry for a directory has to be in a specific format. This is also the first character that you see when you run ls -l on Linux systems. You will either see a hyphen to signify that it is a file or see the letter d to signify that it is a directory.

Table 7.4 presents the options/flags for the DOS attributes. The general syntax for the command is ATTRIB [+R | -R] [+A | -A] [+S | -S] [+H | -H] [+I | -I] [drive:] [path] [filename] [/S [/D] [/L]].

On a Linux file system, a file is represented by an inode. An inode is a kind of serial number containing information about the actual data that makes up the file. At the time a new file is created, it gets a free inode. Every partition on a system has its own set of inodes; thus, throughout a system with multiple partitions, files with the same inode number can exist. In that inode the following information is kept:

- Owner and group owner of the file
- File type (regular, directory, etc.)
- Permissions on the file
- Date and time of creation, last read, and last changed
- Date and time that this information has been changed in the inode
- Number of links to this file
- File size
- An address defining the actual location of the file data

Table 7.4 DOS file attributes

Option/Attribute	Purpose
+	Sets an attribute
-	Clears an attribute
R	Read-only file attribute
A	Archive file attribute
S	System file attribute
H	Hidden file attribute
I	Not content-indexed file attribute
[drive:][path][filename]	Specifies a file or files for attrib to process
/S	Processes matching files in the current folder and all subfolders
/D	Processes folders as well
/L	Works on the attributes of the symbolic link versus the target of the symbolic link

7.5.5 Flash Data Recovery Techniques

7.5.5.1 Fundamental concepts

There are a number of file system analysis tools that are currently available in the market, including Encase, FTK, R-Studio, and TSK. Most of these tools are not completely aware of the physical medium from which an image file originates. To perform advanced data recovery, knowledge of the physical properties is necessary so as to improve the data recovery process. For example, flash file systems often contain different versions of the same data objects. This is because flash memory cannot be erased in small quantities. Particularly for small objects (far smaller than one flash memory block) that have a high update frequency, many old versions might exist outside of the usual high-level file system. The physical characteristics of flash memory offer an opportunity to recover residual data. Residual data is defined as information unintentionally left behind on computer media and can exist at any level of a computer system. Remnant data, on the other hand, is information that can be recovered from a storage medium after new information has been written to the medium. An overwrite operation presents the best opportunity for the recovery of residual data in flash memory.

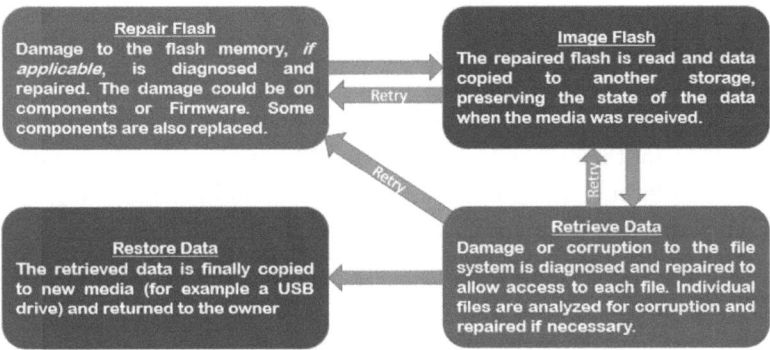

Figure 7.9   Data recovery process.

An overwrite can be due to a user changing data within a file, modifying the metadata associated with a file, or just deleting the file. These operations are all known to cause changes to be made to data at the logical level. However, due to the physical limitations of flash memory, the changed data cannot be rewritten to the same flash page on the flash memory device. Given any directory that was deleted or not yet deleted on a flash device, a common data recovery and analysis tool will show the last version of the directory, probably with some files marked as deleted. However, from the other versions of the directory data, plenty of the user behavior can be reconstructed. The case is similar for other data objects although larger objects (e.g., movie files) are likely to be (partly) overwritten earlier after being deleted because they fully occupy flash blocks, which can be reused immediately after a delete operation. Figure 7.9 shows the general data recovery process.

7.5.5.2 The flash translation layer and flash data recovery

The period for which the old page remains on the flash device without further modification will be dependent upon the implementation of the flash file system or flash translation layer (FTL) and when the system decides it needs to reclaim blocks. The FTL is a software layer built on raw flash memory that performs garbage collection as well as wear leveling strategies and hides the special characteristics of flash memory from upper file systems by emulating a normal block device like magnetic disks [15]. Apart from recovering deleted data, it is possible to use observations of the metadata tags and mapping structures associated with the residual data to recreate previous versions of files that are still existing on the flash medium or, at a bare minimum, indicate whether the current version on the flash device is not the original version. The FTL rewrites the altered data onto a different, clean flash page; marks the old page as dirty; and queues the dirty page for garbage collection.

In accordance to the granularity of the mapping unit, existing FTL schemes can be grouped into four categories: page-level mapping, block-level mapping, hybrid mapping, and variable-length mapping. The FTL disguises the management of flash memory to the file system by masking it as a block device. A file system that is designed uniquely for flash exposes all the details and lets the file system manage wear leveling and block reclamation. The flash file system solution has an advantage in that it can be more efficient since the file system is not hidden underneath layers of mapping used for block device emulation. This situation is ideal for resource-constrained devices such as handheld devices. Examples of flash file systems are JFFS/JFFS2 and Yet Another Flash File System (YAFFS). YAFFS was developed by Aleph One as a file system specifically designed for NAND flash. The YAFFS stores its file data in chunks.

Each YAFFS chunk is equal in size as a NAND flash page, 512 bytes. YAFFS stores metadata, or tags, in the spare area. This metadata contains a file ID number, a chunk number, a write serial number, a tag error correcting code, a page status field, a block status field, and the bytes-in-page used. In YAFFS, the file ID associates a chunk with a file. The chunk number is a count that begins with the number 0, which signifies the file header, and increases by 1 per chunk required to store the whole file. A file ID of 0 signifies that the chunk has been deleted but not yet erased by the flash garbage collector.

The block status attribute signifies whether a block has gone bad. Additionally, the page status field denotes whether the page is valid or discarded. In case 4 bits or more are 0, the page is discarded. A block can be erased in two separate circumstances. The first is in case all pages within a block are dirty, and the second is if there is a page that is still valid. In the second circumstance, the valid page is rewritten to a new block and the current block is then released for collection.The Journaling Flash File System (JFFS) was designed and was released for the Linux Kernel. The JFFS is designed as a log-based file system whereby a circular data structure is used to write all data sequentially as nodes. JFFS writes data linearly on the flash device, in a way in which the oldest node is the head and the newest is the tail. Garbage collection is achieved by observing the node at the head of the log. In case the node is already obsolete, it is skipped and the garbage collection process shifts the head to the next node. In case the node is valid, it is made obsolete by rewriting the data and header information to the tail and the head pointer moves to the next node. This process is performed continuously until JFFS has rendered a complete flash erase block obsolete. Since the garbage collection process rewrites nodes to the tail in order to free a complete flash block, JFFS cannot wait until the entire flash memory has been fully used (occurs when the tail meets the head). Therefore, a slack space is required at all times between the tail and the head of the log. Basically, JFFS2 is a series of linked lists.

After a flash device is filled for the first time, the place that the flash file system or FTL decides to write fresh data is no longer based upon location but will instead depend on variables, for example, the number of dirty pages in a block or a block's erase count. Regardless of the method used, a file's pages will no longer be categorized by spatial locality, causing files to become ever more fragmented as the flash is used. This fragmentation results to two effects. The first effect is that larger files will become increasingly difficult to recover as the flash device is used more. While logically large video and picture files tend to have very little fragmentation, if any at all, physically they will be stored throughout the flash as pages become available. The files will become even more fragmented as they are changed further by the user. The best recovery opportunity is of files that can fit within a single page or files under 512 bytes since shaping files with various fragmentation points is not easy. This could therefore limit recoverable items to smaller text files and lower resolution pictures. The second effect is that fragmentation provides a signature for a flash device that has been used previously. It is possible to wipe clean a hard disk drive and then reformat it with a new file system to make it look much like a drive that has just been shipped by the manufacturer. However, without physical layer access, any attempt by an end user to wipe a flash memory device and then reformat it will still retain the flash device fragmented and might leave blocks that have not been garbage collected.

7.5.5.3 Data recovery for data loss due to a virus attack

We have cases where the flash drive shows no data in the form of folders or files, but we have some shortcuts present there. This is usually caused by a virus attack. Most viruses in the last decade have been written to target the Windows environment. The created shortcut on the virus-infected medium, whether it is a flash drive or a hard disk, can definitely create endless troubles. The file names might be altered by the virus and become shortcuts. This makes it difficult to open or view the files even after eradicating the virus. Sometimes the same problem is noticed when the space used in the flash drive is the same but you cannot access or open the files. Two steps are followed in the process of addressing the problem of inaccessible data due to virus attacks: (i) remove the virus and (ii) recovere the data. To remove the virus from the flash memory, the flash is connected to a computer that has some latest and updated version of the antivirus software and then scanned for viruses after which removal is done. All the threats that are detected by the antivirus software are cleared.

The next step is data recovery. It is possible for you to recover virus-infected files in two ways: (i) using the attrib command and (b) using data recovery software. The following steps are performed when recovering data using the attrib command. The approach will work as long as the flash is not formatted:

1. Connect the flash drive to a computer.
2. Check the drive assigned to the flash (note it down, for example, F).
3. Go to **Start** → **Run**. Type **cmd**.
4. Change directory to the flash (pen drive, memory cards, or mobile phone) directory.
   i. Type **F:** (where F is the drive letter obtained from step 2 above).
   ii. Press the **Enter** key.
5. Delete all link files in the directory (if any).
   i. Type **del *.lnk** (to delete all link files in the directory).
6. Type the following: **attrib -h -r -s /s /d f:*.***

Here "F" is for the flash drive and should therefore be replaced with any appropriate drive letter for the drive as obtained in step 2. It may not be necessary since it is your current working directory due to step 4. One needs to be careful about the spaces when using the command described in step 6; otherwise, the required action will not be successful. Step 6 will not delete any files but only recover the files that had been converted to shortcuts. The commands are presented in Fig. 7.10. There are also a number of data recovery tools (software) that can also be used to recover data. Some of them are provided for free download online. Caution is often important when selecting the software for use so as to verify that it is legitimate software. Most of the tools are often easy to use with an easy-to-learn user interface. One of the tools for recovering data lost due to viruses is EaseUS Data recovery wizard.

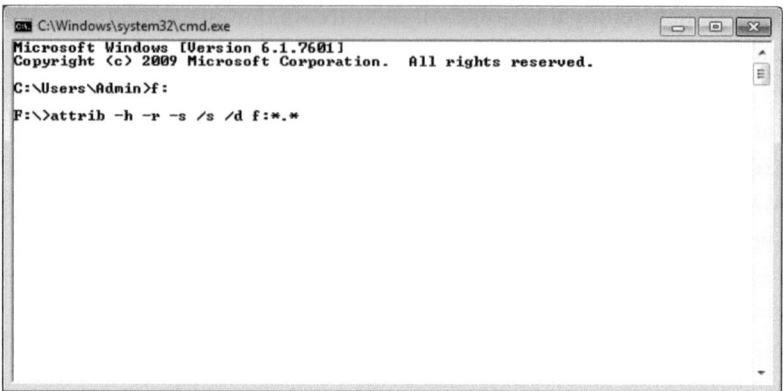

Figure 7.10 Using the attrib command.

7.5.5.4   Data recovery software

The physical characteristics of flash memory give an opportunity to recover residual data. Residual data is that which has been unintentionally left behind on computer media and can exist at any level of a computer system. A number of software tools can be used to recover data from a corrupted or restore data that was deleted from a flash device. These include iCare Data Recovery, Data Rescue for Mac, and Recuva [16, 17]. Recuva (pronounced "recover") is a freely available Windows utility to restore files that have been accidentally deleted. The files could have been emptied from Recycle Bin or deleted by mistake from digital camera memory cards or MP3 players. It will even bring back files that have been deleted from iPods or by bugs, crashes, and viruses. Due to the dangers of overwriting, any data recovery attempt should never be done on the boot volume or by installing the data recovery software on the affected drive. It is always necessary to remove the affected drive and attach it to another computer that is running the data recovery software. Figure 7.11 shows the iCare Data Recovery software window snapshots, while Fig. 7.12 to Fig. 7.16 show snapshots of Recuva in action to recover files.

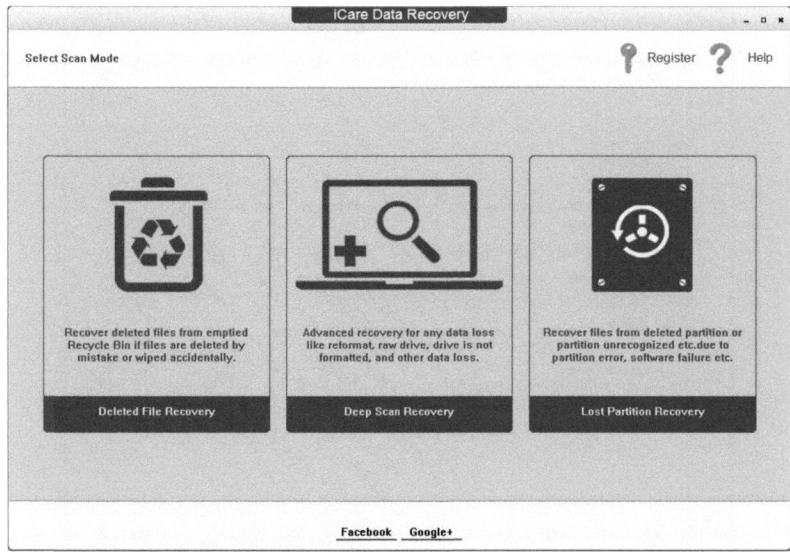

Figure 7.11   iCare Data Recovery.

Figure 7.12   Recuva Wizard: Open the Start page.

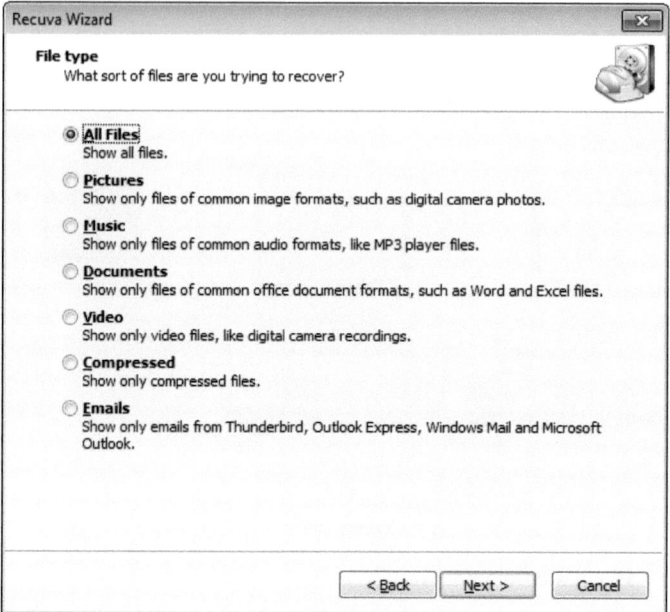

Figure 7.13   Recuva Wizard: Select the file type.

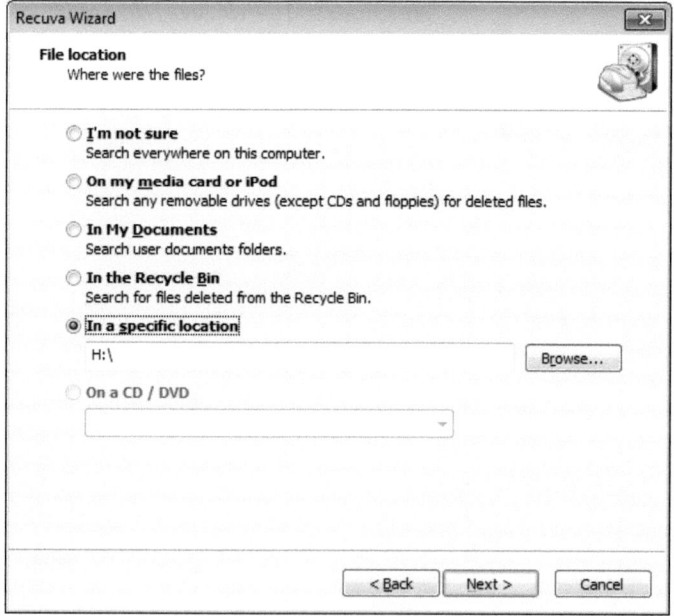

Figure 7.14   Recuva Wizard: Select the file location.

Figure 7.15   Recuva Wizard: Search for files.

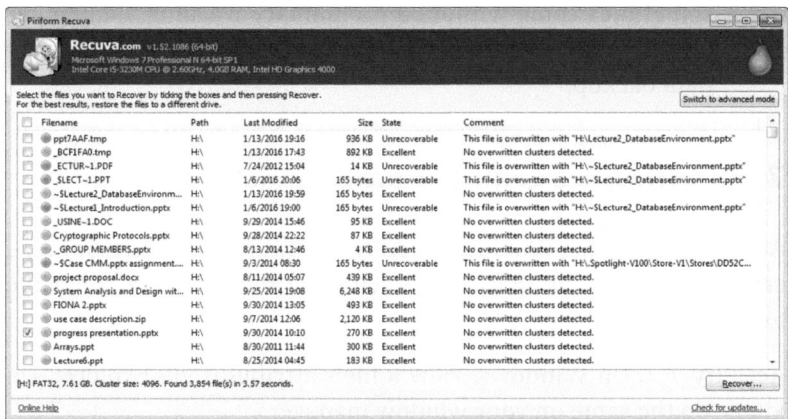

Figure 7.16   Recuva Wizard: Select and recover.

7.5.5.5 Best practice for flash

The best practices while using flash are as follows:

- Do not remove the flash drive when data is being transferred. Always stop it before removing it. You stop it by going to Safely Remove Hardware, which is found in the system tray, and then stopping the USB drive and removing it when it displays a message that it is safe to remove the USB drive.

- Keep a duplicate copy of your data in the flash drive all the time. In case some section of your flash drive is damaged, there is the possibility that the other copy can work.

- Do not double-click a flash drive. Double-clicking it may allow any harmful executable file located in the flash drive to execute itself, which can damage the flash drive or the computer. The best practice is to always right-click the flash drive and then click open or explore to access the flash contents.

- Always scan the flash drive for viruses with good updated antivirus software before opening it. This will minimize the possibility of the computer getting infected by any such viruses.

- Always keep backup of your most valuable data, no matter how much you are following the safety rules. Problems sometimes come without a preliminary warning. Remember that data recovery is often a more complicated process than data backup.

7.6   Windows User Laboratory Activities

Activity 7.1. On the Windows operating system, click **Start** and type **cmd**. Then start the black (DOS) window that appears. Type the following: **attrib /?**

Activity 7.2. On Windows, view a file's attributes via Explorer. On Linux, do the same on your window manager.

Activity 7.3. On the Windows operating system, create a new file and set it to be hidden using the attrib command. Additionally, unhide the file, create a folder, move the file to the folder, and use attrib to hide the folder and all its contents. Unhide it.

Activity 7.4. Get an empty flash medium and format it to FAT32. Analyze the disk, especially in terms of blocks (how are they?). Attempt to read the disk on a machine that has a Linux operating system. Repeat the process for NTFS (format it to NTFS and do the rest).

Activity 7.5. On your flash medium, perform a chkdisk /f.

Activity 7.6. Install Recuva and launch it. Attempt to recover a deleted file from a flash medium.

7.7   Summary

In conclusion, we can say that flash cards are popular because of their low price, low power consumption, lack of noise, shock resistance, increased capacity, speed, and compact size, in addition to the fact that they are a good alternative to network security risks and can be fitted in multiple interfaces (portability), among other reasons. Flash has actually achieved major performance leap over hard disks in the recent years. Over the past decade, CPU speed performance has increased roughly 9-fold, DRAM speed roughly 8-fold, network speed roughly 100-fold, and bus speed roughly 20-fold times. However, disk (storage) speed has only increased roughly 1.5-fold. With flash, storage finally catches up. Cutting-edge NAND technology is complicated and varies significantly by NAND vendor. There were 60 flash card brands in U.S. retail by July 2014. Applications are moving onto flash today, and databases present the main use case for flash (after autotiering). In-memory computing (IMC) accelerates the analysis of big datasets. In as far as people know the importance of backing up, they do not backup. Flash devices can and do fail, or data is often lost due to human error. There is therefore a growing need to continue developing techniques of data recovery on flash media. More work is underway to develop techniques of data recovery from self-encrypting solid-state drives (SSDs).

References

1. Lythos (2011). *elitepcbuilding.*

2. Rouse, M. (2015). http://searchstorage.techtarget.com/definition/flash-memory

3. Vidyabhushan, M. (2010). *Modeling the Physical Characteristics of NAND Flash Memory*, University of Virginia.

4. Crippa, L., Micheloni, R., Motta, I., and Sangalli, M. (2008). *Memories in Wireless Systems,* Springer, Berlin, Heidelberg, pp. 29–53.

5. Dowler, M. (2011). *PCSTATS.*

6. Neal, M., Mathew, A., Jason, H., Shane, C., and Nicholas, W. (2010). *IEEE Petascale Data Storage Workshop (PDSW)*, 1.

7. Wang, Y., Yu, W., Suh, G.E., and Kan, E. (2013). *IEEE Symposium on Security and Privacy*, 271–285.

8. Dario, B. (2015). *Gizmag.* www.newatlas.com/high-capacity-3d-flash-memory/36782/

9. Micron (2013). www.micron.com/~/media/documents/products/product-flyer/nor_nand_flash_guide.pdf

10. Cai, Y., Luo, Y., Haratsch, E.F., Mai, K., and Mutlu, O. (2015). *IEEE 21st International Symposium*, 551–563.

11. Sansurooah, K. (2009). Edith Cowan University, Perth Western Australia, 99–108.

12. Breeuwsma, M., de Jongh, M., Klaver, C., van der Knijff, R., and Roeloffs, M. (2007). *Small Scale Digital Device Forensics Journal*, **1**, 1.

13. Grupp, M.L., Caulfield, A.M., Coburn, J., Swanson, S., Yaakobi, E., Siegel, P.H., and Wolf, J.K. (2009). *42nd Annual IEEE/ACM International Symposium on Microarchitecture (MICRO)*, 24–33.

14. Mittal, S., and Vetter, J. (2015). *IEEE Transactions on Parallel and Distributed Systems (TPDS)*, **27**, 1537–1550.

15. Ma, D., Feng, J., and Li, G. (2011). *SIGMOD*, 1–12.

16. Mah, P. (2013). www.cio.com/article/2389020/disaster-recovery/how-to-recover-data-on-hdd--ssd-and-flash-based-memory.html

17. Yegulalp, S. (2012). *Computerworld.*

Chapter 8

Nanocrystals in Resistive Random Access Memory

Writam Banerjee* and Qi Liu**

Key Laboratory of Microelectronics Devices & Integrated Technology,
Institute of Microelectronics, Chinese Academy of Sciences, No. 3,
BeiTuCheng West Road, ChaoYang District, Beijing 100029, China
University of Chinese Academy of Sciences, Beijing 100049, China
Jiangsu National Synergetic Innovation Center for Advanced Materials (SICAM),
Nanjing 210009, China
writam.banerjee@gmail.com

The term "memory" is very much associated with none other than the brain, a very special part of our living body with receiving, storing, and recalling abilities. Memory allows us to memorize the past events of life. Similar to our body, the memory element is a fundamental need of the storage-class devices and is divided into two broad groups on the basis of the requirement of power to memorize the stored data. One type is volatile memory (VM), which needs constant power supply to remember the state, and the other

---

*Both authors have contributed equally to this work.

Nanocrystals in Nonvolatile Memory
Edited by Writam Banerjee
Copyright © 2018 Pan Stanford Publishing Pte. Ltd.
ISBN 978-981-4774-73-4 (Hardcover), 978-1-351-20327-2 (eBook)
www.panstanford.com

is nonvolatile memory (NVM), which is capable of remembering the data even when no power supply is available. In practice, dynamic random access memory (DRAM) and static random access memory (SRAM) are VM type and flash is NVM type. Due to fundamental limitations associated with the shrinking device size and increased process complexity with miniaturization of baseline NVM devices, emerging NVM devices with exciting architectures are being explored. Because of the simple structure, faster operation, reliable storage capacity, superior scalability, and cost-effective design, resistive random access memory (RRAM) is receiving a huge amount of attention among emerging technologies. Defects like cations or anions dominate the resistive switching (RS) events in the RRAM devices. The performance of the RRAM devices can be manipulated and improved by incorporation of nanocrystals (NCs) in the RS structure. The attractive features of the NCs, especially the ability to enhance the electric field, can effectively enhance the performance of the RRAM devices. In this chapter, we give an overview of the fabrication processes, properties, and improvements in electrical characteristics of NC-based RRAM devices. This chapter will convey a basic understanding of RRAM technology as well as the future scope.

8.1   Introduction

8.1.1   Background

The present generation is depending more and more heavily on advanced electronic equipment, in which different memory technologies—including volatile memory (VM), for example, static random access memory (SRAM) and dynamic random access memory (DRAM), and nonvolatile memory (NVM), that is, flash memory—have been used for specific works. A single type of memory is not enough to do all things together. In general, SRAM is the fastest memory, where the write/erase speed is about 100 ps, but the design of each SRAM cell requires six transistors, which costs a lot of space on a wafer. Therefore, one-transistor-one-capacitor-based compacted DRAM with moderate efficiency is a point of focus. However, its data storage capacity is very limited due to leaky capacitors. One-transistor-based flash memory is an obvious choice because of its nonvolatility and cost effectiveness. The advantages of flash memory are very useful for mass storage applications [1]. Silicon-based flash memory devices hold the biggest share of the semiconductor memory market. But unfortunately, the shrinking of the cell size is creating a lot of issues in SRAM, DRAM, and flash technologies. The miniaturization is increasing the operation power of the VMs. Along with the physical limitation of flash memory device, it has other disadvantages, such as high-voltage and low-speed operation and poor endurance as compared to DRAM. According to a news report from Imec, Leuven, Belgium, [2], it is very clear that

> the scaling of cell size leads to gradual reduction of the number of electrons stored on the floating gate, with a projected number of less than 30 electrons for memorizing a (multilevel) cell state, for an assumed 15-nm feature size.

To meet the requirements of next-generation information technology memory, we are looking for such nonvolatile, scalable, cheap memory technology with ultrafast, low-power, ultrahigh endurance and retention capacities in a single cell. This kind of memory is known as "universal memory" [3]. The need of the time has enhanced the research area for identifying a suitable alternative NVM [4]. For regular industrial adoption, the expectations are really high from any alternative NVM. Compared to the existing memory, a new technology is expected to be a highly scalable one. It should be operable at low power with high speed, have higher endurance and be highly reliable, and be obviously cheaper in price [5]. Some of those promising prototype NVM technologies are ferroelectric random access memory (FeRAM), phase change memory (PCM), and spin-transfer torque magnetic random access memory (STTRAM or STTMRAM). All of those baseline and prototype memory technologies, with their advantages and disadvantages, are summarized in Table 8.1 [4].

Table 8.1 Current baseline and prototype memory technologies

Memory type		Baseline technologies					Prototype technologies		
		DRAM		SRAM	Flash		FeRAM	STTMRAM	PCM
		Standalone	Embedded	6T	NOR	NAND	1T1C	1(2)T-1R	1T(D)-1R
		Volatile memory			Non-volatile memory				
Cell Elements		1T1C		6T	1T		1T1C		
Feature size F, nm	2013	36	65	45	45	16	180	65	45
	2026	9	20	10	25	> 10	65	16	8
Cell Area	2013	$6\,F^2$	$(12\text{-}30)\,F^2$	$140\,F^2$	$10\,F^2$	$4\,F^2$	$22\,F^2$	$20\,F^2$	$4\,F^2$
	2026	$4\,F^2$	$(12\text{-}50)\,F^2$	$140\,F^2$	$10\,F^2$	$4\,F^2$	$12\,F^2$	$8\,F^2$	$4\,F^2$
Read Time	2013	< 10 ns	2 ns	0.2 ns	15 ns	0.1 ms	40 ns	35 ns	12 ns
	2026	< 10 ns	1 ns	70 ps	8 ns	0.1 ms	< 20 ns	< 10 ns	< 10 ns
W/E Time	2013	< 10 ns	2 ns	0.2 ns	1 µs/10 ms	1/0.1ms	65 ns	35 ns	100 ns
	2026	< 10 ns	1 ns	70 ps	1µs/10ms	1/0.1ms	<10ns	<1 ns	<50ns
Retention Time	2013	64 ms	4 ms	-	10 y	10y	10 y	>10y	>10y
	2026	64 ms	1 ms	-	10 y	10 y	10 y	>10y	>10y
Write Cycles	2013	>1E16	>1E16	>1E16	1E5	1E5	1E14	>1E12	1E9
	2026	>1E16	>1E16	>1E16	1E5	1E5	>1E15	>1E15	1E9
Write Voltage (V)	2013	2.5	2.5	1	8-10	15-20	1.3-3.3	1.8	3
	2026	1.5	1.5	0.7	8	15	0.7-1.5	<1	<3
Read Voltage (V)	2013	1.8	1.7	1	4.5	4.5	1.3-3.3	1.8	1.2
	2026	1.5	1.5	0.7	4.5	4.5	0.7-1.5	<1	<1

T: transistor; C: capacitor; D: diode, R: resistor

Source: International Technology Roadmap for Semiconductors (ITRS) [4].

But the adoption of any prototype technology depends on several factors. The performance of the NVM technology must be at least equivalent to if not better than the existing one. In the better-performance list, scalability, density, speed, power consumption, endurance, reliability, and, obviously, the manufacturing cost are matters of concern. A new NVM that can fulfill only one or two expectations is not going to be easily accepted by the industry. Since the emerging NVM devices have to meet the operation speed of SRAM and endurance of DRAM and be beyond NAND flash in terms of scalability and cost effectiveness, the device structure and the basic physics behind the switching have been drastically modified to lead to the development of emerging NVMs, such as FeRAM, PCM, and STTMRAM. All these prototype NVM technologies are going to be introduced briefly.

8.1.2 Prototype NVM Technologies

8.1.2.1 Ferroelectric random access memory

FeRAM is a prototype nonvolatile NVM based on one transistor and one capacitor [6]. The structural design of a FeRAM cell is very similar to that of the DRAM cell. The only difference is in the capacitor material. The nonvolatility of FeRAM mainly depends on its having a ferroelectric layer instead of a conventional dielectric-based capacitor, like traditional DRAM. Lead zirconate titanate $Pb(Zr_xTi_{1-x})O_3$ is the most commonly used dielectric in the FeRAM cell. The general structure of a FeRAM cell is shown in Fig. 8.1a. Due to the formation of a semipermanent electrical dipole, the dielectric constant of a ferroelectric material is much higher than the dielectric constant of the material used in DRAM. Several types of ferroelectric nanocrystals (NCs) have been used to fabricate FeRAM cells. Reports show that there is a huge impact of NCs on standard ferroelectric materials, which can be useful for new applications [7]. Polarization of the ferroelectric capacitor is the basic switching mechanism of a FeRAM device. On applying an external electric field across the ferroelectric layer, the dipoles are aligned according to the field direction. As a result, a small shift in the atomic positions happens. Additionally, in the crystal structure there are shifts in the electronic charge distribution. After the charge is removed, the polarization state is memorized by the dipoles. For FeRAM devices, the stored charge with an applied electric field can produce hysteresis behavior, as shown in Fig. 8.1b.

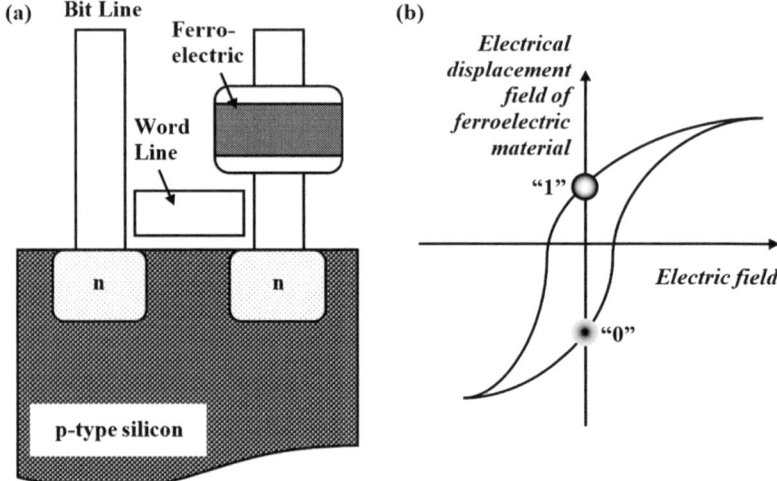

Figure 8.1 (a) Schematic illustration of a one-transistor-one-capacitor-based FeRAM structure. (b) The hysteresis loop.

The possible electric polarization states are denoted in binary 0 and 1. The basic functionality of a FeRAM cell is very similar to that of DRAM, but destructive reading is the major disadvantage of FeRAM. After writing to a cell the access transistor keeps the cell to a particular binary state. Due to the nonvolatile nature the state is unchanged unless a small current pulse is applied to read the state. It is possible to change the binary state by a read pulse. Therefore, the reading of FeRAM is a destructive one.

As compared to DRAM, FeRAM can retain the stored information for 10 years and at the same time FeRAM has good endurability, of >10^{14} cycles. FeRAM has several advantages over flash technology. Firstly, FeRAM is faster than NAND flash. For write operation and read operation NAND flash needs 1 ms and 0.1 ms, respectively, whereas FeRAM needs 65 ns and 40 ns, respectively. Secondly, FeRAM can be operated at a lower voltage than flash. NAND flash needs 15 V to write and 4.5 V to read, but FeRAM can work only at ~3.3 V to write and 1.5 V to read. Although FeRAM has several advantages as compared to the current baseline NVM, the major problem associated with this technology is scalability. A smaller cell size can save a lot of area on a silicon wafer, which will increase the device yield and reduce the product cost. Hence, the cost per bit is much cheaper for flash memory than for FeRAM. Some companies, like Ramtron and Texas Instruments, are producing FeRAM on a large scale and also investing in further research on FeRAM.

8.1.2.2 Phase change memory

PCM is a type of NVM that is mainly based on chalcogenide glass; hence, it is sometimes referred to as CRAM. In chalcogenide films, the presence of two different solid-state phases, crystalline and amorphous, with different electrical resistivity has provided the information storing ability of a PCM cell [8]. The resistivity is lower for a crystalline phase, known as SET, whereas higher resistivity can be observed for the amorphous, known as RESET. The phase transition between crystalline and amorphous phases is a reversal phenomenon. The schematic structure of a simple PCM cell is shown in Fig. 8.2a.

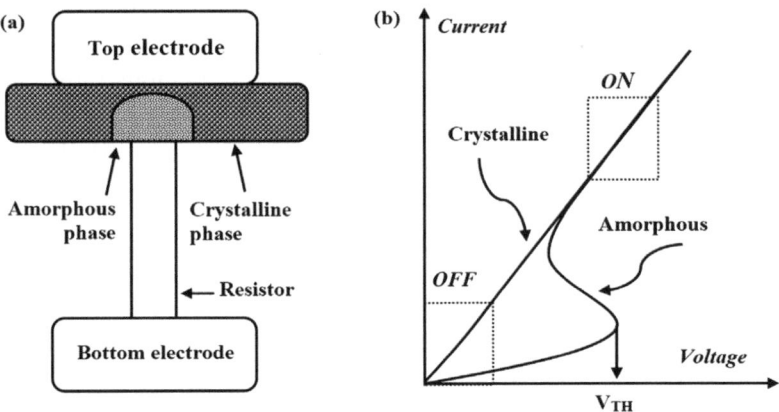

Figure 8.2 (a) Schematic illustration of a basic PCM cell. (b) The *I–V* characteristics.

Generally, after fabrication, a PCM cell is in the crystalline phase with a low resistance state (LRS) due to the high processing temperature. By application of a larger electrical current pulse between the two electrodes for a shorter period of time, the PCM cell can be reset to a high resistance state (HRS), that is, the amorphous phase. To regain the crystalline phase and set the cell, a medium electrical current pulse is applied between the crystallization and melting temperatures for a sufficiently longer period. Figure 8.2b shows the typical current–voltage (I–V) curves of the SET (crystalline) and RESET (amorphous) states. During the SET operation, the current of the cell increases with the increase of applied voltage as the resistivity of the chalcogenide material decreases with an increasing electrical field. Due to the threshold switching (TS) phenomenon (sudden drop in resistivity) the PCM cell during the RESET process shows an S-type differential resistance effect. In the ON region the SET and RESET curves are superimposed whereas a sufficient amount of gap is observed in the OFF region. Hence a small amount of read voltage must be applied to avoid phase transition during read operation.

As compared to flash technology, PCM is much faster in write and read processes [9]. The typical write time of a PCM cell is 100 ns, which is a thousand times faster than the conventional flash technology. A PCM cell needs a lower write or read voltage as compared to a flash device, with better endurance capabilities. Memory giants like IBM, Infineon, Samsung, and Macronix have already demonstrated the prototypes of PCM chips. This technology is further promoted for the mass production of PCM-based 3D crossbar arrays (3D XPoint) by collaborations between Intel and Micron.

8.1.2.3   Spin-transfer torque random access memory

STTRAM based on magnetic tunnel junctions (MTJs) is a type of MRAM [10]. STTRAM basically consists of one transistor and one MTJ, as shown in Fig. 8.3a. The MTJ has a tunnel oxide barrier sandwiched between two ferromagnetic (FM) layers. One FM layer has a fixed magnetic orientation, and the other has a free magnetic orientation. When both FM layers have a parallel magnetic orientation, the cell is in a LRS. When the alignment is antiparallel magnetic, the cell is in a HRS. A typical resistance–voltage (R–V) characteristic of a MTJ is shown in Fig. 8.3b. Depending on the applied voltage polarity the binary states, 1 and 0, can be achieved.

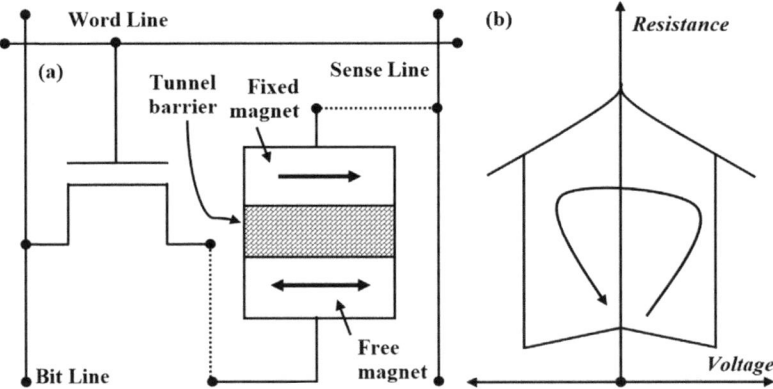

Figure 8.3 (a) Schematic illustration of a STTRAM cell. (b) The resistance–voltage characteristics.

Although STTRAM is a type of MRAM, it has advantages like high scalability, simple architecture, lower power consumption, and faster operation over the conventional MRAM technology. In STTRAM, the writing speed is faster than the writing speeds in flash, FeRAM, and PCM technology. Data retention and endurance are the two most important properties for being a strong contender in storage class memory systems. The endurance of STTRAM is much better than that of flash and PCM, with good data retention properties. Everspin Technologies is actively taking part in the production of STTRAM. Recently, Avalanche claimed that a STTRAM chip can be used for embedded and standalone devices.

It is clear that due to higher expectation, within a few years traditional DRAM, SRAM, and flash technologies are going to suffer a lot. Several properties of prototype NVM technologies make them useful for the next-generation devices. Although several prototype NVMs are challenging the baseline NVM technologies, the quest for the "universal memory" is still on. In the emerging NVM section we are going to discuss the new technologies. The key advantages of emerging NVM are summarized in Table 8.2 [4].

Table 8.2 Key features of emerging NVM technologies

Memory type	Emerging Ferroelectric		Carbon	Mott	Macro-molecular	Molecular
Subclass	*FeFET*	*FTJ*				
Storage mechanism	Remnant polarization on a ferroelectric dielectric	Giant tunnel electro-resistance	Multiple mechanisms	Multiple mechanisms	Multiple mechanisms	Multiple mechanisms
Cell Elements	1T	1T1R or 1D1R	1T1R or 1D1R	1T1R or 1D1R	1T1R or 1D1R	1T1R or 1D1R
Device type	FET with FE gate insulator	M-FJT-M-FJT-Semiconductor	Nanotube, amorphous carbon, graphene	Mott transition	M-I-M(NC)-I-M	Bi-stable switch
Advantages	Excellent endurance. Scalability.	Large ON/OFF ratio. High speed, low power	Scalability.	High speed switching. Low write energy.	Control over structure. Simple processing.	High scalability. Low switching energy.

Source: International Technology Roadmap for Semiconductors (ITRS) [4].

8.1.3 Emerging NVM Technologies

8.1.3.1 Emerging FeRAM

The emerging FeRAM is subdivided into two categories: ferroelectric field-effect transistor (FeFET) [11] and ferroelectric tunnel junction (FTJ) [12, 13]. The structural design of FeFET is a simple one-transistor type, as shown in Fig. 8.4a. In a FeFET cell, the gate oxide is a ferroelectric material. On application of a positive pulse on the top electrode (TE), the polarization will be downward in direction and the channel will be under inversion mode, leading to the low resistance and the ON state. On application of a negative pulse, upward polarization will lead to the OFF state, as the channel will be in depletion mode and the resistance will be high. The scalability potential and endurance are good, but the retention behavior, write/erase disturbs, and complementary metal-oxide semiconductor (CMOS) process integrations are the key challenges for FeFET technology.

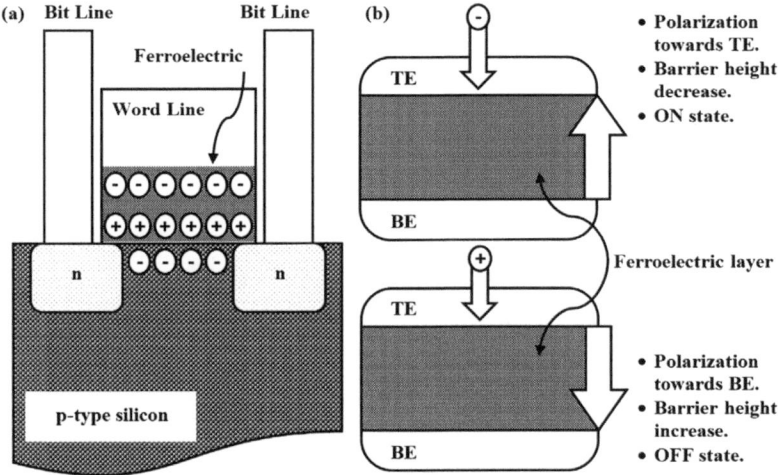

Figure 8.4 Schematic illustration of (a) FeFET and (b) FTJ.

The basic structure of a FTJ is shown in Fig. 8.4b. The ferroelectric layer is not polarized without the application of an external electric field. A negative voltage on top will direct the polarization toward the TE. In this condition the average barrier height will decrease; therefore, the current can easily pass through it, resulting in an ON state. On the contrary, a positive voltage on the TE will change the direction of the polarization. Hence the average height of the barrier will increase, which will block the current flow, resulting in an OFF state. The nondestructive readout is a major advantage of the FTJ over the normal FeRAM. However, endurance and retention are still problematic for the FTJ.

8.1.3.2 Carbon memory

Carbon memory, sometimes referred to as nano–random access memory (NRAM), is an emerging NVM based on carbon nanotubes (CNTs), amorphous carbon, and graphene. NRAM was first proposed by Nantero [14]. The cell of NRAM consists of one transistor and one resistor (Fig. 8.5a) or one diode and one resistor. The basic idea is very simple. Under suitable conditions, if the CNTs are in contact, then the current will flow through them and the device will be in the ON state. But if the CNTs are not in contact, then the resistance will be higher and the device will be in the OFF state (Fig. 8.5b). The physics behind this technology is not yet understood properly.

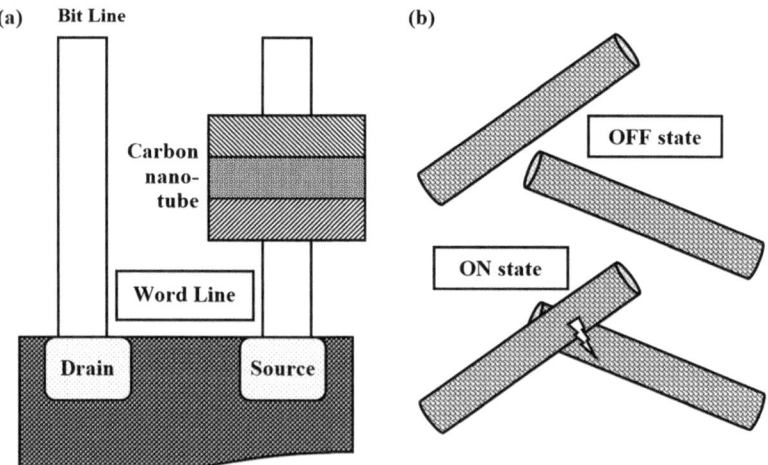

Figure 8.5 (a) Schematic illustration of a carbon-nanotube-based emerging NVM device. (b) The variation of the OFF state and ON state.

8.1.3.3 Mott memory

Mott memory is an emerging NVM technology based on the principal applications of Mott insulators, especially those materials that can go through metal-to-insulator transitions [15]. In the complex oxide thin films, the electronic-structural phase changes can develop the memory phenomenon. During writing and reading operations of the Mott memory devices, the working principle can be understood on the basis of Gibbs free energy, as illustrated in Fig. 8.6.

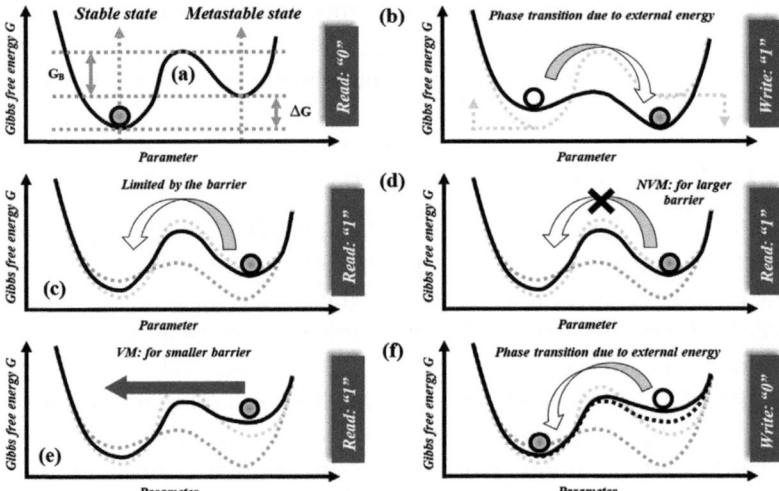

Figure 8.6 Gibbs free energy variation of the stable phase (state 0) and the metastable phase (state 1) of the typical Mott memory under writing and reading operations. (a) The phase is stable in the equilibrium system at a constant temperature. ΔG is the free energy difference between state 0 and state 1. (b) The phase transition from state 0 to state 1 due to external energy. (c) After the writing signal removal an energy barrier (G_B) will be faced by the system in state 1. Before the transition from 1 to 0, (d) the system will behave as NVM if the energy barrier is large or (e) the system can behave like VM if the energy barrier is smaller. (f) Another external signal is necessary for the transition from 1 to 0.

The initial stable phase, state 0, can be broken by an external stimulation, and the system can go through a phase transition to the metastable phase, that is, state 1. For simplicity, one can consider that the resistivity of the system can change from an insulating phase to a metallic phase. The kinetics of the phase transition dominate the stability of the state. In the case of a higher kinetic energy barrier than the thermodynamic driving force, the metastable state 1 is stable and the memory can behave as nonvolatile (Fig. 8.6d). The volatile switch occurs with a smaller kinetic energy barrier (Fig. 8.6e). In the case of NVM, another external stimulus is needed to switch from 1 to 0. Depending on temperatures it is also possible to realize VM and NVM operations within a single material system.

Compared to DRAM and SRAM, Mott memory can be designed with a two-terminal cross-point array with a $4F^2$ cell area size (F is the minimum chip feature size). Mott transitions are faster than flash transitions. In Mott memory, the demonstrated write energy per transition is sub-100 fJ, which could be further scaled down with smaller device dimensions.

8.1.3.4   Macromolecular memory

NVM devices can also be made by using macromolecular materials, such as polymers. This type of memory is generally known as macromolecular memory or organic memory. A macromolecule is a very large molecule typically composed of 10^2 to 10^3 atoms or more. In this category several materials are available, such as synthetic and biological polymers and polyelectrolytes. CNTs and graphene can also be considered as macromolecules. In macromolecules, mainly carbon atoms (sometimes silicon) are connected in a chain. The chemical structure can be modified by incorporating hydrogen or other heteroatoms, such as oxygen, nitrogen, and sulfur. Macromolecular memory can be fabricated using different structural designs, as shown in Fig. 8.7. Generally, this type of memory can be fabricated using a printing technique, which reduces the fabrication cost. Macromolecular memory is mechanically very flexible and has the potential of device scaling.

8.1.3.5   Molecular memory

In general, molecular memory consists of a TE-molecule-BE structure (BE is the bottom electrode). Especially redox-active molecules are grabbing a lot of attention due to their easily understood redox behavior. However, the data storage mechanism can vary with the structure design, such as redox-active molecular memory, solid-state molecular memory, and nanowire- or nanotube-based molecular memory.

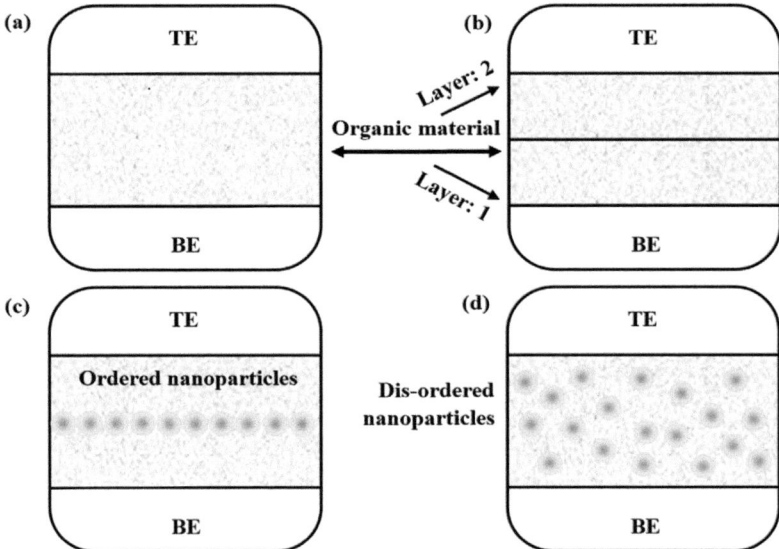

Figure 8.7 Schematic illustration of (a) single-layer macromolecular memory, (b) bilayer structure, (c) with ordered NC, and (d) with disordered NC.

Although several emerging devices are driving the research, the adoption of new memory depends on several factors, such as scalability potential, lower operating power, reliability, and cost effectiveness. Among the emerging devices, several features establish resistive random access memory (RRAM) as the most promising emerging NVM for next-generation electronic devices. The performances of prototype and emerging NVM are summarized in Table 8.3 [16].

8.1.3.6 Resistive random access memory

The emerging RRAM is one of the most promising technologies for the next-generation memory applications. In RRAM, one can change the basic resistance property of the device repeatedly across the insulating material by applying a bias. This emerging resistive switching (RS) technology has advantages over other technologies. The high endurance, $>10^{12}$ cycles, makes RRAM a potential alternative to the DRAM technology [17, 18].

Table 8.3   Performance comparison of emerging NVM technologies

Parameters	Prototypical Memories			Emerging Memories					
	Fe-RAM	STT-MRAM	PC-RAM	Emerging ferroelectric	Nano-mechanical Memory	Redox Memory	Mott Memory	Macromolecular Memory	Molecular Memory
Scalability	☹	☺	☻	☺	☹	☻	❓	❓	☻
MLC	☹	☹	☻	☹	☹	☻	❓	☺	☹
3D Integration	☹	☺	☻	☹	☹	☺	❓	☻	☹
Fabrication cost	☺	☺	☻	☺	☹	☺	❓	☻	❓
Endurance	☻	☻	☺	☻	☹	☻	☺	☹	❓

Symbol	Scalability	MLC	3D Integration	Fabrication cost	Endurance
☹	$F_{min} > 45\ nm$	Difficult	Difficult	High	< 1E5
☺	$F_{min} = 10\text{-}45\ nm$	Feasible	Feasible	Medium	< 1E10
☻	$F_{min} < 10\ nm$	Solution anticipated	Difficult	Potentially low	>1E10

Source: International Technology Roadmap for Semiconductors (ITRS) [16].

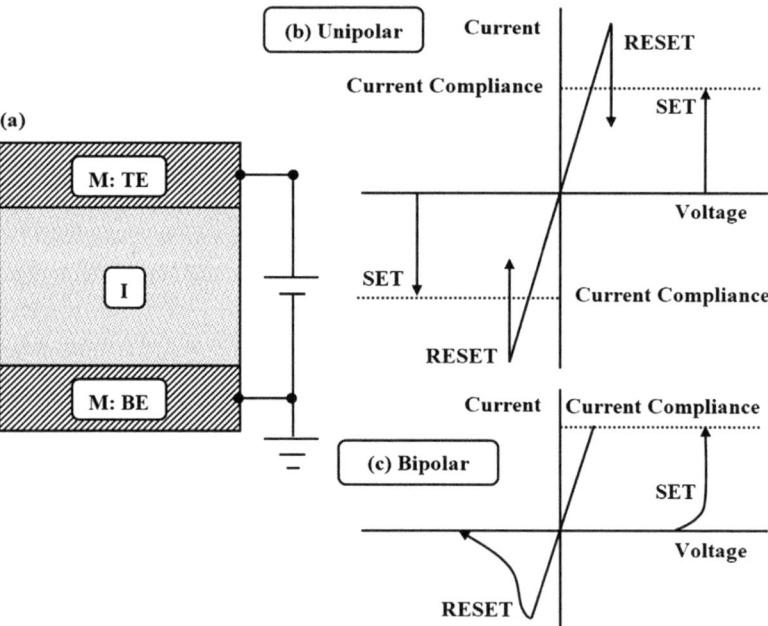

Figure 8.8   (a) Illustration of the simple M-I-M memory structure, which is based on the resistance change of the switching layer by an external bias. Basically, it shows two types of switching behaviors (b) unipolar and (c) bipolar.

The operation speed of RRAM is <1 ns [19]. As compared to the higher operating voltage of flash devices RRAM can work with ~1 V [20]. The F of RRAM have been demonstrated down to <5 nm [21]. The resistance state of RRAM can be varied from a HRS to a significantly LRS or vice versa. The structure of a RRAM device is based on a simple metal-insulator-metal (M-I-M) stack. The two metal layers of RRAM are the TE and the BE. Depending on the design and requirements, the metal layer can be an inert or active material. Many insulating materials are being explored as the RS material, such as binary/multinary oxide [22–26], chalcogenides [27, 28], and organic compounds [29, 30]. The basic structure of RRAM is shown in Fig. 8.8a. The simple design of RRAM provides integration facility in a 2D crossbar array with a small size of $4F^2$ [31, 32]. The density of the array can be further increased by stacking layers with 3D architectures [33, 34]. Therefore, the size can be further reduced to $4F^2/n$, where n is the number of stacking layers.

Normally, an as-prepared RRAM cell maintains a HRS. Therefore, before operating the cell an initial formation process is necessary. Applying a bias with suitable current compliance (I_{CC}) can break the initial HRS of the cell. After the initial process the device can be operated with a lower voltage and current. The change of resistance states from HRS to LRS is known as SET, that is, the ON state. The change of resistance from LRS to HRS is known as RESET, that is, the OFF state. RRAM is basically of two types in terms of the $I–V$ characteristic, unipolar and bipolar. Figure 8.8b shows the $I–V$ curve of typical unipolar switching (US). In US, SET and RESET occur under the same direction by thermochemical-effect-induced conductive filament (CF) formation and dissolution. The advantage of US is the high resistance ratio (HRS/LRS). However, some challenges of US still need to be solved, such as the high RESET current, poor uniformity, and low reliability. Figure 8.8c shows the $I–V$ curve of typical bipolar switching (BS), where SET and RESET occur with the opposite bias polarity. BS can be filamentary or nonfilamentary. In filamentary BS, the filament can form due to the nanoionic redox effect. RRAM is a CMOS-compatible process and highly scalable. Higher endurance ($>10^{12}$) and higher operation speed (<1 ns) are demonstrated as compared to US. In nonfilamentary BS, the resistive switching process takes place due to the change in the tunneling mechanism near the interface of metal and insulator. Generally, a forming process is not required for this type of switching. Nonlinearity can be observed at HRS and LRS with a very uniform switching behavior. The device nonlinearity factor can be defined as the ratio of the current at read voltage V_R to the current at half of V_R, that is nonlinearity $= \dfrac{I @ V_R}{I @ \frac{1}{2} V_R}$. Device nonlinearity factor is an important parameter to achieve high-density memory. Although nonlinearity is good for the nonfilamentary BS RRAM, the retention behavior has to be improved in this case.

8.1.3.7 History of RRAM

The RS process was first introduced in 1962 by Hickmott [35]. He observed a large negative differential resistance in five anodic oxide materials: SiO_x, Al_2O_3, Ta_2O_5, ZrO_2, and TiO_2. Similar phenomena have been observed in 1964 [36, 37]. In 1967, the possible application of RS in memory technology was indicated by Simmons and Verderber [38] and Varker and Juleff [39]. Later on, many materials were studied to perform RS and to understand the physics behind the switching. Development of RS up to the mid-1980s was reviewed by Pagnia and Sotnik [40]. In the search for an alternative to Si-based memory, in late 1990s, RS technology became an attractive area of research. In 1998, the first patent of RS was published by Kozicki and West [41]. Although from the late 1960s to the beginning of the 21st century, several reports showed the possible application of RS in the memory area [42], the first practical application was made by Zhuang et al. in 2002 [43]. They fabricated a one-transistor-one-resistor (1T1R) 64-bit RRAM array based on $Pr_{0.7}Ca_{0.3}MnO_3$ using a 0.5 μm CMOS process. In the process of developing RRAM technology, binary transition-metal-oxide-based RRAM integrated with a 0.18 μm CMOS process was reported by Samsung Electronics in 2004 [22]. The device was capable of functioning even at 30°C and had good SET/RESET cycles of 10^6 and read cycles of 10^{12}. In the same year, Sakamoto et al. [44] reported RRAM technology with the crossbar architecture. The attractive features of the device came as a boost for RRAM technology. Scientists made efforts not only to improve the device performance but also to understand the basic device physics. In 2006, a group of researchers from Forschungszentrum Julich, Germany [45], published a detailed study of a $SrTiO_3$-material-based RRAM device. In 2008, high-temperature multilevel operation in HfO_x-based RS with a 1T1R memory cell integrated with 0.18 μm CMOS technology was reported by the Industrial Technology Research Institute, Taiwan [46]. Later on, new structure designs, such as 3D architecture and cross-point architecture, put RRAM technology on a new level. Because of the simple two-terminal structure of RRAM, there is a huge opportunity to invent novel and advanced architectures. Crossbar is a useful architecture that is basically the fourth fundamental passive circuit element, called a memristor (memory-resistor). The memristor was invented in 1971, by Chua [47]. In 2008, a group of scientists from HP Labs, California, US, researched the existence of RS behavior in a simple $Pt/TiO_2/Pt$-based memristor structure [48]. This is a most promising structure due to its inherent $4F^2$ cell size and 3D integration possibilities for mass storage devices. In 2009, Yoon et al. [49] reported ultrahigh-density vertical RRAM. Figure 8.9 shows the evolution of RRAM technology. In the same year, high-dielectric-constant (high-κ)-Ta_2O_5-based RRAM with an ultralow-current operation, of 5 pA, was reported by Maikap et al. [50]. In 2010, using advanced nanoinjection lithography technique, the scalability potential of substoichiometric WO_x-based RRAM below 10 nm was addressed by the National Nano Device Laboratory, Taiwan [51]. Along with the technological development, theoretical understanding has seen equal development [52–59].

The technological development of RRAM devices is very dependent on the proper understanding of the composition, structure, and dimensions of switching filaments. Universal SET/RESET characteristics [60], effect of switching parameters [61], switching mechanism [62], and filament structure and growth [63] were studied in detail. In 2010, Kwon et al. [64] reported the atomic structure of the conductive nanofilaments in TiO_2-based RS devices.

In the very next year, 2011, the RESET current was scaled down to 23 nA for a nitrogen-doped AlO_x RRAM device with a 1T1R structure [65]. In the same year, the endurance of RRAM hit over 10^{12} program/erase cycles, reported by Samsung Electronics [18]. The RS device is based on a bilayer TaO_x material with a 30×30 nm^2 crossbar structure (Fig. 8.10a) and is equally capable of switching with a 10 ns RESET and SET pulses (Fig. 8.10b). In January 2012, Elpida Memory announced a RRAM prototype using a 50 nm manufacturing process technology with a capacity of 64 Mbits [66]. In May 2012, Panasonic launched its Ta_2O_5-based 1T1R RRAM cell integrated with a 0.18 μm CMOS technology [67]. On the basis of the development of basic device physics and the development of technology, RRAM devices have demonstrated the ultimate scalability potential with a feature size of 5 nm [68]. In 2015, 3-bit-per-cell storage was also achieved for TaO_x-based RRAM devices [69]. Supernonlinear HfO_2/CuGeS RRAM with ultralow power for 3D vertical nanocrossbar arrays was reported by Luo et al. in 2016 [33]. The RRAM technology is suitable for the new area of application like brain-inspired computing. Furthermore, FM, optical, and superconducting properties combined with RRAM have been reported [70–72].

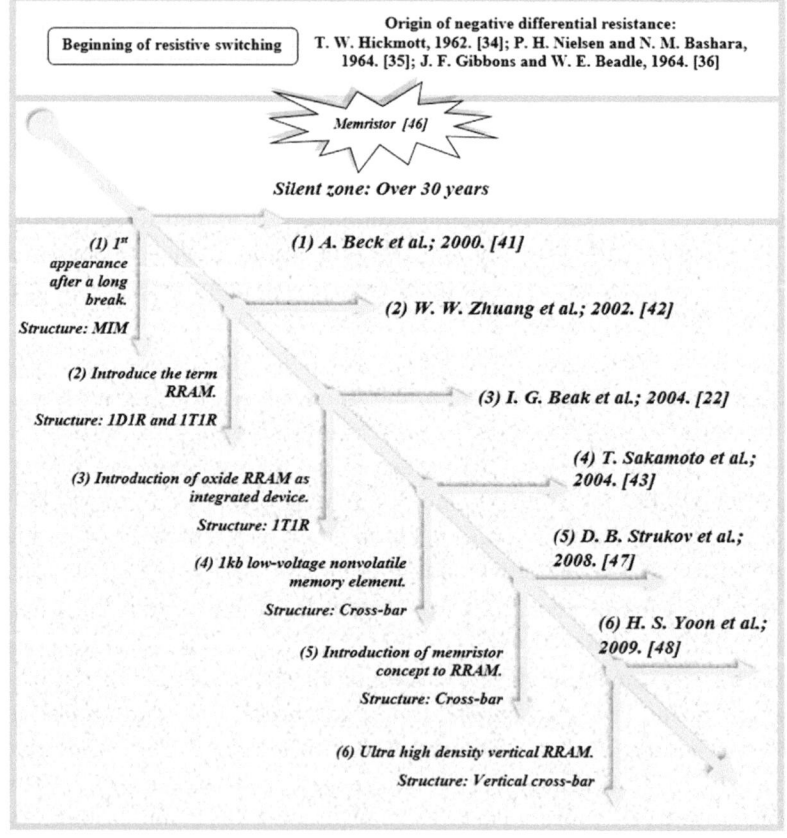

Figure 8.9 Evolution of RRAM technology.

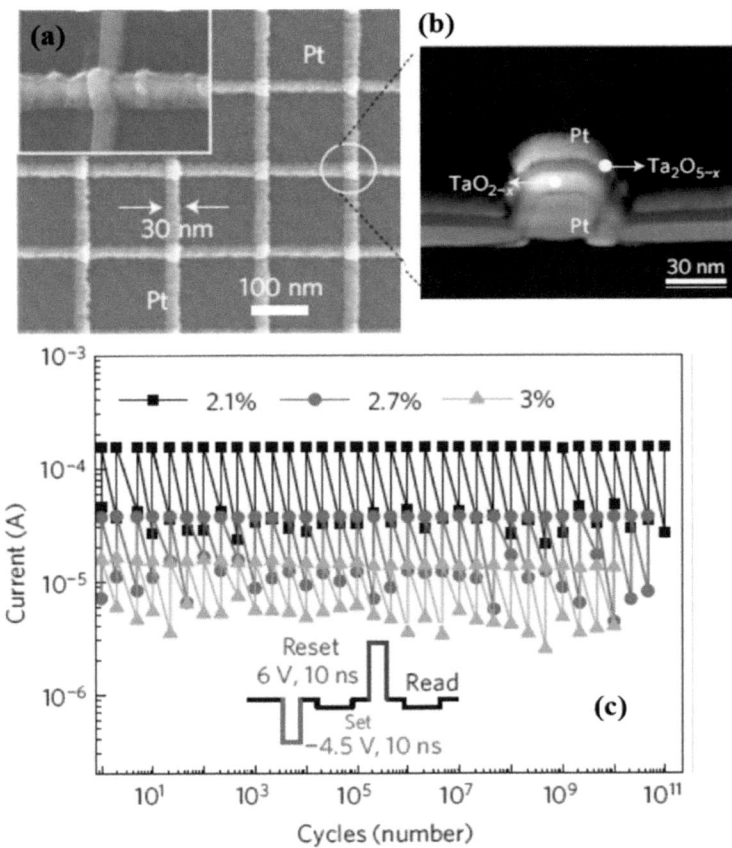

Figure 8.10 (a) 30 nm cross-point array of TaO$_{2-x}$/Ta$_2$O$_{5-x}$ bilayer RRAM by using e-beam lithography technique. (b) Devices showing excellent endurance with a 10 ns pulse. Reprinted by permission from Macmillan Publishers Ltd: *Nature Materials* (Ref. [18]), copyright (2011).

8.2 Mechanisms and Materials in RRAM

8.2.1 Resistive Switching Mechanisms

The RS process in the RRAM devices can be categorized into several groups on the basis of their switching mechanisms. A variety of physical phenomena lead to RS in RRAM devices [73]. The classification of RS is shown in Fig. 8.11. The physics behind

RS not only depends on the materials but also depends on the device fabrication process and device operation. The common switching mechanisms can be classified into three major groups: electrochemical metallization (ECM) type, valence change memory (VCM) type, and thermochemical reaction type. Because of low power consumption, superior scalability up to the atomic level, and a simple fabrication process, ECM and VCM devices are the center of attraction of research.

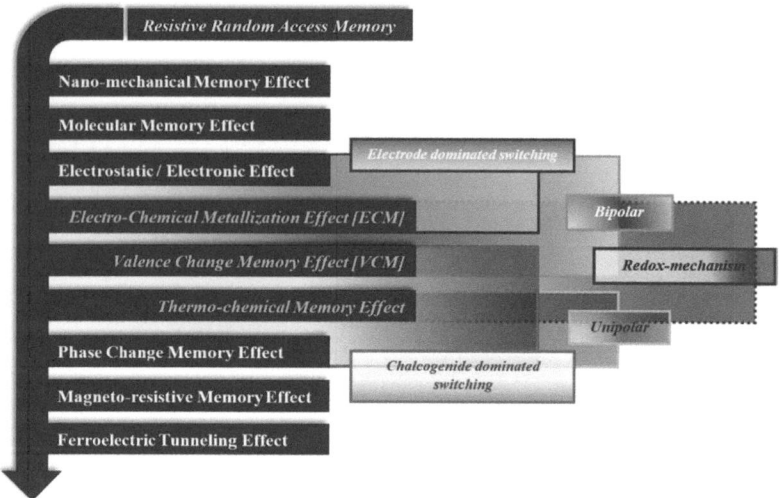

Figure 8.11   Classification of RS processes.

8.2.1.1   Electrochemical metallization type

Generally, the electrodes of ECM devices are defined as active electrodes (AEs) and inert or counter electrodes (CEs). The AE is typically Cu and Ag. Plenty of options are available for CEs. All of the useful electrodes for RRAM devices are shown in Table 8.4. The switching principle of ECM devices is based on the formation and rupture of the CF by the cation migration process. The mobile ions, like Cu^{2+} or Ag^+, from the AE directly participate in the RS event. The CF forms via the electrochemical dissolution process from the AE and final redeposition on the CE; this kind of RRAM is usually known as an ECM type of memory. Due to the formation of a metallic bridge between the electrodes, it is also known as conductive bridging random access memory (CBRAM), a programmable metallization cell (PMC), or a gapless-type atomic switch [74–78]. The optical microscopic evidence of ECM switching was initially reported in 1976 by Y. Hirose and Hirose [79]. The formation of a metallic bridge is shown in Fig. 8.12.

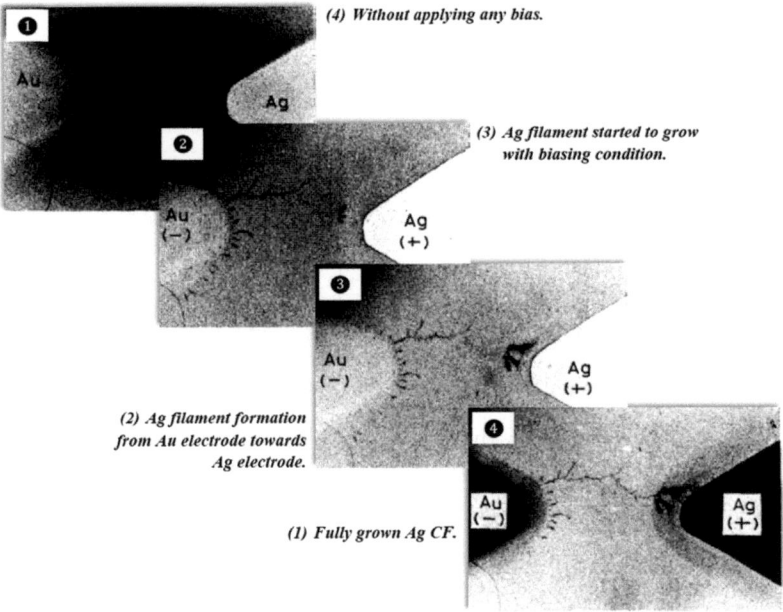

Figure 8.12 Ag metallic bridge grows from the Au electrode, that is, toward the positively biased Ag electrode. Reprinted from Ref. [79], with the permission of AIP Publishing.

Later on the operation principle is rigorously reviewed. ECM switching mainly depends on the AE materials. Although Cu and Ag are the standard AEs for ECM, there are several other options available, such as Ni [80], Al [81], Ti [82], Zn [83], Nb [84], and Au [85]. Ag and Cu are among the most commonly used AEs due to their physical advantages. The standard electrode potentials for Ag^+ (0.8 V) and Cu^{2+} (0.34 V) are much smaller as compared to those of the others, like Pt^{2+} (1.19 V) and Au^+ (1.83 V), due to which Ag^+ and Cu^{2+} can electrochemically dissolve very easily. The standard Gibbs free energy of the formation of oxides for Ag and Cu is much lower than that of other metals, like Ir, Pt, and Ni.

Table 8.4 Useful metals as electrode material in RRAM technology

H																	He
Li	Be											B	C	N	O	F	Ne
Na	Mg											Al	Si	P	S	Cl	Ar
K	Ca	Sc	Ti	V	Cr	Mn	Fe	Co	Ni	Cu	Zn	Ga	Ge	As	Se	Br	Kr
Rb	Sr	Y	Zr	Nb	Mo	Tc	Ru	Rh	Pd	Ag	Cd	In	Sn	Sb	Te	I	Xe
Cs	Ba	LA	Hf	Ta	W	Re	Os	Ir	Pt	Au	Hg	Tl	Pb	Bi	Po	At	Rn
Fr	Ra	AC	Rf	Db	Sg	Bh	Hs	Mt	Ds	Rg	Cn		Fl		Lv		

LA	La	Ce	Pr	Nd	Pm	Sm	Eu	Gd	Tb	Dy	Ho	Er	Tm	Yb	Lu
AC	Ac	Th	Pa	U	Np	Pu	Am	Cm	Bk	Cf	Es	Fm	Md	No	Lr

Lanthanide Series   Actinide Series

AE of ECM   AE of VCM   CE   AE/CE

8.2.1.1.1 *Growth of a metallic filament*

According to the ECM theory, the growth process of a metallic filament depends on three steps. Considering M as metal atoms and M^{z+} as metal ions:

1. Oxidation reaction at the AE, that is, the anode ($M \rightarrow M^{z+} + Ze^-$).

2. Electromigration of M^{z+} ions from anode to cathode (CE) direction

3. The reduction of M^{z+} ions to metal atoms ($M^{z+} + Ze^- \rightarrow M$) at the cathode

Finally, there is the formation of a CF between the two electrodes, which can conduct the RS event. However, for different material systems, the growth direction of the CF can be different.

8.2.1.1.2 *Solid electrolyte-based ECM*

The classical theory of ECM is based on water (H_2O) and solid electrolyte materials. Those systems are dominated by ionic conductivity. Review of literatures show that ECM devices can be designed on the basis of H_2O [86, 87], Ag-Ge-Se [88, 89], Ag_2S [90], GeTe [91], GeS [92], etc. In those mediums, the cations can be generated from the AE and can migrate through the solid electrolyte material for the final reduction on the CE. In this case, the CF growth direction is from the CE to the AE. The *I–V* characteristics of an ECM device based on a Ag/Ag-Ge-Se/Pt structure is shown in Fig. 8.13. Initially the device is in the OFF state. On application of a positive bias on the AE, the oxidation process will take place and the Ag⁺ ion will start to move toward the CE. The reduction process will deposit the Ag atoms on the CE. As a result, the CF will be formed from the CE to the AE, as shown in Fig. 8.13b. At some point the Ag CF will connect the AE and the CE, resulting in an ON state on the device, as shown in Fig. 8.13d. To break the CF, a negative bias is applied on the AE, which will break the filament and reset the device, as shown in Fig. 8.13e.

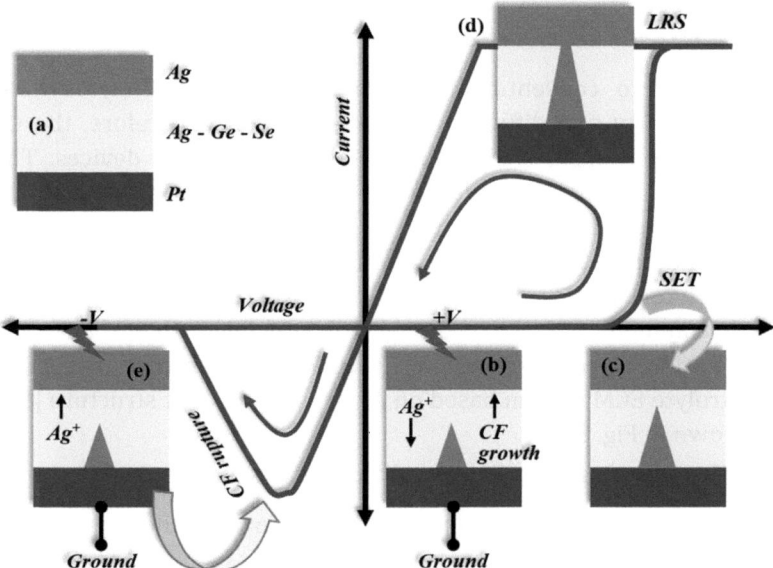

Figure 8.13 Schematic *I–V* characteristics of an ECM device with a switching mechanism.

The growth direction of the CF in a solid-electrolyte-based medium has been observed in a Cu/Cu-GeTe/Pt-Ir-based ECM structure using scanning transmission electron microscopic (STEM) measurements [91]. As shown in Fig. 8.14, the HRS was broken and a Cu filament was formed by applying a negative voltage on the Pt-Ir

TE and the filament was ruptured was using a positive voltage on top.

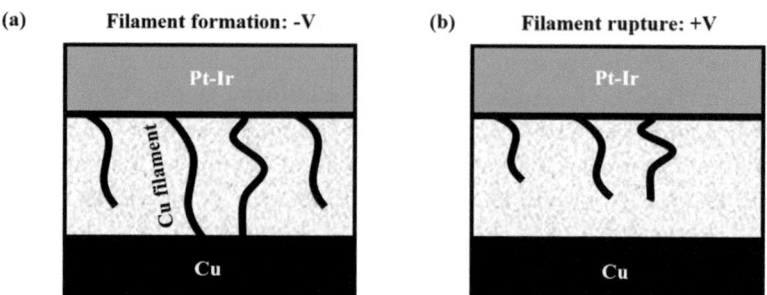

Figure 8.14 Schematic growth of a Cu filament.

8.2.1.1.3 *Oxide electrolyte-based ECM*

Compared to conventional electrolytes, oxide electrolytes have low solubility and diffusion coefficients of M^{z+}; therefore, the CF growth mechanism is different in oxide-based ECM devices. The ionic conductivity depends on the metal ion flux in a particular electrolyte system. For example, Cu ion flux in an oxide-electrolyte-based system is 10 orders of magnitude lower than the conventional electrolyte systems (i.e., Cu/CuS/Pt). Therefore, generally, the ionic conductivity in an oxide electrolyte is lower than that in the conventional electrolyte systems. The CF growth process in an oxide electrolyte ECM system based on Ag or a $Cu/ZrO_2/Pt$ structure [58] is shown in Fig. 8.15.

The growth of the CF can be divided into several steps:

1. A positive bias is applied to the AE to oxidize the metal atoms to M^{z+} ions.

2. Under the applied electric field, the mobile M^{z+} ions move toward the surface of the AE.

3. At a lower migration speed M^{z+} ions combine with electrons and reduce back to M on the surface of the AE.

4. The growth of the extended virtual electrode on the AE/oxide interface continues due to the continuous oxidation/reduction processes.

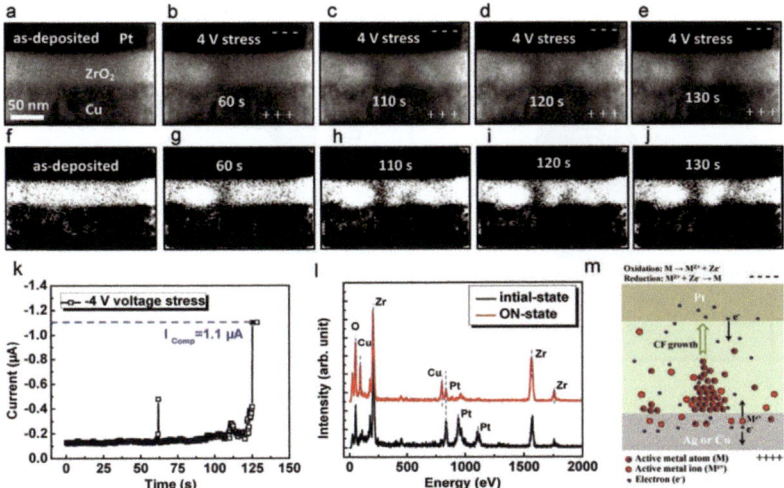

Figure 8.15 (a–e) Transmission electron microscopic images of the dynamics of CF growth processes in the Cu/ZrO$_2$/Pt ECM device. (f–j) The CF is highlighted in black-and-white contrast images of the images in (a–e). (k) A typical current-time characteristic under a bias of −4 V. (l) The energy dispersive spectrum for the initial and ON states. (m) Schematic illustration of the CF growth mechanism. From Ref. [58]. Copyright © 2012. Reproduced with permission of John Wiley & Sons.

The process continues until the CF reaches the CE. Therefore, the growth direction of the CF in an oxide electrolyte system is from AE to CE.

8.2.1.1.4 *Organic electrolyte-based ECM*

A similar growth direction of the CF can be observed in the organic-electrolyte-based systems. Figure 8.16 shows CF growth and rupture processes in an Ag/WPF-BT-FEO/heavily doped p-type poly-Si organic RRAM device [93]. The WPF-BT-FEO stands for poly[(9,9-bis(6′-(*N,N,N*-trimethylammonium)hexyl)-2,7-fluorene)-co-(9,9-bis(2-(2-methoxyethoxy)ethyl)-fluorene)-co-(2,1,3-benzothiadiazole)] dibromide. The material of the CF was verified by the transmission electron microscopic observation and energy-dispersive X-ray spectroscopy analysis. A Ag metallic bridge exists between the Ag electrode and heavily doped p-type single-crystalline Si BE. A negative voltage was applied on the Ag electrode to break the connection near the bottom side.

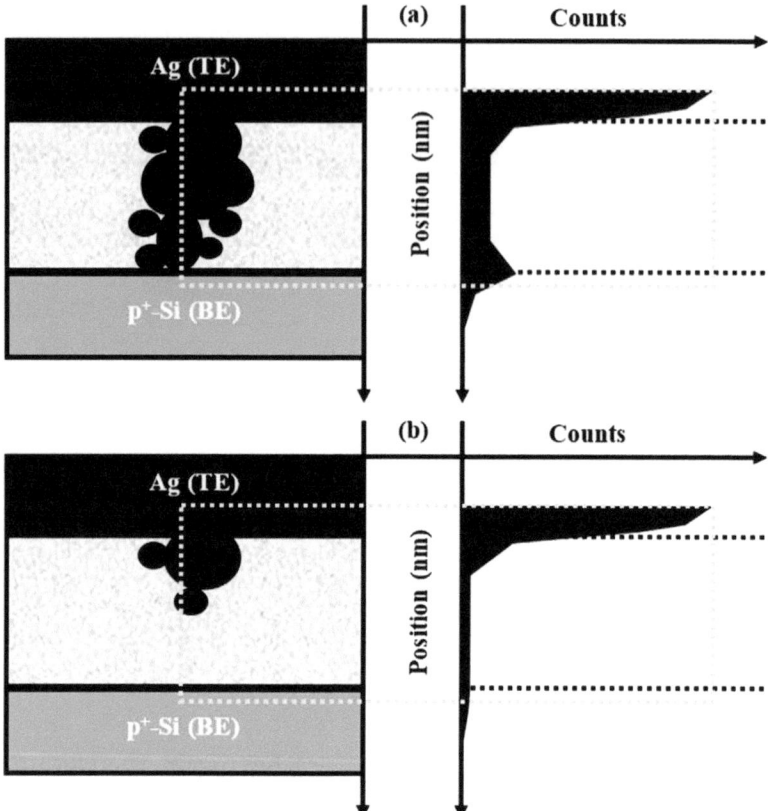

Figure 8.16 Schematic illustration of a Ag filament and its material after (a) formation and (b) rupture in the Ag/WPF-BT-FEO/heavily doped p-type single-crystalline Si RRAM cell.

8.2.1.2 Valence change memory type

Generally, transition-based RRAM with an inert type of metal electrode is also able to show the RS behavior on the basis of the oxygen vacancy (V_O), like the anion migration process. In a VCM-type system, the electrode materials such as Pt, Au, and Ir are not easily oxidized or the oxidized form can't easily convert back to the metal, for example, in the case of Al, Ti, and Nb. Among many insulating materials, TiO_x [94], NiO_x [95], HfO_x [96], TaO_x [97], AlO_x [98, 99], and WO_x [100] have been studied rigorously. Nitrides such as AlN

[101] and NiN [102] have also been studied for VCM-type devices. This is because, in these systems, the valency of oxides or nitrides will be changed due to the migration of positively charged V_O or nitrogen vacancies *(V_N)* under an external electric field. Hence, the anion-migration-based systems are usually known as VCM. The VCM can be categorized into two broad groups, filamentary switching and interface switching.

8.2.1.2.1 *Filamentary switching*

In VCM devices, the RS event takes place due to the CF formation by the movement of vacancies. Figure 8.17 shows a schematic illustration of the filamentary switching mechanism in a VCM cell. On applying a positive voltage in the TE, the oxygen ions drift toward the top interface. The nonlattice oxygen ions accumulate near the TE if the TE is made of an inert metal, like Pt. If the TE is an oxidizable metal like Ti, then the nonlattice oxygen ions will form an interfacial layer between the TE and the insulating layer. In any of the cases the TE/insulator interface will behave like an oxygen reservoir. This process will create the initial CF formation in the VCM cell, resulting in a soft breakdown phenomenon due to high electric field. Generally, after the formation of the initial CF, the next RS events take smaller voltages. Therefore, the SET process can be achieved with a lower positive voltage as compared to the device formation process. Once the CF forms, the current can easily flow through it. The size and thickness of the CF depend on the applied current compliances. The higher the current compliance, the thicker the CF, as shown in Fig. 8.18.

The CF can be either metallic or semiconductor. A simple temperature-dependent study of the resistance state can give an idea about the CF. Temperature-dependent resistance change of the CF is shown in Fig. 8.19. With an increasing temperature, the resistance will increase for the metallic CF and decrease for the semiconducting CF. Due to the abundant number of mobile carriers, metals are usually highly conductive. Therefore, at a low temperature or at room temperature, the resistance is low.

But with an increase of temperature, the increasing number of vibrations of lattice atoms constantly increase the interference with the transportation path of mobile carriers. Hence, at a higher temperature, the mobile carriers are not free as they were at a lower temperature, which will increase the resistance of metallic CF at a higher temperature. The situation is different in the case of semiconductors. At a normal temperature there are fewer mobile carriers, resulting in a higher resistance at a low temperature. With an increasing temperature, the resistance will decrease and the lightly bound carrier will be free. Once all of the mobile carriers become free the semiconductor will behave like a metal. Coming back to the RESET process of VCM devices, a negative voltage on the TE can do the work. The oxygen ions move back to the insulating layer by recombining with the V_O sites or by oxidizing the interfacial layer. The backward movements of oxygen ions will break the CF and the device will be in a HRS.

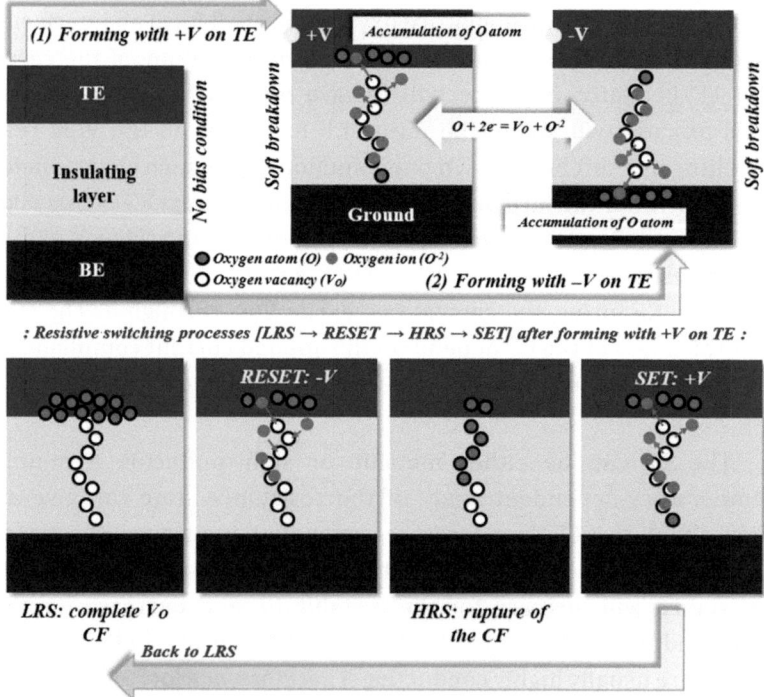

Figure 8.17 Schematic switching mechanism of typical VCM devices.

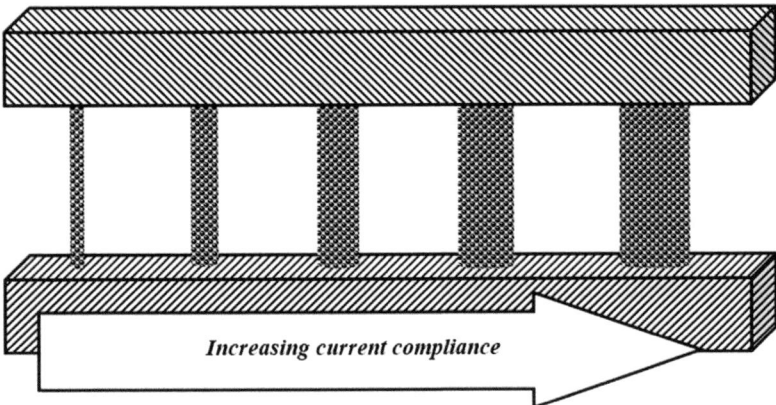

Figure 8.18 The CF diameter will increase with I_{CC}.

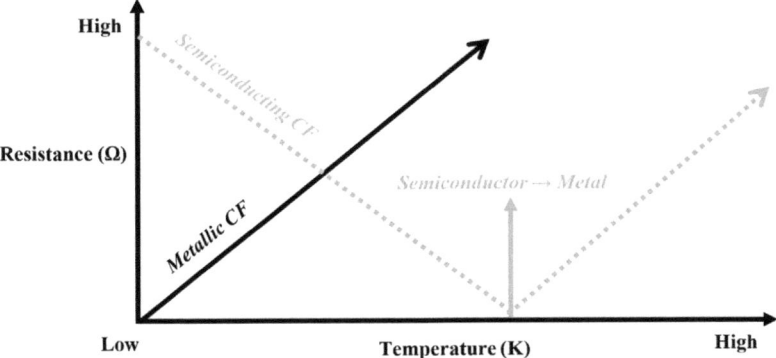

Figure 8.19 CF resistance variation with temperature. The resistance will increase with temperature for a metallic CF and will decrease for a semiconducting CF. When all carriers are free to move, the semiconductor behaves like a metal and a further increase in temperature will increase the resistance of the filament.

8.2.1.2.2 *Interface switching*

In VCM devices, the RS phenomena also may happen in the Schottky junction due to the charge-trapping/detrapping process [103, 104]. The applied bias has a huge impact to determine the trapping states in the junction. The charge carriers trapped near the interface will form a space charge region and will be the modulating barrier of the interface. It is reported that for an Au/Nb:STO junction, the VCM can be in the SET (LRS) state when Au is positively biased and the device can be switched to RESET (HRS) when Au is negatively biased [104]. In a more generalized form, under a reverse bias the charge-trapping process will be in progress at the interface. As a result, the barrier will increase and the device will be in a HRS. On the other hand, under a forward bias condition the charge detrapping process takes place, which will reduce the barrier and lead to the LRS process. The barrier modulation process is shown in Fig. 8.20.

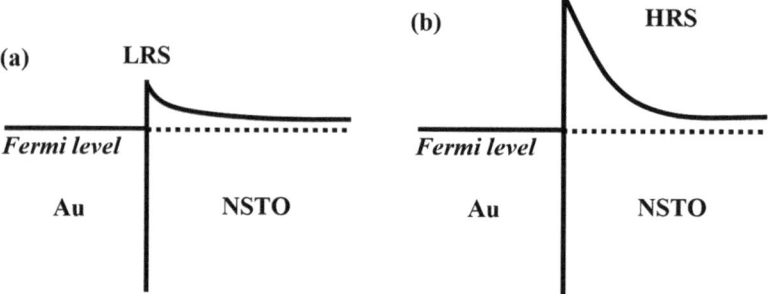

Figure 8.20 Barrier height modulation under forward and reverse bias conditions in a Au/NSTO structure. (a) The charge detrapping process takes place under a forward bias, and the lower barrier height will switch the device to a LRS. (b) A reverse bias will detrap the charges; as a result, the device will switch to a HRS.

8.2.1.3 Thermochemical reaction type

Several material systems show unipolar characteristics with a strong thermal effect during the RS process. Most importantly, PCM devices are also a prominent type in this category. In here, we are only discussing thermochemical-reaction-type RRAM devices. Several transition metal oxides can show this kind of switching. NiO is a vital candidate in this class. The switching mechanism of devices of this kind has been studied rigorously. Figure 8.21 shows the schematic illustration of the formation and rupture of the CF in an Pt/ZnO/Pt structure [105]. The RS mechanism can be explained in the following way: with the application of a bias, V_O are created as the oxygen ions are moving in the bulk ZnO film. During the SET process, Zn-rich CF is formed inside the ZnO film and the device switches to the ON state, as shown in Fig. 8.21a. The RESET operation is dominated by the joule heating process. Joule heating provides energy sufficient to thermochemically break the Zn-rich CF and put the device back to the OFF state. The CF rupture by the joule heating process is shown in Fig. 8.21b. The basic switching mechanisms of any type of RRAM devices are dependent on the RS/electrode materials, device structure, and fabrication process.

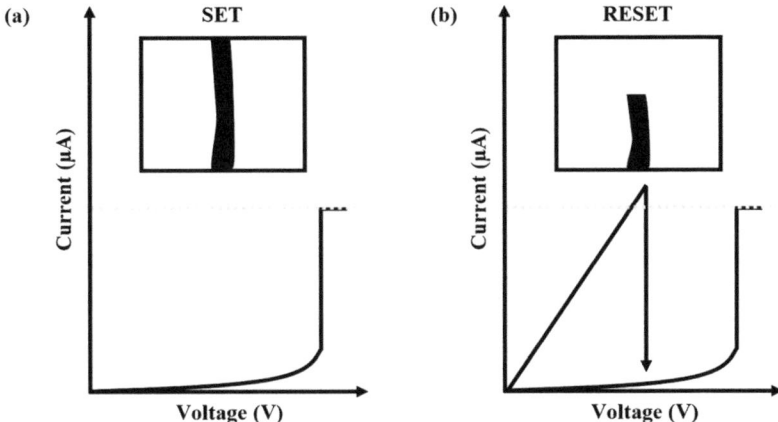

Figure 8.21 Schematic switching mechanism observation in the Pt/ZnO/Pt RRAM device. (a) CF formation and switching to a LRS. (b) CF rupture process by joule heating, resulting in a HRS transition.

8.2.2 Materials in RRAM

A simple RS stack is usually composed of an insulating layer sandwiched between two metal electrodes. The switching process can be equally affected by the insulating layer and also by the metal layers. Therefore, selection of the proper material is a major concern of RRAM technology. The following sections are going to discuss separately the materials involved in a metal electrode layer and an insulating layer in RRAM.

8.2.2.1 Metal electrode layer

The metal electrode material is one of the vital parts to design a RRAM stack. The metal layer can have a significant effect on the RS process and also on the switching mechanism. In general, the metal electrodes can be subdivided into three categories: active metal, inert metal, and oxidizable metal. A large number of metals used in RRAM are identified in the section of the periodic table shown in Table 8.4 (see Section 8.2.1).

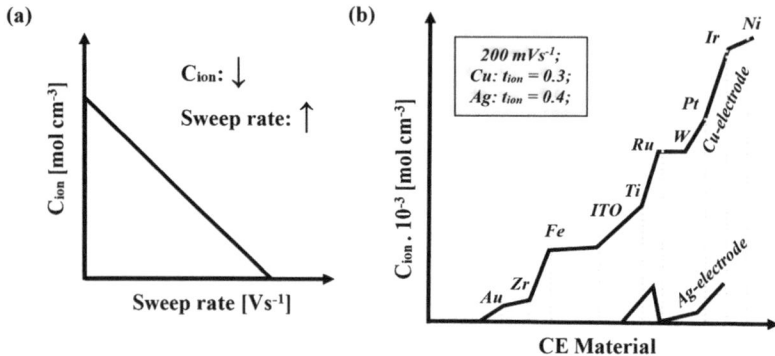

Figure 8.22 Variation in the ion concentration depending on (a) the sweep rate and (b) the material.

RRAM based on active metals, such as Cu and Ag, is usually of the ECM type. RRAM characteristics can vary with variation in the ion concentration (C_{ion}) in the CE metal for different Cu- or Ag-based SiO_2 RRAM cells [106]. The C_{ion} on a CE can be calculated from a simple cyclic voltammetry measurement. The C_{ion} is dependent on the cyclic voltammetry measurement sweep rates as shown in Fig. 8.22a. The C_{ion} always decreases with increasing sweep rates because of the lesser reaction time at the AE with a higher sweep rate. The C_{ion} is also dependent on the AE material. Generally, as Fig. 8.22b indicates, the RRAM device based on the Ag AE shows lower C_{ion} compared to the RRAM based on the Cu AE. However, a lower voltage amplitude was applied during the measurement for the Ag-based ECM devices [106].

In the case of VCM-type RRAM devices, inert metals such as Pt and Ir and oxidizable metals, such as Ti, can be used. Along with them heavily doped silicon, graphene, and CNTs; conductive oxides such as indium tin oxide (ITO) and iridium oxide (IrO_x) [107]; and nitrides such as TiN and TaN are also used as electrode materials.

Table 8.5   Inorganic materials in RRAM

Storage medium	Switching mode	ON/OFF ratio	Operation speed	Endurance (cycles)
		Binary oxides		
MgO_x	Unipolar, bipolar	$>10^5$	-	$>4 \times 10^2$
AlO_x	Unipolar, bipolar	$>10^6$	<10ns; <10ns	$>10^4$
SiO_x	Unipolar, bipolar	$>10^7$	<100ps; <100ps	$>10^8$
TiO_x	Unipolar, bipolar	$>10^5$	<5ns: <5ns	$>2 \times 10^6$
CrO_x	Bipolar	$>10^2$	< 4 μ.s; < 5 μ.s	$>6 \times 10^4$
MnO_x	Unipolar, bipolar	$>10^4$	< 100ns; < 200ns	$>10^5$
FeO_x	Bipolar	$>10^2$	<10ns; <10ns	$>6 \times 10^4$
CoO_x	Unipolar, bipolar	$>5 \times 10^3$	<20ns; <20ns	$>10^3$
NiO_x	Unipolar, bipolar	$>10^6$	<10ns; <20ns	$>10^6$
CuO_x	Unipolar, bipolar	$>10^5$	<50ns; <50ns	$>1.2 \times 10^4$
ZnO_x	Unipolar, bipolar	$>10^7$	<5ns; <5ns	$>10^6$
GaO_x	Bipolar	$>10^2$	<400ns; < 600 ns	$>10^4$
GeO_x	Unipolar, bipolar	$>10^9$	<20ns; <20ns	$>10^7$
ZrO_x	Unipolar, bipolar	$>10^6$	10ns; <10ns	$>10^6$
NbO_x	Unipolar, bipolar	$>10^8$	<100ns; <100ns	$>10^7$
MoO_x	Unipolar, bipolar	>10	<1 μs: <1 μs	$>10^6$
HfO_x	Unipolar, bipolar	$>10^5$	<300ps; <300ps	$>10^{10}$
TaO_x	Unipolar, bipolar	$>10^9$	<105 ps; <120ps	$>10^{12}$

(Continued)

Table 8.5 (*Continued*)

Storage medium	Switching mode	ON/OFF ratio	Operation speed	Endurance (cycles)
WO_x	Unipolar, bipolar	$>10^4$	<300ns; <50ns	$>10^8$
CeO_x	Unipolar, bipolar	$>10^5$	<1µs; <200ns	$>10^4$
GdO_x	Unipolar, bipolar	$>5 \times 10^5$	<1 ns; <1 ns	$>10^7$
YbO_x	Unipolar, bipolar	$>10^5$	-	$>10^5$
LuO_x	Unipolar, bipolar	$>10^4$	<10ns; <30ns	$>8 \times 10^2$
Ternary and more complex oxides				
$LaAlO_3$	Bipolar	$>10^4$	-	$>10^2$
$SrTiO_3$	Bipolar	$>10^5$	<5ns; <5ns	$>10^6$
$BaTiO_3$	Unipolar, bipolar	$>10^4$	<10ns; <70ns	$>10^5$
LC(or S)MO	Bipolar	$>10^3$	<25ns; <25ns	$>10^3$
PCMO	Bipolar	$>10^3$	<8ns; <8ns	$>10^{10}$
$BiFeO_3$	Unipolar, bipolar	$>10^5$	<50ns; <100µs	$>10^3$
Chalcogenides				
Cu_2S	Bipolar	$>10^6$	<100µs; <100µs	$>10^5$
GeS_x	Bipolar	$>10^5$	<50 ns; <50ns	$>7.5 \times 10^6$
Ag_2S	Bipolar	$>10^6$	<200ns; <200ns	-
Ge_xSe_y	Bipolar	$>10^6$	<100ns; < 100ns	$>3.2 \times 10^{10}$
Nitrides				
AlN	Unipolar, bipolar	$>10^3$	<10 ns; <10ns	$>10^8$
SiN	Unipolar, bipolar	$>10^7$	<100ns; <100ns	$>10^9$
Others				
a-C	Unipolar, bipolar	$>3 \times 10^2$	<50 ns; <10ns	$>10^3$
a-Si	Bipolar	$>10^7$	<5ns; <10ns	$>10^8$
AgI	Bipolar	$>10^6$	<50 ns; <150ns	$>4 \times 10^5$

Source: Reprinted from Ref. [111], Copyright (2014), with permission from Elsevier.

8.2.2.2 Insulating layer

Plenty of materials have been researched for potential application as the insulating layer in RRAM. The RS phenomena have been observed in various material systems, such as chalcogenides, transition metal oxides, perovskites, solid-state electrolytes, and organic compounds. For simple classification, the insulating materials are categorized into inorganic and organic insulating layers. Both have their own advantages and disadvantages. RRAM with an inorganic insulating layer has stable RS properties as compared to RRAM with an organic one.

RRAM with an organic insulating layer has a very simple, cost-effective fabrication process and it is easy to make flexible devices. Binary oxide materials are extensively investigated for RRAM technology, such as AlO_x, SiO_x, TiO_x, CoO_x, NiO_x, CuO_x, ZnO_x, ZrO_x, HfO_x, TaO_x, TiO_x, and WO_x. Complex oxides such as $SrTiO_3$ [108], $La_{0.7}Sr_{0.3}MnO_3$ [109], and $BiFeO_3$ [110] are equally studied as potential materials in RRAM technology. The design of the RRAM stack is not limited to only a single insulating layer; bilayer structures also show promising characteristics. A large number of available inorganic materials are summarized in Table 8.5. Organic-insulating-layer-based RRAM technology attracts a lot of attention due to the simple low-cost fabrication processes. Organic insulating layer are generally polymers and small molecule systems. Useful organic materials for RRAM technology are summarized in Table 8.6.

Table 8.6 Organic materials in RRAM

Storage medium	Switching mode	ON/OFF ratio	Operation speed	Endurance cycle
Small molecules				
AIDCN	Bipolar	$>10^4$	<10ns; <40ns	-
Alq$_3$	Unipolar, bipolar	$>10^6$	-	$>2 \times 10^4$
Cu:TCNQ	Bipolar	$>10^4$	<10ns; -	$>10^4$
NPB	Bipolar	$>10^2$	-	$>10^6$
Rose bengal	Bipolar	$>10^5$	< 500 ns; <50ms	$>7.2 \times 10^7$

(Continued)

Table 8.6   (*Continued*)

Polymers				
MEH-PPV	Unipolar, bipolar	$>10^4$	<l µs; -	$>5 \times 10^1$
P3HT	Unipolar, bipolar	$>10^4$	<300 ns; < 300 ns	$>3 \times 10^4$
PARA	Unipolar, bipolar	$>10^5$	<80 ns; <1 µs	$>10^3$
Parylene-C	Bipolar	$>10^7$	<15ns; <15ns	$>2.5 \times 10^2$
PEDOT:PSS	Unipolar, Bipolar	$>10^5$	< 500 ns; < 500 ns	$>10^4$
PFN-C	Unipolar	$>10^7$	< 300 ns; < 500 ns	$>4 \times 10^3$
PFO	Unipolar	$>10^3$	-	$>5 \times 10^3$
PI	Unipolar, bipolar	$>10^8$	< 200 ns; < 200 ns	$>10^5$
PMMA	Bipolar	$>10^7$	-	$>1.5 \times 10^5$
PS	Unipolar, bipolar	$>10^4$	<100ns; <100ns	$>10^2$
PVA	Bipolar	$>10^7$	-	$>10^2$
PVK	Unipolar, bipolar	$>5 \times 10^9$	<5ms; <30ms	$>10^3$
PVP	Bipolar	$>4 \times 10^3$	<10ns; <10ns	-
WPF-oxy-F	Bipolar	$>10^5$	<10 µs; <l00 µs	$>10^4$
Others				
Graphene oxide	Bipolar	$>10^4$	<5ns; <5ns	$>10^3$

Source: Reprinted from Ref. [111], Copyright (2014), with permission from Elsevier.

8.2.2.3   Defect-related improvement of RRAM performance

Defect engineering is an effective way to improve the performance of a RRAM device [112]. For example, the RS properties of Al_2O_3 have been improved by Cu doping [113] and nitrogen (N) doping [65]. N-doped Al_2O_3 RRAM has achieved formation-free switching. Improvement in the uniformity of RS cycles has been achieved by N-doping in a $Ta/TaO_x/Pt$ RRAM device [114]. Switching characteristics of the undoped and N-doped RRAM are shown in

Fig. 8.23. Major improvements by N doping have been achieved in the variability of switching parameters and also in device nonlinearity. Device nonlinearity can be defined as the ratio of the current at a particular LRS voltage to the current at half of that voltage. However, control over the percentage of N doping is very important. For N-doped RRAM devices, the percentage of N doping was varied from 0% to 9%. The RS characteristics varied accordingly. As shown in Fig. 8.23d, good RS property has been achieved with a 3%–6% N-doped TaO$_x$ RRAM. The device is suitable for a 3-bit multilevel cell (MLC) storage application. Moreover, oxygen migration can be confined by the excess N in the insulating layer. During biasing condition, the N can make a strong bond with the oxygen ions with a higher bonding energy. The process confines the CF formation to the localized region and hence the stability of the switching improves.

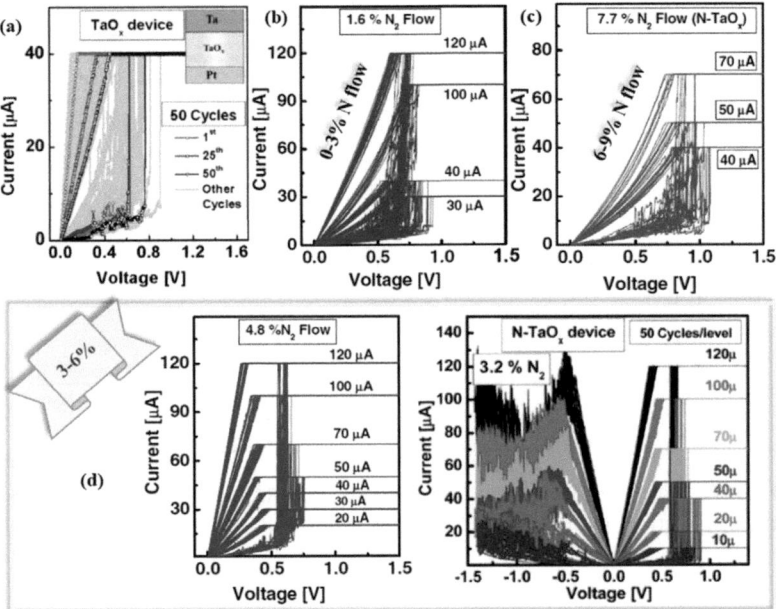

Figure 8.23 RS characteristics of (a) an undoped TaO$_x$-insulating-layer-based RRAM. Improvement of RS with (b) a 1.6% N-doped TaO$_x$ insulating layer, (c) a 7.7% N-doped TaO$_x$ insulating layer, and (d) a 3%–6% N-doped TaO$_x$ insulating layer. 3%–6% N-doped TaO$_x$ RRAM devices are suitable for MLC storage applications. Reprinted with permission from Ref. [114], licensed under CC-BY 4.0 (http://creativecommons.org/licenses/by/4.0/).

Liu et al. [115] reported a Ti-doped improvement of the Cu/ZrO_2:Ti/Pt RRAM structure. A narrow distribution of the SET/RESET voltages and also in HRS/LRS resistances is visible in Fig. 8.24. A large number of metal dopants have been reported so far [116]. Both the single-layer and bilayer RRAM can be equally affected by the doping processes. Inserting NC into the insulating layer is another process to increase the controllable defects in RRAM devices. RRAM devices based on Ru NC [117], IrO_x NC [99], TiO_2 NC [118], CdS NC [119], and Au NC [120] have shown an improvement over the controlled one. The roles of NCs in RS properties are going to be discussed in detail in the following section.

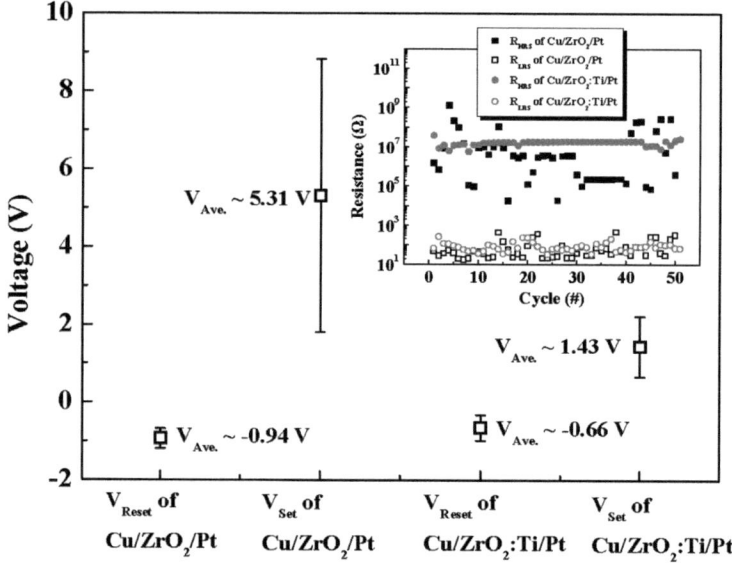

Figure 8.24 By incorporating Ti in a ZrO_2-based RRAM device improvement in the distribution of (a) SET/RESET voltages and (b) HRS/LRS states.

8.3 Applications of Nanocrystals in RRAM

In RRAM devices, NCs can be used for various functions. NCs can be deposited during the fabrication of the RS stack, or they can be generated while operating the RRAM devices [120]. In any situation, NCs can greatly affect the RS properties. In the deposited format, several materials, such as Ru NC and IrO_x NC, have been reported.

8.3.1 Improvement of Electrical Performance

The electrical behavior of RRAM devices can be affected by the presence of NC material. Several deposition techniques have been reported for fabricating NC materials. IrO_x NCs were deposited by sputtering in an $IrO_x/AlO_x/IrO_x$-NC$/AlO_x/WO_x/W$ structure [121], and Ru NCs were deposited using atomic layer deposition (ALD) in a $Pt/TiO_2/Ru$-NC$/TiO_2/Pt$ structure [122]. In a $Pt/TiO_2/Pt$-NC$/TiO_2/Pt$ structure [123], Pt NCs were grown using a reduction process from an ultrathin PtO_x layer. In spite of different fabrication processes, all these NC RRAM devices were showing improvements in the electrical properties.

8.3.1.1 Forming process

The forming process is the basic requirement of RRAM devices. It is widely accepted that the RS processes in RRAM devices are mainly conducted by the formation of filament. Hence, the device forming process has a great impact on the electrical performance and reliability issues. Especially when the device is scaled down to subnanometer range and only a single filament governs the complete RS event, the geometry of the filament has a special impact on the repeated RS operation. The forming process is dependent on many parameters, such as operating conditions, temperatures, insulating layer thickness, and capping metal layer thickness. The effect of insulator and capping metal thickness in the case of HfO_2/Hf RRAM is shown in Fig. 8.25 [124]. A thicker Hf capping layer will reduce the forming voltage. Moreover, forming free RRAM is also possible with scaling of an insulating layer below 2.5 nm. Along with several other conditions, the device-to-device forming process can be improved by incorporating of NC in the insulating layer.

   The importance of NCs can be easily understood using the RRAM stacks as shown in Fig. 8.26a. A negative voltage on top in IrO_x NC RRAM can break the purity of the stack with a soft breakdown phenomenon. Improvement in device forming as compared to pure Al_2O_3 RRAM is shown in Fig. 8.26b [99]. A device with NCs has a lower forming voltage due to a higher charge-trapping density as compared to the device without NCs. A similar improvement in the formation process has been reported for TaN/Al_2O_3:Ru NCs:$Al_2O_3/$ Pt RRAM by Chen et al. [117].

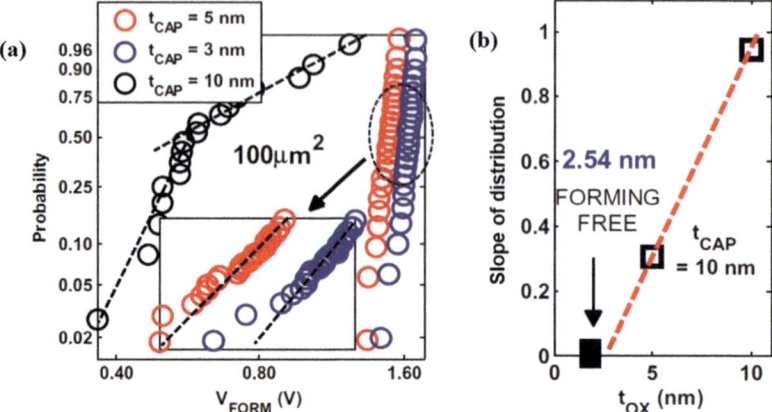

Figure 8.25 (a) Statistical distribution of HfO$_x$ RRAM with different Hf capping layer thicknesses. (b) The forming voltages decrease with decreasing insulating oxide layer thickness. A very thin layer (~2.5 nm) of HfO$_x$ RRAM can show a forming-free nature. Reprinted from Ref. [124], Copyright (2013), with permission from Elsevier.

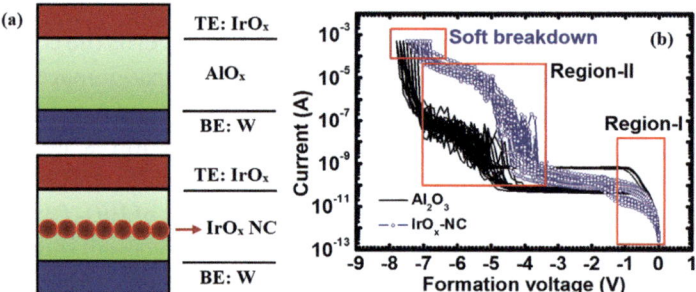

Figure 8.26 (a) Schematic design of AlO$_x$ RRAM and IrO$_x$ NC incorporated AlO$_x$ RRAM. (b) Improvement in device forming [99].

8.3.1.2   SET/RESET operation

A comparison of *I–V* switching with and without NC-based RRAM [99] is shown in Fig. 8.27a. The presence of NC in RRAM may decrease the SET/RESET voltages, or it may not, depending on the material and structure design. For example, IrO$_x$ NCs are less effective in decreasing the SET/RESET voltages as compared to the Al$_2$O$_3$ RRAM. But Ru NCs can effectively decrease the SET/RESET voltages [117]. The ratio of HRS to LRS can be varied according to the NC material and the RRAM structure. The presence of Pt NCs in TiO_2/Pt-NC/ TiO_2 RRAM does not affect the resistance ratio as compared to their presence in TiO_2 RRAM [122]. On the other hand, the resistance ratio decreases in Al_2O_3/IrO_x-NC/Al_2O_3 RRAM and increases in $Al_2O_3/$ Ru-NC/Al_2O_3 RRAM as compared to the Al_2O_3 devices. Therefore, the SET/RESET voltages and resistance ratio may or may not be influenced by the presence of NCs in the RRAM. But in all of these NC-based RRAM devices the RS variabilities have improved a lot. Highly uniform distributions of SET/RESET voltages (Fig. 8.27b) and HRS/LRS resistances (Fig. 8.27c) have been achieved for IrO_x-NC-based RRAM.

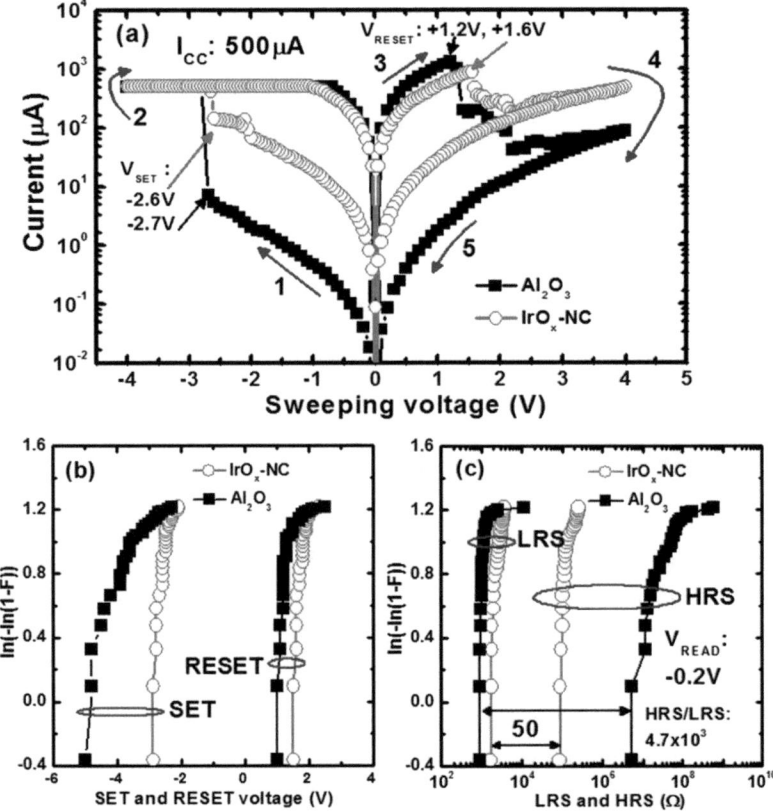

Figure 8.27 Improved characteristics of IrO_x NC RRAM. Improvement has been achieved for (a) stable RS operation, (b) device-to-device SET/RESET voltage distribution, and (c) HRS/LRS states [99].

8.3.1.3 Reliability of RRAM devices

For application as NVM, endurance and retention properties are the most important reliability parameters, which potentially can define the quality of a memory device, whether it is useful or not. Endurance is the sufferance ability of the device with longer program/erase capability. Retention is the ability to retain the stored information for a certain amount of time. Improvement of device endurance property has been reported with the use IrO_x NCs in RRAM, as shown in Fig. 8.28a. The presence of the NC material in-between the insulating layer can confine the filament in a particular position, which shows a much stable cycle-to-cycle uniformity as compared to pure Al_2O_3 RRAM. Poor retention of the TiO_2 RRAM has been reported in the literature [123]. But remarkable improvement has been achieved by embedding Pt NCs in between TiO_2 layers, as shown in Fig. 8.28b. The CF formed by the local enhancement of the electrical field by embedding Pt NC can keep the stored information for longer time.

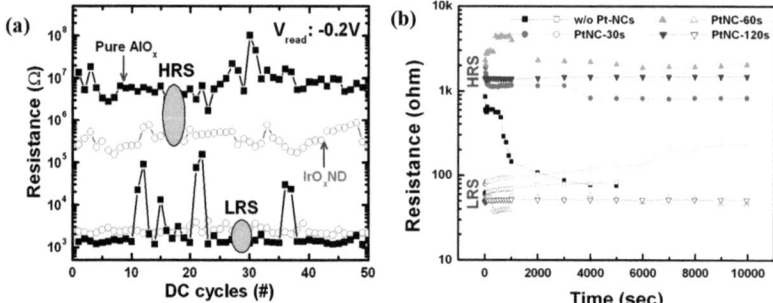

Figure 8.28 Reliability improvement in the NC RRAM devices in (a) endurance [99] and in (b) retention (Reprinted from Ref. [123], with the permission of AIP Publishing).

8.3.2 Conductive Filament Formation Based on Nanocrystal Migration

NCs can be generated in RRAM devices during the RS process under biased conditions. Ag NC formation during the RS process has been observed in several RRAM devices. Xu et al. [90] have reported that in an Ag/Ag$_2$S/W structure, an ionic electronic conductive channel wss formed by both the Ag$_2$S argentite phase and the Ag NCs. Before the RS process took place, the Ag$_2$S maintained an acanthite nonconductive phase. When the Ag electrode was positively biased, the nonconductive acanthite phase transformed to a conducting argentite phase and Ag$^+$ cations started to migrate toward the W electrode (Fig. 8.29). The Ag$^+$ cations mainly reduced at the cathode interface. Therefore, Ag NCs started to grow from the cathode.

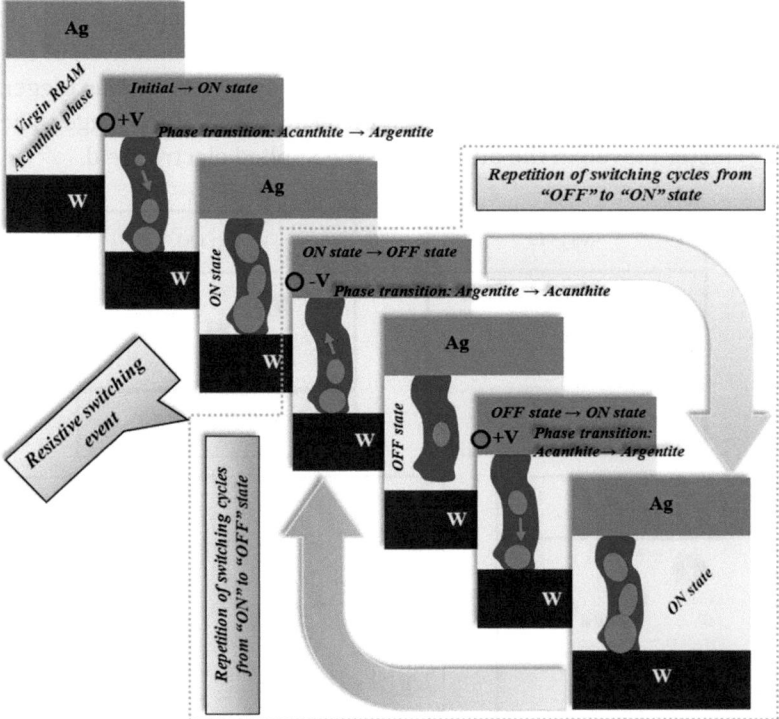

Figure 8.29 Schematic illustration of the switching mechanism based on Ag NC migration in Ag/Ag$_2$S/W RRAM.

Once the complete conductive channel is connected between two electrodes the resistance will drop. Applying a negative bias on the Ag electrode, the argentite phase transforms back to the acanthite phase and the Ag$^+$ cations move toward the anode. At some point the complete device is turned off. Since the device is in the OFF state now, Ag$^+$ cations stop moving and some of the Ag NCs stay inside, in a locally Ag-rich environment. Tian et al. [125] reported the formation of Ag-NC-based CF in the Ag/SiO$_2$/p$^+$-Si structure.

The formation of the Ag-NC-based CF was in the direction of the cathode from the anode. However, originally solid-electrolyte-based ECM devices usually show the CF from cathode to anode. For a better understanding of the driving force of the electrochemical reaction, another structure with Ag NCs in a SiO$_2$ matrix was fabricated. The Ag mass transfer process can be easily understood by this method, as shown in Fig. 8.30. The formation of an Ag cluster in the SiO$_2$ matrix always maintains a nanogap between neighboring clusters. The time-dependent measurement shows a size modification of the Ag cluster. Initially the Ag cluster near the Ag electrode was bigger, but during the process the Ag cluster closer to p$^+$-Si grew bigger. The Ag cluster moved in the direction of the applied electric field.

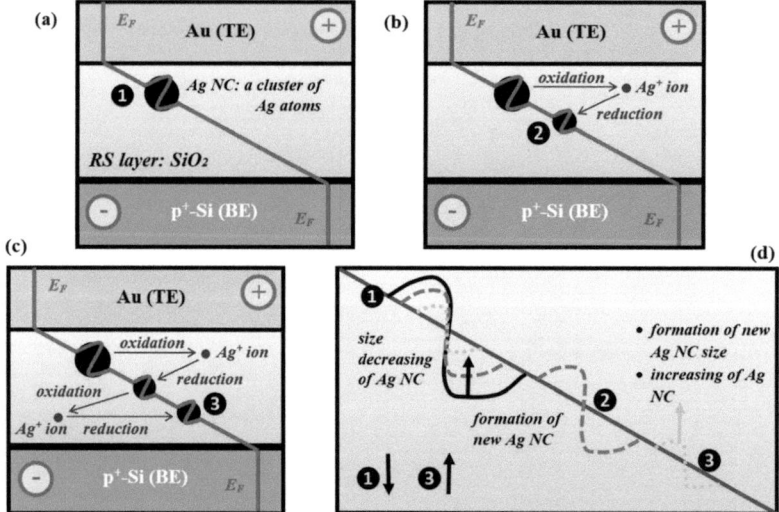

Figure 8.30 (a) The schematic image of the Au NC in the Ag/SiO$_2$/p$^+$-Si structure. Schematic illustration of the formation of (b) first new cluster ① and (c) the second cluster ③ from ② by a mass transfer process. (d) The schematic Fermi level illustration of the formation of new Ag NC and its growth by mass transfer. The increasing size of ③ will decrease the size of ①.

The formation of a Ag NC chain has also been reported for the organic RRAM devices. Gao et al. [126] reported Ag NC CFs in a Ag/ poly(3,4-ethylene-dioxythiophene):   poly(styrene-sulfonate)/Pt planar device. In this case several Ag-NC-based CFs formed during the RS process. Most interestingly, the nucleation of CFs starts from a middle region of the organic layer. In the four different samples (prepared in the same conditions with the same structure), a constant voltage of +1 V was applied for different periods (2 s, 4 s, 10 s, and 130 s). The corresponding currents for the four different times are shown in Figs. 8.31a–8.31d. The device resistance gradually decreases with the increase of time. The scanning electron microscopy (SEM) images corresponding to Figs. 8.31a–8.31d are shown in Figs. 8.31e–8.31h. From the SEM images it can be easily observed that the Ag NCs initially formed in the middle region of the organic material. With the increasing voltage stress time, more and more Ag NCs can emerge together and increase the size of the chain. Depending on the time the Ag NC chain will make a conducting path between Ag and Pt electrodes. Once the path becomes stronger, then the current will flow through it and the device will switch from a HRS to a LRS.

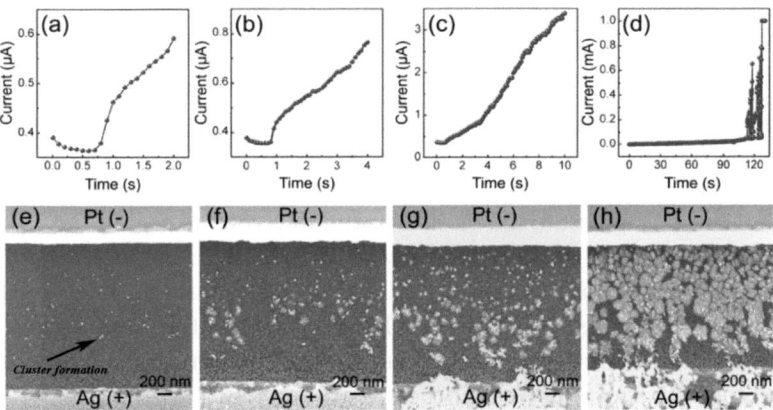

Figure 8.31 (a–d) Current-time characteristics of the pristine Ag/PEDOT:PSS/ Pt RRAM devices (applied voltage of 1 V at different periods of time). (e–h) The corresponding scanning electron microscopic images. Reprinted from Ref. [126], with the permission of AIP Publishing.

8.3.3 Charge Trapping Using NC

The presence of NCs in the RS stack can enhance the charge trapping during the switching process. The NC materials can be located in the middle of the insulating layer or be distributed (Fig. 8.7). The insulating materials can be an inorganic form, such as ZnO [127], Al_2O_3 [128], or TiO_2, or an organic form, such as Alq_3 [129]. The most common NCs used in RRAM devices are Ru, IrO_x, Cu, and Al. Generally during the fabrication process a nanolayer 2–10 nm thick is deposited. During the fabrication process the nanolayer will be transformed into NCs. The Cu-NC-based RS behavior in a Cu/ZnO/Cu/ZnO/Pt RRAM cell [127] shows typical bipolar *I–V* switching, as shown in Fig. 8.32a. The conduction behavior and the switching process can be easily understood by replotting the positive part on a double-logarithmic scale, as shown in Fig. 8.32b.

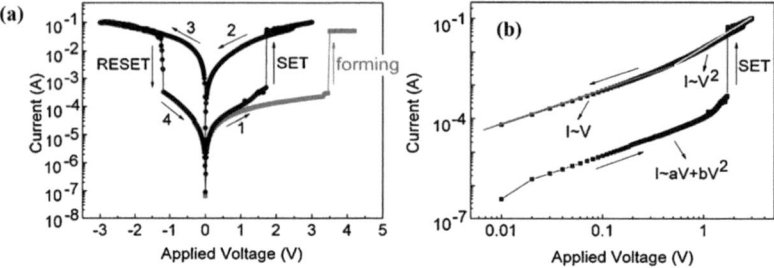

Figure 8.32 (a) RS characteristics and (b) conduction mechanism of Cu/ZnO/Cu/ZnO/Pt RRAM devices. Reprinted from Ref. [127], with the permission of AIP Publishing.

The fitting result shows that in the low-voltage region the HRS follows a linear ohmic nature, with an additional quadratic term at a higher voltage, that is $I \approx aV + bV^2$. This indicates the typical nature of an insulating material, with shallow traps and space-charge-limited current (SCLC) injection. However, the conduction in the LRS shows two distinct regions. The first part fits well with the ohmic principle $(I \approx V)$, followed by a second part, which fits with Child's law region $(I \approx V^2)$, which can be explained by the trap-filled SCLC. Therefore, the bipolar switching can be explained by the trap-controlled SCLC model. During positive voltage application the charge-trapping process leads to the SET operation, and under a negative bias the charge detrapping process leads to the RESET operation. The charge-trapping/detrapping process can be enhanced by random incorporation of NCs such as metal NCs (e.g.,

Au and Ag), semiconductor NCs (e.g., ZnO, ZnS, and CdSe NCs), and graphene quantum dots [130]. Most of the cases show a RS nature similar to what we have discussed earlier. But the charge-trapping/detrapping process in randomly distributed atom-level traps can significantly change the device switching. One type of atom-level traps is intrinsic crystal defects such as vacancies. This type of RRAM shows a conventional bipolar behavior, which can be well explained by the trap-controlled SCLC model [42, 131].

In 1958, Anderson [132] distinguished random insulators and conductors on the basis of the "diffusion" distance ξ for electrons at 0 K. Usually insulators have a finite value of ξ and conductors have an infinite value of ξ. The nanometallic films used by Chen et al. [133] were a mixed solution of insulators doped with electronic conductors. The insulator:conductor solutions were SiO_2:Pt and Si_3N_4:Pt. Other solid solutions were prepared using one of two excellent insulators, $LaAlO_3$ (LAO) and $CaZrO_3$ (CZO), and doped with small amounts of one of two electron conductors, $LaNiO_3$ (LNO) or $SrRuO_3$ (SRO). The combination gave four different solutions: $LaAlO_3$:$LaNiO_3$, $LaAlO_3$:$SrRuO_3$, $CaZrO_3$:$LaNiO_3$, and $CaZrO_3$:$SrRuO_3$. For simplicity, we focused more on the SiO_2:Pt system. Because the lateral percolation limitation of the SiO_2:Pt is $f \approx 0.38$ (f: molar fraction of Pt in the SiO_2:Pt), the SiO_2:0.2Pt must have a finite ξ. The value can be determined by measuring the resistance of the film with different thicknesses δ. If the film thickness $\delta << \xi$, it is metallic across the thickness, and if $\delta >> \xi$, the film is insulating. Figure 8.33a shows the initial resistance–voltage characteristics of the Pt/SiO_2:0.2Pt/Mo RRAM cells with various layer thicknesses. It is easily understood that the conductivity of the structure is dependent on the SiO_2:0.2Pt thickness. A thinner film shows a more metallic nature compared to a thicker film. The film becomes nonohmic when the thickness reaches 16 nm. Therefore, ξ is between 16 nm and 21 nm. Within the thickness range of 7 nm to 16 nm, a peculiar bipolar RS can be observed where the resistance states are not as stable as when the film is 21 nm thick. Ultraviolet (UV) irradiation was able to switch the Pt/SiO_2:Pt/Mo cell from a HRS to a LRS, as shown in Fig. 8.33b. All of the phenomena suggest a charge-trapping/detrapping mechanism.

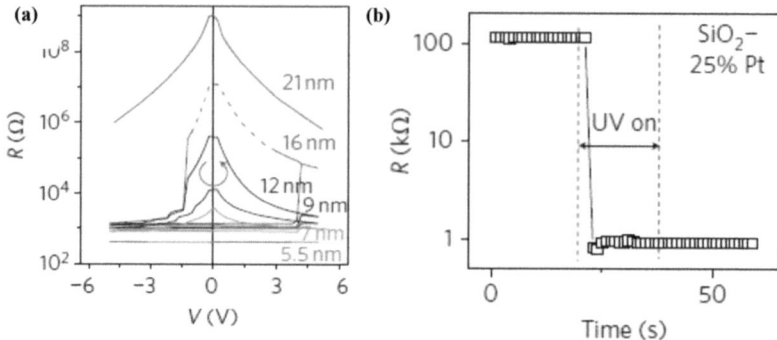

Figure 8.33 (a) The resistance–voltage characteristics of a Pt/insulator:conductor/Mo RRAM structure. The SiO_2:0.2 Pt was used for the insulator:conductor layer of various thicknesses. When the diffusion distance ≈ the film thickness, the RRAM shows a NVM nature. (b) The UV light effect on resistance states for 20 nm SiO_2:0.25 Pt RRAM devices. Reprinted by permission from Macmillan Publishers Ltd: *Nature Nanotechnology* (Ref. [133]), copyright (2011).

During the fabrication, the structure was designed in such way that the TE (Pt) should maintain a higher work function value as compared to the BE. Hence the electrode with a lower work function will inject electrons by FN tunneling and the transition from a LRS to a HRS will take place. Firstly, the injected electrons are trapped at sites near the original conducting pathways. Secondly, the trapped electrons provide Coulomb repulsion force to block further electron transport along the same. Due to a thin SiO_2:0.2Pt layer (5 nm) in a Pt/SiO_2:0.2Pt(5 nm)/Mo RRAM cell, the number of trapped electrons are insufficient to block further electron transport through the same pathway. Under a reverse bias condition, successful detrapping of those trapped electrons can generate a series of intermediate states, which is very useful for multilevel storage applications.

8.3.4 Threshold Switching to Memory Switching

Volatile threshold switching (TS) and nonvolatile memory switching (MS) are two types of RS phenomena in RRAM devices.

In TS, the LRS cannot be maintained after the voltage is removed. Therefore, the resistance behavior in TS mode is of the volatile type. Typical $I-V$ characteristics of positive TS and negative TS are shown in Fig. 8.34. The voltage below the threshold voltage (V_{Th}) always maintains the device in the HRS, but once the voltage exceeds V_{Th} the HRS will switch to a LRS. A sufficient amount of voltage is necessary to hold the LRS position. The hold voltage (V_{hold}) can vary according to the material and structure design. Interestingly, coexistence of TS and MS has been observed in several RRAM devices. Let us discuss the existence of TS and MS in a VCM-type $IrO_x/AlO_x/TiN$ RRAM cell [97].

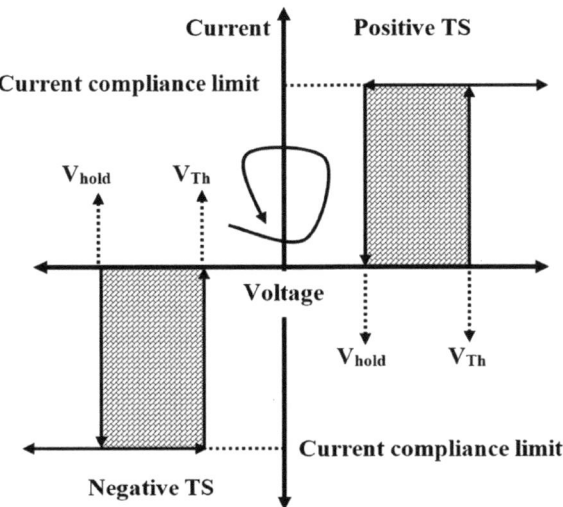

Figure 8.34 Schematic *I–V* characteristics of TS in a RRAM device.

When the device is in MS mode, a SET operation with a I_{CC} of 100 µA changes the HRS (MS-OFF) to a LRS (MS-ON) when a positive voltage is applied on top of IrO_x. A reverse polar bias can reset the device and make a transition from MS-ON to MS-OFF. In this case, MS-ON fits well with the SCLC model. In the same structure, TS was also possible at a I_{CC} of 10 nA. The device was switched from TS-OFF to TS-ON with a voltage change from $0 \rightarrow +V$. But a $V_{hold} > 2$ V was necessary to hold the TS-ON. The difference between MS and TS can be easily understood by dividing the resistance into four parts:

- Initial resistance of the device
- Resistance after the forming process
- Resistance after the first RESET process
- Resistance after the normal SET operation

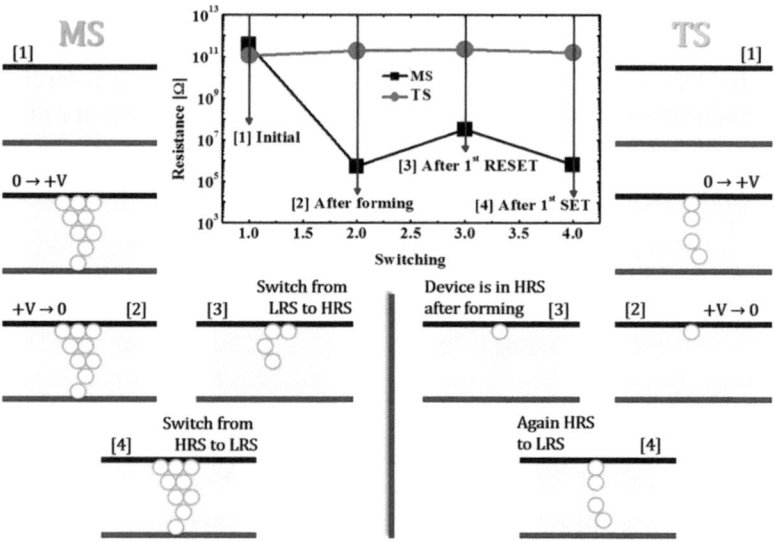

Figure 8.35 Change in the resistance during MS and TS operation in IrO$_x$/AlO$_x$/ TiN RRAM. ① → ④ shows LRS and HRS switching mechanisms for both.

Figure 8.35 shows the variation in resistances under MS mode and TS mode. To understand the differences between MS and TS, one should measure the resistances at a read voltage that is lower than V_{hold} for TS. If the read voltage is greater than V_{hold} then the device under both MS mode and TS mode will be in a LRS. The clear formation of a continuous V_O filament will conduct the RS process (MS ① → ④) at a high I_{CC} of 100 μA, as shown in Fig. 8.35. With a positive bias applied, that is, $0 \rightarrow +V \rightarrow 0$, a filament will be formed (① → ②). With the application of a negative bias, that is, $0 \rightarrow -V \rightarrow 0$, the device will switch from a LRS to a HRS ③. The next RS event will continue with the application bias, that is, $0 \rightarrow +V \rightarrow 0 \rightarrow 0 \rightarrow -V \rightarrow 0$. A completely different scenario will develop when the same structure is measured at 10 nA (TS ① → ④). With applied bias from $0 \rightarrow +V$ TS-ON transition is possible when the $+V$ reaches $+V_{Th}$. Before the TS-ON transition, at such a low current the device displays Schottky conduction. A high electric field (E) > 10^8 V/m guides the tunneling process and TS-ON displays FN tunneling. The high field generates point defects or trap sites in the AlO$_x$ matrix. In the direction $+V \rightarrow 0$ ②, the flow of electron will be stopped when the decreasing E increases the height of the potential barrier. Therefore, if $V < V_{hold}$ TS-ON will switch to TS-OFF. The number of trap sites will increase with increasing I_{CC}; hence TS is dependent on I_{CC} and it is more stable at a lower I_{CC} than at a higher I_{CC} (Figs. 8.36a and 8.36b).

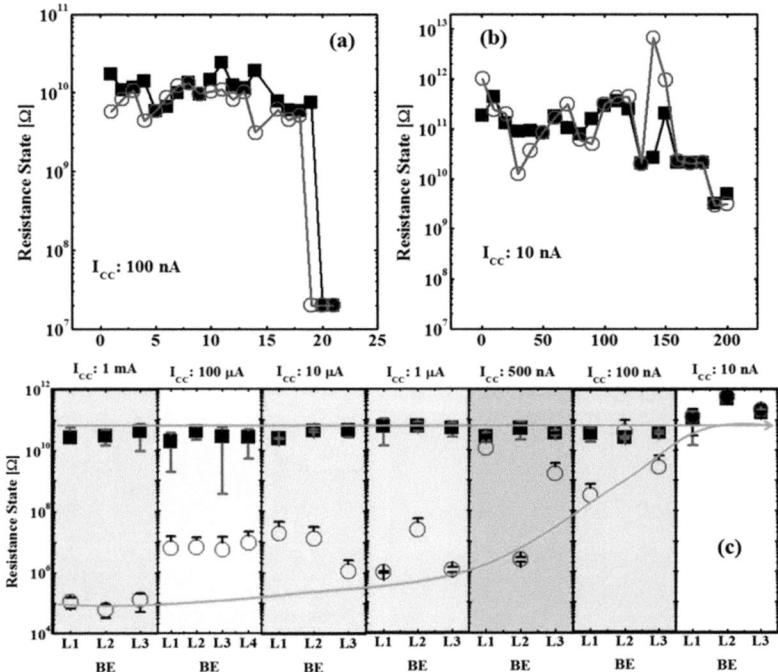

Figure 8.36 The I_{CC}-dependent endurance at (a) 100 nA and (b) 10 nA. The TS shows a better performance at low compliances. (c) Change of the LRS with I_{CC}. The lower the I_{CC}, the higher the resistance of the LRS. The switching is unidirectional, from TS to MS. The metal layers of 3D vertically stacked RRAM are represented by L1, L2, L3, and L4.

However, once a sufficient amount of trap sites is generated within the matrix then it is sometime hard to obtain TS after a successful MS in the same device. As shown in Fig. 8.36c, for AlO_x VCM devices the conversion between TS and MS is a unidirectional process. Fortunately, for ECM RRAM devices, the transformation of TS to MS can be controlled by controlling NC generation. The experimental results represent TS to MS transformation of a 3D vertical RRAM device. The device was fabricated with four-layer-stacked metal electrodes. All devices of the 3D stack show similar results.

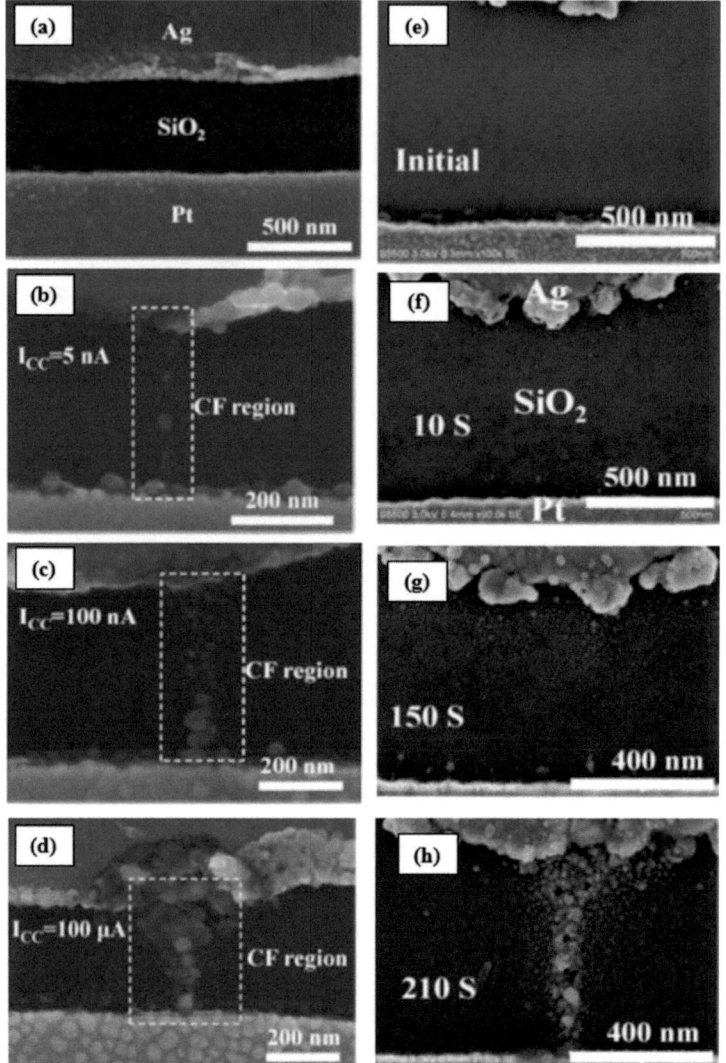

Figure 8.37 The I_{CC}-dependent CF observed by TEM in (a) a Ag/SiO$_2$/Pt RRAM structure with (b) 5 nA, (c) 100 nA, and (d) 100 μA. SEM images of the CF growth dynamic in (a) fresh device and after (b) 10 s, (c) 150 s, and (d) 210 s (V = +30 V; I_{CC} = 10 μA). From Ref. [134]. Copyright © 2014. Reproduced with permission of John Wiley & Sons.

The conversion between TS and MS was proved by Sun et al. [134] with direct experimental evidence. The device was a Ag/SiO$_2$/

Pt-based ECM cell. The RS mechanism and also the conversion from TS to MS are based on the formation of Ag NCs. In a similar way to the VCM devices, I_{CC}-dependent transition from TS to MS has been achieved for Ag/SiO$_2$/Pt ECM devices. A lower I_{CC} shows TS, while a higher I_{CC} shows MS. The formation of a Ag-NC-based CF is shown in Fig. 8.37.

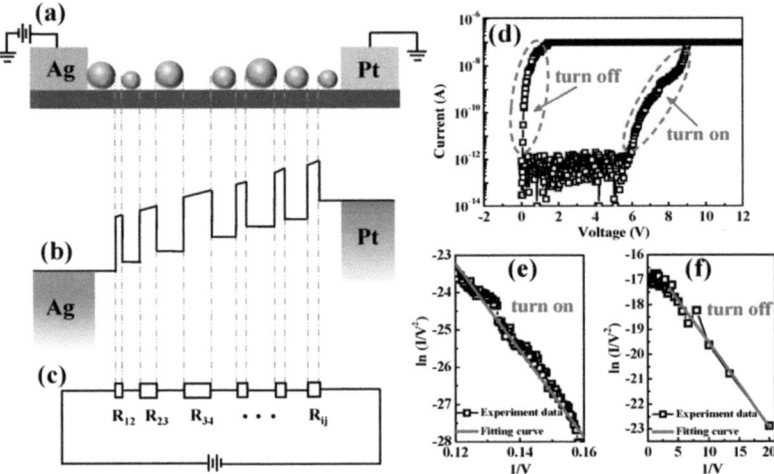

Figure 8.38 Schematic illustration of (a) a CF consisting of isolated Ag NCs, (b) the corresponding band diagram, and (c) the equivalent circuit. (d) A typical *I–V* curve in TS mode. The tunneling mechanism is fitted well in the (e) turn-on and (f) turn-off regions. From Ref. [134]. Copyright © 2014. Reproduced with permission of John Wiley & Sons.

Isolated Ag NCs were formed within the interelectrode region with a low I_{CC} (<5 nA). Increasing the I_{CC} from 5 nA to 100 nA widened the Ag NC region and decreased the gap between NCs. On increasing the I_{CC} to 100 μA, a continuous CF was formed between both electrodes. A constant voltage stress method was used to understand the dynamic growth of the Ag NC chain. A constant voltage of +30 V with a I_{CC} of 10 μA was applied with a variation of time. The growth dynamics are shown in Fig. 8.37e–h. At 10 s some Ag NCs were generated near the Ag electrode, but the formation of NCs was much clearer at 150 s. Finally, a continuous Ag NC chain between two electrodes was formed after 210 s. Although TS and MS are based on the formation of a Ag NC chain the MS mode and

TS mode have different conduction mechanisms. The LRS in MS mode perfectly displays linear ohmic conduction, which indicates that the current transportation was through a metal-like continuous conductive path formation. Instead of a continuous CF, physically isolated Ag NCs are present in the TS mode. Figure 8.38a–c shows the schematic illustration of the physically isolated Ag NC chain with a corresponding energy diagram and an equivalent circuit. The tunnel junction was formed between the gap of each pair of adjacent Ag NCs. If you apply enough voltage, the charge carriers can tunnel through those junctions. As shown in Figs. 8.38d to 8.38f, both TS-ON and TS-OFF states fit the tunneling mechanism. As the equivalent circuit shows, the total resistance (R) of the device can be described as $R = \Sigma R_{ij} = V/I$, where R_{ij} can be defined as the tunneling resistance of the corresponding tunneling junction.

8.4 Nanocrystals as the Seed Layer in RRAM

Although the performance of RRAM devices improves a lot, the new technology suffers from high randomness in the formation and rupture of CFs and a large distribution of switching parameters such as RS voltages and different resistance states. Several methods have been explored to improve device performance, such as structure modification, optimization of operating conditions, and interface engineering. Nanostructures are also involved in this development. NCs, nanowires, and nanopyramids in RRAM devices can improve the RS performance in the real sense. Nanostructures can improve the controllability of the CFs and hence a performance improvement can be observed in RRAM devices.

8.4.1 Effect of Nanocrystals in the Resistive Switching Layer

The presence of NCs in the RS layer can significantly influence device characteristics. In RRAM devices NCs can serve as a complete colloidal NC switching layer. They may be present in some places in the RS layer, such as near the TE, near the BE, or as a middle layer, and NCs may also be distributed randomly in the RS layer (as illustrated in Fig. 8.7). The electrode materials also have great influence on NC-based RRAM devices.

8.4.1.1 Colloidal nanocrystals as the switching layer

Several studies show that the assembly of colloidal NCs exhibits distinguishable RS characteristics. Due to the solution-based fabrication process the colloidal NC layer is very cost effective. For flexible applications RRAM based on solution-processed colloidal NCs is in demand. By changing the chemical and synthesis process, the size, density, and structure of NCs can be modified. As an example, a RRAM device structure based on a solution-processed colloidal γ-Fe$_2$O$_3$ NC layer is shown in Fig. 8.39 [135]. The chemical synthesis of a γ-Fe$_2$O$_3$ NC layer was done by adding Fe(CO)$_5$ (0.6 ml) in a mixture of oleic acid (4.2 ml) and octadecene (30 ml). The solution of oleic acid and octadecene was prepared at room temperature. The added Fe(CO)$_5$ was stirred for 10 min. at 100°C. The residual oxidation in octadecene thermal decomposition of Fe(CO)$_5$ formed γ-Fe$_2$O$_3$ NCs. The surface of the NCs was encapsulated by oleic acid, which prevents the aggregation of NCs in the solution. Afterward, the solution was cooled to room temperature and washed with ethanol and the NCs were dispersed in hexane. The γ-Fe$_2$O$_3$ NCs are hydrophobic and well dispersed in a nonpolar solvent, such as hexane. The thickness of the γ-Fe$_2$O$_3$ NC assembly was found to be 150 nm. However, the switching mechanism of this kind of devices is not clear yet.

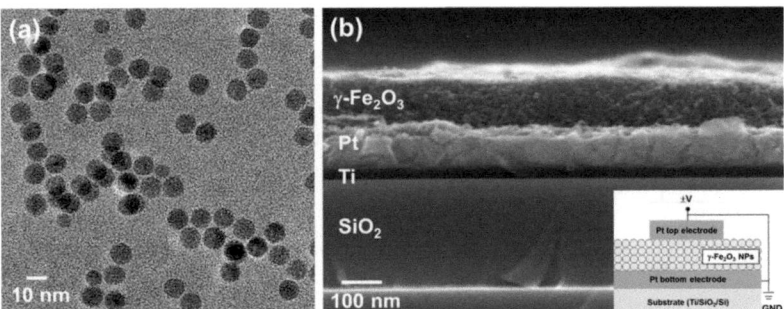

Figure 8.39 (a) Observation of γ-Fe$_2$O$_3$ NCs by TEM analysis and (b) a fabricated RRAM device with a colloidal NC layer. Reprinted from Ref. [135], with the permission of AIP Publishing.

One possible explanation is that the switching mechanism is due to the transition between the higher-resistive γ-Fe$_2$O$_3$ (resistivity > 10^2 $\Omega\cdot$cm) and lower-resistive Fe$_3$O$_4$ (resistivity $\approx$ 6.5 m$\Omega\cdot$cm) by the redox process (3γ-Fe$_2$O$_3$ + $2e^-$ $\leftrightarrow$ 2Fe$_3$O$_4$ + O^{2-}). The γ-Fe$_2$O$_3$ NC layer has a porous structure with a large interface area between adjacent NCs. The pores can act as a reservoir of atoms (ions) and vacancies. In devices of this kind, the RS process depends on several factors, such as stoichiometry, microstructure, vacancy concentration, and defect state. Therefore, depending on the film preparation the nature of the RS process may change. This kind of RRAM structure is also affected by the selection of the electrode material. For example, Au/Ni-Au NC/ITO and Au/Ni-Au NC/NSTO devices have been reported by Zhong et al. [136]. As Ni can be easily oxidized when deposited in air, hence the surface oxidation will create an interface of an NiO$_x$ layer and, consequently, some oxygen vacancies will be created. Due to the presence of Ni^{2+} vacancies, NiO$_x$ is an intrinsic p-type semiconductor and has a complex, highly correlated electronic structure with a bandgap of 3.6 eV and a work function of 5.0 eV. As the work function of Au is 5.1 eV, the contact of Au/NiO$_x$ is ohmic. At the same time, both Nb-doped SrTiO$_3$ (NSTO) and ITO are heavily doped n-type semiconductors, whose energy gaps are approximately 3.2 eV and 3.5 eV, respectively, and the work functions of NSTO and ITO are 4.2 eV and 4.7 eV, respectively. An illustration of the band diagrams is shown in Fig. 8.40.

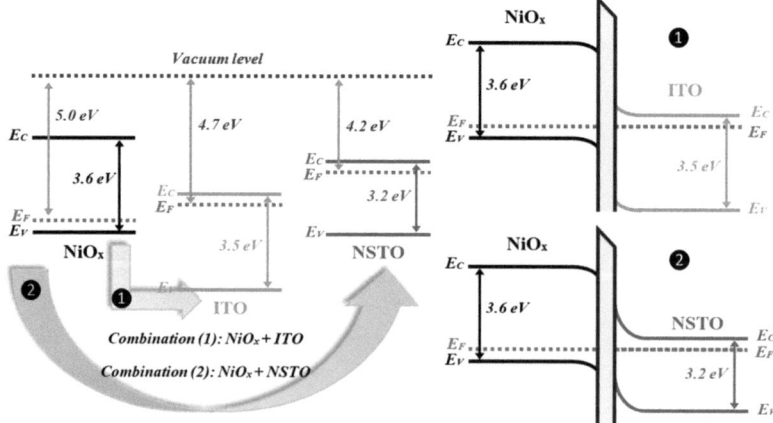

Figure 8.40 Schematic band diagrams for ① a NiO$_x$ + ITO system and ② a NiO$_x$ + NSTO system.

It is clear from ② that the higher diffusion potential at the Ni-Au NC/NSTO interface will introduce rectifying properties whereas the lower diffusion potential in the Ni-Au NC/ITO interface at ① is beneficial to the formation of a CF. The different interface structures and defects using different electrode materials will result in different trapping effects, which will directly affect the electrical properties of the devices.

8.4.1.2 Local electric field enhancement with nanocrystals

In the NC-based RRAM devices, the position of the NC layer, the size of the NCs, and the deposition technique are also very important. To give a clear picture about those parameters we will discuss a Ru-NC-based RRAM device with a Pt/TiO$_2$/Ru-NC/TiO$_2$/Pt structure [137]. The relation between the HRS current, the SET voltage, the RESET power, and the RESET voltage for Ru NC RRAM deposited using ALD is very dependent on the position of the NC layer and the ALD cycles, as shown in Fig. 8.41.

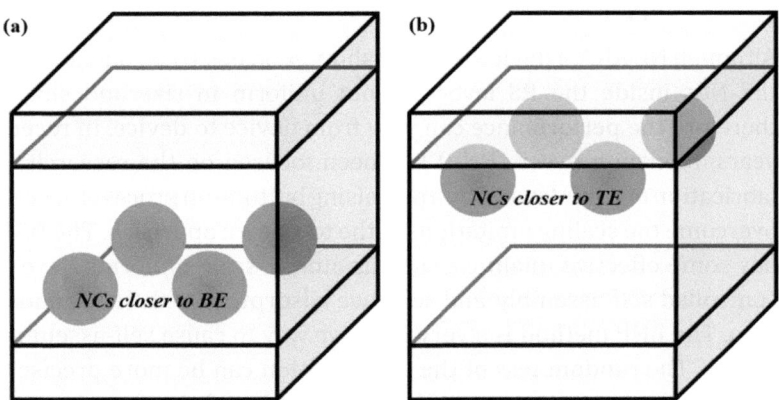

Figure 8.41 Schematic illustration of (a) bottom-Ru-NC-based TiO$_2$ RRAM and (b) top-Ru-NC-based TiO$_2$ RRAM.

The device parameters can show a large variation for the simple Pt/TiO$_2$/Pt structure. But the distribution can be controlled well by introducing Ru NCs in the TiO$_2$ layer. One hundred fifty ALD cycles are better for the fabrication of the bottom Ru NC layer, and 250 ALD cycles are better for the fabrication of the top Ru NC layer. Hence, an opposite trend can be found at two different Ru NC positions. For the bottom Ru NC RRAM the parameter uniformity is better with NCs of a smaller size; however, for the top Ru NC RRAM uniformity is much better with NCs of a larger size. In these RRAM devices, CFs were formed due to the electric field enhancement by the Ru NCs. A major factor that controls the CF formation is the carrier injection at the BE interface. For the bottom Ru NCs the higher field concentration is limited to the region where the Ru NCs reside. The nominal applied electric field across devices was ~1 MV/cm. The electric field for the bottom Ru NC RRAM was ~10 MV/cm, whereas the maximum field for the top Ru NC RRAM was almost 10 times lower. In the case of nonuniformly distributed NCs, the CF grows in the denser NC area. Electric field enhancement is the major difference between RRAM with Ru NCs and RRAM without Ru NCs. An excess electric field due to NCs will form a stable CF as compared to the CF present in the non-Ru-NC RRAM devices. As a result, the performance will be improved by NCs.

8.4.1.3 Formation of homogeneous NCs for RRAM applications

Although NC RRAM devices usually show good electrical phenomena the NCs inside the RS layer are not uniform in size and shape, therefore the performance can vary from device to device. In recent years, bionanoprocess (BNP) has been focused on the research of fabrication of nanodevices by a promising bottom-up process that can overcome the scaling limitations of the top-down approach. The BNP has some effective qualities, such as atomic scale uniformity, well-controlled self-assembly, and selective adsorption to the designated area. The BNP method is a very effective way to cause self-assembly of NCs. The randomness of the CF formation can be more precisely controlled by homogeneous NCs, which can be fabricated using this method. Here, we are going to discuss the fabrication of the Pt NCs in the cavity of cage-shaped ferritin protein. Ferritin protein has a spherical cage structure with an inner diameter of 7 nm and an outer diameter of 12 nm, as shown in Fig. 8.42a.

By the biomineralization method, a variety of inorganic NCs, like Fe oxide, Co oxide, and Ni oxide can be crystallized inside their vacant cavities using ferritin. The best biomineralization condition for the PtS NCs in the ferritin cavity is possible by applying a slow chemical reaction system. As the ferritin cavity works in a spatially restrictive chemical synthesis chamber, the PtS NCs synthesized in the cavity are the same size as the cavity.

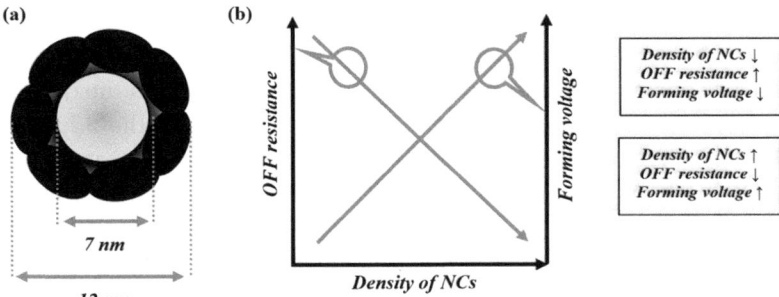

(a)

7 nm

12 nm

(b)

OFF resistance

Forming voltage

Density of NCs

Density of NCs ↓
OFF resistance ↑
Forming voltage ↓

Density of NCs ↑
OFF resistance ↓
Forming voltage ↑

Figure 8.42 (a) Schematic image of ferritin with a PtS core. (b) With the increase of NC density the forming voltage will increase and the OFF resistance will decrease.

In brief, the PtS core within the ferritin cavity was synthesized using a solution of 0.5 mg ml^{-1} apoferritin, 2 mM K_2PtCl_4, 2.6 mM $(NH_2)2CS$, and 100 mM sodium phosphate buffer (SPB) pH 8.0. A RRAM device with uniform BNP Pt NCs was designed within the Pt/NiO/Pt stack [138]. A higher electric field can be observed below and above the Pt NC, whereas the region on the side is weakened. During the fabrication of the RRAM stack, a 3 nm thick NiO film was deposited by electron beam evaporation onto the substrate. Then, ferritin with a PtS core (concentration of 5 mg ml^{-1} in pure water) was dropped onto the NiO layer and kept for 10 min. After that, in order to remove the protein shell and convert PtS NCs into Pt NCs, UV/ozone treatment was carried out at 115°C for 50 min. The Pt NCs were densified on the NiO surface with a density of 10^3 μm^{-2}. To understand the effect of the Pt NCs on the RRAM performance, devices with different densities of Pt NCs were fabricated. The density of NCs was controlled by modifying the concentration of the ferritin solution on the NiO surface. After the fabrication process of the Pt NCs, the final 6 nm thick NiO layer was deposited. Although taking advantage of the BNP, controlled NC density has been achieved by the ferritin concentration, the distribution of the Pt NCs over the larger area was not uniform. The electrodes used in this case were large enough (20 × 20 μm^2) to identify the switching parameter dependence on the NC density. As shown in Fig. 8.42b, the forming voltage increases and the OFF resistance decreases with the increase of Pt NC density. The electric field around the NCs will affect the migration of V_O in the NiO film during the forming process. Since a lower density of NCs allows easier V_O migration over a longer distance and increases the concentration of V_O on the negative side of the NCs, the forming voltage is lower for RRAM devices with the lower density of Pt NCs. A different situation occurs with the denser NC system. Due to the effect of the electric field on the adjacent NCs, it becomes more difficult to increase the local concentration of V_O with an increasing density of NCs. A higher forming voltage is necessary to attract sufficient V_O from the vicinity of the NCs. It is easily understood that when the density of Pt NCs is high, many forming sites exist in the NiO film, which make the forming process more difficult.

The formation of NCs for a RRAM application is also possible by using porous templates. HfO_2 nanodots in $Au/HfO_2/Pt$ RRAM have been fabricated using a self-organized nanoporous anodized aluminum oxide (AAO) template [139]. Using nanoporous templates NCs can be easily designed by a chemical method or a physical method.

The porous structure of the AAO template was designed with pores of various sizes, from 25 nm to 95 nm, and the sizes were confirmed by field emission scanning electron microscopy (FESEM) measurements, as shown in Fig. 8.43a. Generally, templates of this kind can be designed with various geometries, such as templates with different pore sizes, different pore densities, and different lengths. As shown in Fig. 8.43b, the NC size and density are dependent on the nanoporous AAO template; hence the controllability of the NCs or nanodots is much better. The complete device fabrication process using an AAO template is shown in Fig. 8.43c.

Figure 8.44a shows an SEM image of the controlled distribution of HfO_2 nanodots. The size of the nanodots has been confirmed by the 2D view and 3D view, as shown in Fig. 8.44b and Fig. 8.44c, respectively, using atomic force microscopic (AFM). The thickness of the HfO_2 nanodots can be calculated from the thickness profile (Fig. 8.44d) along the AB axis. It is clear that using templates NCs of a uniform size distribution can be easily fabricated, which can play a vital role in RRAM stability. Therefore, in general, the performances of NC-based RRAM devices depend on several factors, but electric field enhancement is a common effect of using NCs in the structure.

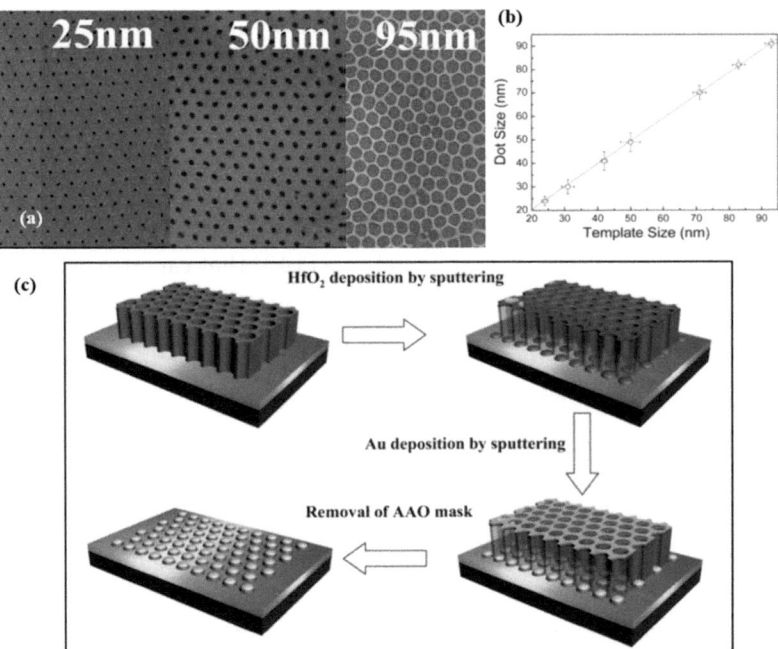

Figure 8.43 (a) Plan view of the FESEM images with pore sizes from 25 nm to 95 nm of the pore-size-controlled AAO nanotemplate. (b) The HfO$_2$ nanodot size is dependent on the pore size of the template. (c) The RRAM device fabrication process using the template. Reproduced from Ref. [139] with permission of The Royal Society of Chemistry.

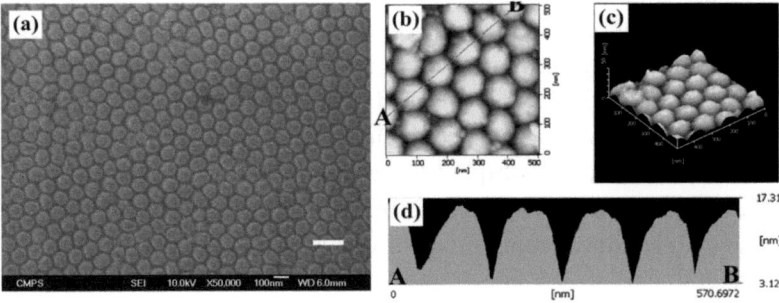

Figure 8.44 The SEM image of the array of (a) HfO$_2$ nanodots. The (b) 2D and (c) 3D AFM images of the HfO$_2$ nanodots, with (d) the thickness profile along AB. Reproduced from Ref. [139] with permission of The Royal Society of Chemistry.

8.4.2 Bottom Electrode Modification

The applications of NCs are not limited to the RS layer but have been extended to modify the BE structure also. The major advantage of the local electric field enhancement by the NC can be utilized well by using a NC-covered BE. Apart from the applications of NCs in the BE, there are nanopyramid BEs [140], nanopeak BEs [141], arc-shaped BEs [142], and several other structures [143] that can also improve the RS performance by enhancing the localized electric field on the surface of the electrode.

8.4.2.1 Nanocrystal-based bottom electrode

To manipulate the location and orientation of the controlled CF, NC BE RRAM has been studied [144]. The Cu-NC-based BE structure for a Ag/ZrO$_2$/Cu-NC/Pt RRAM device is shown in Fig. 8.45a. The NC-controlled CF growth process is very effective due to several reasons, for example, it reduces the randomness of the CF growth process, improves the device performance, and enables a deeper understanding of the microscopic mechanisms of the CF formation process. A dark, nanometer-sized CF region can be observed as a conducting bridge between the Ag TE and the Pt BE across the ZrO$_2$ layer, as shown in Fig. 8.45b, and a high-angle annular dark field image under scanning TEM mode in Fig. 8.45c. As an energy dispersive spectroscopy (EDS) analysis reveals the CF is directly connected to the Cu NC, as shown in Fig. 8.45d, which can be identified by the rightmost region in the relative atomic concentration plot. The shape of the nanobridge region is nearly cylindrical, which can be observed in solid-electrolyte-based RRAM. Therefore, it is clear that switching phenomena of such devices are dominated by formation and annihilation of multiple CFs. A composition analysis of the CF shows that Ag is more concentrated in the CF region and Ag is the primary elemental component of the Ag/ZrO$_2$/Cu-NC/Pt memory device.

The impact of the Cu NC on the electric field distribution in the ZrO$_2$ layer can be investigated by using the Matlab partial differential equation (PDE) tool and technology computer-aided design (TCAD). Cu NCs on a BE play an important role in strengthening the electric field around the NC tip in the ZrO_2 layer due to the larger curvature of the semisphere NC BE than the planar BE. As shown in Fig. 8.46, applying a positive voltage on the Ag TE, the electric field is directed toward the Cu NC and the intensity of electric field (length of line of red arrows) is also enhanced around the NC location.

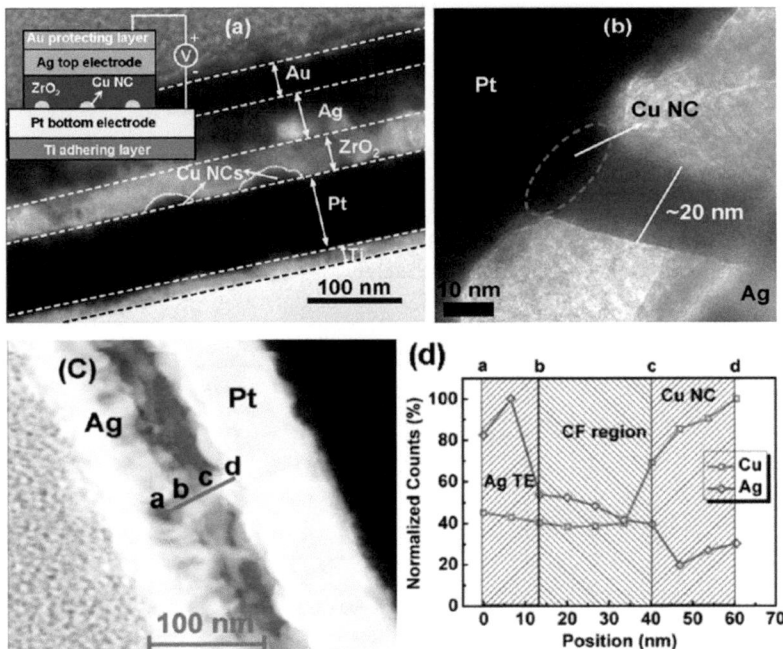

Figure 8.45 (a) The Cu-NC-based Pt BE structure, (b) the structure in HRTEM mode, and (c) a high-angle annular dark field image under scanning TEM mode. (d) An EDS analysis of the CF. Reprinted with permission from Ref. [144]. Copyright (2010) American Chemical Society.

The simulation results show that the intensity of the electric field on the NC tip is almost double as compared to the electric field intensity in the planar region for a given NC radius of 8 nm with a ZrO_2 thickness of 40 nm, as shown in Fig. 8.47a. Outside the column region above the Cu NC, the electrical field intensity decays rapidly. Therefore, it is expected that the CF growth might be guided toward the column region during the formation process. Figure 8.47b indicates the growth of the electric field with NC size.

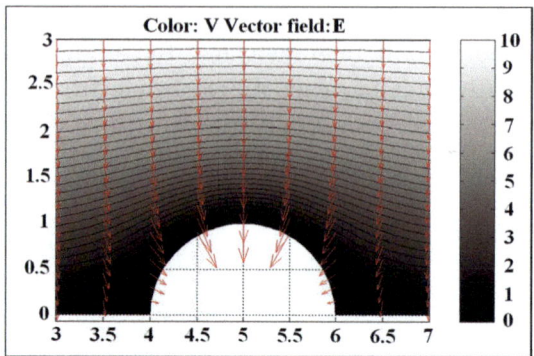

Figure 8.46 The simulated electric field distribution on a Cu NC. The direction of the electric field is represented by red arrows. The size of the arrows defines the strength of the electric field. Reprinted with permission from Ref. [144]. Copyright (2010) American Chemical Society.

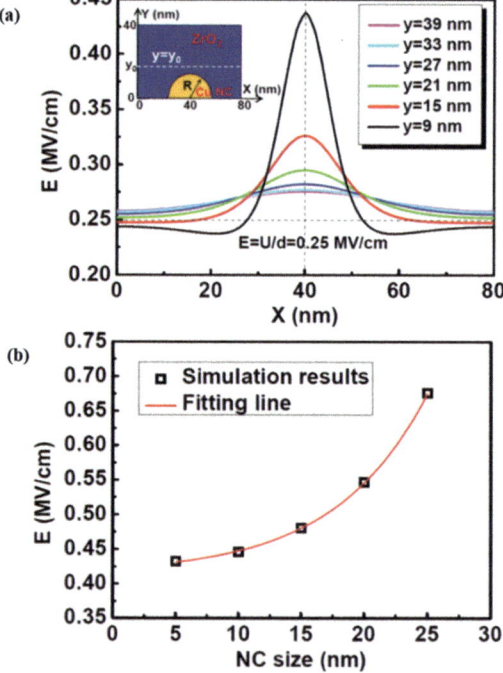

Figure 8.47 (a) A TCAD simulation of the intensity of the electrical field. (b) The electrical field intensity dependence on the size of the NC. Reprinted with permission from Ref. [144]. Copyright (2010) American Chemical Society.

The results clearly indicate that a large NC size or a large ratio of NC size and insulator thickness will enhance the NC-controlled CF growth effect. A Cu NC surrounded by a higher electric field attracts more Ag^+ ions than other regions of the Pt surface, which greatly increases the nucleation probability of Ag atoms at the NC sites. Due to the higher electric field, after the nucleation of Ag atoms the electrochemical deposition rate on the tip of the Ag nucleus will increase. This process will lead to the growth of the Ag CF on a Cu NC BE. In the reverse process, by applying a negative voltage, a reverse electrochemical reaction or a thermally assisted electrochemical reaction will lead to the dissolution or rupture of the Ag CF. Although the device will be reset, on top of the NC a sufficient number of Ag elements remain, which could serve to promote CF formation during the next SET process by a positive voltage of the Ag TE. Therefore, on the basis of the NC-enhanced electrical field analysis, it can be understood that a CF is easily formed, ruptured, and reformed along the same path in repetitive switching cycles.

8.4.2.2 Nanopyramid-shaped bottom electrode

Just like a NC-based BE, a nanopyramid-shaped BE will enhance the localization of the electric field (Fig. 8.48a) via the electrode structure [140]. Huang et al. [145] reported a nanopyramid-Cu-BE-based RRAM, as shown in the schematic illustration of Fig. 8.48b. The fabrication was started with p-type (100)-oriented crystalline silicon wafers dipped into hydrofluoric acid to remove the native oxide and rinsed in deionized water. Afterward, the pyramid structure was formed in a mixture of 1 wt% KOH and 30 vol.% isopropyl alcohol buffer solution heated at 80°C for 10 min. The design of the RRAM stack was started by the direct current (DC) sputtering deposition of 100 nm thick Cu as the BE material. As the RS layer, a 200 nm of TiO_2 layer was reactively sputtered with a Ti target in the following conditions: 190 W power; 5×10^3 torr total pressure; and 22.4 and 9.6 sccm total pressure for Ar and O_2 gases, respectively. As the TE metal, a 50 nm thick Pt layer was deposited through a shadow mask, followed by the 1 μm thick Al layer as the disk-shaped contact metal pad. The pyramid formed using this method is randomly distributed on the surface. As the base size of those pyramids ranges from 1 to

2 μm, the minimum suitable cell size of the pyramid BE samples is 5 μm × 5 μm. In comparison with flat BE devices a pyramid BE shows superior performances in lower operating parameters like SET/RESET. The electric field distribution of the pyramid BE can be simulated by an ISE TCAD 10.0 simulator. According to the model of Mott and Gurney for ionic transport, the ionic current density (J) driven by the electric field can be described by the following [145]:

$$J = 2ZqN_i af \exp(-E_a/kT)\sinh(ZqEa/2kT), \tag{8.1}$$

where q is the charge, Z is the number of the ions charged, N_i is the concentration of mobile cations, a is the jump distance of the cations, f is the attempt-to-escape frequency, E_a is the activation energy, E is the electric field, and kT is the thermal energy.

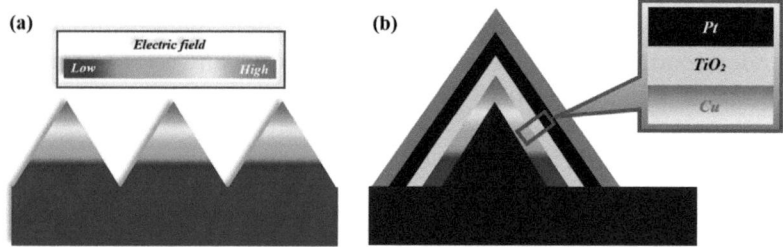

Figure 8.48 Schematic illustration of (a) the electric field distribution with a nanopyramid structure and (b) the Al/Pt/TiO$_2$/Cu/Si-based RRAM design with a Cu nanopyramid BE. The electric field will be maximum at the tip of the nanopyramid.

From Fig. 8.48, it is clear that the electric field at the tip of the pyramid is very high as compared to the other places. During the SET/RESET operation, the local enhancement of the electric field by the pyramid BE structure will enhance the current density through the RS layer. The enhanced electric field will also improve the localization process and fasten the growth process of the CF, as the ionic conductivity will be higher at the tip as compared to the other areas. In terms of reliability, narrow distributions of the operating current and voltage as well as good cycling and retention properties were observed in the nanopyramid BE RRAM as compared to the conventional flat BE RRAM.

8.4.2.3 Arc-shaped bottom electrode

A similar kind of electric field enhancement is also possible with an arc-shaped BE, as shown in Fig. 8.49a. The arc-shaped BE can be fabricated by using a polystyrene (PS) nanosphere template [142]. A RRAM structure based on Pt/InGaZnO/Al is shown in Fig. 8.49b. A high electric field on the surface of the arc improves performance relative to a high electric field on a BE without the arc shape. To fabricate an arc-shaped BE, a SiO_2/Si substrate was chosen on which an ordered monolayer of PS nanospheres of 50 nm diameter was coated by the self-assembly method. A 20 nm thick Al as the BE was deposited on the PS nanosphere monolayer. Then, amorphous InGaZnO as an RS layer of 60 nm thickness and Pt as the TE were deposited. In Pt/InGaZnO/Al RRAM devices the electric field near the BE would point to the center of the arc. A preferential growth of CFs is possible along the electric field direction around the arc-shaped Al BE. In the case of a flat BE RRAM, the randomness in CF formation/rupture processes will increase, which can be controlled well by the arc-shaped structure. CFs can also be locally enhanced and become more concentrated on the arc-shaped BE. Local and controlled growth of CFs contributes to the narrow HRS/LRS distribution and a stable RS process.

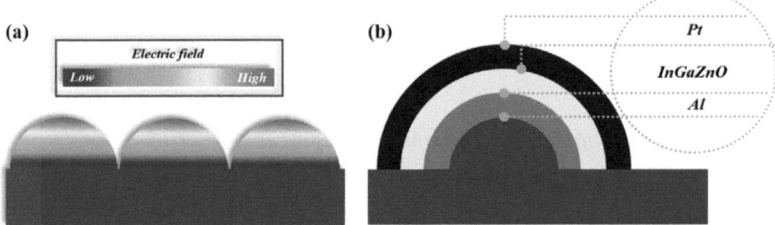

Figure 8.49 Schematic illustration of (a) the electric field distribution with an arc-shaped BE structure and (b) the Pt/InGaZnO/Al-based RRAM design with an arc-shaped BE.

Therefore, all BEs engineered by NCs, those in the shape of nanopyramids, or arc-shaped BEs enhance the effective electric field at the tip of the structure. The higher electric field helps to localize the formation of CFs and improves the RS performance in RRAM devices.

8.5   Summary and Future Scope

Scaling limitation is one of the major hindrances for conventional baseline memory technologies, that is, DRAM, SRAM, and flash. New technologies are being explored to identify a suitable replacement for the baseline devices or one that can act as a standalone universal one. RRAM devices have the potential for superior scalability, higher density, higher nonlinearity, higher endurance as compared to flash, higher speed, lower power operation, good retention, good compatibility with standard CMOS processes, and good control over the economic budget. Hence, this emerging technology has plenty of opportunities to become a useful NVM in the near future. Among RRAM devices, ECM, VCM, and thermochemical reaction type of memory attract a lot of attention in the research division.

Table 8.7   Key research challenges and future scope of ECM- and VCM-type RRAM

ECM
- ✓ *Process compatibility with CMOS technology*
- ✓ *Design of the selector device*
- ✓ *Development of a suitable model for operation and reliability*
- ✓ *Random telegraph noise*

VCM

Filamentary
- ✓ *Understanding of physics behind the RS process*
- ✓ *Improvement of switching uniformity and device reliability*
- ✓ *Random telegraph noise*

Nonfilamentary/interface
- ✓ *Understanding of physics behind the RS process*
- ✓ *Improvement of switching uniformity and device reliability*

Due to the migration of cations like Cu^+ and Ag^+ the RS mechanism can be clearly understood in the solid-electrolyte or oxide-electrolyte-based ECM devices. Generally, in the solid electrolyte ECM, the CF growth direction is from the CE to the AE side and in the oxide electrolyte ECM, the CF growth direction is from the AE to the CE side. In VCM-type RRAM devices the RS mechanism is mainly based on the V_O-type anion migration process. Under the applied bias condition, the oxygen ions can be pulled out from the interstitial sites and the resultant V_O will make a connection between the electrodes. Advancements in RRAM technology have been extended to graphene-based design of the RS stack. Liu et al. [146] reported a recently developed RRAM design using graphene as a diffusion barrier layer. The attractive features of RRAM technology have been extended to transparent electronics [147] and to new areas of research such as neuromorphic computing [148] and high-density memory [149]. Although several positive aspects confirm the footprint of RRAM technology, it is still not a matured technology. The key research challenges and future scope of investigation are summarized Table 8.7.

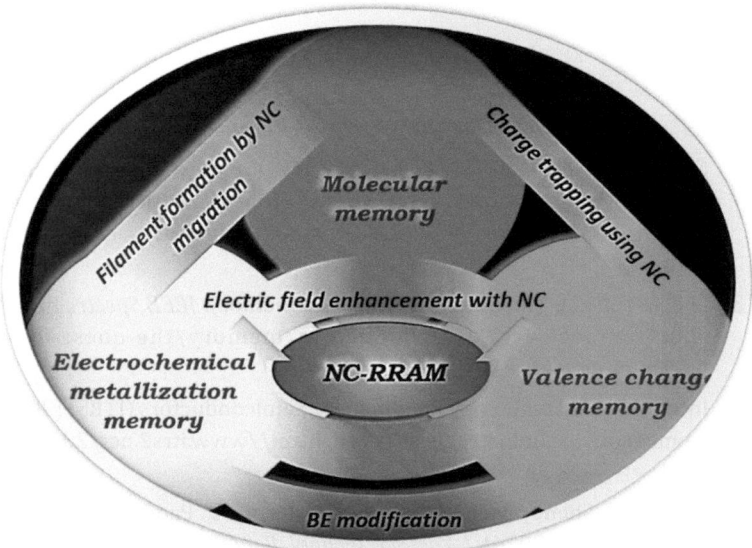

Figure 8.50 Applications of NCs in RRAM devices.

NCs have been proved to be useful in the electrical performance of RRAM devices. NCs can be used for ECM-type, VCM-type, and macromolecular memory design. As shown in Fig. 8.50, the applications of NCs in RRAM are nicely summarized by Banerjee et al. [150]. As defect engineering can play a vital role in the development of RRAM devices, NCs in the insulating layer can provide the necessary defect sites in the design, which may be or may not be orderly. In any case, the major advantage of using NCs in RRAM is the enhancement of localized electric field around the NCs. The improved field can effectively control the formation of the filament through a localized position and result in an improvement in switching uniformity of RRAM. However, several challenges need to be resolved, such as identifying a suitable and compatible material for NCs, ensuring proper tuning between the size and density of NCs, and investigating the RS mechanism of NC-based RRAM devices because the mechanism is dependent on the NC material. If these problems could be solved, RRAM devices could achieve supreme performance, which is necessary in near the future, and the improved RRAM may act as a universal memory.

References

1. Kahng, D., and Sze, S.M. (1967). *Bell Syst. Technol. J.*, **46**, 1288–1295.

2. Govoreanu, B. (2012). Resistive RAM for next-generation non-volatile memory. http://www.eetimes.com/design/memory-design/4237923/Resistive-RAM-for-next-generation-nonvolatile-memory

3. Lee, G. (2012). The quest for universal memory. *IEEE Spectr.*, http://spectrum.ieee.org/semiconductors/memory/the-quest-for-a-universal-memory

4. International Technology Roadmap for Semiconductors (ITRS) (2013). Semiconductor Industry Association. http://www.itrs2.net/2013-itrs.html.

5. Burr, G.W., Shenoy, R.S., Virwani, K., Narayana, P., Padilla, A., Kurdi, B., and Hwang, H. (2014). *J. Vac. Sci. Technol. B*, 040802.

6. Ishiwara, H. (2012). *J. Nanosci. Nanotechnol.*, **12**, 7619–7627.

7. Nishida, T., Asahi, K., Miura, Y., Lu, L., Echizen, M., Yoneda, Y., Kimura, H., Ishikawa, Y., and Uraoka, Y. (2011). *IEEE Conf.*, 106.

8. Raoux, S., Wełnic, W., and Ielmini, D. (2010). *Chem. Rev.*, **110**(1), 240–267.

9. Statt, N. (2016). IBM's phase-change memory is faster than flash and more reliable than RAM. http://www.theverge.com/2016/5/17/11693054/ibm-phase-change-memory-breakthrough-ram-flash-storage

10. Kawahara, T., Ito, K., Takemura, R., and Ohno, H. (2012). *Microelectron. Reliab.*, **52**(4), 613–627.

11. Hoffman, J., Pan, X., Reiner, J.W., Walker, F.J., Han, J.P., Ahsn, C.H., and Ma, T.P. (2010). *Adv. Mater.*, **22**, 2957–2961.

12. Lonescu, A.M. (2012). *Nat. Nanotechnol.*, **7**, 83–85.

13. Garcia, V., and Bibes, M. (2014). *Nat. Commun.*, **5**, 4289.

14. Mellow, C. (2012). Flash-killer nanotube memory firm teams with Belgians to try again: 3 yrs late, and counting - but now moving 'even faster'. http://www.theregister.co.uk/2012/11/06/ nantero_cnt/.

15. Zhou, Y., and Ramanathan, S. (2015). *Proc. IEEE*, **103**(8), 1289–1310.

16. International Technology Roadmap for Semiconductors (ITRS) (2011). Semiconductor Industry Association. http://www.itrs2.net/2011-itrs.html

17. Kim, Y.-B., Lee, S.R., Lee, D., Lee, C.B., Chang, M., Hur, J.H., Lee, M.-J., Park, G.-S., Kim, C.J., Chung, U.-I., Yoo, I.-K., and Kim, K. (2011). *IEEE VLSI Technology Symposium (VLSI) Technical Digest*, 52–53.

18. Lee, M.-J., Lee, C.B., Lee, D., Lee, S.R., Chang, M., Hur, J.H., Kim, Y.-B., Kim, C.-J., Seo, D.H., Seo, S., Chung, U.-I., Yoo, I.-K., and Kim, K. (2011). *Nat. Mater.*, **10**, 625–630.

19. Torrezan, A.C., Strachan, J.P., Medeiros-Ribeiro, G., and Williams, R.S. (2011). *Nanotechnology*, **22**, 485203.

20. Gilbert, N., Zhang, Y., Dinh, J., Calhoun, B., and Hollmer, S. (2013). *IEEE Symposium on VLSI Circuits (VLSIC)*, C204–C205.

21. Tsai, C.-L., Xiong, F., Pop, E., and Shim, M. (2013). *ACS Nano*, **7**, 5360–5366.

22. Hasan, M., Dong, R., Lee, D.S., Seong, D.J., Choi, H.J., Pyun, M.B., and Hwang, H. (2008). *J. Semicond. Technol. Sci.*, **8**, 66–79.

23. Baek, I.G., Lee, M.S., Seo, S., Lee, M.J., Seo, D.H., Suh, D.-S., Park, J.C., Park, S.O., Kim, H.S., Yoo, I.K., Chung, U.-In., and Moon, J.T. (2004). *IEDM Technical Digest. IEEE International Electron Devices Meeting*, 587–590.

24. Kim, K.M., Jeong, D.S., and Hwang, C.S. (2011). *Nanotechnology*, **22**, 254002.

25. Sawa, A. (2008). *Mater. Today*, **11**, 28–36.

26. Lee, M.-J., Han, S., Jeon, S.H., Park, B.H., Kang, B.S., Ahn, S.-E., Kim, K.H., Lee, C.B., Kim, C.J., Yoo, I.-K., Seo, D.H., Li, X.-S., Park, J.-B., Lee, J.-H., and Park, Y. (2009). *Nano Lett.*, **9**, 1476–1481.

27. Kozicki, M.N., Yun, M., Hilt, L., and Singh, A. (1999). *Pennington NJ USA: Electrochemical Society* 99-13, 298–309.

28. Redaelli, A., Pirovano, A., Pellizzer, F., Lacaita, A.L., Ielmini, D., and Bez, R. (2004). *IEEE Electron Device Lett.*, **25**, 684–686.

29. Verbakel, F., Meskers, S.C.J., de Leeuw, D.M., and Janssen, R.A.J. (2008). *J. Phys. Chem. C*, **112**, 5254–5257.

30. Scott, J.C., and Bozano, L.D. (2007). *Adv. Mater.*, **19**, 1452–1463.

31. Banerjee, W., Rahaman, S.Z., and Maikap, S. (2012). *Jpn. J. Appl. Phys.*, **51**, 04DD10.

32. Prakash, A., Maikap, S., Banerjee, W., Jana, D., and Lai, C.S. (2013). *Nanoscale Res. Lett.*, **8**, 379.

33. Luo, Q., Xu, X., Liu, H., Lv, H., Gong, T., Long, S., Liu, Q., Sun, H., Banerjee, W., Li, L., Gao, J., Lu, N., and Liu, M. (2016). *Nanoscale*, **8**(34), 15629–15636.

34. Luo, Q., Xu, X., Liu, H., Lv, H., Gong, T., Long, S., Liu, Q., Sun, H., Banerjee, W., Li, L., Gao, J., Lu, N., Chung, S.S., Li, J., and Liu, M. (2015). *IEDM Technical Digest. IEEE International Electron Devices Meeting*, 10.2.1–10.2.4.

35. Hickmott, T.W. (1962). *J. Appl. Phys.*, **154**, 2669–2682.

36. Nielsen, P.H., and Bashara, N.M. (1964). *IEEE Trans. Electron Devices*, **11**, 243.

37. *Gibbons, J.F., and Beadle, W.E. (1964). Solid-State Electron.*, **7**, 785.

38. Simmons, J.G., and Verderber, R.R. (1967). *Proc. R. Soc. A*, **301**, 77–102.

39. Varker, C.J., and Juleff, E.M. (1967). *Proc. IEEE*, **55**, 728–729.

40. Pagnia, H., and Sotnik, N. (1988). *Phys. Status Solidi*, **108**, 11–65.

41. Kozicki, M.N., and West, W.C. (1998). Programmable metallization cell structure and method of making. US Patent US005761115A/5,761,115.

42. Beck, A., Bednorz, J.G., Gerber, Ch., Rossel, C., and Widmer, D. (2000). *Appl. Phys. Lett.*, **77**, 139.

43. Zhuang, W.W., Pan, W., Ulrich, B.D., Lee, J.J., Stecker, L., Burmaster, A., Evans, D.R., Hsu, Tajiri, S.T.M., Shimaoka, A., Inoue, K., Naka, T., Awaya, N., Sakiyama, K., Wang, Y., Liu, S.Q., Wu, N.J., and Iganatiev, A. (2002). *IEDM Technical Digest. IEEE International Electron Devices Meeting*, 193–196.

44. Sakamoto, T., Kaeriyama, S., Sunamura, H., Mizuno, M., Kawaura, H., Hasegawa, T., Terabe, K., Nakayama, T., and Aono, M. (2004). *IEEE International Solid-State Circuits Conference (ISSCC)*, 16.3.

45. Szot, K., Speier, W., Bihlmayer, G., and Waser, R. (2006). *Nat. Mater.*, **5**, 312–320.

46. Lee, H.Y., Chen, P.S., Wu, T.Y., Chen, Y.S., Wang, C.C., Tzeng, P.J., Lin, C.H., Chen, F., Lien, C.H., and Tsai, M.-J. (2008). *IEDM Technical Digest. IEEE International Electron Devices Meeting*, 297–300.

47. Chua, L.O. (1971). *IEEE Trans. Circuit Theory*, **CT-18**, 507–519.

48. Strukov, D.B., Snider, G.S., Stewart, D.R., and Williams, R.S. (2008). *Nat. Lett.*, **453**, 80–83.

49. Yoon, H.S., Beak, I.-G., Zhao, J., Sim, H., Park, M.Y., Lee, H., Oh, G.-H., Shin, J.C., Yeo, I.-S., and Chung, U.-I. (2009). *Symposium on VLSI Technology Digest of Technical Papers (VLSI)*, 26-7.

50. Maikap, S., Rahaman, S.Z., Wu, T.Y., Chen, F., Kao, M.-J., and Tsai, M.-J. (2010). *IEEE Proceedings of the European Solid-State Device Research Conference (ESSDERC)*, 978.

51. Ho, C.H., Hsu, C.-L., Chen, C.-C., Liu, J.-T., Wu, C.-S., Huang, C.-C., Hu, C., and Yang, F.-L. (2010). *IEDM Technical Digest. IEEE International Electron Devices Meeting*, 436–439.

52. Rozenberg, M.J., Inoue, I.H., and Sanchez, M.J. (2004). *Phys. Rev. Lett.*, **92**, 178302.

53. Yamauchi, T., Yang, M.Y., Kamiya, K., Shiraishi, K., and Nakayama, T. (2012). *Appl. Phys. Lett.*, **110**, 203506.

54. Sheridan, P., Kim, K.-H., Gaba, S., Chang, T., Chen, L., and Lu, W. (2011). *Nanoscale*, **3**, 3833–3840.

55. Park, S.G., Köpe, B.M., and Nishi, Y. (2011). *Symposium on VLSI Technology Digest of Technical Papers*, 46–47.

56. Bersuker, G., Gilmer, D.C., Veksler, D., Kirsch, P., Vandelli, L., Padovani, A., Larcher, L., McKenna, K., Shluger, A., Iglesias, V., Porti, M., and Nafría, M. (2011). *J. Appl. Phys.*, **110**, 124518.

57. Köpe, B.M., Tendulkar, M., Park, S.-G., Lee, H.D., and Nishi, Y. (2011). *Nanotechnology*, **22**, 254029.

58. Liu, Q., Sun, J., Lv, H., Long, S., Yin, K., Wan, N., Li, Y., Sun, L., and Liu, M. (2012). *Adv. Mater.*, **24**, 1844–1849.

59. Raghavan, N., Pey, K.L., Wu, X., Liu, W., and Bosman, M. (2012). *IEEE Electron Device Lett.*, **33**, 712–714.

60. Ielmini, D. (2011). *IEEE Trans. Electron Devices*, **58**, 4309–4317.

61. Guan, X., Yu, S., and Wong, H.-S.P. (2012). *IEEE Trans. Electron Devices*, **59**, 1172–1182.

62. Kamiya, K., Yang, M.Y., Park, S.-G., Köpe, B.M., Nishi, Y., Niwa, M., and Shiraishi, K. (2012). *Appl. Phys. Lett.*, **100**, 073502.

63. Stefano, F.D., Houssa, M., Kittl, J.A., Jurczak, M., Afanas'ev, V.V., and Stesmans, A. (2012). *Appl. Phys. Lett.*, **100**, 142102.

64. Kwon, D.-H., Kim, K.M., Jang, J.H., Jeon, J.M., Lee, M.H., Kim, G.H., Li, X.-S., Park, G.-S., Lee, B., Han, S., Kim, M., and Hwang, C.S. (2010). *Nat. Nanotechnol.*, **5**, 148–153.

65. Kim, W., Park, S.I., Zhang, Z., Liauw, Y.Y., Sekar, D., Wong, H.-S. P., and Wong, S.S. (2011). *Symposium on VLSI Technology Digest of Technical Papers*, 22–23.

66. Clarke, P. (2012). Elpida announces ReRAM chip, aims to enter market 2013. http://www.eetimes.com/electronics-news/4235304/ Elpida-announces-ReRAM-chip

67. Panasonic (2017). Panasonic Japan prepares to release ReRAM nonvolatile memory with microprocessor evaluation kit. http://www.memristor.org/electronics/807/panasonic-reram-nonvolatile-memory-microprocessor-kit

68. Zhirnov, V.V., Meade, R., Cavin, R.K., and Sandhu, G. (2011). *Nanotechnology*, **22**, 254027.

69. Prakash, A., Park, J., Song, J., Woo, J., Cha, E.-J., and Hwang, H. (2015). *IEEE Electron Device Lett.*, **36**(1), 32–34.

70. Chen, G., Song, C., Chen, C., Gao, S., Zeng, F., and Pan, F. (2012). *Adv. Mater.*, **24**, 3515.

71. Gao, S., Song, C., Chen, C., Zeng, F., and Pan, F. (2012). *J. Phys. Chem. C*, **116**, 17955.

72. Zhu, X., Su, W., Liu, Y., Hu, B., Pan, L., Lu, W., Zhang, J., and Li, R.-W. (2012). *Adv. Mater.*, **24**, 3941.

73. Waser, R., Dittmann, R., Staikov, G., and Szot, K. (2009). *Adv. Mater.*, **21**, 2632–2663.

74. Lu, W., Jeong, D.S., Kozicki, M., and Waser, R. (2012). *Mater. Res. Soc. Bull.*, **37**, 124.

75. Hasegawa, T., Terabe, K., Tsuruoka, T., and Aono, M. (2012). *Adv. Mater.*, **24**, 252.

76. Yang, Y., and Lu, W. (2013). *Nanoscale*, **5**, 10076.

77. Lv, H., Xu, X., Sun, P., Liu, H., Luo, Q., Liu, Q., Banerjee, W., Sun, H., Long, S., Li, L., and Liu, M. (2015). *Sci. Rep.*, **5**, 13311.

78. Lv, H., Xu, X., Liu, H., Liu, R., Liu, Q., Banerjee, W., Sun, H., Long, S., Li, L., and Liu, M. (2015). *Sci. Rep.*, **5**, 7764.

79. Hirose, Y., and Hirose, H. (1976). *J. Appl. Phys.*, **47**, 2767.

80. Sun, J., Liu, Q., Xie, H., Wu, X., Xu, F., Xu, T., Long, S., Lv, Li, H., Y., Sun, L., and Liu, M. (2013). *Appl. Phys. Lett.*, **102**, 053502.

81. Pearson, C., Bowen, L., Lee, M.-W., Fisher, A.L., Linton, K.E., Bryce, M.R., and Petty, M.C. (2013). *Appl. Phys. Lett.*, **102**, 213301.

82. Peng, P., Xie, D., Yang, Y., Zang, Y., Gao, X., Zhou, C., Feng, T., Tian, H., Ren, T., and Zhang, X. (2012). *J. Appl. Phys.*, **111**, 084501.

83. Wang, Z., Griffin, P.B., McVittie, J., Wong, S., McIntyre, P.C., and Nishi, Y. (2007). *IEEE Electron Device Lett.*, **28**, 14.

84. Zhu, X., Su, W., Liu, Y., Hu, B., Pan, L., Lu, W., Zhang, J., and Li, R.-W. (2012). *Adv. Mater.*, **24**, 3941.

85. Peng, C.N., Wang, C.W., Chan, T.C., Chang, W.Y., Wang, Y.C., Tsai, H.W., Wu, W.W., Chen, L.J., and Chueh, Y.L. (2012). *Nanoscale Res. Lett.*, **7**, 559.

86. Guo, X., Schindler, C., Menzel, S., and Waser, R. (2007). *Appl. Phys. Lett.*, **91**, 133513.

87. Pan, F., Yin, S., and Subramanian, V. (2011). *IEEE Electron Device Lett.*, **32**, 949.

88. Kozicki, M.N., and Mitkova, M. (2006). *J. Non-Cryst. Solids*, **352**, 567.

89. Kozicki, M.N., Ratnakumar, C., and Mitkova, M. (2006). *Proceedings of Non-Volatile Memory Technology Symposium*, 111.

90. Xu, Z., Bando, Y., Wang, W., Bai, X., and Golberg, D. (2010). *ACS Nano*, **4**, 2515.

91. Choi, S.J., Park, G.S., Kim, K.H., Cho, S., Yang, W.Y., Li, X.S., Moon, J.H., Lee, K.J., and Kim, K. (2011). *Adv. Mater.*, **23**, 3272.

92. Fujii, T., Arita, M., Takahashi, Y., and Fujiwara, I. (2011). *Appl. Phys. Lett.*, **98**, 212104.

93. Cho, B., Yun, J.-M., Song, S., Ji, Y., Kim, D.-Y., and Lee, T. (2011). *Adv. Funct. Mater.*, **21**(20), 3976–3981.

94. Yang, J.J., Pickett, M.D., Li, X., Ohlberg, D.A., Stewart, D.R., and Williams, R.S. (2008). *Nat. Nanotechnol.*, **3**, 429.

95. Yoshida, C., Kinoshita, K., Yamasaki, T., and Sugiyama, Y. (2008). *Appl. Phys. Lett.*, **93**, 042106.

96. Lin, Y.S., Zeng, F., Tang, S.G., Liu, H.Y., Chen, C., Gao, S., Wang, Y.G., and Pan, F. (2013). *J. Appl. Phys.*, **113**, 064510.

97. Chen, C., Song, C., Yang, J., Zeng, F., and Pan, F. (2012). *Appl. Phys. Lett.*, **100**, 253509.

98. Banerjee, W., Xu, X., Liu, H., Lv, H., Liu, Q., Sun, H., Long, S., and Liu, M. (2015). *IEEE Electron Device Lett.*, **36**(4), 333–335.

99. Banerjee, W., Maikap, S., Rahaman, S.Z., Prakash, A., Tien, T.-C., Li, W.-C., and Yang, J.-R. (2012). *J. Electrochem. Soc.*, **159**, H177–H182.

100. Banerjee, W., Rahaman, S.Z., Prakash, A., and Maikap., S. (2011). *Jpn. J. Appl. Phys.*, **50**, 10PH01.

101. Chen, C., Pan, F., Wang, Z.S., and Zeng, F. (2011). *J. Nanosci. Lett.*, **1**, 102.

102. Kim, H.-D., An, H.-M., and Kim, T.G. (2012). *IEEE Trans. Electron Devices*, **59**, 2302.

103. Chen, X.G., Ma, X.B., Yang, Y.B., Chen, L.P., Xiong, G.C., Lian, G.J., Yang, Y.C., and Yang, J.B. (2011). *Appl. Phys. Lett.*, **98**, 122102.

104. Yan, Z.B., and Liu, J.-M. (2013). *Sci. Rep.*, **3**, 2482.

105. Chen, J.-Y., Hsin, C.-L., Huang, C.-W., Chiu, C.-H., Huang, Y.-T., Lin, S.-J., Wu, W.-W., and Chen, L.-J. (2013). *Nano Lett.*, **13**, 3671–3677.

106. Tappertzhofen, S., Waser, R., and Valov, I. (2014). *ChemElectroChem*, **1**, 1287–1292.

107. Banerjee, W., Maikap, S., Tien, T.-C., Li, W.-C., and Yang, J.-R. (2011). *J. Appl. Phys.*, **110**, 074309.

108. Zhang, X.T., Yu, Q.X., Yao, Y.P., and Li, X.G. (2010). *Appl. Phys. Lett.*, **97**, 222117.

109. Hasan, M., Dong, R., Choi, H., Lee, D., Seong, D.-J., Pyun, M., and Hwang, H. (2008). *Appl. Phys. Lett.*, **92**, 202102.

110. Liu, L., Zhang, S., Luo, Y., Yuan, G., Liu, J., Yin, J., and Liu, Z. (2012). *J. Appl. Phys.*, **111**, 104103.

111. Pan, F., Gao, S., Chen, C., Song, C., and Zeng, F. (2014). *Mater. Sci. Eng. R*, **83**, 1–59.

112. Waser, R., Dittmann, R., Salinga, M., and Wuttig, M. (2010). *Solid-State Electron.*, **54**, 830–840.

113. Lee, D., Seong, D.-J., Choi, H.J., Jo, I., Dong, R., Xiang, W., Oh, S., Pyun, M., Seo, S.-O., Heo, S., Jo, M., Hwang, D.-K., Park, H.K., Chang, M., Hasan, M., and Hwang, H. (2006). *IEDM Technical Digest, IEEE International Electron Devices Meeting*, San Francisco, CA, p. 1.

114. Misha, S.H., Tamanna, N., Woo, J., Lee, S., Song, J., Park, J., Lim, S., Park, J., and Hwang, H. (2015). *ECS Solid State Lett.*, **4**(3), P25–P28.

115. Liu, Q., Long, S., Wang, W., Zuo, Q., Zhang, S., Chen, J., and Liu, M. (2009). *IEEE Electron Device Lett.*, **30**(12), 1335–1337.

116. Yuanyang, Z., Jiayu, W., Jianbin, X., Fei, Y., Qi, L., and Yuehua, D. (2014). *J. Semicond.*, **35**(4), 042002.

117. Chen, L., Gou, H.-Y., Sun, Q.-Q., Zhou, P., Lu, H.-L., Wang, P.-F., Ding, S.-J., and Zhang, D.W. (2011). *IEEE Electron Device Lett.*, **32**, 794–796.

118. Cheng, C.H., Chen, P.C., Wu, Y.H., Yeh, F.S., and Chin, A. (2011). *IEEE Electron Device Lett.*, **32**, 1749–1751.

119. Ju, Y.C., Kim, S., Seong, T.G., Nahm, S., Chung, H., Hong, K., and Kim, W. (2012). *Small*, **8**(18), 2849–2855.

120. Guan, W., Long, S., Jia, R., and Liu, M. (2007). *Appl. Phys. Lett.*, **91**, 062111.

121. Banerjee, W. (2012). Nonvolatile memories using nanoscale IrO_x metal nanocrystals. PhD Thesis.

122. Yoon, J.H., Kim, K.M., Lee, M.H., Kim, S.K., Kim, G.H., Song, S.J., Seok, J.Y., and Hwang, C.S. (2010). *Appl. Phys. Lett.*, **97**, 232904.

123. Chang, W.-Y., Cheng, K.-J., Tsai, J.-M., Chen, H.-J., Chen, F., Tsai, M.-J., and Wu, T.-B. (2009). *Appl. Phys. Lett.*, **95**, 042104.

124. Raghavan, N., Fantini, A., Degraeve, R., Roussel, P.J., Goux, L., Govoreanu, B., Wouters, D.J., Groeseneken, G., and Jurczak, M. (2013). *Microelectron. Eng.*, **109**, 177–181.

125. Tian, X., Yang, S., Zeng, M., Wang, L., Wei, J., Xu, Z., Wang, W., and Bai, X. (2014). *Adv. Funct. Mater.*, **26**, 3649–3654.

126. Gao, S., Song, C., Chen, C., Zeng, F., and Pan, F. (2013). *Appl. Phys. Lett.*, **102**, 141606.

127. Yang, Y.C., Pan, F., Zeng, F., and Liu, M. (2009). *J. Appl. Phys.*, **106**, 123705.

128. Banerjee, W., Maikap, S., Lai, C.S., Chen, Y.Y., Tien, T.C., Lee, H.Y., Chen, W.S., Chen, F.T., Kao, M.J., Tsai, M.J., and Yang, J.-R. (2012). *Nanoscale Res. Lett.*, **7**, 194.

129. Bozano, L.D., Kean, B.W., Deline, V.R., Salem, J.R., and Scott, J.C. (2004). *Appl. Phys. Lett.*, **84**, 607.

130. Kou, L., Li, F., Chen, W., and Guo, T. (2013). *Org. Electron.*, **14**, 1447.

131. Odagawa, A., Sato, H., Inoue, I.H., Akoh, H., Kawasaki, M., and Tokura, Y. (2004). *Phys. Rev. B*, **70**, 224403.

132. Anderson, P.W. (1958). *Phys. Rev.*, **109**, 1492–1507.

133. Chen, A.B.K., Kim, S.G., Wang, Y., Tung, W.-S., and Chen, I.-W. (2011). *Nat. Nanotechnol.*, **6**, 237.

134. Sun, H., Liu, Q., Li, C., Long, S., Lv, H., Bi, C., Huo, Z., Li, L., and Liu, M. (2014). *Adv. Funct. Mater.*, **24**, 5679–5686.

135. Kim, J.-D., Baek, Y.-J., Choi, Y.J., Kang, C.J., Lee, H.H., Kim, H.-M., Kim, K.-B., and Yoon, T.-S. (2013). *J. Appl. Phys.*, **114**, 224505.

136. Zhong, S., Duan S., and Cui, Y. (2014). *RSC Adv.*, **4**, 40924.

137. Yoon, J.H., Han, J.H., Jung, J.S., Jeon, W., Kim, G.H., Song, S.J., Seok, J.Y., Yoon, K.J., Lee, M.H., and Hwang, C.S. (2013). *Adv. Mater.*, **25**, 1987–1992.

138. Uenuma, M., Kawano, K., Zheng, B., Okamoto, N., Horita, M., Yoshii, S., Yamashita, I., and Uraoka, Y. (2011). *Nanotechnology*, **22**, 215201.

139. Lyu, S.-H., and Lee, J.-S. (2011). *J. Mater. Chem.*, **22**, 1852.

140. Kim, H.-D., Yun, M.J., Hong, S.M., and Kim, T.G. (2014). *Nanotechnology*, **25**, 125201.

141. Otsuka, S., Shimizu, T., Shingubara, S., Makihara, K., Miyazaki, S., Yamasaki, A., Tanimoto, Y., and Takase, K. (2014). *AIP Adv.*, **4**, 087110.

142. Wang, Z., Zhao, K., Xu, H., Zhang, L., Ma, J., and Liu, Y. (2015). *Appl. Phys. Express*, **8**, 014101.

143. Shin, H.W., Park, J.H., Chung, H.Y., Kim, K.H., Kim, H.-D., and Kim, T.G. (2014). *Appl. Phys. Express*, **7**, 024202.

144. Liu, Q., Long, S., Lv, H., Wang, W., Niu, J., Huo, Z., Chen J., and Liu, M. (2010). *ACS Nano*, **4**(10), 6162.

145. Huang, Y.-C., Tsai, W.-L., Chou, C.-H., Wan, C.-Y., Hsiao, C., and Cheng, H.-C. (2013). *IEEE Electron Device Lett.*, **34**, 1244–1246.

146. Liu, S., Lu, N., Zhao, X., Xu, H., Banerjee, W., Lv, H., Long, S., Li, Q., Liu, Q., and Liu, M. (2016). *Adv. Mater.*, **28**, 10623–10629.

147. Yan, X.B., Hao, H., Chen, Y.F., Li, Y.C., and Banerjee, W. (2014). *Appl. Phys. Lett.*, **105**(9), 093502.

148. Wang, W., Li, Y., Wang, M., Wang, L., Liu, Q., Banerjee, W., Li, L., and Liu, M. (2016). *IEEE Silicon Nanoelectronics Workshop (SNW)*, 50–51.

149. Luo, Q., Xu, X., Liu, H., Lv, H., Gong, T., Long, S., Liu, Q., Sun, H., Banerjee, W., Li, L., Lu, N., and Liu, M. (2015). *IEEE International Electron Devices Meeting (IEDM)*, 10.4.1–10.4.4.

150. Banerjee, W., Liu, Q., Lv, H., Long, S., and Liu, M. (2017). *J. Phys. D: Appl. Phys.*, **50**, 303002.

Chapter 9

Measurement Aspects of Nonvolatile Memory

Alberto Campisi

Xcerra, Via Paracelso 22, 20864 Agrate Brianza, Italy
alberto.campisi@mail.polimi.it

Testing integrated circuits is an activity that does not always receive the attention it deserves when describing the overall process of chip manufacturing. Many persons, including some of those directly involved in this business, think that once a chip is properly designed and manufactured using a stable technology, it should work, period. Well, that is unfortunately not true. Each and every unit must be carefully checked, often multiple times, before it can reach its final user. And this activity prevents many issues, ranging from inconsequential to catastrophic, that may otherwise emerge later, once the unit sits in a system. Testing also compensates for small differences in behavior of different units, making the underneath technology look more uniform than what it is in reality. Those who recognize the importance of testing often turn completely upside

Nanocrystals in Nonvolatile Memory
Edited by Writam Banerjee
Copyright © 2018 Pan Stanford Publishing Pte. Ltd.
ISBN 978-981-4774-73-4 (Hardcover), 978-1-351-20327-2 (eBook)
www.panstanford.com

down the common understanding mentioned earlier, stating that the yield is determined by a good testing methodology. Whether this is true or not, a lot of technology and money are involved in this activity, along with a lot of effort from a number of engineers, who often play a hidden role unlike their design or process colleagues. Without any wish to be exhaustive, this chapter aims to give an overview of the instruments and techniques used to test a nonvolatile memory chip.

9.1 Introduction

Flash memory is ubiquitous. Every PC, tablet, and smartphone contains a flash memory integrated circuit (IC). The continuous drop in flash memory cost per bit is transforming the hard disk market by replacing traditional drives with faster and less-power-consuming solid-state storage devices, or solid-state drives (SDDs). At the same time, since flash memory manufacturing is so unique and has such a high cost and more complex tuning compared to other IC types, only a small number of companies worldwide can afford to do it. This means that only a handful of companies produce and test a tremendous quantity of wafers to meet the market's growing demand, while simultaneously working to achieve advanced levels of technology in order to maintain a competitive advantage over the other players.

In recent years NAND has overcome NOR architecture in flash memory products (Fig. 9.1), and this is in in large part due to a more efficient use of the silicon real estate. While the floating gate technology in production today is close to reaching its limits, companies are trying to store more bits in each memory cell (2–3 bits per cell) and simultaneously conducting research on 3D technology solutions. The drawback of storing more bits per cell is poorer data retention and endurance. A strong dependence on an external controller to cope with the cells' degraded performance has nowadays become the rule.

Memory is so different compared to other types of ICs that any company that wasn't born exclusively as a memory maker (Intel, AMD, ST, Siemens, etc.) at some point has had to spin out new companies focused on memory. This chapter is not focused on flash memory technology or design (a complete introduction to the subject can be found in Refs. [1, 2]) but specifically on what is needed to test it, both in terms of equipment and in terms of internal resources.

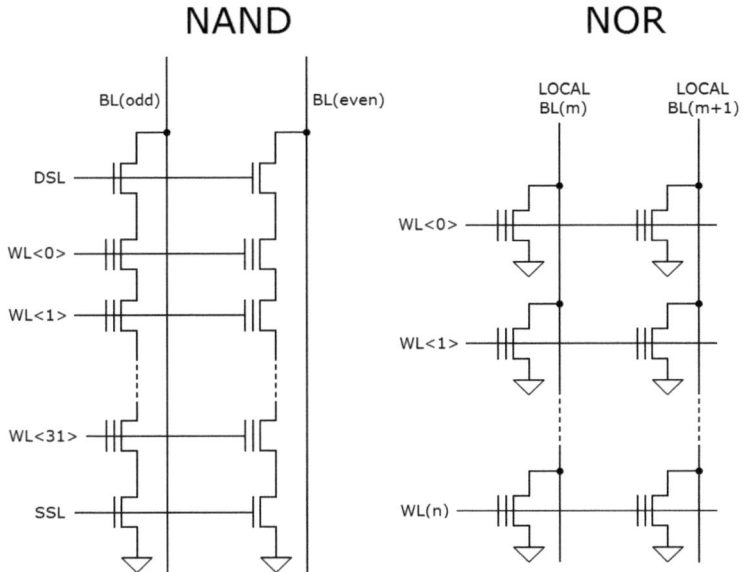

Figure 9.1 NAND versus NOR array arrangement.

9.2 Testing Memory

The main reasons why testing flash memory differs from testing any other silicon chip are:

- **Extremely fast lifecycle:** Test program and test hardware must be developed in a timely manner in order to be ready by the time the silicon is available. Any program or test setup debug delays must be avoided because immediate feedback on the product functionality from an electrical analysis is key to addressing as many issues as possible with the first mask respin (if required).

- **Large production (and testing) volumes:** Many different test fixtures (interface boards and probe cards) must be designed quite early in the product development cycle. These are very complex pieces of equipment that take a long time to be manufactured and require exact knowledge not only of the pad positions on a die but also of the dies' exact number and position (layout) on the wafer. While working on a new product, designers are reluctant to declare the exact locations of the pads too soon, as this poses a severe constraint on any circuit layout change that may be necessary later on.

Today, the design phase of the chip must be over[a] before the evaluation can start of how many chips fit into the wafer and what the exact position is of each of them in the wafer. The process team provides this information, and according to corporate rules that can vary across manufacturers. These rules derive from the equipment being used. However, in a more integrated design flow, this information could be available immediately at the time the die size is set.

- **Profit per unit:** The profit per unit is strongly dependent on the test time per unit.

- **Long test time:** The last three points emphasize the need for achieving a high degree of parallelism in testing. Due to the large size of flash memory, relatively few die can fit on a wafer in comparison to other silicon products. It is not uncommon, however, that the entire wafer is tested in a single shot (single touchdown). This means that thousands of dice are tested in parallel.

9.2.1   Memory Tester

The tester is the primary tool used to validate the electrical functionality of any IC, including flash memory. Figure 9.2 shows the key tester blocks.

In principle, a tester is a set of electrical resources (channels) that can be connected to the pins of the device to be tested, also known as a device under test (DUT). In memory testers, there are two kinds of channels: pin electronics (PE) and programmable power suppliers (PPSs). Both can force voltage but vary in terms of power, resolution, precision, accuracy, and speed, just to mention a few key parameters.

---

[a]This occurrence is often referred to as "tape out."

Another resource that can be connected to the DUT's pin when a precise measurement is required is the parametric measurement unit (PMU). A PMU can force current or voltage and measure voltage or current. Channels communicate with the DUTs through one or more highly packed connectors on the tester test head. Some additional resources are also available through the connectors to power and control any component present on the load board, or the device interface board (DIB), although typically minimal to zero additional circuitry is required for flash memory testing other than wires. All the channels refer to a common ground, and there is a ground pin for each channel in the connector to guarantee a solid ground reference.

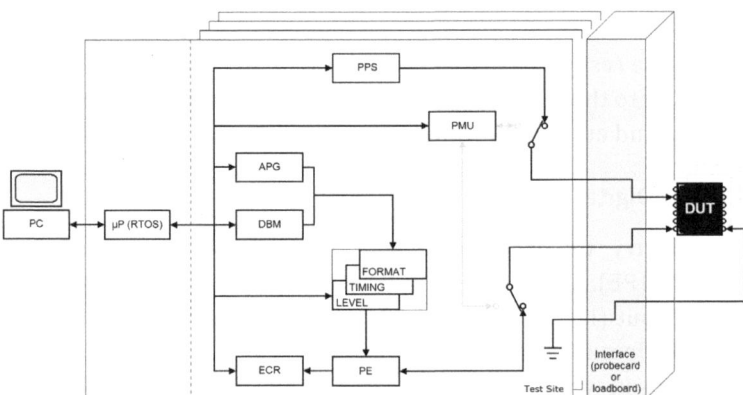

Figure 9.2 Flash memory tester basic blocks.

A user programmable real-time operative system (RTOS) central processor coordinates channel operations. The test environment that runs on a PC (or a workstation) allows one to load the test program into the tester and collect data generated during test execution. It also allows, typically in the debug phase, to pause program execution and proceed by steps to verify that the configuration of the resources and parameters is correct.

A flash memory tester has a given number of direct current (DC) programmable power supplier (PPS) channels, capable of a relatively high current (500 mA to 1.0A), that connect to the IC power pins (V_{dd}[b]

---

[b]V_{dd} owes its name to the field-effect transistor (FET) drain (d), once afferent to the power supplier node. The "d" is doubled to indicate a power supply voltage.

[5] and $V_{ddq}{}^c$). It has many digital channels, which can supply a small current (say 10 mA) but can quickly switch between predefined high and low analog values and can be grouped into buses that operate in a coordinated fashion under the control of the central processor. The number of channels of these two types of resources depends on the flash specifications. It is important to note that the NOR type of flash has a larger number of digital pins compared to the NAND type. This is because NOR flash has always had an address bus and a data bus (which causes the number of pins to change with each memory size increase), while NAND flash contains a common bus for addresses and data. The common ground is connected to the $V_{ss}{}^d$ pin of the IC.

A precision voltage/current meter is required to measure continuity, leakage, operative, and standby power consumption and a few other characteristics (e.g., bandgap and I_{ref}). There is usually one of these resources per site as its usage can be serialized without any impact to the test time. Test-time-heavy operations are the read, program, and erase operations.

9.2.1.1 Digital channel

The circuitry underneath each digital channel is called a pin electronic (PE). Digital channels are connected to the IC control and input/output (I/O) pins. Figure 9.3 shows the simplified structure of a PE. Different sections get activated for different operations. The driver is used when we need to force a logical value to a DUT's pin, V_{IH} is the voltage value that the driver is forcing on the pin when it is set to a logical high, and V_{IL} is the voltage value forced on the pin when it is set to a logical low. When we need to read from a DUT's pin acting as an output, the driver is set to high-impedance mode so as not to interfere with the read mode, which is achieved by two comparators that detect when the voltage at the pin is a logical high or a logical low. The DUT's output is assigned a logical high when its voltage exceeds the V_{OH} value and a logical low when it is lower than V_{OL}. When V_{OH} and V_{OL} are not equal and the output voltage lies in this midband, one can calculate how long it takes for a DUT's I/O to switch from a forcing condition to a high-impedance state.

---

[c] V_{ddq} is the power pin dedicated to the IC I/O bus. It can have the same voltage value as the V_{dd} or lower.

[d] V_{ss} owes its name to the FET source (s), once afferent to the ground node. As in the case of V_{dd}, the "s" is doubled to indicate a power supply voltage.

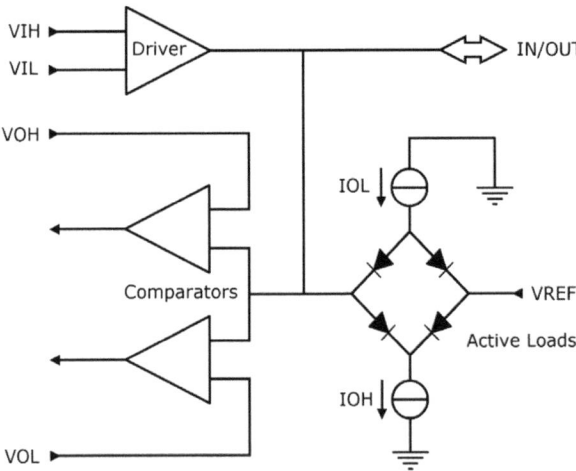

Figure 9.3 PE simplified architecture.

The value will actually be read by averaging its value in a time window or at a specific time (strobe time). The structure including the diodes is referred to as active load. The purpose of this circuitry is to load the DUT's output to verify its current sourcing/drawing capacity. Disabled active load gives a best-case read condition as the comparators have high-impedance inputs and could make the DUT's output appear to switch faster than when placed in a real application, in which other components are connected at the DUT's output. V_{ref} sets a voltage threshold: if the DUT's output voltage is higher than that, a current I_{OH} is drawn from it, while if it is lower, a current I_{OL} is injected in it. Conventionally I_{OH} is set as a negative current and I_{OL} as a positive current. This circuitry can be disabled by setting negligible current values. As previously said, the driver circuitry and the receiver circuitry aren't activated simultaneously: that at least is true most of the time, unless, for instance, you want to use the 50 Ω driver output impedance to terminate a transmission line while you are reading through the high-impedance comparators.

Exercise 9.1 Can you figure out a way to measure a capacitance using a digital channel like the one shown in Fig. 9.3 ?

Exercise 9.1a If in Exercise 9.1 you can sample the logic state of the PE at a 100 MHz frequency, how many samples will be necessary to measure a 30 pF capacitance (initially not charged) if I_{OL} = 100 uA and V_{OL} = 1 V?

At what level should V_{ref} be set?

Exercise 9.2   Can you verify, using the digital channel like the one shown in Fig. 9.3, if a DUT's I/O pin is in a high-impedance condition? What should be the expected condition? Which parameters must be set, and which are not important to set? What are meaningful values for these parameters?

Solution 9.1   Set your digital channel to drive mode low (V_{IL} = 0 V) to discharge the capacitor first, then switch to the receive mode expected high ($V_{OH} = V_x$, $I_{OL} = I_x$, and $V_{ref} > V_{OH}$). From the definition of capacitance $C = dQ/dV \rightarrow C = I \times dt/dV$, which becomes $C = I_x \times (t_1 - t_0)/(V_x - 0 \text{ V})$, t_0 being the drive-to-receive switch time and t_1 the time at which the PE strobe returns the expected condition. If $t_0 = 0$ then $C = (I_x/V_x)t_1$. The same result could be obtained by precharging the capacitor to a different voltage level, for example, V_{IH} = 5 V, and using I_{OH} to discharge the capacitor. What should be the expected logic value in this case? And how should V_{OL} and V_{ref} be set?

Solution 9.1a $C = (I_x/V_x) \times t_1 \rightarrow$ 30 pF = (100 uA/1 V) $\times k \times$ 10 ns $\rightarrow$

30×10^{-12} = $100 \times 10^{-6} \times 10 \times 10^{-9} \times k \rightarrow$

$30 = 100 \times 10 \times 10^{-3} \times k \rightarrow k = 30$.

Solution 9.2 Set the digital channel to the receive mode expected midband, which is a value between V_{OH} and V_{OL}. V_{ref} has to be placed between those values.

I_{OH} and I_{OL} should be high enough to discharge the pin in a reasonable time.

For digital channels, timing also has to be considered. A system calibration procedure guarantees that, at the connector pins, all channel edges and strobes relevant delays respect exactly those set through the test program, with an error tolerance of a few hundred picoseconds.

However, when a DIB is docked to the test-head connector, a second calibration is necessary to evaluate each signal's additional delay caused by the length of the DIB trace connecting each connector pin to the relevant DUT pin. In fact, if a signal transition is expected to occur at the DUT input pin exactly at time T2 (see Fig. 9.4), the tester actually has to switch a bit earlier (T1), to allow the signal enough time to propagate from the hardware source to the DUT pin through the connectors and the DIB trace. For the same reason, if the event at T2 causes the DUT output pin to commutate at T3, this transition is not detectable at the tester connector pin until a bit later, at T4. The target of this second calibration (DIB calibration) is to evaluate, per each digital channel, the delay associated with the additional trace that goes from the connector pin to the relevant DUT pin: T2–T1 and T4–T3 in the example considered. Once evaluated, those delays are taken into account by the tester and automatically added or subtracted to the target values so that the user does not have to handle them himself or herself. The user, while writing the test program, focuses on events occurring at the DUT pins only (T2 and T3). The tester issues a signal to DUT in advance (T1), such that it occurs at DUT's input pin when the user needs it (T2). Similarly, the tester strobes DUT's output signals with some delay (T4) compared to the time expected for a transition to occur at DUT's output pin (T3). Each signal trace on the DIB has known characteristic impedance, and DIB calibration can be obtained in two ways, direct or indirect. In direct (measured) DIB calibration, before contacting any DUT, each channel uses its driver to generate a probe transition and its receiver to measure the signal round-trip time along the attached DIB trace, time that is proportional to the length of the trace. For an open-end trace like this, while no DUT is contacted, the impedance seen by the PE driving source equals the trace characteristic impedance as long as it takes to the traveling transition to reach the open end and return to the source (round-trip time), then grows to infinite value. In other words, the round-trip interval starts when the driver issues the step at the source end of the trace and stops when the receiver captures the impedance value change at the same source end.

Delay values, half the measured round-trip time, are stored in a file or in a small serial memory IC[e] placed on the DIB, from which

---

[e]This serial access electrically erasable memory (DIB IDI is a chip mounted on the DIB with a unique identifier information programmed, which gets read at the beginning of the test to prevent accidental usage of this hardware with test programs of different products.

they are retrieved at the next test-head power-on. At test-program runtime these delays are added to or subtracted from each edge according to the channel operation (drive or sense).

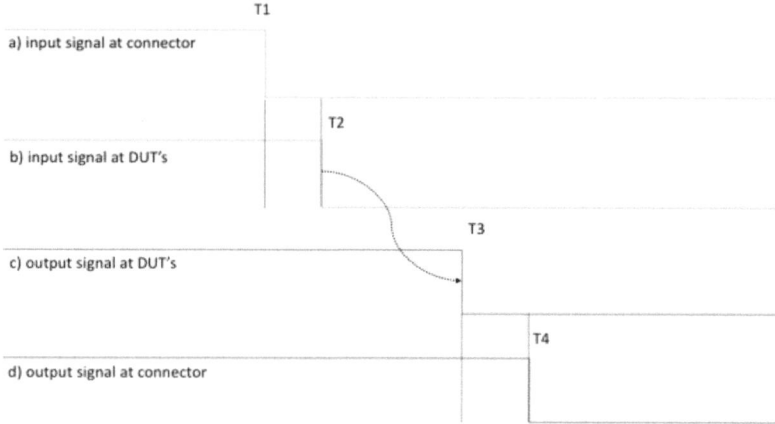

Figure 9.4 Signal transition at the connector and at the DUT's pins.

In the indirect method the delays, rather than being measured, are worked out from the list of trace lengths extracted by the DIB layout and provided by the DIB manufacturer. In a typical medium for printed circuit board (PCB), the speed of the electrical signal in a strip line[f] is 0.466 times the speed of light in vacuum (c), and the characteristic impedance is 55 Ω.

9.2.1.2 PMU

A precise meter known as PMU is another important part of a memory tester. A small number of PMUs is present in a memory tester, typically one per site. Each PE output can be temporarily disconnected from its DUT's pin and a PMU can be connected in its place, whenever a precise force and measurement is required. A PMU is used to force voltage and measure current (FVMI) or force current and measure voltage (FIMV).

A PMU has different ranges that can be set both for the voltage and for the current, and the digital-to-analog converter (DAC) underneath adapts to the range selected; so if a current range is set

---

[f]A strip line is a trace on the PCB surrounded on all sides by solid metal planes.

to ±1 mA and the DAC is 10 bits wide, the two closest values you can determine are $2\text{ mA}/2^{10}$ apart, that is ~2 µA.

Figure 9.5 shows the conceptual schematic of a PMU in the FIMV configuration. For the resistance values reported, the current forced into the load resistor is $(V_1 - V_2)/(2R_x)$. Changing R_x does change the range of the forced current.

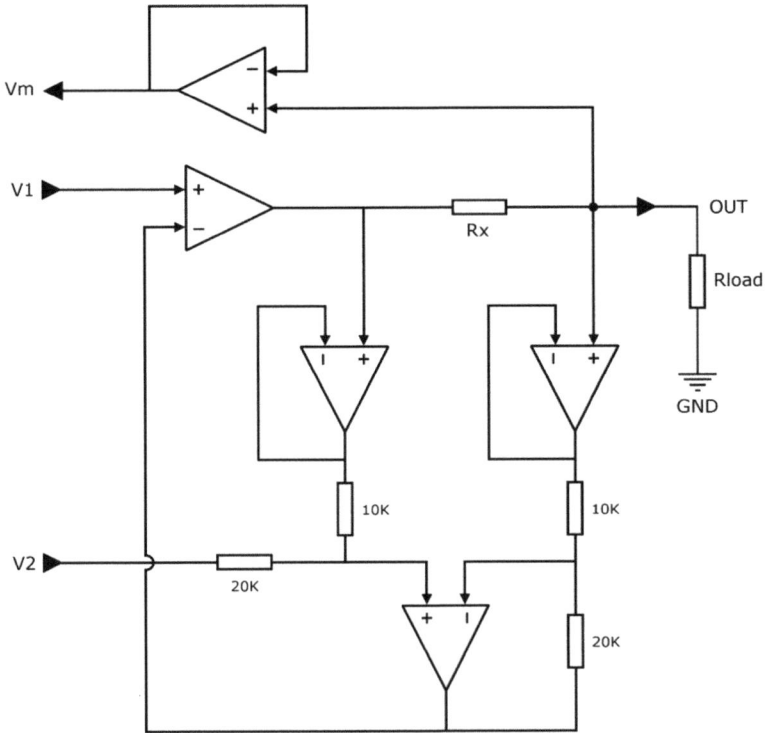

Figure 9.5 FIMV configuration of a PMU.

Exercise 9.3 Assuming the operational amplifier (OPAMP) behavior ideal, prove that for the simplified PMU reported in Fig. 9.5, the current forced through R_{load} is $(V_1 - V_2)/(2 \times R_x)$. How should 20 kΩ resistors' value be modified to make the forced current through R_{load} equal to $(V_1 - V_2)/(4 \times R_x)$?

Solution 9.3   Let's name V_a the voltage at the R_x left end and V_b the voltage at its right end. Considering the OPAMP ideal, the current drown at its inputs is zero, its gain is infinite, and the voltage at the inverting input (−) equals that at the non-inverting input (+).

The current flowing through R_{load} therefore equals that flowing through R_x, that is:

$$I_{Rload} = I_{rx} = (V_a - V_b)/R_x.$$

Voltage at bottom OPAMP non-inverting input (+) $= V_2 + (V_a - V_2) \times 2/3$.

Similarly, voltage at its inverting-input (−) $= V_1 + (V_b - V_1) \times 2/3$.

The fact that these values must be equal implies that $V_a - V_b = (V_1 - V_2)/2$.

Current through $R_{load} = I_{Rload} = (V_a - V_b)/R_x = (V_1 - V_2)/(2 \times R_x)$.

Doubling 20 kΩ resistors and repeating the above calculation leads to:

$$I_{rx} = I_{Rload} = (V_a - V_b)/R_x \rightarrow \underline{(V_1 - V_2)/(4 \times R_x)}.$$

9.2.1.3   Device power supply

The structure of a device power supply (DPS) is, in principle, even simpler. Normally, the DPS voltage is changed via macrocode a few times in a test flow. Making the DUT's power supply voltage vary with too complex a profile is an unusual task. Furthermore, a quite large capacitor is placed on the testing fixture near the DUT power pad to prevent a sudden current surge, and such a capacitor would filter complex waveforms. The possibility to control the slew rate is normally enough for flash testing purpose, and not all the testers have the ability to do that. The DPS has a characteristic not present in PE channels: the possibility to modify the voltage at its output to compensate voltage drops along the trace connecting to the DUT's pin. While a PE is expected to supply a small current, thereby suffering a small voltage drop on its way to the DUT's pin, being the relevant resistance fractions of ohms, such a drop is not negligible for currents typically drawn by a V_{dd} pin of a memory running operative modes. So more pins at the tester connector are reserved to each DPS channel: two or more are related to the force branch (which is a tester output) and one to the sense (a tester input).

Separate branches are routed from the DPS source and sense pins at the tester interface connector to the DUT's pin and get shorted only in its proximity, possibly at the pin itself, by means of a dedicated socket with multiple contacts per pin or multiple probes per pad in the case of wafer testing. This allows the sense branch to measure the voltage present at the DUT's pin and the DPS circuitry to operate the necessary modification at the force output to keep the voltage at the DUT's pin on target; however, the current may vary on the force branch. This is possible because the DPS sense input has high impedance and the current flowing in the sense branch is negligible, as are the voltage drops along the sense branch. This is known as a Kelvin contact. The common ground may vary as well on the basis of the current flowing on ground traces, so the ground also has a sense pin to keep the ground value stable.

9.2.1.4   Control unit

Tester channels operate under the control of a dedicated central processing unit (CPU). The source code is compiled[g] to generate an executable file that controls the resources as a whole to generate the waveforms necessary to dialog with a flash command user interface (CUI).

Once a one-to-one correspondence is established between a tester digital channel and a DUT's pin, either the pin or the tester's channel connected to it is indicated with the same name.

A typical flash tester source code is structured into two levels: a high-level code (macrocode), used to set all the digital channels' static parameters (voltages, timings, and formats); and a low-level code (microcode), to run at speed patterns that comply with the above-defined parameters.

Voltages are all those listed in the above description of the PE: V_{IL}, V_{IH}, V_{OL}, V_{OH}, I_{OL}, I_{OH}, and V_{ref}.

Timings are the duration of a microcode instruction (period) and the instant (edges) of events occurring within the instruction (rise and fall times at input pins and the strobe times at output pins).

Formats specify the nature of events occurring at edges and period ends, for example, G2H (go to high), G2L (go to low), G2D (go

---

[g]Other testers interpret the source file rather than compile it.

to data, where data is an argument of the microcode instruction), and ED (expected data).

These parameters can be assigned to individual pins or pingroups.[h]

A macrocode source is much like the standard C code (sometimes it is C code). Besides libraries of functions to handle files (standard I/O) or perform math, a specific library of specific functions to interface tester resources[i] is available. Some not-time-critical measurements or pins' preconditioning[j] can be performed on the DUT at this level; the division between the two levels of coding is not always sharp. However, the time taken by a macrocode instruction to be executed is long, making its use inefficient to talk to the DUT for most of user operations and making it subject to interrupts, and therefore not repeatable.

Microcode, on the other hand, runs on a dedicated processor, which runs operations at speed and whose repeatability depends only on the precision of the clock that controls such a processor.

The microcode source looks more like an assembler language. In principle each source row contains a user-defined instruction that affects the DUT, a timeset and some operands to control the microcode execution sequence, and a set of available registers. The instruction can be a read, a write, a nop,[k] a set-to-high-impedance, or an arbitrary combination of these, each one addressed to a specific pin or pingroup (group on pins). A write instruction can be a composed write, if at each instruction call, a constant logic pattern is repeatedly issued to a pingroup (control pins), while a cycle-dependent pattern is issued to another pingroup (DUT's I/O pins). This last pattern is built combining counters that can be increased or decreased at each write instruction execution, according to user-defined rules. The existence of multiple counters allows the generation of complex patterns that can span the entire memory array with only a few lines of microcode coding the rule. A typical write-read combined instruction is when some DUT's pins (e.g., the control pins) are driven to a logical state while others (e.g., DUT's

---

[h]A set of pins allowed to be assigned a common property.
[i]Set and readback.
[j]"Preconditioning" here means apply a given voltage to a pin.
[k]Or a noop, meaning "no operation."

I/O pins) are read. This ability to generate a complex pattern in a few lines of microcode is a prerogative of the memory tester algorithmic pattern generator (APG). It is possible to keep microcode sources short (a microcode bank size is in the range of a few kilo instructions) because at testing, the different blocks of the memory utilize the same patterns. When each block requires a specific pattern to be programmed in (or read against), another resource is necessary, known as reference memory. Reference memory is mandatory for flash programmer equipment that customers use to duplicate a target pattern (e.g., a piece of firmware) into many blank flash units, but production testers may have none. Units, in fact, are sold with no pattern programmed, but fully erased (all one [all1]).

9.2.1.5   Capture memory: data buffer memory

Another essential resource in a memory tester is the error capture random access memory (ECR), a bank of fast-access-time dynamic random access memory (DRAM) cells.

A correspondence is established between each bit of the ECR and each flash cell in the array being tested. This correspondence can be one-to-one, but that is not necessarily the case.

Whenever the flash memory cells array is read during the test flow, error-free cells return a known pattern. For instance, in the case of fully erased memory, the expected pattern is a solid 0xFF, while in the case of memory entirely programmed (every flash cell in the array is programmed) the expected pattern is a solid 0x00. A known pattern is also read from an only partially programmed array, as a physical checkerboard (CK),[l] for instance. A physical CK is a pattern used to verify floating gate shorts between adjacent cells. If a bit is read as 0 while expected to be 1 (or vice versa), the corresponding bit in the ECR is marked. That is, the EXOR of a flash cell expected and actual read values are stored in the capture memory at the physical address of the relevant cell in the array. It is the programmer's responsibility to handle ECR address counters such that they reproduce exactly the flash memory array form factor. Fail bits can be accumulated on the ECR while different read operations, each one aimed to screen a different type of defectiveness in the memory, are run in sequence.

---

[l]As the name suggests, any cell (with the exception of edge ones) is surrounded by an equal number of programmed and erased cells.

When observed with a dedicated graphical viewer, the cumulative ECR content can reveal some characteristics of the failures that would not be evident otherwise. In an erased NOR memory sector, a couple of adjacent bits read as 0 could be ascribed to an open contact, while in a NAND memory this would cause a segment of consecutive vertical cells[m] to fail. In a production test, ECR is used to identify the elements to repair through the redundancy analysis. ECR is an expensive resource, and often it is not necessary to have a one-to-one correspondence between the memory under test and the ECR size. If the NAND memory under test can only repair defective blocks of pages or strings, it's useless to know the exact coordinate of the page within the block, so a page-size area of ECR, for repair purpose, is enough to describe each block. This already drastically reduces the amount of ECR needed by a 16 or 32 factor, depending upon how many cells wide a string is. Multiple bits failing on the same string cause, in this case, a single bit to be marked on the ECR area associated to a block. Data buffer memory (DBM) has characteristics similar to ECR, but rather than being used to store failing locations, it is used as a reference memory to program arbitrary content into memory under test[n] when necessary. DBM arbitrary content must be loaded from a file, typically at test program load, as at tester power-on it does contain a meaningless random combination of 1s and 0s. Certain tests may require the error pattern stored into the ECR to be written into dedicated locations of the DUT's array; therefore, it must be possible to switch memory bank usage from ECR to DBM in the same test program (dynamically). Wherever the pattern content can be expressed as a function of the addresses or a set of counters, which is quite common at test, the APG is used rather than the DBM.

A physical CK can be taken as an example of how the APG works: a word at a given address (characterized by a column and a row counter) is built by taking the column counter lsb xor'd with the row counter lsb and making each word's bit equal to the result. In a diagonal pattern, which is used to verify that array decoders work properly, each word's bit is set to the result of the bytewise AND of the least significant part of the column and row address.

---

[m]In a NAND flash, a set of consecutive array cells connected in series and delimited by two pass transistors (one connecting to a contact to a global bitline and the other to a sourceline) is called a string.

[n]Or read against it.

For other digital products one important resource to have is the vector memory or scan memory, to input long scan patterns on selected pins (scan_in) and read results from selected outputs (scan_out); this is not the case of flash memory.

9.2.1.6 Redundancy analysis processor

Throughout a test flow, consecutive readings may have marked a number of bits in the ECR (fail cells). At some point in the test, the program must understand if the DUT has enough spare resources to repair the defective locations spotted, and if so, it must find out the coordinates of the broken structures.[o] These coordinates must be programmed into the DUT dedicated nonvolatile cells to make the repair permanent.[p] Returning the fail coordinated is a task accomplished by the redundancy analysis processor, previously instructed with a set of rules that implement the repair constraints of the specific product. Some repair characteristics/constraints are how many bitlines, wordline couples, and blocks can be repaired per DUT; whether spare elements can be used individually or grouped; and whether each spare element can replace any fail, whatever its position is in the array. ECR scan, although executed by a dedicated processor, requires a not-negligible test time, even when the ECR size is only a fraction of the memory under test. So, normally, the redundancy analysis is run a few times per test flow (say one to three times) within a first wafer sort flow.

9.2.1.7 Other remarks on flash testers

The huge total available flash market is of utmost interest to corporations operating in the automated test equipment (ATE) business. Yet the peculiar requirements in terms of resources and high parallelism, in particular, make flash production test using testers that are not specifically targeted to flash virtually impossible. The high parallelism, necessary to compensate for the long test time due to intrinsically slow program and erase operations, requires these testers to be extremely compact. Tester compactness clashes

---

[o]Cells or strings or coupled wordlines or blocks.
[p]At least as long as the charge injected into the nonvolatile cell floating gate is retained. In the past, as an alternative to nonvolatile cells, dedicated traces were laser-cut after electrical test or broken with a high current at test (fuses).

with other testers' concept of modularity. In fact, non–flash testers, with parallelism limited to a few dice, are assembled by gathering on a chassis different boards (modules), each one containing one specific electrical channel[q] required by the product to be tested. So virtually there is one specific tester configuration for each product, although in practice a family of similar products adopts the same configuration, to allow a more flexible assignment of available testers to different products, as per the corporate planning directives. If electrical resources in the tester are enough to test more than a single DUT, for example, two or more, then they can be concurrently tested, provided special care is reserved for the usage of the arithmetic logic unit (ALU) and other resources that are unique per tester. This approach is known as multiDUT and is normally assisted by a dedicated library of functions.

Flash testers overturn this modularity concept. A site is the elementary unit that contains an ALU and each and every electrical resource required to test one DUT. A multiDUT approach is valid for flash testers too. In this case it applies to each site rather than to the whole tester.

The module is the physical board that contains one or more sites $(2, 4, \ldots 2^N)$. The definition "tester on a board" well describes this concept. A possible configuration for a flash tester could be a chassis populated with say 32 boards, each one with 4 sites, each site able to test 4 DUTs in a multiDUT approach, for $32 \times 4 \times 4 = 512$ DUTs tested in parallel. With such modularity, if a higher level of parallelism is required, sometimes a wider chassis and more powerful transformers can suffice. On the opposite end, a single board can be hosted in a small chassis to have a fully equipped engineering station, where production programs can be developed that can be later moved to high-parallelism production systems with minimum or no modification.

This type of modularity also simplifies flash testers' maintenance and dramatically cuts the tester downtime,[r] as once a problem is

---

[q]Or a set of equal channels.
[r]Down time (DT) is one of the metrices to evaluate a tester's productivity. It is expressed in time percentage and represents the time a system is not operative for issues or maintenance intervention versus the tester lifetime. The other metric is mean time between failures (MTBF), that is, the estimated elapsed time between intrinsic failures in a system operating continuously.

encountered, for example, a consistent fail condition on a site, the board with the suspect site is replaced with a spare one[s] to let production restart immediately after the new board calibration. Since the board contains most of the components that are more likely to fail in production, a small inventory of a single type of boards guarantees almost uninterrupted production.

Not all the silicon integrated device manufacturers (IDMs) decided to pursue the memory market, because of its peculiarity, its unpredictable short-term market behavior, and the huge investments associated with it. Similarly, not all ATE corporations decided to develop memory testers, and some decided to quit this market to refocus their resources on other, more stable markets.

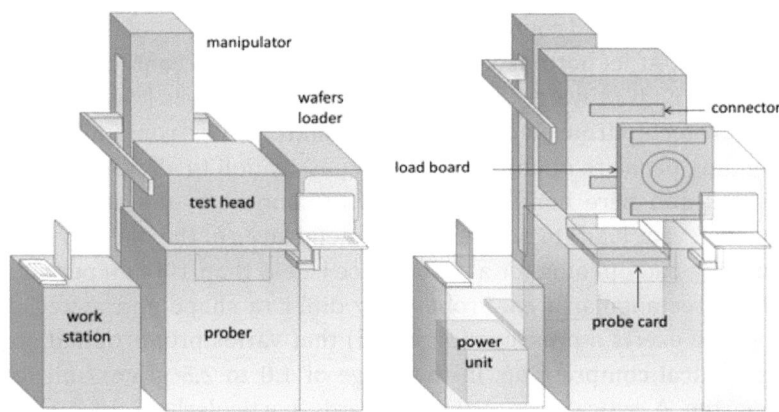

Figure 9.6 Electrical wafer-testing cell.

Figure 9.6 shows a complete wafer test cell. A workstation (or a PC) runs the program that communicates with the tester, and the spare room in the case can host the voltage transformers to power it. The workstation also stores test programs and data collected during a test session, which can be accessed remotely via a network connection. A bundle of cables connects the workstation to the test head. The test head is the heart of the tester. It needs to be positioned as close as possible to the units under test not to degrade significantly the signals involved. In the shown configuration a manipulator—a

---

[s]Often boards can be replaced on flash testers from a back site, so the system can remain docked, making this operation even faster.

servo-assisted pneumatic tool—helps the operators to quickly dock the test head to the prober or lift and rotate it with a small effort to replace the load board when a different product needs to be tested.

The load board and the probe card are two product-specific printed circuit boards (PCBs): they form the interface between the tester and the DUT. The load board docks to the tester test head's connectors, while the probe card is loaded into the prober through a frontal door and automatically placed and fastened in its place. The electrical connection between these two boards, once the test head is docked on top of the prober, is due to an interposed cylinder ring (pogo tower), built in the prober, containing enough vertical spring contacts (pogo pins) to assure the two facing boards' electrical continuity.

The probe card routes the tester signals to the pads of the dies in the wafer. At its bottom a probe card has probes to contact the die-bonding pads. A wafer is covered by a dielectric layer (passivation) that protects it from mechanical friction and chemical contamination. Bonding pads are openings etched in passivation to allow electrical probes and wire bonds (later used to connect to the assembled unit relevant load) to access the top metal layer of the die. A typical bonding pad opening for a flash device is less than 100 μm per side of the rectangular area. Probes may differ in shape and material. A probe exerts a pressure (on a pad) that varies proportionally to its vertical compression, in the range of 1.0 to 2.5 grams/mil for tungsten. A critical aspect of probe cards, particularly in the case of high parallelism, is to guarantee the planarity of all the probe tips, that is, to guarantee that all the probe tips lie in the narrow layer defined by two ideal close planes parallel to the card (and the wafer). The closer the distance between these planes, the more uniform is the value of the pressure mutually exerted between probe tips and wafer pads. Since perfect planarity is only an ideal goal, the probe's pressure variation with vertical compression cannot be too sharp. Technologies able to guarantee good planarity can tolerate minor pin flexibility compared to others, for an equal number of contacts. To give a reference number the planarity should not exceed 25 mm (1.0 mil). A probe ends with a sharp-edged tip to scrub a bit of the oxide present on the pad surface, thus ensuring a good electrical contact.

The leading-edge probe card industry took advantage of the latest findings in MEMS technology to develop a whole wafer test solution (Fig. 9.7).

Figure 9.7 A MEMS whole wafer probe card. Figure courtesy of SoulBrain MemSys.

A prober is a wafer robotic manipulator that picks one wafer at the time from a cassette (typically a 25-wafer-capable carrier) inserted in the wafer loader, places it on a chuck (a metallic cylinder whose diameter is slightly bigger than the wafer), and keeps it firmly in place by means of vacuum. The chuck, which can move along the three axes with a high degree of precision (fractions of micron) by means of brushless linear motors, places the wafer exactly below the probe card, letting each probe tip contact the relevant pad on the wafer die, before giving to the tester the green light to start. The tester also, once its test flow is completed, sends to the prober the green light to move the wafer to the next position, until the entire wafer test is completed. A robotic arm picks up tested wafers from the chuck and puts them in their original slots of the cassette.

Once the probe card is fastened in its housing, the position in the space of probe tips is measured with high precision by a video camera. Similarly, the position and angle of every wafer to be tested, once secured to the chuck, are accurately adjusted, via a preselected visual target recognition iterative procedure, to guarantee the best coupling between the probe card and the wafer. The chuck can be heated (or cooled) if the test requires it. While heating does not pose major issues, cooling the wafers requires a local nitrogen atmosphere to prevent frost-related issues (humidity present in the air would condense on the wafer surface, causing unwanted leakages, and on the tester circuits, causing serious damage). However, in the case of high parallelism, both hot and cold pose some constraints in the choice of probe card material. In this case, the probe card material's coefficient of thermal expansion must be as close as possible to that of the silicon; otherwise at different temperatures peripheral die pads could be out of reach of their probes, given the small size of the bonding pad opening. To prevent issues derived from accumulation of environmental pollutants on the delicate equipment and material treated, the wafer-level test is run in a clean room, in an environment with a controlled amount of particles in the air. A typical testing area clean room is class 1000 (named after the maximum number of particles permitted per cubic foot, about a thousandth of an urban environment value).

9.2.2 DUT Built-In Test-Oriented Resources

The datasheet is the master document that describes the characteristics of a flash memory product or a family of products. It includes the list of commands to perform standard operations. A manufacturer's restricted-use additional set of commands is available (test modes or reserved modes). These extra commands allow one to measure characteristics hidden from standard users and to modify both volatile and nonvolatile registers that change the way some operations are performed. They also allow stressing the memory array in ways that would be impossible to obtain, or extremely time consuming, using standard operations. Finally, they can correct a given number of cells or blocks showing as failed or underperforming at test (redundancy). Test modes may include a number of other functionalities, for example, product debug or technology endurance evaluation oriented. The only limits to the number of additional functions included are design time, silicon area occupation, and company-consolidated methodology.

Box 9.1   Testchips

When a company is ready to introduce a new, more productive technology step (typically a photolithography shrinking), products in its portfolio are gradually converted to it, starting with those generating the highest income (high runners).

In the case of shrinking, the design effort on derived products is mainly focused on re-aligning subunits impacted by the modifications in the electrical characteristics.

But when the new technology (or a radically different architecture choice) poses too many unknowns, it is not uncommon that companies decide to manufacture pure characterization devices (testchips) rather than proceeding directly with a commercial product.

Besides, of course, the object of the characterization, testchips include only the minimum set of characteristics to make them work. This cuts drastically the time required to design them. When the silicon is available, a quick test is performed at the wafer level just to screen out defective units and units are packaged into open-top packages used only for engineering purposes. Obviously, it's necessary that the results of the new solutions' characterization are available before making crucial decisions for the commercial product that is going to incorporate the new solutions. The number of pins (pincount) of a testchip is not necessarily correlated to the pincount of the product that will derive from it. A testchip may easily have a larger pincount than a product, for example, some of its pins may be nodes that in the product would be internal, controlled by circuitry not present in the testchip; in this case the tester may help to identify the best sequence of some operations' algorithm. Often more testchips targeted to different analyses can be placed on the same lithographic step (the elementary rectangular frame that gets repeated to cover the entire wafer surface), to evaluate possible alternative solutions to the same problem and to maximize the amount of information to collect at the same mask set cost.

Despite the fact that flash products of different brands run standard operations (almost) identically, and it is hard to distinguish them from outside other than for the manufacturer code, the internal structure may actually be radically different.

The brain of the product is a microcontroller (μC) or a programmable logic array (PLA). The μC can be a general-purpose type (in this case the flash manufacturer purchases the intellectual property [IP] from the company owning the μC rights and embeds it into the product) or designed from scratch by the manufacturer itself. Let us call the first generic µC and the latter custom µC.

Besides the µC, the other key blocks of a circuitry are page buffers in the case of NAND flash (or sense amplifiers in the case of NOR), decoders, and voltage regulators; we can refer to them as peripheral units.

A µC coordinates the activity of a peripheral unit both directly, through digital signals, and indirectly, through the analog voltages at its inputs (generated themselves from other µC-controlled peripheral units).

The instructions of the µC are stored in a dedicated ROM bank. A custom µC can communicate with specific peripheral units faster than a generic µC can do, having been designed to this purpose. If µC (or PLA) communication with peripheral subunits is too slow to have direct low-level control of their functionality, each peripheral subunit needs a local status machine to control the local process at speed, exchanging with a µC only high-level signals. A µC able to exert direct control on peripheral units at a low level offers higher flexibility than a generic µC, making it possible to operate deep changes to the operation's algorithms by modifying relevant read-only memory (ROM) instructions rather than modifying the local status machine's circuitry as well. In this case the cost modifying a complete set of instructions (ROM bank change) is limited to a mask respin of a via 2 mask (in a three-metal process).

A µC stores runtime variables to and retrieves them from latch registers.

Furthermore, at DUT power-on, the content of a bank of reserved flash cells is copied into specific latch registers that contain the parameters[t] used by algorithms and redundancy information. This reserved flash bank can be programmed at the wafer level in two (or more) steps. While algorithm parameters can be programmed early in the testing flow, redundancy information can only be programmed after all the tests pertaining to the memory array have been run. This reserved flash is generically named content addressable memory (CAM) after the architecture of flash cells used to store redundancy information in earlier products: when the user pointed to a failed location, its address matched a CAM address field content,

---

[t]We can call them parameter latches.

making its data field[u] select one of the available spare elements in replacement of the one originally pointed to. Today CAM is the name used to identify not only the reserved flash where redundancy information is stored but also those where the configuration bits are stored. If CAMs are not programmed, parameter latches get loaded with default values at power-on, which should guarantee acceptable functioning, though not optimized functioning. CAMs, like any other sector (block) of flash cells, can be erased in test mode.

At test, dedicated test modes allow writing directly into the parameter latches, overruling what is read from the relevant CAM at power-on. This makes it possible to evaluate different operations' behaviors before choosing the best one by programming the CAM.

Architectural choices, personal preferences, and a company's previous choices influence the way test modes are conceived and realized. A manufacturer may decide to have rigidly predefined library test modes, such as drain stress, gate stress, erase stress, direct memory access (DMA), and redundancy program.

A slightly more flexible approach is the possibility to choose, per each one of these tests, from among a limited set of voltage and duration configurations or to pass critical voltages through selected pins and control critical durations by logic values applied to selected pins.

Another manufacturer may decide to grant a privileged user low-level control of peripheral units so the user can assemble test modes with the maximum flexibility. In this scenario, previously mentioned test modes, like stresses, are still doable but on providing a combination of low-level settings, parameter latches, and standard user mode (UM)[v] commands (whose action is influenced by concurrent settings) rather than a single predefined reserved command. This approach also makes it possible to build custom test modes to investigate unexpected issues that may emerge during a product's early life and were not necessarily simulated and validated in the design phase. For example, in the case of high current

---

[u]Address and data are two fields of a redundancy word, which gets programmed at test time. A third, single-bit field of this word specifies whether it is active or not (default case), that is, whether the spare element whose address is coded into the data field actually replaces the element whose address coincides with the address field.

[v]Any command listed in the datasheet (document addressed to the common user) goes under this name.

consumption, it could be possible to disable different peripheral units one at a time to narrow the analysis to a DUT-restricted circuitry only. The cost of this higher flexibility is a less readable program that requires good communication between the design and the test (and product) engineering teams.

Test modes require nonstandard commands (out of specification) to be issued. In the elder flash NOR CUI it was rather common to have a high voltage[w] steadily applied to a control pin for the entire duration of the test mode (or alternatively pulsed at test mode entry). New products became capable of generating, through internal charge pumps, the high voltage necessary to the program and erase operations, making the pin V_{PP}[x] obsolete. That made the product fully operative with a single supply, a very convenient feature in the case of portable devices. However, that allowed accidental changing of the flash content if signals applied to the CUI were wrongly interpreted as a test-mode modify commands! For such products, it became crucial to adopt special command patterns to prevent accidental modification of memory content, and unlock sequences (made of 0xAA and 0x55 intermittently applied to the address and data buses) became a common practice. This made it possible to design safe test mode entry, making the use of third-level voltage obsolete.[y] This remark applies to the NAND CUI as well.

9.2.2.1 DMA

DMA is a test mode that allows one to bypass the read circuitry normally connected at one end of the bitline to which the selected flash cell is linked and to route this node directly to a pin, so that the current drawn by the cell can be measured with an external meter (the tester PMU).

In conjunction with test mode allowing one to apply an arbitrary voltage value to the selected flash cell control gate (wordline),[z] the DMA can collect enough measures to draw the cell I_{ds}[aa] versus V_{gs}[ab] transcharacteristic. Figure 9.8 shows three cells' current plots: the green one is a reference cell; cells on its left (erased)

---

[w]This high voltage is called third level, in contrast to logical levels (high and low).
[x]High-voltage pin to input high voltage originally necessary in program and erase operations.
[y]With the additional advantage of having the same ESD protection on all inputs.
[z]And with the additional assumption of keeping V_d fixed to 1.0 V typically.
[aa]Drain–source current.
[ab]Gate–source voltage.

are conventionally given the logic value 1, while those on its right (programmed) are given the value 0. Incidentally, two cells' voltage threshold difference ΔV_{th} is proportional to the difference of charge (number of electrons) stored in their respective floating gates according to the formula

$$\Delta V_{th} = -Q/C_{tot} \tag{9.1}$$

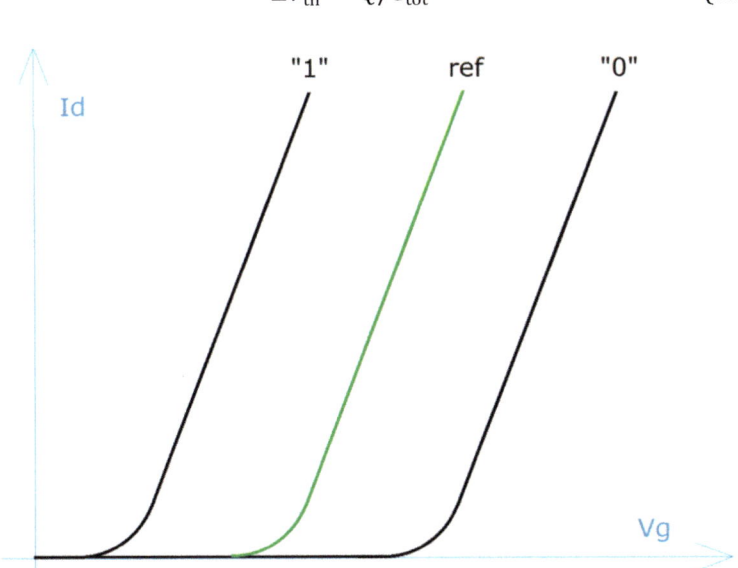

Figure 9.8 Flash cell transcharacteristic.

DMA is a key tool for NOR flash, where the I_{ds} of a cell is easily in the range of microamperes or tens of microamperes but becomes hardly usable in the case of NAND memory, where the serial structure of a string makes this current close to the resolution limits of the tester PMU. However, even in the case of NAND the I_{ds}, despite its small value, can be accumulated for a fixed time interval on a dedicated internal capacitor whose voltage is sampled and buffered to an output pin, to let the PMU measure a voltage proportional to the selected cell current.

9.2.2.2 Threshold distribution

A V_{th} distribution is a common visual representation of a population of cells obtained from repeated readings at different control gate voltages and a fixed reference current. Without going too much into details, it is worth mentioning how reading internally works in flash. NOR and NAND adopt different reading mechanisms.

In read, the current drawn by a NOR flash cell whose control gate is biased at V_{read}[ac] is compared to the current drawn by a reference cell whose threshold has been programmed to a precise threshold[ad] value at the wafer-level test. The reference control gate is also biased at V_{read}, and its current is I_{read}. If the cell current is lower than I_{read}, then the cell is read as programmed (0); otherwise it is read as erased (1).

On a NAND read, a source charges to a target voltage value the selected cell bitline and then switches to high impedance. From that time on, if the flash cell whose control gate is biased at V_{read}[ae] does not draw any current (programmed), the bitline keeps its precharged voltage value; otherwise its value is free to fall with a slope that is proportional to the current drawn by the cell. In fact, the bitline acts as a capacitor that integrates a cell's[af] current. After a fixed evaluation time the bitline voltage is digitized into a logic status by a latch and stored there for as long as necessary.

If we repeatedly read an ideal population of same-voltage-threshold[ag] NOR cells, increasing V_{gs} of a Δ each time, but this time keeping I_{read} constant,[ah] we would initially read a solid all-zero (all0) pattern and then suddenly a solid all1 pattern as soon as we cross the common threshold value. If we plot on a graph the number of cells read as one against V_{read}, we obtain a cumulative V_{th} distribution.[ai]

However, a more widely used plot is [ones(V_{gs}) – ones(V_{gs} – ΔV_{gs})] against V_{gs}, which for the ideal population considered is equal to the total number of cells for $V_{gs} = V_{th}$ and 0 elsewhere. A real

---

[ac]The control gate of any other cell connected to the same local bitline is grounded to not interfere (Fig. 9.1).

[ad]The reference cell is programmed at EWS to a threshold that guarantees enough current when biased at V_{read} (enough "overdrive" $V_{OV} = V_{gs} - V_{th}$).

[ae]The control gate of any other cell of the same string is biased at a high enough voltage V_{pass} to not interfere (Fig. 9.1).

[af]It is smaller than in NOR; that's why the read process is longer on NAND.

[ag]And the same in terms of $I_{ds} - V_{gs}$ characteristic slope.

[ah]The test mode that allows fixed Iref readings, necessary to build the V_{th} distribution, can be also called fast DMA, as it makes it possible, like DMA does, to build per points a cell $I_{ds} - V_{gs}$ characteristic. However, it uses the sense amplifier to this purpose and reads logical values from the DUT rather than measuring small currents with an external meter. Therefore, it is much faster than DMA.

[ai]A step plot in this example.

population distribution is more similar to the orange bell-shaped curve in Fig. 9.9. A double-y-axis figure is used to show the relation existing between V_{th} distribution and $I_{ds} - V_{gs}$ plots of the population minimum (a), typical (b), and maximum (c). If I_{ref} was negligible,[aj] the extremes of the distribution would coincide with the knee of curves (a) and (b). In a real case I_{ref} must be measurable[ak] by the product-specific circuitry and cannot be too small. The higher the I_{ref}, the more shifted to the right is the V_{th} distribution. This means that each V_{th} distribution has as an implicit parameter a reference current.

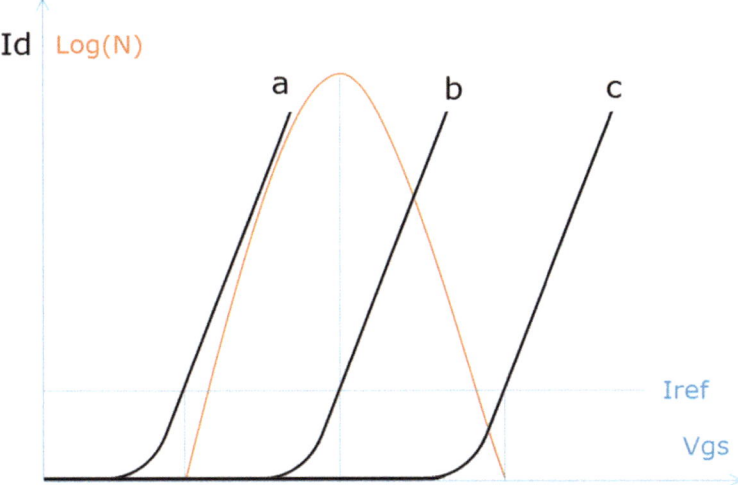

Figure 9.9 V_{th} distribution.

The NAND read mechanism does not compare cell I_{ds} to an external current: the implicit parameter in this case is the evaluation time duration, which is carefully identified once at product debug and from that point on is set equal for all units tested, making the V_{th} distributions comparison meaningful.

The plot's y axis is commonly expressed on a logarithmic scale to magnify the impact of the tails on which the analysis activity normally focuses.

[aj]Excluding the subthreshold area.
[ak]And must be read in a reasonable time.

Figure 9.10 shows two NAND product threshold distributions. The arrow indicates the effect of programming and erasing. The erased distribution is dotted because it is not observable, as V_{gs} cannot assume negative values. A NAND cell's population can be erased to negative V_{th}, and actually the erase verify (EV) V_{th} often does coincide with the 0, causing every cell to be pushed below 0 by the UM erase algorithm. During read, every cell in the string is turned on when a sufficiently high V_{pass} is applied to its control gate, except the one being read, which is biased at V_{read}. However, this works if the programmed cell's threshold is lower than V_{pass}, so it is necessary that programming operation does not push cells too high. It is enough that one cell in the string has a higher V_{th} value than V_{pass}, to cause any other cell in the string to be read as programmed.

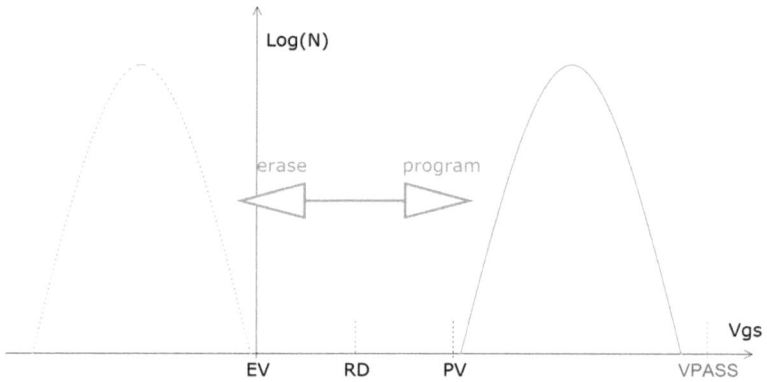

Figure 9.10   NAND programmed and erased V_{th}.

Conversely, in NOR (Fig. 9.11), there is no particular issue pertaining to the high-voltage-threshold cell but negative V_{th} cells are not tolerated as a cell with V_{th} below 0 (depleted) draws current when its control gate is grounded, therefore adding an offset to the current drawn by a cell intentionally selected and connected to the same local bitline, making the selected bit appear more erased than what it really is. During a UM program, a cell would consistently fail all program verifies following program pulses if a cell connected to its local bitline is depleted. That is why during user-mode erase command execution, in NOR flash, an additional step (reprogram or soft-program) follows the real erase phase. The goal of this further step is to move all negative V_{th} cells to the right enough to overpass a reference depletion verify (Fig. 9.10), having a positive V_{th} but very close to the origin of the axis. Sometimes this soft-program may cause some bits to overpass again EV, but that is one of the reasons why RD (read reference V_{th}) is placed at a safe distance from EV.

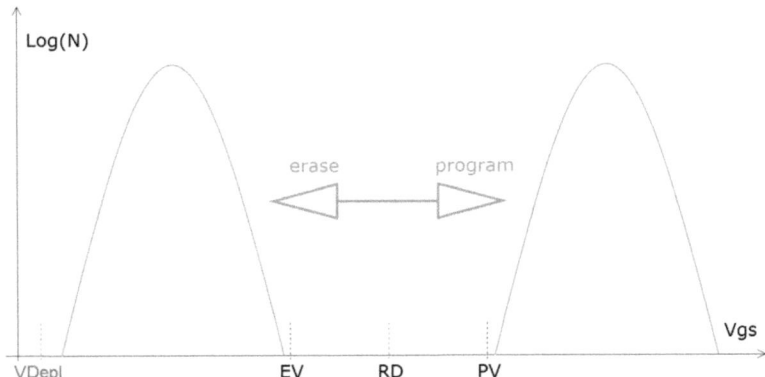

Figure 9.11   NOR programmed and erased V_{th} distributions.

Threshold distribution is an essential characterization tool to describe a population of cells and the effects induced on a population of cells by several causes: electrical stresses, poor retention, read and program/erase cycling, and read artifacts caused by near cells coupling, just to mention the common ones.

It is used to evaluate the effectiveness of erase and program algorithms and identify the best parameters for these operations. It allows to quickly identify tail bits of the distribution, whose coordinates can be found with bitmap viewers, to identify signatures that may help in the debug activity.

However, V_{th} distribution is rarely taken in production programs (or is limited to small samples data collection). The reason is that it implies multiple readings and failure counting at each step. What is rather common though is to identify the minimum or maximum V_{th} (tail voltages).

Let us assume we want to find the maximum V_{th} (distribution upper tail): starting from a V_{gs} that we know to be lower than the upper tail, we repeatedly run a reading against all1, interrupting each read at first fail encountered and restarting at increased V_{gs}

from the coordinates at which failure occurred during the previous iteration.[al] In this way we can get the wanted information with a single read (completed in multiple partial steps) of the array under analysis.

9.3   Test Flow

Once a wafer manufacturing phase is completed, it must be electrically tested before it can be sold. The products can reach the customer in the form of tested wafer[am] or packaged units, each one containing a die that ended successfully the electric test phase. The primary purpose of the electrical test is to detect and reject defective units, to trim some of their characteristics to make electrical behavior more uniform among units, and to repair as many defective memory cells as spare elements available for that purpose (redundancy). On other chips, trim and repair require blowing fuses or laser cuts, while on flash, the availability of nonvolatile memory cells makes it convenient to use a group of them for this purpose. Another target of testing is collecting data to constantly monitor the quality of the manufacturing process, to catch early drifts, and to allow timely corrective actions. It is very important to summarize the huge amount of information generated at this stage in a few figures easy to understand, share, and correlate. The market offers many software tools to extract meaningful charts from huge amounts of rough data.

The overall test flow consists of different steps (insertions). Figure 9.12 shows a scheme of the overall flow. Occasionally there may be more insertions than the four shown in the figure, or between the steps, units may undergo temperature stress. The reason why there are two separate insertions as the wafer level (electrical wafer sort 1 [EWS1] and EWS2) rather than a single one are different:

- EWS1 and EWS2 run at different temperatures.
- Units are heated between the two insertions to accelerate data loss phenomena.

[al]This is actually a best case, as in reading a failure at maximum speed can trigger an operation abort after a given time on a real tester (pipeline handling). However, the additional test time burden is negligible.
[am]Bare die or known good die (KGD).

- Different numbers of pins are contacted in the two insertions.

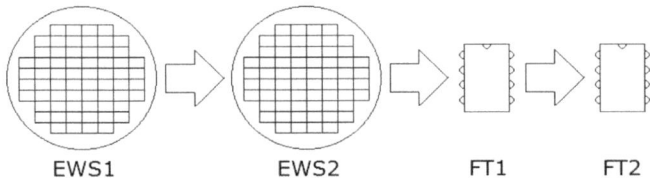

$$\text{EWS1} \qquad \text{EWS2} \qquad \text{FT1} \qquad \text{FT2}$$

Figure 9.12   Conceptual test flow.

A careful analysis of the results of different test insertions can sometimes reveal unexpected correlations between different classes of rejection and can allow removing redundant tests in the overall flow; the target is to have a specific defectiveness captured by a specific test.

9.3.1   Wafer Sort

The EWS[an] is a key step in the flash testing process (Fig. 9.13). A defective DUT should be rejected as early as possible at the wafer level, as the cost of the assembly should be reserved to good units. Furthermore, some products (stacked) contain in the same package more units of the same die, and chipmakers cannot afford throwing good units as they share the package with a bad one. In the past dies were tested one at a time; at that time the common practice was to reject each unit as early as possible in the flow, to spare test time. Nowadays this is not required as multiple dies are tested in parallel and the odds that all of them will fail to complete the entire test flow are negligible; and a single die still testing prevents the tester from giving the prober the green light to step to the next set of dies. As it is no longer useful to interrupt a bad die test as early as possible, it is possible to gather information regarding failing units too, with the limitation that after the failure is detected units cannot run a different flow with respect to the ones still good. Figure 9.14 reports a wafer sort flow. The column on the left represents the first wafer level insertion and the column on the right the second. The purpose is to provide a general idea of a test flow, and a lot of details have been omitted. Furthermore, to avoid a too-fragmented description

---

[an]Electrical wafer sort is also known as wafer sort or probe test.

derived from different terminologies and characteristics of NOR and NAND flash, this example is focused on a NAND product.

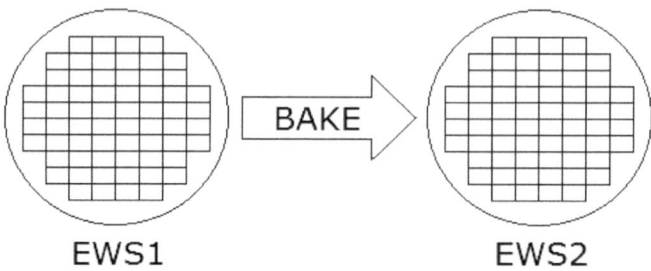

Figure 9.13   Wafer sort steps.

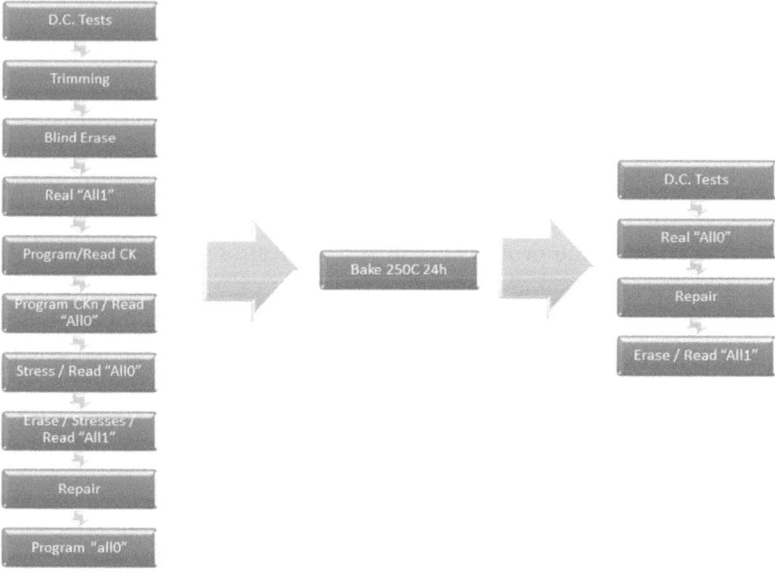

Figure 9.14   Wafer flow overall test.

- **D.C. Test:** Check the electrical continuity of each pin, their lack of leakages, and current drawn by power pins (V_{dd} and V_{ddq}). Failures at these tests are unrecoverable; units failing them have to be rejected. Bitline leakage is measured by keeping all the wordlines deselected while reading all0 a wordline per block. If test modes exist to detect shorts between adjacent wordlines and between adjacent bitlines, this check can be also performed at this point.

- **Trimming:** By design, all memory chips in a wafer should behave identically. In reality random defectiveness and small differences cumulated during the wafer manufacturing process make each die different from any other in many aspects. As a consequence, any analog measurement, collected on a population of dice, follows a statistical distribution. The point is that the values of some parameters acting as internal reference in product operations must be close to their nominal values with a small deviation admitted.

One or more register bits, both volatile (latches) and not volatile (one-time programmable [OTP][ao]), are associated to each one of these parameters. Each register bit modifies the parameter values of a known quantity, so a careful setting of these registers can recover off-target parameters, or at least most of them. Typical parameters to be adjusted on a die-to-die basis are bandgap, I_{ref}, µC, and charge pumps clocks. The search for the best values uses volatile latches, and only once the set of best values is complete, results are programmed into the configuration OTP to make the changes permanent, together with other information found during product debug and characterization. Therefore, at the next power-on, the die circuitry will work at its best.

- **Blind Erase:** Early flash could be erased by exposing wafers to UV light, as erasable programmable read-only memory (EPROM) did. This generated a narrow V_{th} distribution and allowed an easy detection of the outlier bits.

Today that is no longer possible due to tight technology and a denser circuitry. So an electrical erase is run first, to recover those cells spuriously programmed during the wafer manufacture process.

In UM erase, on a fully tested unit, erase pulses are repeated on a block until it gets fully erased. To verify that, after each erase pulse, the block is verified to allow a gap between the

---

[ao]Any flash cell can function as a OTP.

upper tail of the distribution and V_{read} to tolerate a small bit of threshold unwanted shift without impairing the read ability to distinguish 1s from 0s.

But at this stage, the presence of unrepaired bits stuck at a high voltage, or extremely low to erase, would cause the erase procedure to continue to its timeout on many blocks of the DUT, generating different distributions on blocks. To have similar block distributions, the erase time duration must be the same for all the blocks, and to that purpose it's sufficient to skip the verify phase after the erase, hence the term "blind erase."

- **Read All1:** A read all1, extended to the entire DUT, at a $V_{gs}{}^{ap}$ dependent on the blind erase used, is sufficient to identify open string-bitline contacts and slow-to-erase bits that, although not the immediate cause of failure, if not repaired could cause a constant overerase in the field to the block, affecting its ability to sustain repeated write-erase cycles.

  This test is preceded by a page buffer integrity check, obtained by setting and resetting each page buffer through proper datain patterns and comparing each pattern dataout to datain.

  It is worth noticing that in NAND architecture, a read all1 can be performed in an alternative, and faster, way: rather than reading one page of the block at the time, with the wordline of the page being read biased at $V_{eraseverify}$ and others at $V_{passread}$, all the wordlines in the block can be biased to the same V_{th} value. This concurrent reading will behave as a single-cell-equivalent reading. A single bit in a string not canceled enough will cause the UM erase to issue another erase pulse. However, it is not trivial to find a concurrent reading V_{th} that correlates to the $V_{eraseverify}$.

- **Program/Read CK:** At this point array cells are programmed. Bits that do not program at all or are simply too slow to program must be identified and replaced. Too many reiterated attempts to program a page, caused by one of more slow-to-program bit's resistance to overcome the program_verify level, would cause too much program disturb to the page sharing the same

---

$^{ap}V_{gs}$ applied to the control gates.

wordline in a UM program command. Therefore, as already done earlier for the erase, a single program pulse is applied blindly to all the pages. A CK^{aq} pattern is programmed to allow the following reading to detect shorts between close bits. In fact, if two cells have the floating gates shorted, programming one will cause both to appear partially programmed at a following reading. Programming a solid all0 throughout the whole DUT would mask this issue.

In NAND flash architecture two adjacent pages are often interlaced parts of the same wordline; if this is the case, programming a physical CK actually means programming a whole page with a solid all0 pattern, leaving the following page erased, doing the opposite for the next two pages, and so on.

It is actually the CK reading that follows the program, run at a V_{gs} value tailored to the characteristics of the blind program used, to detect shorted and slow-to-program bits. If two pages are interlaced on the same wordline as mentioned above, programmed pages will be read all0 and those left untouched will be read all1: at the array level this means reading a CK in two steps. On selected blocks only, before programming the CK, a diagonal pattern can be programmed to verify that decoding circuitry works properly, as a CK pattern because of its symmetric nature does not allow the detection of enough decode-related issues.

- **Program CKn/Read All0:** Cells left erased at the previous program are now programmed as well with a program CKn (a checkerboard complement). Then read all0 stores additional bits detected for slow program to the error capture ram for later repair. A test that would make sense to run at this point is a read all1 at a V_{th} set between V_{read} and $V_{passread}$ values; in fact, in reading, all wordlines are biased at $V_{passread}$ but the one containing the page to be read. If the program causes a cell V_{th} to overpass $V_{passread}$, the cell will always be off and no cell of that string will be readable.

[aq]A checkerboard means that each programmed cell is surrounded horizontally and vertically by erased cells and diagonally by programmed cells (and vice versa).

- **Stress/Read All0:** Now that all the cells are programmed above a known threshold, we can apply to the cells those electrical stresses whose expected result is a charge loss. One of the stress tests that could be placed here is the bulk stress. On blocks unselected (inhibited) during an erase command, capacitive coupling cause the wordlines, left floating, to follow the rising bulk potential, preventing these blocks' erasure. The existence of a sufficient voltage gap between the wordline and the bulk triggers, in fact, the Fowler–Nordheim tunneling mechanism, erasing the cells. However, in case the inhibition does not work properly on some blocks due to local issues, a spurious erase will take place, shifting the block distribution to the left enough to determine a massive failure at the subsequent read all0, run at previous test conditions.

- **Erase/Stresses/Read All1:** A single repair is scheduled at a later step in this hypothetical flow. If a first repair would occur at this stage, it would suddenly get rid of all failing bits cumulated so far in the flow, including those slow-erase bits preventing the UM erase, leaving to the test engineer the option of launching such erase. However, UM erase would not leave the best starting condition for subsequent stresses to run on erased because UM erase provides enough erase pulses to get the slowest-to-erase bit in the block erased. These bits vary across different blocks, so different blocks' V_{th} distributions may look not aligned after erase. Therefore stress, whose effect is a spurious programming of poor cells, may appear more severe on blocks with average V_{th} higher than others. A stress effect detection criteria setup is a mixture of theoretic work and experimental evidence.

Electrical stresses applied to the memory cells at test mimic those that cells will be subject to during their operative life, considering the worst case plus a further margin. The worst case can be easily found for operations that have a computable maximum number of times they can be repeated before conditions get reset. Programming is a good example of that, as once all the pages in a block have been programmed, the block in object will only be read or erased but none of its cells will endure any further program stress before the block is erased again. The datasheet states[ar] that each page programming can be fractioned in a maximum number of partial (NOP)[as] programming and reports what this number is. It also reports the number of pages per block. What the datasheet does not report is that two pages may actually be on the same wordline and what the maximum amount of program pulses is that the program operation can issue before reaching its timeout. Knowing these allows one to define a worst-case condition for the test, both for the V_{pass} stress and for the program stress.

- **Repair:** Redundancy analysis should be run at this point at least, but this does not mean it must be run at this point only. If a multi-repair-step strategy is adopted, wordline couples may have been repaired earlier in the flow, decoupling these two types of failures and allowing only single bit or bitline detection in the rest of the flow, making the redundancy analysis easier at this point of the flow.

  To activate the spare elements, CAMs need to be programmed. The activation of spare element poses some issues that each manufacture may handle differently. Spare parts of the array, in fact, should go through the same tests and stresses as the rest of the array, and when replaced, they should be in the same status (programmed/erased) as the rest of the array, unless the next operation is a modify, which brings all the array cells in a common condition.

  Testing may be cumbersome if you need special test modes to get to the spare area[at] rather than accessing it as an extension of the standard array. In this case, in fact, you basically need to duplicate most program and read tests. Fails encountered in the spare area before repair is enabled should not generate an immediate rejection of the DUT, because the defective spare element, in the end, may possibly not even be used. Yet, handling the spare part's failures in a highly optimized way adds a not-negligible level of complexity to the redundancy analysis.

[ar]At least in the case of single-level NAND flash.

[as]Each partial part of the page that can be programmed is often referred to as a "chunk."

[at]The same may be addressed for both a standard location and a spare location. If this is the case, a specific test mode allows switching between the two.

- **Program Read All0:** The DUT is powered-off and powered-on to make the changes effective. And the entire array is programmed as to charge all the cells' floating gates in preparation of the thermic step, which will follow. This program can be run in UM, with the DUT powered by a voltage that does not need to be nominal. As already mentioned, all the possible corners should be checked in the overall flow. The UM program programs the cells above the program_verify V_{th}, but the gap between this reference and worse programmed cell V_{th} is unknown. Optionally, a test could be added at this point to track after program distribution low-tail V_{th}, with respect to the program_verify (margin versus programmed [MVP] test).

- **D.C. Test:** The same tests are repeated as at the beginning of the first insertion. The main goal is to verify continuity to allow a prompt intervention in the case of wrong probe card Z overdrive, but another goal is to make sure that none of the tests at the first insertion caused any damage to the DUT.

- **Read All0:** The goal of retention bake applied between the two wafer-level test insertions is to accelerate the leakage of electrons from weak oxide cells' floating gates by increasing their average speed. Such leakage causes the affected cell to shift its voltage threshold to the left.

  Therefore, an all0 reading run at this point captures those bits whose voltage threshold has shifted too much to the left. If the reading uses as reference V_{th} the MVP stored at the end of the first insertion (minus a voltage margin of physiological loss and reading repeatability) this reading somehow adapts to the DUT process characteristics. An all0 distribution has a certain width, so capturing those cells that emerge from the left tail is especially severe for those cells that already were in the left part of the distribution after programming but before bake; a cell starting from the right of the distribution after all0 programming (let's name it A) may lose the entire distribution width as a consequence of the bake without been captured. You may object that we should not worry about A because, regardless of how much is lost, it is still read as programmed; however, in the customer case, A may have been the only one to be programmed surrounded by cells left canceled. The lack of their coupled charge would cause A V_{th} to appear much lower then what it was in an all0 distribution, causing it to fail a reading after enough time from the time of programming to reproduce the effect of the bake. Unfortunately, there is not much that can be done in this case. Repeating multiple retention bake on DUT programmed with different patterns is not realistic as it would imply too many insertions to make it not sustainable in terms of test cost. By the way, temperatures like those reached at retention bake can only be applied at the wafer level. Standard packages would not tolerate them.

- **Repair:** Redundancy analysis is repeated. Information from such analysis must be integrated with the redundancy information already present in the CAMs after the first repair occurred in the first insertion. Otherwise the information present in the CAMs may be used to mark the error capture random access memory (ECR) to allow previous all0 fail bits to be accumulated on it. The way CAMs have been designed may or may not allow an incremental programming of new encountered failures. If CAMs were placed in a block on flash organized as any other user block, the number of programming would be limited by the NOP, and it may be necessary to read the block's original content, store its image into the ECR, erase the bank, and then reprogram on it the saved image, updated with the new failures. This is not a trivial task, but it's not something the testing engineer cannot deal with.

- **Erase/Read All1:** After the wafers finish the two flows reported here, they are sewed to single chips to be placed in their final package, after the bad units are thrown away. Sewing (dicing) and packaging are stressful operations to the units, whose cells may undergo unwanted soft program or charge loss due to the stress experienced. So the first operation performed at the package test will normally be an electrical erase. Therefore, ending the wafer sort with an electrical erase (and relevant real all FF) may be debatable. However, this step necessary to those units that will not undergo any package test, because they are sold in the form of wafers to the final user (bare die or known good die). This concludes the wafer test.

Not too long ago, units rejected at the wafer test were ink-marked. Then a sticky foil was attached to the back of the wafer before this was sawed along the scribe lines. The sticky foil itself wasn't cut by the wafer sawing, and the ink-marked units were identified through an optical check and detached from the foil before moving them to the next step. More recently runtime-collected PASS/FAIL wafer-maps were directly passed to the sawing machine, avoiding the use of the ink since ink doesn't fit a clean area all. Furthermore, after wafer sort, wafers were back-grinded to reach the desired thickness for the assembly phase that followed and the dried-ink-layer thickness would cause mechanical stress to the units given the extremely low thickness reached at this point.

9.3.2   Final Test

The final test (FT)[au] is the electrical test that follows the unit packaging. While at probe test most of the steps use test modes, at this stage DUTs are tested mainly through UMs. It is important that considering the entire testing flow, unit functionality at all critical corners be checked. "Corner" is a jargon indicating an extreme bound of a validity range reported in the datasheet, be it a voltage, a temperature, or a current.

Since each unit sold is guaranteed to work at any corner, test conditions must include a bit more extreme values than those admitted by the datasheet.

The FT is expected to have a much higher yield than the EWS. Among the fails we expect to find at this stage, we count those related to bad bonding or data losses induced by temperatures encountered during the assembly process. It is not rare to retest units rejected at the FT when the results collected on a meaningful sample of units show an abnormally high rejection rate on a specific test site, suggesting a bad contact between the DUTs and the socket they are plugged into, while this is rarely an option at the wafer level. The test program must be designed to admit this possibility.

[au]The final test is also known as the factory test or the class test.

With few exceptions, units are sold in the erased form. The last step of the overall test flow is an all-array erase.

Box 9.2 User Mode Program

Programming flash cells relies on two different physic phenomena, NAND and NOR.

In a NOR flash cell, electrons from the source are accelerated by a lateral field to reach the drain area with enough kinetic energy to cross the thin tunnel oxide and reach the floating gate under a proper vertical field condition (channel hot electron injection [CHEI]).

In NAND the programming responsible mechanism is Fowler–Nordheim tunneling (the kind of tunneling occurring through the insulator triangular area in a band diagram), also responsible for the erase for both NAND and NOR.

In UM, the program command not only issues to the selected cells a programming pulse but also checks that each of them overpassed a predefined V_{th} value (PV = program verify); if that is not the case, another program pulse is applied to those cells still not enough programmed, while others are inhibited (with methods that are different for NOR and NAND). This sequence goes on until the slowest cell to program has passed the PV or the maximum number of trials is reached (whatever comes first). This iterative procedure, which keeps programming slow cells only, has the effect of making the program distribution tighter than what it would be if all the cells were given the number of pulses necessary to program the slowest one.

There are several reasons why a program sequence is used where pulses are not repeated on slow bits at the same voltage but at a voltage that grows with the number of pulses (incremental step pulse program [ISPP]). This strategy is mandatory on multilevel flash; in this case several distributions (three in the case of 2 bits per cell and seven in the case of 3 bits per cell) must be positioned with extreme precision along the V_{th} axis, and this approach guarantees each distribution width to be as large as the voltage increase (ΔV) in consecutive program pulses. However, even standard NOR flash benefits from this approach, not at the test phase but later, when the device has already been erased and reprogrammed multiple times. In fact, with cycling the working window, that is the distance between the worst-erased cell V_{th} and the worst-programmed cell V_{th}, becomes narrower, requiring more pulses to program effectively. At test, the ISPP maximum voltage theoretically reachable must be tried to prevent the unfortunate situation of this condition to be encountered for the first time in a DUT's life once installed on a customer board.

This is a general rule to observe in testing: any maximum voltage reached by every adaptive algorithm, plus a safety margin, must be tried at test first. The safety margin is to guarantee that conditions at test are worse than those ever encountered in each DUT life span. The yield loss at test for oxide break is largely preferable to later customer returns that adversely affect the manufacturer's image.

Incidentally, erase uses the same approach as program (incremental step pulse erase [ISPE]). The same precautions mentioned above for the ISPP at test extend to ISPE as well. While the program voltage grows slowly, in the case of NOR only, and is otherwise utilized to control V_{th} distribution width in multilevel products, erase ages quickly with write/erase cycling required a much faster voltage growth, and ISPE is essential to keep the overall erase time within the datasheet limits for this operation.

9.4   Brief History of Flash

The basic characteristic of flash memory is to keep the information stored in it when the power is turned off. Saying that flash is a nonvolatile type of memory means exactly that. EPROM, a flash ancestor, was invented by Dov Frohman in 1971. It could be electrically programmed one word at a time but needed to be exposed to ultraviolet (UV) light to be erased. This was a serious limitation to many applications, together with the high cost of the package, which needed a small window for the UV erase, yet this product had enormous success as it allowed a new methodologic approach to system design as code could be adjusted to get quickly to the target functionality with minimum hardware rework needed. In 1977 electrically erasable programmable read-only memory (EEPROM) was developed by Eli Harari. The first "E," in the acronym, for "electrically," indicates that the added feature to the EPROM is the capability to be electrically erased, although erase is on a word basis and the cost is a per cell decoding architecture that doesn't allow huge memory size products. EEPROM is still used in certain applications, while EPROM mainly exists in the legacy niche. However, those limitations were overcome when Fujio Masuoka invented the flash in the early 1980s.

9.5   Redundancy

In a memory die, the area taken by the array of cells is by far the widest. The ratio of the array area over entire chip is a key parameter in recognizing a properly designed flash memory chip. It is known as "array efficiency." The technology used for flash is very tight, intrinsically affected by process defectiveness. That is why a set of extra flash cells, normally extra columns and rows of cells, is present in the chip to replace nonfunctional cells. Without these extra elements it would be virtually impossible to have any yield. It's not a trivial task to foresee the proper number and nature of spare elements that should be part of a new product, particularly when it will be manufactured using a new technology. Adding too many spare structures impacts the array efficiency and makes the design and test more complex, yet underestimating defectivity may negatively impact the yield.

9.6   Cycling

Flash can be electrically programmed on a per-word (or per-page) basis and electrically erased on a per-sector basis (sector being a set of pages). Yet electrical erasure and programming (the modify operative modes) cannot be repeated indefinitely. The physical phenomena responsible for the erase and program operations cause stress that in the long run damages the cell structure, progressively impairing its capability to perform operations within datasheet time limits. The nature of such phenomena is behind this article's scope. From the testing standpoint, it must be verified on a statistically meaningful sample of DUTs that at the end of a cycling trial, operations are still performed in the specified time. Dedicated testers (cyclers) are normally used for this purpose. Cyclers are rather simple drivers that control many DUTs in parallel placed on a dedicated board. The cyclers' target is to provide a long sequence of program and erase commands, verifying that such commands are accepted and executed by the DUTs. DUTs that fail to perform one operation at a given cycle are marked as bad and disabled. A cycler can provide an indication of how program and erase time changes with cycling per DUT. More sophisticated cyclers can deal with the management of bad blocks in the case of NAND flash products. Normally finer electrical characterizations on cycled units are brought on using a tester.

It's always debatable whether the fast cycling trial operated by a cycler can actually mimic what in a real application spans the DUT's entire life. That's why it has become more and more relevant to agree on a usage model of a flash. Such a usage model depends on the application a flash product is going to be used for.

9.7 Retention

Although flash memory used in different applications may require the information to be retained for different amounts of time, the bottom line is that it should store uncorrupted information for as long as possible.

Once taken, a digital picture will be stored unchanged in a photo camera's secure digital (SD) card possibly for years, without any further refresh, while the parameter portion of a photo camera solid-state drive (SSD) will be updated quite often as long as the PC is used. Characterizing a product capability to retain uncorrupted information is of upmost importance. The typical data retention time reported in a datasheet is 10 years, and assessment of this capability must clearly rely on a mechanism to accelerate the process. Arrhenius law provides an expression to calculate how the temperature accelerates data loss, provided the activation energy E_a for this phenomenon. It allows expression of the accelerating factor (AF), $t_{T\text{usage}}/t_{T\text{stress}}$]), index of how slower data loss occurs at temperature T_{usage} than at T_{stress}, with the formula AF= $\exp([E_a/k] \times [1/T_{\text{usage}} - 1/T_{\text{stress}}])$.[av]

Temperature must be expressed in Kelvin (K). Assuming T_{usage} = 25°C (room temperature), we need to find T_{stress}, which, in time, compatible with the process qualification, induces the same data loss that we tolerate in 10 years (87,600 h) at T_{usage}.

[av]k is Boltzmann's constant = 8.617×10^{-5} eV/K.

A reasonable amount of time may for the stress is 6 weeks (1008 h), and assuming the activation energy E_a to be 0.8 eV, we must find the stress temperature related to AF = 10 years/6 weeks = 87,600 h/1008 h = 86.9. The formula returns T_{stress} = 74.9°C.

Exercise 9.4   For any new technology it is necessary to estimate the value of E_a.

On programmed samples, the first readout to show an unrecoverable amount of fails occurs after a 168 h bake at 120°C or after a 35 h bake at 150°C.

Can you provide an estimation of E_a?

Exercise 9.5   On programmed samples, the first readout to show an unrecoverable amount of fails occurs after a 240 h bake at 110°C or after a 24 h bake at 150°C.

How long should a bake at 100°C last to mimic a 10-year data loss at room temperature?

Solution 9.4

AF = t_1/t_2 = $\exp([E_a/k] \times [1/T_1 - 1/T_2]) \to \ln(t_1/t_2) = (E_a/k) \times (1/T_1 - 1/T_2) \to$

$E_a/k = \ln(t_1/t_2)/(1/T_1 - 1/T_2) \to E_a = k \times \ln(t_1/t_2)/(1/T_1 - 1/T_2) \to$

$E_a = 8.617 \times 10^{-5} \times \ln(168/35)/(1/[273.16 + 120] - 1/[273.16 + 150]) \to$

$E_a = 0.75$ eV.

As you may expect, the result is the same whatever you consider to be conditions 1 and 2.

Solution 9.5   Using the same approach as the previous exercise we start with finding E_a first:

AF = t_1/t_2 = $\exp([E_a/k] \times [1/ T_1 - 1/T_2]) \to$

$E_a = 8.617 \times 10^{-5} \times \ln(240/24)/(1/[273.16 + 110] - 1/[273.16 + 150]) \to$

$E_a = 0.805$ eV.

Then applying once again the formula, we calculate the duration of the trial at 100°C, observing that AF = 10 years/$t_{100°C}$, where $t_{100°C}$ is the duration of the trial at 100°C.

$E_a = k \times \ln(AF)/(1/T_{usage} - 1/T_{stress}) \rightarrow$

$0.805 = 8.617 \times 10^{-5} \times \ln(10 \times 365 \times 24/t_{100°C})/(1/[273.16 + 25] - 1/[273.16 + 100]) \rightarrow$

$\ln(10 \times 365 \times 24/X) = (1/[273.16 + 25] - 1/[273.16 + 100]) \times 0.805/[8.617 \times 10^{-5}] \rightarrow$

raising e to both terms $\rightarrow$

$3650 \times 24/t_{100°C} = \exp([1/(298.16) - 1/(373.16)] \times 0.805/[8.617 \times 10^{-5}]) \rightarrow$

$t_{100°C} = 3650 \times 24/\exp([1/(298.16) - 1/(373.16)] \times 0.805/[8.617 \times 10^{-5}]) \approx \mathbf{161\ h}.$

9.8   Silicon Debug/Design Validation

An activity that is not often ascribed to testing is silicon debug. However, this is often the first time a new product is interfaced with automated test equipment (ATE).

Normally, a production probe card is ready and debugged by the time the first fully processed wafer is handed over to the product engineer, as well as a test program, but it is pretty rare that any component involved in the process will work fine at the first trial. The bug can be in the test hardware, the test program, or the product itself. A well-planned design of experiment (DoE) and the experience of the product engineer are key factors in expedite the solution.

The production environment is not the best place to conduct a debug session.

9.9   Testing Readiness

As for any other high-tech productive environment, things must be planned well in advance to be prepared to the silicon out (SO). Many factors must concur to avoid dead time and to prevent a bottleneck.

A high-parallelism probe card, as that required to test flash memory, takes a long time to be designed, checked, and manufactured. So do the FT interface fixture and the boards to run qualification.

9.10   Characterization

Once a flash product reaches a given level of maturity, it must be verified against the datasheet, so a conventional number of units is tested at maximum and minimum voltages and temperature corners and must fulfill the datasheet requirements in such extreme conditions, adding a "safety" margin to the nominal limits. A dedicated test program is normally coded to this purpose, and the results are summarized in the form of tables for quick reference. This analysis is incompatible with the wafer-testing environment, as many of the parameters to assess signal transition timings are particularly disturbed by parasitic elements of the bulky interfaces used. So this analysis is done on a limited sample of assembled units using load boards that minimize disturbs and are easily interfaced to airflow temperature controllers.

9.10.1   Shmoo Plot

Some software tools' usage is limited to specific phases of the product life only. The usage of a shmoo plot, for instance, pertains to the characterization phase. This tool, curiously enough named after a cartoon strip character Al Capp created in 1948, repeats a given operation (typically a read of a portion of the cell array) varying two parameters of interest, say a voltage and a timing, between two boundaries with a fixed step. The result of each operation run (a point in a shmoo plot) is pass or fail, conventionally represented as a dot "." and an asterisk "*," respectively. The outcome is a picture as the one shown below.

The picture is an example, and it is not indicative of the nature of the characteristic being swept. The timing could be an access time (e.g., data valid after chip enabled) and the voltage V_{dd}.

```
 5  10  15  20  [ns]
----|----|----|----|
3.3  ********.............
3.2  *********...........
3.1  **********..........
3.0  ***********.........
2.9  ***********.........
2.8  ***********........
2.7  ***********........

[V]
```

To an expert eye the shmoo plot provides at a glance valuable information of how the product working window and its margins toward the minimum/maximum admitted values are declared in the datasheet. Besides being run at room temperature, a shmoo is repeated on units that have been heated (cooled) corner temperatures in order to guarantee they correctly operate in a wider range of conditions. Running shmoos is a time-consuming activity, so it is a common practice that long lists of parameters are combined in long batch files to be later run sequentially in automatic mode. In this case the tester must also control the heater/cooler unit (thermostream), alternating the temperature with special care to prevent the load board (and potentially the test head) from overheating or overcooling. Overcooling is particularly dangerous as it causes the environment air humidity to condense on the load board's backside and freeze, causing short circuits and potentially severe damage to the tester.

What is just described is an example of a standard shmoo. Often sets of related variables are changed rather than a single one: while increasing V_{cc} by a 100 mV step, one may want to concurrently increase V_{IH} by the same amount or V_{th} by half the step width.

To prevent misinterpretation of the results, it is the programmer's responsibility to guarantee that the result of the operation is affected exclusively by the parameters being varied within the shmoo, setting any other parameter to a value that does not cause a failure, whatever is the value assumed by the parameter that we are going to sweep.

Another feature shmoo may offer is to show how consistent is a given result for a given couple of parameters. The picture below refers to the same setup already considered earlier; provided the measure has been repeated 6 times for each point of the shmoo plot, the number indicates the occurrence of pass results. Some uncertainty is expected at the edge between the pass area and the fail area. The picture indicates an issue may have been passed unnoticed had a conventional shmoo been run. Can you spot it?

```
5 10 15 20 [NS]
----|----|----|----|
3.3 *******1466666666666
3.2 *********56665666666
3.1 *********16666666666
3.0 ***********666666666
2.9 ***********666666666
2.8 ***********666666666
2.7 ***********566666666
[V]
```

A shmoo plot can be enriched in many different ways. Each point in it could be the result of a measurement rather than a simple pass or fail flag, or an additional parameter could be swept, making the plot 3D. The shmoo plot could be part of an interactive graphic user interface (GUI) rather than a datalog made of a sequence of ASCII characters.

In that case the tool may offer additional features, like the possibility of repeating run for a given position of the shmoo plot, pointed out by the mouse cursor, without necessarily running all and every combination of parameters swept by the plot.

Exercise 9.6 Would it be possible to use the shmoo tool to visualize output pin transition? How would you do that? What are the limitations of this technique, if any?

Exercise 9.7 Could a shmoo tool be used to characterize operations other than read? Would it work to investigate erase efficiency dependency on V_{dd}?

Solution 9.6 Restrict the analysis to a single pin, unless you want to capture the envelope of multiple output pin transitions. Read a specific address's word, unless you want to capture the envelope of nearby cells' readings. Set expected data valid as format.[aw] Select

---

[aw]Such format returns pass whenever the output pin level is higher than VOH or lower than VOL.

$V_{OH} = V_{OL}$ (=V_{th} if active loads are on) and T_{strobe} as variables to be swept. The limitation is that the shape of the transition gets reconstructed as a result of multiple readings, rather than a single one as it happens for an oscilloscope. The concept adopted here is similar to that of the sampling oscilloscope.

Solution 9.7   Virtually, any operation that can return a PASS or FAIL condition can be used to build a shmoo plot. The PASS or FAIL condition can pertain to an entire test or a subset of it only (e.g., a pattern run). However, an erase run is largely influenced by the previous operation run on the block/sector under investigation, by its previous status (erased/programmed), and by many other factors (including the relaxation time allowed between sequential erases). Therefore shmoo is rarely used to characterize modify (program/erase) operations.

9.11   Qualification

Every company operating in flash has its own process to evaluate product robustness, a process known as qualification. The aim of qualification is to catch each and every issue that could otherwise emerge on customer application. A statistically meaningful sample of units has to pass different trials, ranging from package to electrical stress. Guidelines of the qualification process have been coded by the organization JEDEC.

9.12   Datasheet

The datasheet is the master document describing in detail how a product, or a family of products, works.

It reports any information useful to the final user. For example, it provides:

- A general description
- A pin name and function
- A product block diagram
- The command set

- DC characteristics
- AC characteristics
- Endurance characteristics
- Package details
- Order code

9.12.1   Product General Description

This section includes a brief product description. It indicates the size of the memory, its nominal voltage, and the size of individually erasable blocks and individually programmable pages. It also describes the functionality supported, cycling performances, endurance data, and special features if present.

9.12.2   Pin Name and Function

This section describes each pin. Figure 9.15 shows the pinout of two ideal flash products, one parallel NOR and one NAND type. Pins in common to both types are power-related ones, some of the controls, ready-busy, and an I/O bus.

Parallel NOR uses a dedicated input bus (address) to set the address of the word to be read or programmed and a different bidirectional (I/O) bus to exchange data relevant to such address. NAND, on the other hand, shares the I/O bus to set addresses and exchange data in sequential serial steps.

The size (number of bits) increases on a flash product according to Moore's law, and the number of bits necessary to pass a word address grows as well. For this reason the NOR flash number of pins rapidly grew very large; and when the I/O bus increased from 8 bits to 16 bits, the pin count became even larger. NAND products are different: a wider address or word side just causes an operation to be serialized in a larger number of serial steps.

Sharing the same bus presents many advantages in terms of testing hardware: it needs fewer tester physical resources, which means higher possible parallelism with the same amount of resources, and the possibility to reuse equipment for different generations of products, as long as the size of the package doesn't change. Yet the tester may need an upgrade in terms of memory, where to store the image of the fail location at the wafer sort, and SW capabilities to perform a faster redundancy analysis.

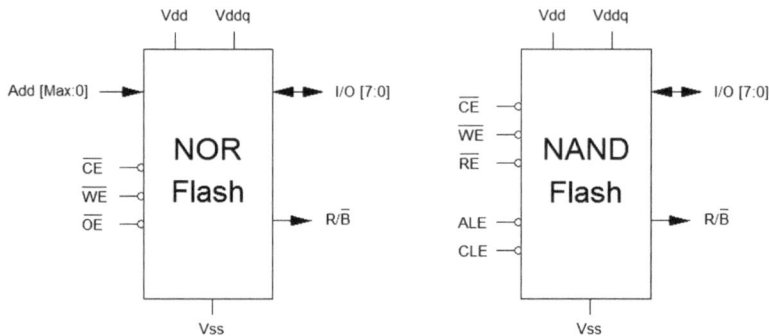

Figure 9.15 NOR and NAND pinout.

Another type of flash exists (serial), which has a low pin count and quite a different interface.

9.12.3 Product Conceptual Schematic

This is a block diagram that shows a basic level of details of the product architecture. It is an introductive figure that does not provide any operative instruction but helps to introduce the product to new users.

9.12.4 Command Set

This section reports the logical values to be applied to the DUT input pins and what the DUT is expected to put at output pins in return. Some pins can actually work as both input and output (I/O pins); in this case their function in a particular moment derives from the operation, or phase of the operation, being executed. Both NAND and NOR flash have a set of I/O pins, used to pass a command or write/ read data to be programmed into the array (I/O bus).

However, a NOR product has distinct sets of pins for addresses and data/commands while a NAND product exploits this characteristic even further, passing the addresses as well through the I/O bus. Each pin in a bus has a certain weight, so each set can be represented in hexadecimal format rather than in binary. Most of the commands are common in the market, so there it is relatively straightforward to replace a component from a vendor with one of another vendor.

9.12.5   DC Characteristics

This section reports the minimum voltage recognized as a logical high and the maximum voltage recognized as a logical low at the DUT's input. It also reports the DUT's output logic high-state minimum voltage when driving spec. current I_{OH} and logic low maximum voltage when sinking spec. current I_{OL}. The maximum current the unit sinks while performing main operations (I_{dd} and I_{ddq}) is listed here as well. Values reported in this section and in the next must result far enough from those causing units to fail and require a huge validation phase of the product on wide statistics collected at various temperatures. Parameters here reported can be easily compared with competitor products in terms of performance, and a manufacturer is largely measured on them. So one of the main aims of multiple masks respin and process recipe tuning is to get to values.

9.12.6   AC Characteristics

This section reports the timing prescriptions' constraints that signals must comply with to effectively dialog with the DUT. For each available operation (read, write, program, erase, etc.) information is reported both in the form of a table and in the form of a time diagram to capture any mutual dependency between signals at a glance. An anacronym marks each critical interval separating two events in a diagram. Events can pertain to different signals/buses or the same one, as for instance AVAV (JEDEC for address valid to next address valid). The table reports a validity range per each critical interval (identified by its acronyms in the related diagram).

Most of the critical intervals are of the setup and hold type.

The setup time indicates how long in advance a given condition must be reliably established on a signal/bus to let the event occurring on another signal/bus cause a successful operation. In other words, it is the minimum interval between an arm and a trigger condition involving two distinct signals (or buses) that play an active role in the execution of a certain operation.

Similarly, the hold time indicates for how long a given signal/ bus must be kept in its state after an event has occurred on another signal/bus, to produce a successful operation.

The intervals mentioned above show a huge variation with the temperature.

Therefore, although the tables do not state it clearly, the interval values are those guaranteed by the manufacturer throughout the entire product validity temperature range. Practically they are a collection of worst cases.

9.12.7   Endurance Characteristics

Endurance characteristics pertain mainly how many time blocks can be programmed/erased or read before their behavior is no longer guaranteed to respect the specification parameters. PE stresses flash cells differently than reading; yet the way these operations degrade are related.

A block undergoing multiple program/erase cycles is more prone to the read stress effect than a fresh one, as, incidentally, it's more prone to charge loss. Multilevel cells, as well as 3D technology architecture and tighter technology nodes, all have a negative impact on the ability to sustain program/erase cycles and data robustness. Rather than mentioning a maximum value per single endurance parameter, recent datasheets, rather, include figures that show the relationship existing between these parameters.

9.12.8   Package Dimensions

This section reports the detailed size information of the product. This is a key info not only for the final user who is going to assemble units on applications but also to build the sockets to test them and design the trays used to store them.

9.12.9   Order Code

Information reported here does not have much technical importance. But for the purchase department of the company that is going to use the units in its devices or the retailer that is going to buy units for its store, this information is key to ordering the wanted product in the proper package.

What the datasheet does not cover is the set of reserved commands that each manufacture uses at test to screen defectiveness and repair and configure each die in the least time possible.

9.13   Datasheet Gray Areas

The datasheet is the product master document. It should contain any information the customer needs to use on the applications. That is the reason this document receives particular care right from its first draft. Were other pieces of device-relevant information written with the same care, life would be much easier for testing and product engineers while developing and debugging a test program. Unfortunately, design communication to test engineering (and vice versa) flow is not likewise well planned, but that's another story. Each statement in a datasheet can cause endless discussion afterward with the customer. One may think that each behavior of the device could be exactly interpreted after a careful read of the datasheet. Actually, this is not completely true. Two datasheets from two different manufacturers may look exactly the same, but that doesn't necessarily mean that the two products behave the same way. The datasheet simply cannot describe any conceivable combination of conditions the DUT may experience, and whatever is not explicitly described in detail in the datasheet could actually be different. This should not be a problem in theory, as any sequence not described in that document is by definition not supported by the manufacturer. However, it is not uncommon that if you use two devices from two different manufacturers on a third-party application, this won't work with one of the two. How is that possible? The reason is that the third party may have run some operation a bit out of specification, although obviously not on purpose. The third party priority is to debug the application enough to make it work as expected, no matter if the specification is strictly respected in every minimal detail. But if another manufacturer tries to sell its product as a one-to-one replacement of that currently used by the third party, it will have a hard time convincing the potential customer that the device doesn't work on the application as something not supported was implemented. It is more advisable to align as much as possible to the "out-of-specification" behaviors as well. This implies that in the design phase, unless your product is the first its kind in the market, it is useful to have competition units to characterize to align as much as possible with the behavior not explicitly treated by the datasheet.

This is not an easy task though, because it is difficult to figure out the number of possible out-of-specification configurations to include in your design. The testing phase often starts simultaneously with the design phase, on units available on the market of competitor manufactures.

9.14   Error Correction Code

Besides the array defectiveness that is found at wafer test, which can be repaired through redundancy, failures can occur during the operative life of a product, either when conditions unverified at test are encountered or because of the unavoidable flash cell performance degradation (aging). In the first case, provided the failure mechanism gets well understood and a procedure able to catch it can be coded on the tester, the procedure is added to the test program, extending its coverage and preventing other units affected by this bug to reach the market. Also failures induced by flash cell aging, occurring in program/erase, can be handled to some extent, but this requires the circuit connected to the flash in the application to take care of the issue by tracking the position of failures to cease using them from that point onward.

Erratic behavior of a cell, or a particular coupling with nearby cells (which is particularly bad when the cell stores more than 1 bit), can make a cell read fail. Redundancy, of course, cannot cope with such a problem. Another data integrity preserving mechanism must be adopted in this case. One such method is the error correction code (ECC).

In the past the objective of the production test flow was to select zero-defect memory units, which is an essential requirement when storing operative system and, in general, any information that is not data loss tolerant.

More tight technology nodes, lower current architectures (NAND flash having replaced NOR flash for its better bits/area figure), 3D architecture, and multiple bits storage per cell made a certain density of failures in read unavoidable. With the introduction of NAND products, datasheets started to mention the maximum number of read defects tolerated for a given page, reflected by the bit error rate (BER), that is the number of fail bits in a sample over the sample size, in other words the fail density.

This made it impossible to use raw NAND to store sensitive data and opened the way to two possible countermeasures: embedding an ECC controller in the product and/or using an external controller. The first option implies the flash manufacturer needs to develop expertise in a field that has nothing in common with what was done previously. The second boosted the activity of semiconductor companies that were not in memory business earlier but could join the race to develop such controllers. Both approaches were actually pursued. Nowadays, the second approach is more widely adopted. The availability of controllers that could manage different levels of defectiveness at the cost of a variable amount of spare cells made the memory maker reconsider approaches previously not usable due to their high BER, as in the case of multilevel cells.

We may cut here the ECC description, because with a few exceptions they took a different path and became an external appendix of the memory, often sharing the same package. But it is interesting to mention at least the basics of this approach.

You are probably familiar with the concept of parity check. Assume you (sender) want to transmit a message, say 7 bits wide, to a receiver, through a noisy channel; you can count how many 1s your message contains and set an additional bit (parity) to 1 if the count is odd and 0 if even. So this "extended" message (code) will now be 8 bits wide instead of 7 and by construction it will have an even number of 1s. Such code is less prone to transmission channel issues than the original message. In fact, the receiver expects a code with an even number of 1s. Were any (single) bit flipped in the transmission process, the result would be a spoiled code with an odd number of 1s. If the receiver has the chance to ask the sender to resend the defective code, then the receiver wouldn't care to identify the position of the flipped bit.

I have used the terms "transmit" and "message" as the frame theory involved mainly pertains to communication through noisy channels. In the case of flash memory "You are transmitting through *time* rather than *space*" [3]. In fact, in flash, ECC mostly copes with data retention issues, meaning that bits can be flipped when the number of electrons stored in a floating gate drops below a certain number, causing that cell to be read as a 1 rather than a 0.[ax]

How does ECC work, in short? You may see it as an extension of the parity check. Assume you have a 4-bit-wide message that you want to send through a noisy channel to a receiver. How many additional bits should you add to it to build a code tolerant to 1-bit loss? The answer is 3, and there is an intuitive visual representation of this that shows clearly how this concept can be seen as an extension of the parity check.

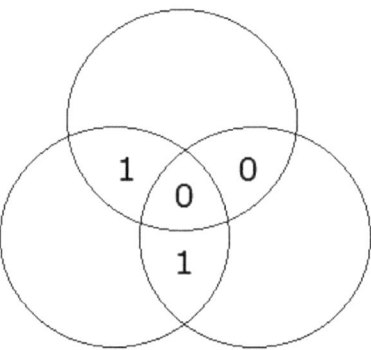

Figure 9.16   A 4-bit message.

The 1s and 0s in our message (Fig. 9.16) are spread over three partially superimposed circles.

If you write in the empty part of each circle the parity bit of the submessage made of the bits already present in that circle, you have generated a code.

You can conventionally name each portion of the picture as in Figure 9.17. Assuming d3 is the most significant bit (msb) of a message *u* and d0 its least significant bit (lsb), and so on so forth,

---

[ax]Conventionally, a programmed flash cell is given the value 0 and an erased cell the value 1.

and p3 takes the leftmost position in u, Fig. 9.16 content can be read as 0101.

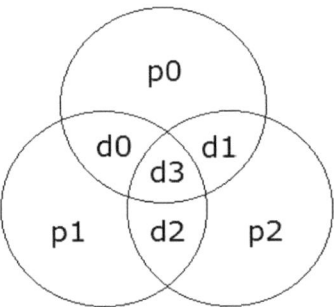

Figure 9.17 Message (d1 . . . d4) and parity (p1 . . . p3) bits.

The same convention applied to parity brings 101 as a result of the parity. Finally, if you combine message u and parity you obtain the code x 0101101. You can easily build a code table by giving the message all possible values and working out the resulting parity.

For those familiar with matrix algebra, code x can be derived from each message u by multiplying u (considered as a single-row array) for a matrix G, also known as a *generator* matrix.

$$G = \begin{vmatrix} 1 & 0 & 0 & 0 & 1 & 1 & 1 \\ 0 & 1 & 0 & 0 & 1 & 1 & 0 \\ 0 & 0 & 1 & 0 & 1 & 0 & 1 \\ 0 & 0 & 0 & 1 & 0 & 1 & 1 \end{vmatrix}$$

The outcome of the product is a row, and the result (code) is its module 2.

$$x = (u \times G) \bmod (2).$$

There isn't a single G. However, the one reported above has the peculiarity that the resulting code contains in its four most significant digits the message itself:

$$\left(\begin{vmatrix} 0 & 0 & 1 & 1 \end{vmatrix} \cdot \begin{vmatrix} 1 & 0 & 0 & 0 & 1 & 1 & 1 \\ 0 & 1 & 0 & 0 & 1 & 1 & 0 \\ 0 & 0 & 1 & 0 & 1 & 0 & 1 \\ 0 & 0 & 0 & 1 & 0 & 1 & 1 \end{vmatrix} \right) \bmod (2) = \begin{vmatrix} 0 & 0 & 1 & 1 & 1 & 1 & 0 \end{vmatrix}$$

Another array H (parity matrix) can be derived from G. Note that G is composed of an identity matrix (in black) and a matrix (in red).

Matrix *H* is made of the same red submatrix superimposed to an identity matrix (in black).

$$H = \begin{vmatrix} 1 & 1 & 1 \\ 1 & 1 & 0 \\ 1 & 0 & 1 \\ 0 & 1 & 1 \\ 1 & 0 & 0 \\ 0 & 1 & 0 \\ 0 & 0 & 1 \end{vmatrix}$$

This matrix *H* is important to decode. Every code in Table 9.1, if multiplied for matrix *H* and (again) considered module 2, returns a null row vector, of three digits. However, if you flip a single cypher in a code and repeat the operation, the result will indicate the position of the flipped cypher. Let's consider the code found above and change a 1 into a 0 (in green):

$$\left(|\,0 \ 0 \ 1 \ 1 \ 1 \ 0 \ 0\,| \cdot \begin{vmatrix} 1 & 1 & 1 \\ 1 & 1 & 0 \\ 1 & 0 & 1 \\ 0 & 1 & 1 \\ 1 & 0 & 0 \\ 0 & 1 & 0 \\ 0 & 0 & 1 \end{vmatrix} \right) \mod (2) = |\,0 \ 1 \ 0\,|$$

Table 9.1 Messages and relevant parity bits

Message *u*	Parity	Message *u*	Parity
0000	000	1000	111
0001	011	1001	100
0010	101	1010	010
0011	110	1011	001
0100	110	1100	001
0101	101	1101	010
0110	011	1110	100
0111	000	1111	111

The result in binary is 2, and in fact the cypher at that position has been flipped. The reason is that the array | 0 0 1 1 1 0 0 | is the sum of a legal code | 0 0 1 1 1 1 0 | and the "error" | 0 0 0 0 0 1 0 |, or more exactly module 2 of that sum. We have mentioned already that each code multiplied by *H* gives all 0s and the linear behavior of the product so defined implies that only the error reveals itself. You can easily verify that the result does not change if you multiply the error by *H*. The product is

$$S = x \times H \bmod (2).$$

The result is known as a "syndrome." This ECC was the first one to be developed by Hamming and corrects a single bit. More complex schemes can correct multiple bits and generally do adopt a recursive algorithm. This is nowadays a very active field of research [4]; the goal is to move closer to the spare cells theoretic minimum number set by Shannon [5], through a minimum area circuitry and maximum speed.

References

1. Marzio, B., and Natasha, S. (2008). *Errori e parità*. Centro Ricerche e Innovazione Tecnologica della RAI.

2. Campardo, G. (2000). *Progettazione di Memorie Non Volatili*. Franco Angeli.

3. Paolo, C., Carla, G., Patrizio, O., and Enrico, Z. (1999). *Flash Memories*. Kluver Academic.

4. MacKay, D.J. (1995). *Information Theory, Inference, and Learning Algorithms*, Cambridge University Press, Cambridge.

5. Shannon, C.E. (1948). *Bell Syst. Tech. J.*, **27**, 379–423, 623–656.

Index

absorption, 39, 46–48, 50
AC characteristics, 501
activation energy, 148, 185, 191, 436, 494–95
active electrode (AE), 108, 390–95, 402, 439, 476
address bus, 454
address field, 473
AE, *see* active electrode
AFM, *see* atomic force microscopy
extended, 36
topographic, 31
Ag nanocrystals, 111–14, 116–17, 119–20, 412–15, 423–24
Al_2O_3 RRAM, 409–10, 412
ALD, *see* atomic layer deposition
algorithmic pattern generator (APG), 463–64
algorithms, 41, 128, 132, 349–51, 471–72
adaptive, 492
operation's, 472
recursive, 511
smart, 128
all-one (all1), 463, 476, 479, 484–85
all-zero (all0), 476, 485–86, 488–89
all0, *see* all-zero
all1, *see* all-one
ALU, *see* arithmetic logic unit
angle-resolved X-ray photoelectron spectroscopy (AR-XPS), 43
annealing, 21, 24, 180, 182, 199, 201–2, 204–5, 207, 209–11, 214, 216–17, 219–24, 226, 271, 274

high-pressure vapor, 174
high-temperature, 182, 270
hydrogen, 247
anti-Stokes radiations, 49
APCVD, *see* atmospheric pressure chemical vapor deposition
APG, *see* algorithmic pattern generator
architecture, 131–32, 370, 377, 385, 387, 450, 455, 471–72, 507
area density, 152, 182, 249
arithmetic logic unit (ALU), 466
Arrhenius law, 494
Arrhenius plot, 185, 190–91
AR-XPS, *see* angle-resolved X-ray photoelectron spectroscopy
ASCII, 334–35, 499
atmospheric pressure chemical vapor deposition (APCVD), 15
atomic force microscopy (AFM), 12, 15, 33–35, 269–70, 273, 282, 284, 288, 290, 307, 430–31
atomic layer deposition (ALD), 12, 21, 153, 299, 409, 427
atom-level traps, 417
ATR-FTIR, *see* attenuated total reflectance Fourier transform infrared
attenuated total reflectance Fourier transform infrared (ATR-FTIR), 51
Auger-electron spectral lines, 40

bad blocks, 132, 344–45, 349–53, 494

bandgap, 16, 19, 46, 90, 92,
    113–14, 161, 184, 189, 246,
    266–67, 300, 305, 426, 454
bandgaps
    effective, 113
    electronic, 47
    larger energy, 23
    narrow, 14
    non-zero, 267
    optical, 217–18
    width-edge-dependent, 267
    zero, 320
barrier height, 24, 83, 91–92, 94,
    108, 110, 112–13, 115–17,
    159–61, 242, 244, 310–11,
    319, 380, 400
basic input/ output system (BIOS),
    339
BCP, *see* block copolymer
BE, *see* bottom electrode
bias, 34, 77–78, 82, 90, 94, 97, 161,
    183, 187, 220, 222, 383, 385,
    395, 399–400
bionanoprocess (BNP), 428–29
BIOS, *see* basic input/ output
    system
bipolar switching (BS), 13, 385,
    416
bit flipping, 344, 350
bit flips, 350–51
bitlines, 340–41, 343, 464–65, 474,
    476, 478–79, 483, 487
block copolymer (BCP), 39, 281–82
BNP, *see* bionanoprocess
Bohr radius, 200, 216, 305
Boltzmann's constant, 94, 102,
    247, 494
bottom electrode (BE), 432, 437
Brownian motion, 45
BS, *see* bipolar switching
bytes, 130, 334–35, 344–46, 349,
    359–60

CAM, *see* content addressable
    memory
capacitance, 11, 16, 24, 77, 79, 87,
    97, 175, 183–85, 187, 291,
    298, 301, 307, 455–56
capacitors, 76, 192, 299, 309,
    372–73, 456, 460, 475–76
    conventional dielectric-based,
    373
    large, 460
    leaky, 371
    parallel plate, 79, 91
    parasitic, 291
    substrate coupling, 85
carbon nanotube (CNT), 15, 40–41,
    267, 273, 284, 380, 382, 402
cathode-ray tube (CRT), 76
CBE, *see* crested barrier
    engineering
CBF, *see* Coulomb blockade factor
CBRAM, *see* conductive bridging
    random access memory
CDSEM, critical dimension
    scanning electron microscopy
central processing unit (CPU), 367,
    461
CF, *see* conductive filament
CG, *see* control gate
channel hot electron injection
    (CHEI), 80, 130, 229, 491
charge carriers, 48, 77, 116, 135,
    137, 240–41, 244, 266, 297,
    320, 399, 424
charge detrapping, 151, 400, 416
charge injection, 129–30, 140, 142,
    153
charge leakage, 141–42, 144,
    228–29, 237, 312, 321
charge loss (CL), 41, 89, 109–10,
    128, 143, 146–47, 151, 154,
    238, 294, 305, 310–11, 486,
    489, 504

charge retention, 79, 88, 106,
109–10, 127, 133, 136–38,
140–42, 152–53, 241–42,
244–45, 247, 252, 291, 295
charge storage, 87, 89, 133–35,
141, 176, 178–79, 182–83,
228, 230–31, 234–36, 240–47,
252–53, 312, 314, 318–20
charge storage nonvolatile
memory (CS-NVM), 127–30,
133, 136, 142, 145, 149, 152,
159, 161–63
charge transfer, 24, 48, 89, 180
charge trapping, 7, 22, 25, 128,
151, 153–54, 179, 297,
300–301, 304–7, 314, 318,
400, 409, 415–16
charge trapping/detrapping, 399,
416–17
charge-trapping memory (CTM),
297, 300, 309, 319–21
checkerboard (CK), 463, 485
CHEI, *see* channel hot electron
injection
chemical vapor deposition (CVD),
10, 12, 15, 140, 174, 180, 189,
200–202, 268, 273–77,
179–81, 288–89, 292, 297, 308
Child's law region, 416
circuitry, 21, 31, 453–55, 471–72,
474, 477, 483, 511
CK, *see* checkerboard
CL, *see* charge loss
CMOS, *see* complementary-metal-
oxide semiconductor
CNT, *see* carbon nanotube
complementary-metal-oxide
semiconductor (CMOS), 129,
131, 144, 175, 298, 319–20,
379, 385, 438
composite gate dielectrics, 90–92,
95, 111–14, 119

conduction band, 24, 84, 92, 98,
100–101, 113, 138, 159–60,
177, 185, 189, 193, 245–47,
250, 252, 266
conductive bridging random access
memory (CBRAM), 391
conductive filament (CF), 24, 35,
385, 414–15, 424, 428, 432,
437
conductive nanofilaments, 387
contact potential difference (CPD),
304, 306
content addressable memory
(CAM), 472–73, 487, 489
control dielectric, 107, 119,
139–40, 149, 152–54, 162–63
control gate (CG), 85, 108, 135–36,
152, 162, 177–78, 181,
228–29, 245–46, 250, 291–92,
340–43, 474–76, 478, 484
control oxide, 77, 88, 98, 105, 116,
137, 140–41, 178, 186, 307
Coulomb blockade, 84–85, 87–89,
106, 137–39, 146, 176, 191,
229
Coulomb blockade factor (CBF), 85
Coulomb repulsion, 97, 418
CPD, *see* contact potential
difference
CPU, *see* central processing unit
crested barrier engineering (CBE),
133, 152, 155, 159–62
critical dimension scanning
electron microscopy (CDSEM),
30
CRT, *see* cathode-ray tube
crystallization, 12, 16, 24, 186,
190, 208, 215, 376
CS-NVM, *see* charge storage
nonvolatile memory
CS-NVM devices, 128–30, 133–34,
141–42, 145–47, 149, 151–53,
155–57, 159–63, 165

CTM, *see* charge-trapping memory
Cu NCs, 432–35
CVD, *see* chemical vapor deposition
  low-pressure, 202, 249, 277
  metal-organic, 242
  plasma-enhanced, 288

DAC, *see* digital-to-analog
  converter
dangling bonds, 156, 190, 222,
  234–35
data buffer memory (DBM),
  463–64
data loss, 327–28, 331, 345–46,
  348, 360, 480, 490, 494–95,
  506
data recovery, 327–28, 331, 345,
  347, 357–58, 360–63, 366–67
data retention, 6, 89, 128, 139–40,
  143–44, 151, 157, 159,
  179–80, 193, 296, 312, 349,
  351, 377
datasheets, 348–49, 470, 473, 487,
  490, 494, 497–98, 500, 504–7
data storage, 129, 131, 134, 180,
  312, 317, 329–30, 335, 337,
  343–44, 353, 371, 382
DBM, *see* data buffer memory
DC, *see* direct current
DC characteristics, 501
DDLS, *see* depolarized dynamic
  light scattering
deep-level transient spectroscopy
  (DLTS), 174–75, 184–92, 252
defects, 47, 49, 51, 134–37, 141,
  176, 178, 215–16, 218–20,
  252, 273, 275, 305–6, 406, 408
density functional theory (DFT),
  252, 267
depolarized dynamic light
  scattering (DDLS), 45–46
design of experiment (DoE), 496
detrapping, 184, 418

device interface board (DIB), 453,
  456–58
device power supply (DPS),
  460–61
devices, 76, 130–31, 175–76,
  228–45, 297–98, 314–16,
  331–34, 343–48, 350–51, 383,
  400–401, 409–10, 419–22,
  424–29, 505
device under test (DUT), 452–58,
  460–62, 464–66, 468, 470,
  472, 474, 476, 481, 484–85,
  487–94, 502–3, 505
DFG, *see* double-floating-gate
DFT, *see* density functional theory
DIB, *see* device interface board
dielectric constant, 16, 23–24, 77,
  79, 90–91, 93, 98, 112, 114,
  117, 133, 161, 373
dielectric layer, 17, 134–37, 152,
  161, 179, 229, 300, 468
dielectrics, 22, 78–79, 82, 86–87,
  90–94, 104, 106, 110–13,
  115–16, 121, 134, 162, 300,
  373, 378
dielectric stack, 140, 159, 161,
  178–79, 213, 249
diffraction, 32, 34, 36, 38, 68,
  215–16
diffusion, 152–53, 157, 201, 207,
  209, 213, 237–38, 394,
  417–18, 427, 439
digital channels, 454–57, 461
digital-to-analog converter (DAC),
  458
dimethyl formaldehyde (DMF), 16,
  316
dip-pen nanolithography (DPN),
  12, 15
Dirac distribution function, 251
Dirac electrons, 266
Dirac equation, 264
Dirac fermions, 264, 266
Dirac material graphene, 267

Dirac point, 264–65
direct current (DC), 21–22, 83,
    435, 453
direct memory access (DMA),
    473–76
direct tunneling, 80, 89–91, 93–94,
    102, 108–9, 112, 115–17, 140,
    149–51, 200, 229, 246–47,
    294, 302
disks, 77, 203, 223, 331, 333,
    335–36, 343, 345, 354–55,
    358, 360, 367
DLS, *see* dynamic light scattering
DLTS, *see* deep-level transient
    spectroscopy
DMA, *see* direct memory access
DMF, *see* dimethyl formaldehyde
    16, 316
DoE, *see* design of experiment
doping, 151, 267, 279, 293, 304,
    316, 319, 338, 406–7
double-floating-gate (DFG), 296
double-stack nanocrystals,
    154–55, 165
DPN, *see* dip-pen nanolithography
DPS, *see* device power supply
DRAM, *see* dynamic random access
    memory
DUT, *see* device under test
dynamic light scattering (DLS), 45
dynamic random access memory
    (DRAM), 6, 129, 192, 331, 367,
    370–74, 377, 382–83, 438, 463

EB, *see* electron beam
ECC, *see* error code correction
ECM, *see* electrochemical
    metallization
ECR, *see* error capture random
    access memory
ECD, *see* error code detection 350
EDX, *see* energy dispersive X-ray
    31

EELS, *see* electron energy loss
    spectroscopy
EEPROM, *see* electrically erasable
    programmable read-only
    memory
EFM, *see* electric force microscopy
EFTEM, energy-filtered
    transmission electron
    microscopy
EL, *see* electroluminescence
electrically erasable programmable
    read-only memory (EEPROM),
    129, 331, 337, 339, 492
electrical wafer sort (EWS), 476,
    480, 490
electric field, 78, 80, 91, 93, 109,
    134, 137–38, 158, 160–61,
    245–46, 370, 373, 427–30,
    432–37, 440
electric force microscopy (EFM),
    12
electrochemical metallization
    (ECM), 390–94, 402, 414, 421,
    438–39
electroluminescence (EL), 200,
    219–22, 253
electron beam (EB), 10, 15, 21, 186
electron beam evaporation, 429
electron beam lithography, 140
electron energy loss spectroscopy
    (EELS), 31
electronic devices, 173–74, 193,
    199, 383
    flexible, 320
    high-performance, 267
    modern, 338
    optical, 2
    portable, 79
electron microscopy, 27–28, 31, 67,
    204, 269, 430
electron tunneling, 80, 84, 110,
    112, 155, 246, 310
energy dispersive X-ray (EDX), 31,
    395

energy-filtered transmission
electron microscopy (EFTEM),
28
energy levels, 5, 34, 89, 101, 103,
114, 222, 252, 263, 291, 297
EOT, *see* equivalent oxide thickness
EPROM, *see* erasable
programmable read-only
memory
equivalent oxide thickness (EOT),
159, 161–62, 242–44
erasable programmable read-only
memory (EPROM), 129, 331,
337, 483, 492
erase cycles, 350–53, 387
erase operations, 80, 87, 100, 104,
142, 229, 309, 331, 345, 454,
465, 474
erase state, 129, 143, 154–55
erase verify (EV), 149, 478–79
erasing, 79, 89, 145, 180, 295, 309,
330, 349, 478, 486
error capture random access
memory (ECR), 463–65, 489
error code correction (ECC),
344–45, 350, 353, 506–8, 511
error code detection (ECD), 350
escape frequency, 84
etching, 25, 223–24, 278–79,
281–82, 305–6, 314
EV, *see* erase verify
EWS, *see* electrical wafer sort
EXAFS, *see* extended X-ray
absorption fine structure
EXOR, 463
extended X-ray absorption fine
structure (EXAFS), 43

fabrication, 10–11, 51–52, 136,
141–42, 144, 153, 281–82,
292, 298–300, 314–15,
320–21, 405, 408–9, 416,
427–30

FeFET, ferroelectric field-effect
transistor
FeRAM, ferroelectric random
access memory
Fermi energies, 98
Fermi level, 9, 94, 102, 107, 153,
251, 265–66, 296, 304, 319,
414
ferroelectric field-effect transistor
(FeFET), 16, 379
ferroelectric random access
memory (FeRAM), 6, 25, 175,
179, 193, 371, 373–75, 377,
380
ferroelectric tunnel junction (FTJ),
378–80
FESEM, *see* field emission scanning
electron microscopy
FETEM, *see* field-emission
transmission electron
microscopy 28
FET, *see* field-effect transistor
few-layer graphene, 266, 275
few-layer nanoribbons, 267
FG, *see* floating gate
FG-based CS-NVM, 134, 136–40,
142–43, 145, 147, 149
field-effect transistor (FET), 6, 16,
77, 194, 279–80, 319, 378,
453–54
field emission scanning electron
microscopy (FESEM), 430–31
field-emission transmission
electron microscopy (FETEM),
28
FIMV, *see* force current and
measure voltage 458–59
flash cells, 350–51, 463, 472–76,
483, 491, 493, 506, 508
flash devices, 79, 309, 357–60,
362, 367, 376, 468
flash drive, 360–61, 366
flash media, 327, 345, 348, 350,
358, 367

flash memory, 77–78, 130–32,
173–77, 179–83, 192–93,
291–92, 294–95, 314, 327–28,
330–32, 334–50, 352–54,
356–60, 362, 450–53
all-organic, 317
ambipolar, 194
charge-trap, 291
charge-trapping, 297
conventional, 297
first-reported graphene, 294
flexible, 318
floating-gate, 291
graphene-based flexible, 315
graphene floating-gate, 296
graphene nanocrystal, 298
graphene nanosheet, 313
graphite nanocrystal, 303
insulated floating-gate, 177
low-voltage, 296
molecular, 194
nanocrystal, 298, 305
nanocrystal-based, 76
nonvolatile, 263
polysilicon, 294
raw, 358
single-layer-graphene, 293
standard FG, 144
ultradense NAND, 144
flash memory devices, 130–31,
135, 182, 196, 297, 306, 309,
320, 349, 357, 360, 371
flash nonvolatile memory, 79
flash translation layer (FTL),
358–59
flat-band voltage, 86, 183, 235–36,
241, 302, 309, 312
floating gate (FG), 6–7, 11, 77–79,
82–83, 127, 133–36, 140–41,
143–47, 157–58, 162, 165,
176–78, 181–82, 228–29,
291–92, 294–95, 297, 300,
340–43, 349, 352, 485, 488,
491

floating gate memory, 7, 78, 82,
173, 178, 263, 291–92, 314,
319, 321
fluorine-doped tin oxide (FTO), 14,
17, 20
Fowler–Nordheim (FN) tunneling,
80, 83–84, 87, 89–93, 95–96,
104, 112, 116–17, 119–20,
130, 135, 140, 161, 229,
341–42, 418, 420, 486, 491
force current and measure voltage
(FIMV), 458–59
force voltage and measure current
(FVMI), 458
Fourier analysis, 38
Fourier transform infrared (FTIR)
50–51
FP, *see* Frenkel–Poole
Frenkel–Poole (FP), 146–48, 291
FTIR, *see* Fourier transform
infrared 50–51
FTJ, *see* ferroelectric tunnel
junction
FTL, *see* flash translation layer
FTO, *see* fluorine-doped tin oxide
full width at half maximum
(FWHM), 38, 41, 220
FVMI, *see* force voltage and
measure current
FWHM, *see* full width at half
maximum

gate bias, 89, 102, 116, 118, 148,
234–35, 240–41
gate dielectrics, 79, 87, 90–92, 102,
106, 110–13, 117–20, 177,
300–301
gate oxides, 35, 95, 109–12, 117,
119, 178–79, 192, 253, 343,
349, 379
gates, 77, 84, 86, 90, 95, 112,
115–16, 119, 177–78, 245,
249, 251, 291, 338–40, 473–75

conductive, 316
conventional floating, 176
isolated floating, 176
logic, 331
lower floating, 296–97
nano floating, 186
nonvolatile cell floating, 465
polycrystalline silicon floating, 178
thinner floating, 1
gate voltage, 89, 95, 98, 103, 114, 116, 119, 129, 138, 183–84, 243–44, 273, 280, 293, 302
Gaussian confinement function, 49
Ge nanocrystals, 35, 153, 199–202, 205–10, 212–14, 216–20, 222–27, 229–31, 233–35, 237, 240, 242–43, 245–46, 248, 250, 252–53
Ge NC memory, 245–46, 249, 253
GFM, *see* graphene flash memory
Gibbs free energy, 381, 391
GNC, *see* graphene nanocrystal
GQD, *see* graphene quantum dot
graphene, 13, 15, 34–35, 264–69, 271–79, 281–83, 285, 287–90, 292–94, 304, 306, 312, 314, 317–21, 378
graphene films, 273, 275–76, 278–79, 284
graphene flash memory (GFM), 194, 294, 297
graphene floating-gate memory, 292, 294
graphene nanocrystal (GNC), 283, 298–300, 302, 304, 307, 311, 313
graphene sheet (GS), 266, 285, 295
graphite nanocrystals, 287, 303–4
graphene quantum dot (GQD), 267, 281–84, 286–88, 299, 302–3
GS, *see* graphene sheet

HAADF-STEM, high-angle annular dark-field scanning transmission electron microscopy
hard disk drive (HDD), 6, 360
HDD, *see* hard disk drive
HHI, *see* hot hole injection
high-angle annular dark-field scanning transmission electron microscopy (HAADF-STEM), 31
high-pressure vapor annealing (HPVA), 174, 180, 182
high resistance state (HRS), 376, 385–86, 393, 398, 400–401, 408, 410–11, 415–20, 427
high-resolution transmission electron microscopy (HRTEM), 28, 51, 105, 209, 215, 226, 238, 248, 283, 286, 433
hole injection, 130, 193
hot hole injection (HHI), 130
HPVA, *see* high-pressure vapor annealing
HRS, *see* high resistance state
HRTEM, high-resolution transmission electron microscopy
Hummers's method, 269, 282–83, 287
hysteresis, 11, 13, 17, 24, 183, 230–31, 235–36, 240–41, 243, 293, 299, 301–2, 374

IBS, *see* ion beam synthesis
IC, *see* integrated circuit
IDMs, *see* integrated device manufacturers
IGD, *see* intergate dielectric
IMC, *see* in-memory computing
immunity, 138–39, 141, 143–45, 155, 165
incremental step pulse erase (ISPE), 140, 491–92

indium tin oxide (ITO), 11, 16, 21, 25, 300, 314, 402, 426, 441
indium zinc oxide (IZO), 317
information, 26, 39, 43–47, 49, 75–76, 187, 327–36, 339–40, 345–47, 349, 356–57, 452, 480–81, 489, 503–6
    compositional, 28–29
    device-relevant, 505
    direct, 34, 49
    header, 359
    nanoscale, 37
    quantitative, 41, 47
    redundancy, 472–73, 489
    sensitive, 327, 350
    stored, 310, 340, 374, 412
    structural, 33, 43, 45
    uncorrupted, 494
injection
    hot-carrier, 343
    hot-electron, 341–43
in-memory computing (IMC), 367
input/output (I/O), 454, 462–63, 501–2
instructions, 332–33, 461–62, 472
    kilo, 463
    macrocode, 462
    microcode, 461–62
    operative, 502
    user-defined, 462
    write-read combined, 462
insulating layer, 341, 397–98, 401, 405, 407–9, 412, 416, 440
    inorganic, 405
    single, 405
insulators, 30, 270, 338, 378, 385, 409, 417–18
integrated circuit (IC), 2, 131, 338, 449–50
integrated device manufacturers (IDMs), 467
intergate dielectric (IGD), 162

International Technology Roadmap for Semiconductors (ITRS), 6, 195, 231, 257, 291, 324, 372, 378, 384, 440–41
Internet of Things (IoT), 130, 165
inversion, 84, 86–87, 92, 222, 229, 251, 253
I/O, *see* input/output
ion beam synthesis (IBS), 178–79
IoT, *see* Internet of Things
irradiation, 143, 165, 201, 417
ISPE, *see* incremental step pulse erase
ITO, *see* indium tin oxide
ITRS, *see* International Technology Roadmap for Semiconductors
IZO, *see* indium zinc oxide

JFFS, *see* Journaling Flash File System
joule heating, 401
Journaling Flash File System (JFFS), 358–59
junctions, 15, 184, 194, 376, 399, 424

Kapton substrate, 30
Kelvin contact, 461
Kerkhof–Moulijn model, 42
KGD, *see* known good die
Klein tunneling, 264
known good die (KGD), 480, 490
Kubo theory, 107

lateral tunneling (LT), 150–51
leakage, 1, 110, 112, 117–19, 134, 149–50, 159, 176, 237, 242, 307, 311, 321, 482, 488
leakage current, 1, 110, 112, 114, 117–19, 134, 159, 176, 192, 242, 307, 311, 319

Lee's model for exact verification, 119

LEISS, *see* low-energy ion scattering spectroscopy

Lerf–Klinowski model, 271

Linux, 354–55, 359, 366–67

load boards, 453, 468, 497–98

longitudinal plasma resonance (LPR), 46

Lorentzian functions, 42

low-energy ion scattering spectroscopy (LEISS), 41

low-pressure chemical vapor deposition (LPCVD), 97, 174, 178, 180

low resistance state (LRS), 375–76, 385–86, 400–401, 407, 410, 415–21, 424

low-temperature polycrystalline silicon (LTPS), 181–82

LPCVD, *see* low-pressure chemical vapor deposition

LPR, *see* longitudinal plasma resonance

LRS, *see* low resistance state

LT, *see* lateral tunneling

LTPS, *see* low-temperature polycrystalline silicon

lucky electron model, 80

macromolecular memory, 382–83, 440

magnetic random access memory (MRAM), 175–76, 179, 376–77

magnetic tunnel junction (MTJ), 376

MANGOS, *see* metal/aluminum oxide/nanographene/silicon oxide/silicon structure

Maxwell Garnett theory, 91, 112

MBE, *see* molecular beam epitaxy

mean time between failures (MTBF), 348, 466

memory, 75–76, 129, 145, 182–83, 302–3, 308, 328, 333, 335–36, 369–71, 382, 462–65, 472, 501–2, 507

  computer's, 335

  conventional floating-gate-type, 87

  data buffer, 463–64

  discrete charge trap, 139

  electronic, 333

  erasable, 337, 457

  erased, 463

  external, 333

  fabricated MOS, 308

  fastest, 370

  ferroelectric, 77

  flash, 316

  flexible, 291

  high-density, 386, 439

  metal-based, 106

  metal-nanocrystal-based, 121

  nanocrystal-based, 87, 106

  nanocrystal-embedded, 120

  new, 383

  next-generation information technology, 371

  nonvolatile, 109

  organic, 318, 382

  phase-change, 30, 176, 371, 375

  portable, 336

  primary/temporary, 333

  reference, 463–64

  scan, 465

  semiconductor, 131, 175

  single-transistor, 320

  temporary, 336

  typical, 178

  universal, 371, 377, 440

  valence change, 390

  vector, 465

memory cards, 345, 361–62

memory cells, 89, 131, 134, 141, 143, 145–46, 155, 165, 228, 231, 237, 298, 387, 450, 463

advanced nonvolatile, 80
defective, 480
floating-gate flash, 135
impacted, 145–46
individual, 194, 298, 343
multilevel, 318
nitride flash, 136
nonvolatile, 480
single, 320
memory chips, 175, 336, 343, 348, 483, 493
memory devices, 7, 13, 30–31, 75–76, 79, 109–10, 174–75, 237, 241, 263, 268, 296–97, 307, 310, 320–21
amorphous ZrO2 resistive, 31
conventional FG, 6
conventional floating-gate, 85, 228
dual-layer NC floating-gate, 229
effective, 79
flash, 7, 30, 76, 79, 131, 162, 320
flexible, 13
floating-gate, 199, 228
hybrid DFG, 297
nanocrystal-based, 110, 305
nonvolatile, 4, 76, 117, 120, 298, 320
organic, 14
reliable, 182
scaled, 152
silicon NC, 192
single-Au-nanocrystal nano-floating-gate, 296
trilayer, 242
volatile, 129
memory structures, 21, 104–5, 242, 297, 299, 307–8, 318, 384
memory switching (MS), 419–20, 423–24
memory technologies, 297, 321, 344, 370–71, 386, 438

memory window (MW), 12, 19, 22, 24, 129, 152, 154–57, 293, 295–99, 301–3, 305–7, 311–14, 316, 318–20
functional, 180
large, 24, 297, 306, 320
millivolt, 302
program/erase, 295
reduced, 88
MEMS, 469
metal/aluminum oxide/nanographene/silicon oxide/silicon structure (MANGOS), 302, 311
metal-insulator-metal (M-I-M), 385
metal-insulator-semiconductor (MIS), 29, 219–20, 233–34, 237, 239, 242–43
metal nanocrystals, 51, 76, 81, 104, 107, 110, 114, 117–19, 121, 128, 152–53, 165, 171, 313, 319
metal nanoparticles, 15, 40, 297–98, 318
metal-oxide high-κ-oxide silicon (MOHOS), 23
metal-oxide semiconductor (MOS), 77, 91, 95–96, 109–12, 175, 177–78, 182–84, 186, 189–92, 228, 253, 302
metal-oxide semiconductor, complementary, 175, 379
metal-oxide-semiconductor field-effect transistor (MOSFET), 6, 77, 86, 153, 177–78, 183
M-I-M, *see* metal-insulator-metal
miniaturization, 79, 175, 263, 370–71
MIS, *see* metal-insulator-semiconductor
MLC, *see* multilevel cell
mobile phones, 174–75, 180, 331, 345, 361

MOHOS, *see* metal-oxide high-κ-
oxide silicon
molecular beam epitaxy (MBE), 10,
32, 68
molecular memory, 382
monolayer graphene, 266, 281,
292–93, 295, 314, 319
Moore's law, 127–28, 131, 139–40,
165, 501
MOS, *see* metal-oxide
semiconductor
MOSFET, *see* metal-oxide-
semiconductor field-effect
transistor
MOS nonvolatile memory, 110
Mott insulators, 381
Mott memory, 381–82
Mott transitions, 382
MRAM, *see* magnetic random
access memory
MS, *see* memory switching
MTBF, *see* mean time between
failures
MTJ, *see* magnetic tunnel junction
multilevel cell (MLC), 345, 350,
352–53, 407
MW, *see* memory window

NAND, 131–32, 331, 343–44,
350–51, 367–68, 451, 454,
474–78, 482, 484, 491, 501–2,
507
NAND devices, 343, 350–51
NAND flash, 343–44, 351–53,
358–59, 373–74, 454, 464,
472, 482, 485, 487, 494, 507
NAND flash memory, 161, 327,
343–45, 349, 368
NAND memory, 351, 464, 475
nanocrystal-based CS-NVM, 128,
133, 137–47, 149, 151–55,
164–66

nanocrystal memory (NCM), 5–6,
27, 51, 77, 79, 81, 89, 139,
176–79, 193, 200, 228–29,
231, 237, 245–47, 302
nanocrystal NVM, 7, 140, 144
nanocrystal (NC), 1–2, 4, 6–7, 9,
13–15, 19–21, 27–31, 76,
79–80, 82–92, 96–113, 116,
118–21, 137–45, 149–55,
173–75, 179–94, 199–213,
222–33, 235–41, 243–53, 282,
296–98, 320–21, 408–12,
414–16, 423–30, 439–40
nano-floating-gate memory
(NFGM), 183, 297, 301, 309,
319
nanographene-based flash
memory, 313
nanographene flash memory, 307
nanomaterial, 2–5, 7, 9–10, 12–13,
16, 18, 43, 47, 67, 72, 263
nanoparticle (NP), 5, 8–14, 25–27,
31–47, 49–52, 100, 176, 208,
296, 314
nano–random access memory
(NRAM), 380
nanorod (NR), 14, 17, 19–20, 40,
46–47
nanostructure, 2, 5, 8, 15, 18, 26,
49, 267, 290, 424
nanowire, 8–9, 12, 15–16, 18, 20,
44, 46, 194, 382, 424
NC, *see* nanocrystal
NC-based memory, 228–29, 237,
245, 253
NC-based NVM, 1
NC-based RRAM, 370, 410–11,
425, 427, 430, 440
NC flash memory, 173–75, 195
NC floating-gate flash memory, 174
NC floating-gate MOS memory
device, 228
NC floating gates, 173, 175–77,
180, 182

NCM, *see* nanocrystal memory
NC RRAM devices, 409, 412, 428
NCs in flash memory, 173–75, 177, 179, 183
NCs in RRAM, 410, 440
NDIKW, *see* noise, data, information, knowledge, and wisdom
N-doped RRAM, 406–7
near-edge X-ray absorption fine structure (NEXAFS), 43
near-field scanning optical microscopy (NSOM), 27, 33, 36–37
NEXAFS, *see* near-edge X-ray absorption fine structure
NFGM, *see* nano-floating-gate memory
nitride-based CS-NVM, 133, 135–36, 139, 141–42, 145, 147, 149, 162, 165
noise, data, information, knowledge, and wisdom (NDIKW), 328–29
nonvolatile memory (NVM), 4–6, 121, 370, 447, 449–50, 452, 454, 458, 460, 462, 464, 466, 468, 470
    conventional, 106
    embedded, 90
    floating-gate, 1, 80, 245
    nanocrystal-based, 75, 80, 101, 109, 120
    nanocrystals in, 1, 75, 127, 173, 199, 263, 327, 369, 449
    nitride-based charge storage, 127
    RRAM, 444
    scaling, 87
    semiconductor-based, 106
    silicon-nanocrystal-based, 121
    traditional floating-gate, 109
nonvolatile read-only memory (NROM), 131–32, 136

nonvolatility, 7, 134, 229, 327, 371, 373
NP, *see* nanoparticle
NRAM, nano–random access memory
NROM, *see* nonvolatile read-only memory
NR, *see* nanorod
NSOM, near-field scanning optical microscopy
n-type semiconductors, 14, 426
nucleation, 30, 39, 141, 180, 203, 206–8, 210, 215–16, 225–26, 275, 277, 289, 415, 435
NVM, *see* nonvolatile memory

NVM devices, 1, 5, 9, 16, 23–24, 157, 163, 166–67, 176, 186, 194, 370, 373, 380, 382
one-time programmable (OTP), 483
ONO, *see* oxide-nitride-oxide
OOB, *see* out of band
OPAMP, *see* operational amplifier
operational amplifier (OPAMP), 459–60
orientations
    fixed magnetic, 376
    free magnetic, 376
    parallel magnetic, 376
Ostwald ripening, 19, 203
OTP, *see* one-time programmable
out-diffusion, 216, 225, 231, 239
out of band (OOB), 345, 352–53
oxide layers, 111, 115, 134, 138, 142, 151, 156, 159, 161, 177, 212, 233, 237–39, 242, 340
oxide-nitride-oxide (ONO), 135, 179, 192

PACS, *see* perturbed angular correlation spectroscopy
parametric measurement unit (PMU), 453, 458–59, 475

parity, 507–10
partial differential equation (PDE), 432
PCB, *see* printed circuit board
PCM, *see* phonon confinement model
PCRAM, phase-change random access memory
PDE, *see* partial differential equation
PEALD, *see* plasma-enhanced atomic layer deposition
PECVD, *see* plasma-enhanced chemical vapor deposition
P/E, *see* programming/erasing
perturbed angular correlation spectroscopy (PACS), 252
PFM, *see* piezoresponse force microscopy
phase-change random access memory (PCRAM), 6, 179
phonon confinement model (PCM), 49, 371, 373, 375–77, 400
photolithography, 15, 17, 19, 471
photoluminescence (PL), 47–48, 51, 72, 200, 216–18
physical vapor deposition (PVD), 10
PI, *see* polymer inorganic
piezoresponse force microscopy (PFM), 12
pins, 452, 454, 456–57, 460–62, 465, 471, 473–74, 481–82, 499, 501–2
    connector, 456–57
    digital, 454
    input, 461
    lower, 343
    output, 461, 475, 502
    pogo, 468
    sense, 461
PL, *see* photoluminescence
PLA, *see* programmable logic array

plasma-enhanced atomic layer deposition (PEALD), 21
plasma-enhanced chemical vapor deposition (PECVD), 12, 22, 25, 174, 180–81, 202, 223, 288–89, 299, 305
plasma etching, 281, 306, 308
PLD, *see* pulsed laser deposition
PMC, *see* programmable metallization cell
PMU, *see* parametric measurement unit
Poisson–Schrödinger solver, 98
Poisson's ratio, 227
polymer inorganic (PI), 13, 406
polymers, 22, 34, 51, 279, 301, 314, 382, 405–6
polystyrene (PS), 12–13, 281, 316, 437
power consumption, 5, 80, 194, 297, 367, 373, 377, 390, 454
PPS, *see* programmable power supplier
printed circuit board (PCB), 348, 458, 468
probe cards, 451, 468–70, 488
    high-parallelism, 497
probes, 26, 33, 36, 41, 43, 45, 47, 51, 457, 461, 468–70
programmable logic array (PLA), 471–72
programmable metallization cell (PMC), 391, 442
programmable power supplier (PPS), 452–53
programmable read-only memory (PROM), 129, 331, 337, 483, 492
programming, 76, 79, 89, 104, 107, 138, 159, 161, 291, 295, 473, 478, 485–89, 491, 493
programming/erasing (P/E), 129, 134, 139–40, 143, 145–50, 153–54, 156–58, 183, 295, 313

programming voltages, 117, 140,
180, 183, 193
PROM, *see* programmable read-
only memory
prototype NVM technologies, 371,
373, 377
Pt NCs, 409, 411–12, 428–30
p-type semiconductor, 314, 426
PS, *see* polystyrene
pulse bias, 187, 189–90
pulsed laser deposition (PLD),
14–15, 20–21, 104
PVD, *see* physical vapor deposition

QCE, *see* quantum confinement
effect
QD, *see* quantum dot
QED, *see* quick electron detrapping
quantum confinement, 5, 98, 101,
109, 138, 217, 305, 320
quantum confinement effect (QCE),
176, 267, 298, 309
quantum dot (QD), 2, 5, 8–9, 34,
45–46, 96–97, 189, 192, 287
semiconducting, 2
quantum wells, 5
quantum wires, 5, 7
quasi-elastic light scattering, 45
quasi-equilibrium approximation,
98
quasi-Fermi level, 252
quick electron detrapping (QED),
157

radiation, 44, 144
e-beam, 201
synchrotron, 44
radio frequency (RF), 14, 23, 201
RAM, *see* random access memory
Raman scattering, 51, 304
Raman shift, 49, 226
Raman spectra, 49–50, 207–8, 214,
216, 226, 289–90

Raman spectroscopic mapping,
277
Raman spectroscopy, 47, 49–50,
73, 207
random access memory (RAM), 6,
129, 175, 179, 331–33,
335–36, 370–71, 373, 376,
380, 383, 391, 463, 485, 489
random telegraph noise (RTN),
157, 438
rapid thermal annealing (RTA), 23,
202, 207–9, 216–17, 223–26,
229–31, 235, 238–39, 243,
249, 299
rapid thermal oxide (RTO), 219,
230–31, 233–34, 237, 239,
241, 243
rate windows, 185, 187–90, 192
read-only memory (ROM), 129,
131, 331, 333, 336–37, 472,
483, 492
real-time operative system (RTOS),
453
Recuva, 362–65
reduced graphene oxide (RGO),
194, 268, 271–73, 294, 296,
316
redundancy, 464–65, 470, 473,
480, 487, 489, 493, 502, 506
reflection high-energy electron
diffraction (RHEED), 32–33
reliability, 79–80, 128, 145–46,
153–55, 159, 163–64, 348,
350, 373, 383, 385, 409, 412,
436, 438
research mode, 141
RESET, 376, 387, 398, 401, 416,
419, 427
resistance, 13–14, 20–21, 25,
272–73, 280, 367, 375–77,
379–80, 384–86, 397–99, 411,
413, 417–21, 424, 429

resistive random access memory
(RRAM), 6, 15, 22, 25, 193,
369–70, 372, 374, 376, 378,
380, 382–440, 442, 444, 446,
448
resistive switching (RS), 14, 17–18,
22, 24, 370, 383, 385–87,
389–90, 396, 399–401, 405–9,
412–13, 415–16, 424–26,
438–39
resistivity, 267, 375–76, 381, 426
retention, 9, 17–18, 104, 106, 138,
140, 142–43, 145, 150–51,
153–55, 157, 159–63, 165,
242–46, 249–50, 252, 296–98,
310–13, 316, 319–20, 327,
350–51, 372, 379–80, 386,
412, 488–89, 494–95
reverse bias, 184, 187, 229, 400,
418
RF, *see* radio frequency
RF magnetron sputtering, 21–23
RF sputtering, 15, 21–24, 202, 213,
229
RGO, *see* reduced graphene oxide
RGO-based flash memory, 316
RGO-based flexible flash memory,
312, 316
RGO-based flexible organic flash
memory, 317
RGO-based organic transistor
memory device, 317
RGO flash memory, 294–95
RHEED, *see* reflection high-energy
electron diffraction
ROM, *see* read-only memory
RRAM, *see* resistive random access
memory
bilayer, 389, 408
binary transition-metal-oxide-
based, 386
emerging, 383
flash devices, 385
flexible applications, 425

free, 409
improved, 440
nanopyramid-Cu-BE-based, 435
nonfilamentary BS, 386
solid-electrolyte-based, 432
stacked, 421
transition-based, 396
ultrahigh-density vertical, 387
ultralow-power, 193
undoped TaOx-insulating-layer-
based, 407
RRAM devices, 370, 385, 387–90,
395, 400–402, 406, 408–9,
412, 414, 416, 418–19, 421,
424–25, 428–31, 437–40
RS, *see* resistive switching
RS memory, 19, 24
RTA, *see* rapid thermal annealing
RTN, *see* random telegraph noise
RTO, *see* rapid thermal oxide
RTOS, *see* real-time operative
system
Ru NCs, 28, 408–10, 427–28

SAM, *see* self-assembled monolayer
SAXS, *see* small-angle X-ray
scattering
scalability, 9, 13, 51, 79, 155, 159,
174, 176, 370, 373, 375,
377–79, 383, 387–88, 390
scanning Kelvin probe microscopy
(SKPM), 304, 306
scanning probe microscopy (SPM),
33, 68
scanning transmission electron
microscopy (STEM), 31–32,
67, 393
scanning transmission electron
microscopy–electron energy
loss spectroscopy (STEM-
EELS), 31–32
scanning tunneling microscopy
(STM), 29, 33–35

scanning tunneling spectroscopy
(STS), 34
Scherrer equation, 38
Scherrer formula, 38
Schottky barrier diode, 184
Schottky conduction, 420
Schottky junction, 399
Schrödinger's equation, 101
Schulz's formulation, 87
SCLC, *see* space-charge-limited
current
SD, *see* secure digital
SDD, *see* solid-state drive 450
secondary ion mass spectroscopy
(SIMS), 212, 239
secure digital (SD), 327, 344, 494
self-assembled monolayer (SAM),
11
semiconductor-based
nanocrystals, 263
semiconductor nanocrystals, 2, 6,
9, 16, 109–11, 113, 117, 119,
164, 263, 297–98, 305, 320,
417
semiconductors, 5, 43, 46–47, 51,
71, 76–77, 87, 90, 109, 111,
115–16, 118, 121, 319–20,
397–99
bulk, 5
compound, 47
indirect-bandgap, 19
zero-bandgap, 305
SET/RESET, 24, 386, 408, 410–11,
436
Shirley background subtraction, 42
shmoo, 497–500
Shockley–Read–Hall, 86
signals, 29, 45, 185, 189–90, 330,
337, 351, 381, 456–58, 467,
472, 474, 497, 503–4
SILC, *see* stress-induced leakage
current
silicon-nanocrystal-based CS-NVM,
133, 137, 141, 154

silicon nanocrystals, 48, 83, 85–87,
89–90, 115, 118, 140–41, 143,
153–54, 178, 191–92, 229,
304–6
silicon-oxide-nitride-oxide-silicon
(SONOS), 87, 136, 179
Simmons model, 94, 115
SIMS, *see* secondary ion mass
spectroscopy
Si-nanocrystal-based memory, 82,
90
Si-NC-based floating-gate memory
devices, 174
Si-NC-embedded MOS structure,
175, 184
single-level cell (SLC), 345, 350,
352–53
single-walled carbon nanotube
(SW-CNT, *see* single-walled
carbon nanotube 12), 12
SKPM, *see* scanning Kelvin probe
microscopy
SLC, *see* single-level cell
small-angle scattering, 38
small-angle X-ray scattering
(SAXS), 38–40, 44
sol-gel process, 12, 24
sol-gel reaction, typical, 24
solid-state drive (SSD), 367, 450,
494
SONOS, *see* silicon-oxide-nitride-
oxide-silicon
SONOS nonvolatile memory, 88
space-charge-limited current
(SCLC), 416–17, 419
spectroscopy, 34, 45–47
deep-level transient, 174–75,
184–85, 187, 189, 191, 252
dispersive, 432
electron energy loss, 31
infrared, 50
optical absorption, 46
optical luminescence, 43

perturbed angular correlation, 252
photon correlation, 45
secondary ion mass, 212
transient capacitance, 184
spin coating, 12, 14, 22, 25, 282, 300, 314
spin-transfer torque, 6, 371, 376
spin-transfer torque magnetic random access memory (STTMRAM), 6, 371, 373, 376–77
SPM, *see* scanning probe microscopy
SPP, *see* surface plasmon polariton
SPR, *see* surface plasma resonance
SRAM, *see* static random access memory
SSD, *see* solid-state drive
static random access memory (SRAM), 6, 129, 370–71, 373, 377, 382, 438
STEM, *see* scanning transmission electron microscopy
STEM-EELS, *see* scanning transmission electron microscopy–electron energy loss spectroscopy
STM, *see* scanning tunneling microscopy
Stokes–Einstein equation, 45
storage, 88, 195, 200, 236, 291, 298, 310, 328, 332–33, 336, 339, 349, 353, 367, 378
storage medium, 134, 136, 327, 337, 357, 403–5
storage nodes, 228–29, 297
    discrete charge, 237
    isolated charge, 141
stress-induced leakage current (SILC), 134, 139, 141, 144, 146–47, 149–51, 176, 192
STS, *see* scanning tunneling spectroscopy

STTMRAM, spin-transfer torque magnetic random access memory
surface plasma resonance (SPR), 46
surface plasmon polariton (SPP), 37
SW-CNT, *see* single-walled carbon nanotube

TAT, *see* trap-assisted tunneling
TBE, *see* tunnel barrier engineering
TCAD, *see* technology computer-aided design
TE, *see* top electrode
technology computer-aided design (TCAD), 139, 432, 434
test modes, 470, 473–74, 476, 483, 487, 490
test programs, 453, 456–57, 464, 490, 496–97, 505–6
thermal annealing, 16, 23, 105, 186, 201–2, 213, 281, 299
thermal detrapping, 246–47, 250
threshold switching (TS), 92, 376, 418–24
TiO_2 RRAM, 411–12, 427
TLC, *see* triple-level cell
TMD, *see* transition metal dichalcogenide
TN, *see* top nitridation
top electrode (TE), 17, 23, 25, 38, 292, 299, 379, 397
top nitridation (TN), 155–58
TPR, *see* transverse plasma resonance
transistors, 77–78, 80, 127, 131, 177, 182, 192, 194, 200, 299–300, 337–41, 370, 372–74, 376, 379–80
transition metal dichalcogenide (TMD), 317–18

transmission emission microscopy, 105

transverse plasma resonance (TPR), 46

trap-assisted tunneling (TAT), 146, 148

trapped electrons, 20, 99, 110, 235, 246–47, 250, 299, 304, 352, 418

trapping layer, 22, 24, 86, 178, 300

trap sites, 148–49, 189, 247, 303, 420–21

triple-level cell (TLC), 345

TS, *see* threshold switching

tunnel barrier, 78, 128, 133, 1 59–62, 171, 245–46, 248

tunnel barrier engineering (TBE), 128, 133, 140, 152, 155, 159–60, 171

tunnel dielectrics, 81, 83, 86, 107–9, 133, 139–40, 152, 155, 159–63, 165, 174, 176, 192, 242–43, 245

tunneling, 33–34, 76, 82–84, 89–91, 98, 109–15, 117–19, 121, 138, 140, 147, 238, 310–11, 319, 423–24
  band-to-band, 250
  defect-assisted, 146
  direct/FN, 159
  electrons/holes, 299
  field-assisted, 112
  interdot charge, 152
  lateral, 150–51
  mechanical, 77
  significant dot-to-dot, 182
  trap-assisted, 146
  trap-to-trap, 250

tunneling spectroscopy, 34

tunnel oxide layers, 6–7, 98, 128, 134–35, 137–40, 142–43, 146–49, 151, 155–61, 165, 229, 237–38, 241

tunnel oxides, 77–79, 111, 113, 115–16, 137, 139–41, 144, 146, 149–50, 152, 189, 192–93, 228–29, 237–39, 242–43

ULE-IBS, ultralow-energy ion beam synthesis

ultrahigh vacuum, 40

ultralow-current operation, 387

ultralow-energy ion beam synthesis (ULE-IBS), 179

ultralow-energy ion implantation, 174, 180, 182

ultrasonication, 13

ultrasonic wave, 10

ultraviolet (UV), 26, 45, 181, 282, 417–18, 483, 492

UM, *see* user mode, 490

unipolar trapping, 194

UNIX, 354

USB drive, 366

USB flash drive, 352

user mode (UM), 473, 478, 490–91

UV, *see* ultraviolet

UV/ozone treatment, 429

UV/Vis, 40, 45–46

UV/Vis/IR spectroscopy, 46

UV/Vis spectroscopy, 39, 45–46

vacuum, 11, 14, 16, 19, 64–65, 76, 93, 104, 114, 202, 271, 458, 469

valence band, 155, 162, 179, 185, 266

valence change memory (VCM), 390, 396–400, 402, 419, 423, 438–39

van der Waals forces, 10, 19

van der Waals interaction, 288

vapor deposition, 11, 180
  chemical, 10, 15, 22, 140, 174, 200, 268, 273, 288
  low-pressure chemical, 174

plasma-enhanced chemical, 12, 174, 202
remote-plasma-enhanced chemical, 305
vapor-liquid-solid (VLS), 15
variable oxide thickness (VARIOT), 133, 152, 155, 159–62
variable range hopping (VRH), 273
VARIOT, *see* variable oxide thickness
VCM, *see* valence change memory
virus, 329, 346, 360–62, 366
VLS, *see* vapor-liquid-solid
VM, *see* volatile memory
volatile memory (VM), 128–29, 175, 369–72, 381–82
voltage, 77–78, 80, 91–92, 116–17, 159–60, 183, 250–51, 372, 376, 418–20, 453–54, 458, 460–62, 475–76, 490–91
  analog, 472
  bitline, 476
  critical, 473
  delta, 129
  high, 11, 416, 474, 484
  low, 115, 129, 176, 200, 374, 385
  negative, 341, 379, 393, 395, 398, 409, 435
  nominal, 501
  output, 454
  small, 397, 460
  stress, 22
  tail, 479
voltage threshold, 455, 467, 475, 488
VRH, *see* variable range hopping

wafer level, 471–72, 480–81, 489–90
wafers, 181, 227, 249, 299, 350, 370, 450, 452, 461, 468–70, 483, 490

wafer sort, 465, 480–82, 489–90, 502
WAXS, *see* wide-angle X-ray scattering
wear leveling, 328, 344, 349–50, 352, 358
Wentzel–Kramer–Brillouin (WKB), 83, 89, 91, 115
wet-etch process, 278
wide-angle X-ray scattering (WAXS), 39
WKB, *see* Wentzel–Kramer–Brillouin
wordlines, 340–41, 465, 474, 482–87
work functions, 6, 9, 51, 77, 104, 110, 114, 118, 120, 152–53, 296–97, 316, 418, 426

XANES, *see* X-ray absorption near-edge spectroscopy
XAS, *see* X-ray absorption spectroscopy
XEOL, *see* X-ray-excited optical luminescence
XPS, *see* X-ray photoelectron spectroscopy
X-ray absorption near-edge spectroscopy (XANES), 43–44
X-ray absorption spectroscopy (XAS), 37, 43–44, 71
X-ray crystallographic data, 39
X-ray data, 70
X-ray diffraction (XRD), 37–38, 43–44, 47, 215–16
X-ray-excited optical luminescence (XEOL), 43
X-ray photoelectron spectroscopy (XPS), 30, 37, 40–43, 70, 203, 205, 207, 234
X-rays, 26, 40, 43, 45, 69
X-ray scattering (XRS), 37
XRD, *see* X-ray diffraction

XRS, *see* X-ray scattering

YAFFS, *see* Yet Another Flash File
    System
Yano's method, 87
Yet Another Flash File System
    (YAFFS), 358–59

Young's modules, 227

ZB, *see* zinc-blende
zero-bandgap semimetal, 264
zinc-blende (ZB), 16
ZnO-based memory, 18